VOLUME ONE HUNDRED AND TWENTY SEVEN

CURRENT TOPICS IN DEVELOPMENTAL BIOLOGY

Forkhead FOXO Transcription Factors in Development and Disease

CURRENT TOPICS IN DEVELOPMENTAL BIOLOGY

"A meeting-ground for critical review and discussion of developmental processes"
A.A. Moscona and Alberto Monroy (Volume 1, 1966)

VOLUME ONE HUNDRED AND TWENTY SEVEN

Current Topics in DEVELOPMENTAL BIOLOGY

Forkhead FOXO Transcription Factors in Development and Disease

Edited by

SAGHI GHAFFARI

Department of Cell, Developmental and Regenerative Biology; Black Family Stem Cell Institute; Tisch Cancer Institute, Icahn School of Medicine at Mount Sinai, New York, NY, United States

ELSEVIER

ACADEMIC PRESS

An imprint of Elsevier

Academic Press is an imprint of Elsevier
50 Hampshire Street, 5th Floor, Cambridge, MA 02139, United States
525 B Street, Suite 1800, San Diego, CA 92101-4495, United States
The Boulevard, Langford Lane, Kidlington, Oxford OX5 1GB, United Kingdom
125 London Wall, London, EC2Y 5AS, United Kingdom

First edition 2018

ISBN: 978-0-12-804253-3
ISSN: 0070-2153

For information on all Academic Press publications
visit our website at https://www.elsevier.com/books-and-journals

Publisher: Zoe Kruze
Acquisition Editor: Zoe Kruze
Editorial Project Manager: Shellie Bryant
Production Project Manager: Denny Mansingh
Cover Designer: Mark Rogers

Typeset by SPi Global, India

CONTENTS

CONTRIBUTORS

Richard C. Allsopp
Institute for Biogenesis Research, University of Hawaii, Honolulu, HI, United States

Maria Almeida
Center for Osteoporosis and Metabolic Bone Diseases; University of Arkansas for Medical Sciences, Little Rock, AR, United States

Abigail K. Brown
Brown University, Providence, RI, United States

Boudewijn M.T. Burgering
Molecular Cancer Research, Center Molecular Medicine, Oncode Institute, University Medical Center Utrecht, Utrecht University, Utrecht, The Netherlands

Philip M.C. Davy
Institute for Biogenesis Research, University of Hawaii, Honolulu, HI, United States

Timothy A. Donlon
Honolulu Heart Program/Honolulu-Asia Aging Study, Kuakini Medical Center; Ohana Genetics, Honolulu, HI, United States

Saghi Ghaffari
Department of Cell, Developmental and Regenerative Biology; Black Family Stem Cell Institute; Tisch Cancer Institute, Icahn School of Medicine at Mount Sinai, New York, NY, United States

Srividhya Iyer
University of Arkansas for Medical Sciences, Little Rock, AR, United States

Ha-Neui Kim
Center for Osteoporosis and Metabolic Bone Diseases, University of Arkansas for Medical Sciences, Little Rock, AR, United States

Raymond Liang
Developmental and Stem Cell Biology, Multidisciplinary Training Area, Icahn School of Medicine at Mount Sinai, New York, NY, United States

Brian J. Morris
Honolulu Heart Program/Honolulu-Asia Aging Study, Kuakini Medical Center; John A. Burns School of Medicine, University of Hawaii, Honolulu, HI, United States; School of Medical Sciences and Bosch Institute, University of Sydney, Sydney, NSW, Australia

Jihye Paik
Weill Cornell Medicine, New York, NY, United States

Rebecca Ring
Center for Osteoporosis and Metabolic Bone Diseases, University of Arkansas for Medical Sciences, Little Rock, AR, United States

Evan E. Santo
Weill Cornell Medicine, New York, NY, United States

Heidi A. Tissenbaum
Department of Molecular, Cell and Cancer Biology, University of Massachusetts Medical School, Worcester, MA, United States

Terry G. Unterman
University of Illinois at Chicago College of Medicine; Medical Service, Jesse Brown VA Medical Center, Chicago, IL, United States

Sabina van Doeselaar
Molecular Cancer Research, Center Molecular Medicine, Oncode Institute, University Medical Center Utrecht, Utrecht University, Utrecht, The Netherlands

Ashley E. Webb
Brown University, Providence, RI, United States

Bradley J. Willcox
Honolulu Heart Program/Honolulu-Asia Aging Study, Kuakini Medical Center; John A. Burns School of Medicine, University of Hawaii, Honolulu, HI, United States

Donald Craig Willcox
Honolulu Heart Program/Honolulu-Asia Aging Study, Kuakini Medical Center; John A. Burns School of Medicine, University of Hawaii, Honolulu, HI, United States; Okinawa International University, Ginowan, Okinawa, Japan

PREFACE

Work in many laboratories across the world in the past three decades has identified transcription factors FOXO as an enigmatic family of proteins that maintain health and longevity. FOXO transcription factors (four related proteins in the mammalian system) maintain homeostasis by regulating a wide range of processes during development and in adult tissues. FOXO (FOXO3) are recognized as main proteins that contribute to human longevity. The aberrant disrupted FOXO function is implicated in disease including diabetes, cancer, neurodegenerative, and blood disorders. The chapters of this book bring together an international board of leading experts to review FOXO's functions and regulation in various tissues, both in homeostasis and in disease. In Chapter 1, Heidi Tissenbaum reviews how the identification of FOXO factors in *C. elegans* as a mediator of worm longevity has set the stage for elucidating the importance of these factors in various species and tissues. In Chapter 2, Liang and Ghaffari discuss how FOXO function in stem cells might promote stem cell longevity and potential implications for organismal longevity. They review the current literature on FOXO in different types of embryonic and adult stem cells and the potential implications for disease. In Chapter 3, van Doeselaar and Burgering discuss a paradigm shift in our understanding of FOXO as tumor suppressors and how these proteins may function paradoxically as oncogenes under favorable conditions. In Chapter 4, Santo and Paik discuss FOXO in neural development and stem cells as well as in neural stress response, the neural phenotype generated by loss of FOXO, and disorders promoted by FOXO modulations in neural cells. In Chapter 5, FOXO regulation and function in the control of metabolic processes and the contribution of FOXO deregulation to diabetes are reviewed by Terry Unterman. In Chapter 6, FOXO's effects on osteogenesis along with consequences for benign and malignant bone disorders are reviewed by Almeida and colleagues. In Chapter 7, the complexity by which FOXO factors are regulated in mammalian cells is reviewed by Brown and Webb. And finally in Chapter 8, Willcox and colleagues discuss FOXO genes as determinants of longevity, the relationship of the insulin pathway and FOXO to extended life span, the transcriptional impacts of

FOXO genes, and the extrinsic influences on FOXO-mediated longevity. Altogether, this book presents an in-depth overview of current thinking on FOXO across species and in various tissues as well as FOXO-deregulated mechanisms that contribute to disease.

Saghi Ghaffari
Department of Cell, Developmental and Regenerative Biology; Black Family Stem Cell Institute;
Tisch Cancer Institute,
Icahn School of Medicine at Mount Sinai, New York,
NY, United States

CHAPTER ONE

DAF-16: FOXO in the Context of *C. elegans*

Heidi A. Tissenbaum[1]
Department of Molecular, Cell and Cancer Biology, University of Massachusetts Medical School, Worcester, MA, United States
[1]Corresponding author: e-mail address: heidi.tissenbaum@umassmed.edu

Contents

Abstract

In *Caenorhabditis elegans*, there is a single FOXO transcription factor homolog, encoded by the gene, *daf-16*. As a central regulator for multiple signaling pathways, DAF-16 integrates these signals which results in modulation of several biological processes including longevity, development, fat storage, stress resistance, innate immunity, and reproduction. Using *C. elegans* allows for studies of FOXO in the context of the whole animal. Therefore, manipulating levels or the activity of *daf-16* results in phenotypic changes. Genetic and molecular analysis revealed that similar to other systems, DAF-16 is the downstream target of the conserved insulin/IGF-1 signaling (IIS) pathway;

Current Topics in Developmental Biology, Volume 127
ISSN 0070-2153
https://doi.org/10.1016/bs.ctdb.2017.11.007

a PI 3-kinase/AKT signaling cascade that ultimately controls the regulation of DAF-16 nuclear localization. In this chapter, I will focus on understanding how a single gene *daf-16* can incorporate signals from multiple upstream pathways and in turn modulate different phenotypes as well as the study of FOXO in the context of a whole organism.

1. INTRODUCTION

The roundworm species *Caenorhabditis elegans* or more commonly referred to as *C. elegans* has emerged as a powerful model system to study FOXO transcription factors. Named *Caenorhabditis elegans* (*Caeno* meaning recent; *rhabditis* meaning rod; *elegans* meaning nice) (Ankeny, 2001; Fatt & Dougherty, 1963; Nigon & Dougherty, 1949), in 1963 by the well-known scientist Sydney Brenner, who chose *C. elegans* to serve as a model system for biological research. Brenner believed that future biological advances, particularly, in studies of development and neurobiology required a model system (a nonhuman species in the laboratory to help understand biological processes) that was inexpensive to maintain, had a simple body plan, and could be easily grown in vast quantities (Ankeny, 2001).

C. elegans fit the criteria put forth by Brenner. These animals are free living, 1-mm long, and easily grown and propagated in the laboratory. *C. elegans* are hermaphrodites and bacteriovores that feed on both virulent and avirulent bacteria. Routine growth in the laboratory involves *C. elegans* feeding on lawns of *Escherichia coli* growing on top of a plastic Petri dish filled with agar (Stiernagle, 2006). Over the last 30 years, numerous breakthroughs have been made using this organism and recognized with several Nobel prizes. These include one award for the early work in the Brenner lab, where the position and lineage of all 959 somatic cells of the adult worm were determined, along with the genetics of cell death to Drs. Robert H. Horvitz, John Sulston, and Sydney Brenner in 2002. Additional Nobel prizes for work in *C. elegans* include the discovery of RNA interference (RNAi) to Drs. Craig Mello and Andy Fire in 2006, and fluorescent reporters to visualize expression of genes in live animals to Dr. Martin Chalfie (Chalfie, Tu, Euskirchen, Ward, & Prasher, 1994) in 2008. In addition, Drs. Victor Ambros and Gary Ruvkun have been recognized with numerous awards including the Lasker award for the discovery of microRNAs.

Over the past three decades, *C. elegans* has emerged as a premiere system to study the molecular biology of aging. This is primarily due to (1) a short, reproducible life span (approximately 18 days at 20°C); (2) similar to humans, life span is measured by the number of days; (3) similar to humans, the survival kinetics show an exponential increase in mortality rate over time; (4) aging is postmitotic; (5) the entire genome is sequenced and many molecular and genetic tools are available; and (6) many evolutionarily conserved signaling pathways. Therefore, together, a comprehensive identification of the genes that affect the aging process in *C. elegans* is possible.

Of all the genes identified as part of the aging process, perhaps, the most important has been the single forkhead "O" Box transcription factor (FOXO) homolog *daf-16* (Lin, Dorman, Rodan, & Kenyon, 1997; Ogg et al., 1997). *C. elegans* FOXO/DAF-16 has emerged as a central regulator of the aging process. Therefore, over the last 20 years, research studies on the regulation of life span or the aging process in *C. elegans* include genetic experiments determining whether *daf-16* is involved.

2. IDENTIFICATION OF *DAF-16*

2.1 Genetic

The original identification of *C. elegans* FOXO/DAF-16 is solely based on the life cycle/development of this animal. The life cycle of *C. elegans* begins with an egg and develops through four larval stages (abbreviated L1–L4) each separated by a molt. The final molt results in the formation of a self-fertilizing adult hermaphrodite (Riddle & Albert, 1997). However, when growth conditions become unfavorable, primarily determined by the levels of a continually secreted pheromone, worms can leave the normal development life cycle and form an alternative L3, a dauer larva (Golden & Riddle, 1982; Hu, 2007). Dauers (German for enduring) are a distinct larval state that is developmentally arrested, nonfeeding, and suited for long-term survival until growth conditions become more suitable in favor of reproduction (Fielenbach & Antebi, 2008; Hu, 2007). Interestingly, the dauer larval stage is nonaging. Therefore, the time an animal spends as a dauer does not affect its adult life span.

A genetic screen to identify genes that modulate dauer formation was performed during the early stages of *C. elegans* research (Albert, Brown, & Riddle, 1981; Riddle, 1977). The mutations identified in this screen: were called **daf** (**da**uer **f**ormation) and were divided into two categories: dauer constitutive (*daf*-c) and dauer defective (*daf-d*). Mutants that were *daf-c* would enter dauer even under plentiful growth conditions, while *daf-d* mutants failed to enter dauer even when growth conditions were unfavorable. *daf-16* was isolated in this screen as a dauer defective mutant (Albert et al., 1981).

Early genetic epistasis analysis revealed three pathways for dauer information. These initial observations placed several genes including *daf-2*, *age*-1 (also initially identified as *daf-23*), and *daf-16* in a separate genetic pathway. Both *daf-2* and *age-1* were initially identified as *daf*-c mutants and were suppressed by a reduction of function mutation in *daf-16*. Therefore, both a *daf-16*; *daf-2* or *daf-16*; *age-1* double mutant behaved similar to *daf-16*. Later, phenotypic analysis revealed that *daf-2* and *age-1* mutants also exhibited changes in life span, stress resistance, fat storage, immunity, and reproduction (reviewed in Murphy & Hu, 2013). All of these phenotypes are also suppressed by a reduction-of-function mutation in *daf-16*.

2.2 Molecular

Remarkably, when molecular studies cloned the genes *daf-2*, *daf-16*, and *age-1*, it was revealed that they were part of an insulin/insulin-like growth factor-1 signaling (IIS) pathway. *daf-2* encodes for an insulin/IGF-1 receptor which is equally identical to both the mammalian insulin and insulin-like growth factor-1 (IGF-1) receptors (Kimura, Tissenbaum, Liu, & Ruvkun, 1997). *age-1* encodes for the p110 catalytic subunit of phosphatidylinositol (PI)-3 kinase (Morris, Tissenbaum, & Ruvkun, 1996) and *daf-16* encodes for a FOXO transcription factor (Lin et al., 1997; Ogg et al., 1997). These findings, together with additional *C. elegans* studies, combining genetic and molecular analyses revealed the remaining components of a PI 3-kinase/AKT-signaling cascade including the kinases AKT-1/2 (Paradis & Ruvkun, 1998) and PDK-1 (Paradis, Ailion, Toker, Thomas, & Ruvkun, 1999) and the phosphatase PTEN (Ogg & Ruvkun, 1998). Together, this IIS pathway functions ultimately to negatively regulate the downstream target DAF-16.

In addition, to its original isolation as a gene that affects dauer formation, *daf-16* has emerged as a regulator and main integrator of many different signaling pathways, interacting proteins, modifications, etc. that result in modulation of many biological processes including life span, (discussed later). However, the upstream pathway that is central for *daf-16* function is the IIS pathway.

The *C. elegans* IIS pathway usually begins with insulin-like peptides and the *C. elegans* genome has 40 putative insulins. Although several insulins have been suggested to be the IIS ligand by genetic or expression studies (including *ins*-7, *daf-28*), none had been shown to biochemically bind the DAF-2 receptor (Cornils, Gloeck, Chen, Zhang, & Alcedo, 2011;

Fernandes de Abreu et al., 2014; Li, Kennedy, & Ruvkun, 2003; Mendler et al., 2016; Ritter et al., 2013). According to studies in mammalian cells, following the receptor, there is an insulin receptor substrate gene which is then followed by a briefly: activation of PI 3-kinase which includes a regulatory subunit p55/p85 and a catalytic subunit p110, and this results in conversion of phosphatidylinositol (3,4,5)-bisphosphate (PIP_2) to phosphatidylinositol (3,4,5)-trisphosphate (PIP_3). In mammalian cell culture, newly converted PIP_2 and/or PIP_3 recruit the downstream kinases PDK1 and AKT1 to the plasma membrane where PDK1 phosphorylates and activates AKT. Phosphorylation of AKT-1 (and AKT-2) leads to their direct phosphorylation of FOXO and causing sequestration in the cytosol (Brunet et al., 1999; Brunet, Datta, & Greenberg, 2001; Tang, Nunez, Barr, & Guan, 1999).

In *C. elegans*, additional components include the insulin receptor substrate gene encoded by the gene *irs-1* (Wolkow, Munoz, Riddle, & Ruvkun, 2002) and the regulatory subunit of PI 3-kinase, p55, encoded by the gene *aap-1* (Wolkow et al., 2002). Phosphorylation of AKT-1 (and AKT-2) leads to their direct phosphorylation of DAF-16 causing its sequestration in the cytosol (Hertweck, Gobel, & Baumeister, 2004; Lin, Hsin, Libina, & Kenyon, 2001; Oh et al., 2005). In addition, studies with worm extracts revealed that signaling through PDK-1 activates AKT-1, AKT-2, and SGK-1 which can then phosphorylate DAF-16 (Hertweck et al., 2004; Lin et al., 2001; Oh et al., 2005) similar to mammals.

Genetic analysis consistent with the molecular studies above has shown that reduction-of-function mutations or RNAi of either *daf-2*, *age-1*, *aap-1*, *pdk-1*, or *akt-1/2* results in phenotypic changes such as an increase in life span, increased stress resistance, a dauer constitutive phenotype (reviewed in Murphy & Hu, 2013). *daf-2* RNAi or reduction of function has been also reported to show an increase in fat storage (Kimura et al., 1997) and increased innate immunity (Troemel et al., 2006; Zou, Tu, Niu, Ji, & Zhang, 2013). All of these phenotypic changes in life span, development, metabolism, and/or stress resistance are dependent on *daf-16* (reviewed in Kenyon, 2011; Murphy & Hu, 2013) (Fig. 1).

C. elegans provides the advantage of studying FOXO in the context of the whole organism. Therefore, studies on *C. elegans* have the ability to assess the role of FOXO on an organismal level. As such, genetic manipulations of the gene (increasing or decreasing levels) can be directly measured as a phenotypic consequence in the worm. Routinely, there are four main phenotypes typically examined for DAF-16 function (Fig. 1).

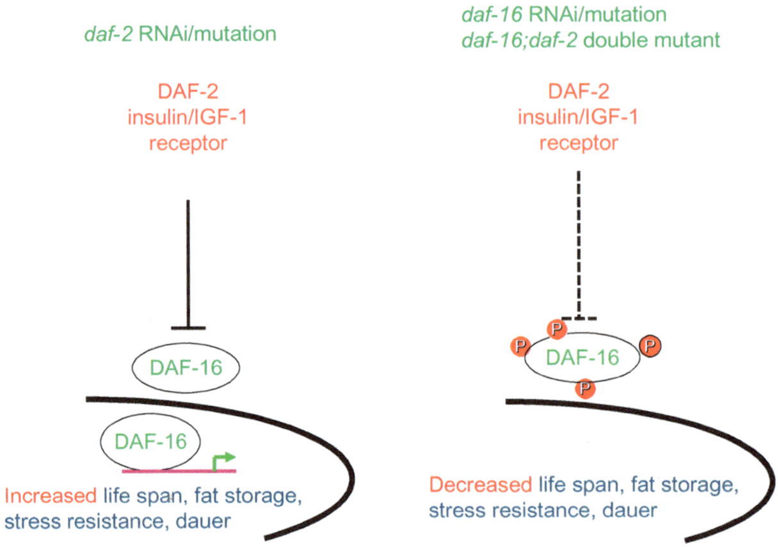

Fig. 1 Schematic showing how the IIS pathway modulates DAF-16 nuclear translocation and the resulting phenotypes. Although *daf-16* modulates additional phenotypes, these are the 4 most often studied in association with this pathway.

As mentioned earlier, DAF-16 signaling is critical for larval development—the decision to enter dauer or grow reproductively. Modulating DAF-16 has been also been shown to cause changes in longevity (assessed directly with a life span assay measuring the number of days the animals lives). Modulating levels of DAF-16 have also been associated stress resistance where changes can be measured (including heat and oxidative stress (reviewed in Kenyon, 2011; Murphy & Hu, 2013). Finally, fat storage which can qualitatively assessed using Oil Red O and Sudan Black staining and quantitatively (Yen et al., 2010; reviewed in Ashrafi, 2007; Barros, Liu, Lemieux, Mullaney, & Ashrafi, 2012; Kenyon, 2011; Murphy & Hu, 2013).

DAF-16 has a high degree of homology to FOXO3a (Lin et al., 1997; Ogg et al., 1997). Mammals have four FOXO genes: FOXO1, FOXO3/FOXO3a, FOXO4, and FOXO6, each with a different function and tissue expression pattern. In contrast, similar to *Drosophila melanogaster*, *C. elegans* has a single FOXO, encoded by the gene *daf-16* and has multiple isoforms from a single open reading frame (discussed later).

The forkhead domain (FHD) was named for its original isolation as a mutant in *D. melanogaster* (Weigel, Jurgens, Kuttner, Seifert, & Jackle, 1989) where mutants displayed a "forked head" appearance. Following this, many genes were identified with a similar FHD domain and together they were named the Fox gene family. The FHD is around 100 bp and contains the signature winged-helix DNA-binding domain. The second classification is based on the second domain the box named A–X. The forkhead box O family members include mammalian AFX (FOXO4), FKHR (FOXO1), and FKHR-L1 (FOXO3a) and FOXO6.

3. ISOFORMS

The initial studies of DAF-16 isoforms focused on three isoforms from the single open reading frame; *daf-16a1*, *daf-16a2*, and *daf-16b*. However, the *C. elegans* genome resource, Wormbase (https://www.wormbase.org), renamed the *daf-16* isoforms *daf-16* (R13H8.1a–h) referring to eight putative isoforms of *daf-16*, with each isoform designated by a lowercase letter alphabetically following the cosmid name. Bansal et al. (2014) investigated these putative isoforms (R13H8.1a–h), first by analyzing the expressed sequence tag (EST) database followed by capped RNA (Cap-Seq) data, total mRNA, and ribosome-associated mRNA deep-sequencing datasets. These analyses together revealed that the existence of only *daf-16*f, *daf-16*d, *daf-16*a1, *daf-16*a2, and *daf-16*b was supported. Only four spliced leader (SL1) sites to the capped 5′ ends of *daf-16*f, *daf-16*d, *daf-16*a1/a2, and *daf-16*b were supported by the Cap-Seq data Bansal et al. (2014). Therefore, five isoforms for DAF-16—DAF-f (R13H8.1f), DAF-16d (R13H8.1d), DAF-16a1 (R13H8.1b), DAF-16a2 (R13H8.1c), and DAF-16b (R13H81.a) have been verified and should be considered moving forward.

RNAi, mutant, and transgenic studies with *daf-16* isoforms (*daf-16a, b, d/f*) (Chen et al., 2015; Kwon, Narasimhan, Yen, & Tissenbaum, 2010; Lee, Hench, & Ruvkun, 2001; Lin et al., 2001) revealed *daf-16b* functions in dauer formation, while *daf-16a* and *daf-16d/f* contribute to life span regulation. More recently, *daf-16*b has been implicated in promoting developmental neurite outgrowth (Christensen, de la Torre-Ubieta, Bonni, & Colon-Ramos, 2011) consistent with its neural expression pattern (Kwon et al., 2010; Lee et al., 2001; Lin et al., 2001). *daf-16b* is unique in that it appears to be toxic at high levels (Kwon, E. & Tissenbaum, H.A., unpublished; Lee et al., 2001).

One mechanism to achieve specificity is modification of the different isoforms by phosphorylation of direct interacting proteins. Although suggested by the work of Kwon et al. (2010) where the DAF-16 translocation data suggested that DAF-16d/f is preferentially regulated by AKT-1 as compared with AKT-2, thus far little is known about isoform-specific phosphorylation. Overall, kinases will phosphorylate DAF-16 to regulate the activity of DAF-16 and ultimately determine the ability of DAF-16 to enter the nucleus to bind its direct targets (Birkenkamp & Coffer, 2003; Van Der Heide, Hoekman, & Smidt, 2004; Vogt, Jiang, & Aoki, 2005; Yen, Narasimhan, & Tissenbaum, 2011).

The contributions of *daf-16a* and *daf-16d/f* have been examined in greater detail. Using animals bearing transgenes with low copy number (Fig. 2), Kwon et al. (2010) revealed that the multiple DAF-16 isoforms enable worms to equilibrate and fine-tune the IIS pathway-mediated processes including life span, dauer formation, stress resistance, and fat storage by virtue of (i) distinct enrichment in specific tissues, (ii) selective preference for upstream kinases, and (iii) specific/overlapping/cooperative regulation of target genes. Additional analysis where the promoters of *daf-16a* and *daf-16d/f* were swapped revealed that the DAF-16 C-terminus is important for nuclear localization, while the DAF-16 N-terminus is important for tissue expression. *daf-16a* and *daf-16d/f* are the major isoforms for modulation of dauer formation and life span (Chen et al., 2015; Kwon et al., 2010). Together, these data reveal that *C. elegans* FoxO/DAF-16 uses different isoforms to modulate multiple biological processes. This is in contrast to mammals, where genes on different chromosomes perform different functions. Thus far, many of the in vivo inferred molecular and biochemical characteristics of DAF-16 are deduced from its similarity with mammalian FOXOs, due to the lack of a *C. elegans* cell culture system similar to *Drosophila* S2 cells or mammalian tissue culture systems.

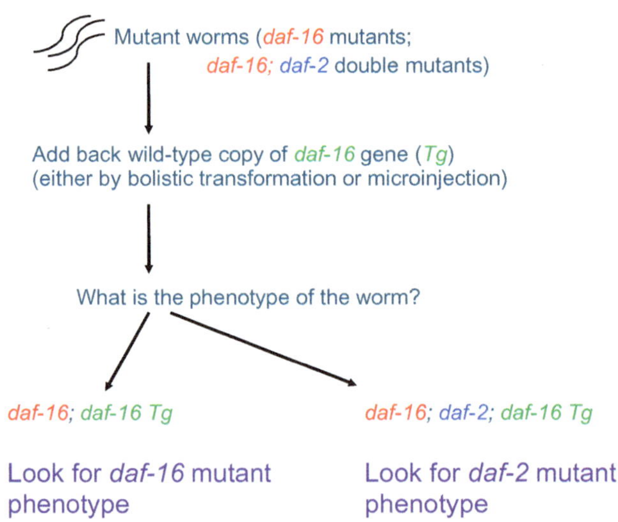

Fig. 2 Outline of transgenic analysis used to study *daf-16*.

4. DIRECT PHOSPHORYLATION

4.1 AKT-1/AKT-2/SGK-1

AKT-1 is reported to phosphorylate DAF-16 on four distinct sites, and three of these are conserved in mammalian FOXO (Lee et al., 2001; Lin et al., 2001; Tran, Brunet, Griffith, & Greenberg, 2003). Studies using *daf-16* mutants bearing a DAF-16::GFP transgene where the four AKT-1 phosphorylation sites were mutated have been used to examine the importance of direct phosphorylation of DAF-16 by AKT. Absence of AKT phosphorylation was sufficient to cause dauer arrest indicating that DAF-16 is the major target of AKT, at least for dauer formation (Lee et al., 2001). However, a second study revealed contrasting results where absence of AKT phosphorylation facilitated DAF-16 entry into the nucleus, but the resulting DAF-16 nuclear localization was insufficient to induce any phenotypes (Lin et al., 2001); perhaps indicating a role for additional proteins that activate DAF-16 (Lin et al., 2001).

The two studies highlight studying FOXO/DAF-16 in the context of a whole organism. As mentioned, the advantage of *C. elegans* is that one can modify an enzymatic site and discover the phenotypic consequences. How/why then would these studies have different results? Oftenlabs use different transgenic strains as in this instance where Lin et al. (2001) used an integrated genomic construct tagged with GFP, while the Lee et al. (2001) used an extrachromosomal cDNA construct. Therefore, it is entirely possible that the expression levels of the DAF-16 transgenes could have been substantially different and account for the results. With the new development of CRISPR/Cas9 genome editing systems (Dickinson, Ward, Reiner, & Goldstein, 2013), some of these expression of differences may not exist.

In mammals (Brunet, Park, et al., 2001) and *C. elegans* serum–glucocorticoid kinase, SGK-1, is suggested to function at the level of AKT-1/2 (Hertweck et al., 2004). AKT-1/AKT-2/SGK-1 have been isolated from whole-worm extracts showing they form a protein complex that transduces the PI 3-kinase signals via PDK-1 to control the localization and activation of DAF-16 by direct phosphorylation (Hertweck et al., 2004).

However, additional studies from other labs indicate that the tissue expression patterns of these kinases show little overlap. AKT-1::GFP and AKT-2::GFP are expressed in the head and tail neurons, pharynx, and spermathecae; SGK-1::GFP is primarily in the intestine (Narasimhan et al., 2011; Padmanabhan et al., 2009). Therefore, in vitro, similar to mammals, *C. elegans* lysates show that AKT-1, AKT-2, and SGK-1 can directly phosphorylate DAF-16 (Hertweck et al., 2004; Lin et al., 2001; Oh et al., 2005), resulting in the translocation of DAF-16 to the nucleus and an increase in dauer formation (Hertweck et al., 2004). Overall, phosphorylations by AKT-1/AKT-2/SGK-1 lead to inactivation of DAF-16.

4.2 Jun Kinase

Another kinase that has been shown to directly phosphorylate DAF-16 is Jun kinase (JNK) and this direct interaction with DAF-16 results in life span modulation (Oh et al., 2005). JNK is part of a large kinase family that is a subgroup of the mitogen-activated protein kinase (MAPK or MAP kinase) superfamily associated with mammalian regulation of multiple functions including development, apoptosis, and cell survival (Davis, 2000).

JNK itself is downstream from a MAPK signaling pathway that responds to changes in stress such as heat stress and oxidative stress. Phenotypically, overexpression of JNK, in worms and flies, extends life span and increases stress resistance, and this life span extension is dependent on DAF-16/dFOXO (respectively, Oh et al., 2005; Wang, Bohmann, & Jasper, 2005). Moreover, in both *C. elegans* and mammalian cell culture, JNK physically interacts with and phosphorylates DAF-16 at sites different from the AKT phosphorylation sites and in *C. elegans* results in enhanced nuclear translocation of DAF-16 that presumably leads to the phenotypic changes in life span and stress resistance (Essers et al., 2004; Oh et al., 2005). Therefore, JNK represents an additional modification of DAF-16 perhaps only under conditions of stress and overall, phosphorylation by JNK-1 leads to activation of DAF-16.

4.3 AMP Kinase

AMP-activated protein kinase (AMPK) has emerged a central integrator of metabolism reviewed inBurkewitz, Weir, and Mair (2016). AMPK has also been shown to directly phosphorylate DAF-16 on at least six different sites which results in activation of DAF-16 (Greer et al., 2007). AMPK, is a heterotrimeric kinase complex with an α (catalytic) subunit and regulatory β and γ subunits. Tullet et al. (2014) investigating the multiple subunits of AMPK showed a positive feedback loop with DAF-16 and AAKG-4 (γ subunit isoform). Therefore, AMPK and DAF-16 show a complex interaction whereby AMPK can directly act on DAF-16 and is also a target (Tullet et al., 2014). Regulation of DAF-16 by AMPK is further complicated since AMPK has inhibitory effects on target of rapamycin (TOR) signaling.

TOR is a highly conserved kinase that integrates many nutritional signals and itself is part of a pathway that converges with DAF-16 (Narasimhan, Yen, & Tissenbaum, 2009). In *C. elegans*, the TORC2 complex activates SGK-1 and inhibition by AMPK should likely result in the activation of DAF-16 signaling (Jones, Greer, Pearce, & Ashrafi, 2009; Soukas, Kane, Carr, Melo, & Ruvkun, 2009). Additionally, a recent study suggested that inhibition of TORC1 results in an increase in expression of DAF-16d/f target genes and hence transcription (Robida-Stubbs et al., 2012) revealing another example of how upstream signaling may contribute to isoform-specific function. Therefore, overall AMPK activates DAF-16 signaling both directly and indirectly.

5. COFACTORS/INTERACTING PROTEINS

5.1 NHL-1

nhl-1, a member of the trim-nhl gene family, was uncovered in a search for cofactors that mediated the function of *daf-16* in stress resistance. Interestingly although *nhl-1* is expressed in chemosensory neurons, it activates DAF-16 in a distal tissue, the intestine to promote stress resistance (Volovik et al., 2014). Therefore, NHL-1 represents regulation by a DAF-16 cofactor involving different tissues.

5.2 SWI/SNF

Riedel et al. (2013) looked for proteins that physically interact with FOXO/DAF-16, coupled with a functional RNAi of the corresponding gene, and identified several members of the SWI/SNF family. Further, genome-wide analysis including RNA-seq and chromatin immunoprecipitation followed by sequencing (ChIP-Seq) revealed the regulatory regions of the target genes were occupied by both DAF-16 and the SWI/SNF complex. Moreover, these target genes were largely related to function including the aging process and stress resistance. Components of the SWI/SNF complex were also identified in an RNAi screen of transcriptional regulators that promote longevity specifically via DAF-16d/f (Bansal et al., 2014). Together, these findings add a new level of regulation of DAF-16 by the SWI/SNF complex.

5.3 14-3-3 Proteins

A major player in mediating mammalian FOXOs and DAF-16 nuclear translocation is the 14-3-3 family of proteins. In cell culture, AKT-1/2/SGK-1 phosphorylation of FOXO results in an increased binding affinity to the 14-3-3 proteins. Once bound to the 14-3-3 proteins, FOXO is released from the DNA and then FOXO relocalizes to the cytosol (Cahill et al., 2001). Bound 14-3-3 will mask the nuclear localization signal on FOXO thereby preventing its reentry into the nucleus (Burgering, 2008; Burgering & Kops, 2002). Therefore, the interaction of 14-3-3 proteins with FOXO contributes to its nuclear/cytoplasmic shuttling regulation.

DAF-16 regulation is also modified by association with 14-3-3 proteins. *C. elegans* have two 14-3-3 proteins (PAR-5 and FTT-2). Complexes isolated from whole-worm extracts suggest that *C. elegans* PAR-5 and FTT-2 interact with DAF-16 and similar to mammals this modulates its nuclear/cytoplasmic distribution (Berdichevsky, Viswanathan, Horvitz, & Guarente, 2006; Li, Tewari, Vidal, & Lee, 2007; Wang et al., 2006).

C. elegans 14-3-3 proteins also function to mediate the direct interaction between SIR-2.1 (*C. elegans* SIRT1) and DAF-16 (Berdichevsky et al., 2006; Wang et al., 2006). Originally identified in *Saccharomyces cerevisiae* as a gene important for gene silencing, *sir2* (Rine & Herskowitz, 1987), known as mammalian SIRT1, similar to DAF-16, has emerged as a major player in life span regulation across phylogeny (Kaeberlein, McVey, & Guarente, 1999; Rogina & Helfand, 2004; Tissenbaum & Guarente, 2001). In *C. elegans*, overexpression of *sir-2.1* (*C. elegans* SIR2 ortholog) results in increased dauer formation and life span extension dependent on *daf-16* (Tissenbaum & Guarente, 2001). It is possible this interaction also adds to DAF-16 specificity. Therefore, the nuclear/cytoplasmic shuttling of DAF-16 is mediated by an interaction with the 14-3-3 proteins and SIR-2.1.

A number of additional proteins have been identified that modulate DAF-16 function. These proteins are listed as cofactors since they have not been shown to directly interact with DAF-16 in whole worms.

5.4 SMK-1

SMK-1 (suppressor of MEK null gene), originally identified in *Dictyostelium discoideum* was identified as a potential cofactor for DAF-16 (Wolff et al., 2006). Interestingly, analyses uncovered that *smk-1* is not needed for heat stress but is necessary for the life span, innate immune, UV, and oxidative stress functions of *daf-16*.

Importantly, this could present at another layer of specificity whereby what direct DAF-16 to particular target genes. However, although SMK-1 possesses all of the requirements for a cofactor of DAF-16, there is no direct biochemical data to confirm this interaction.

5.5 STK4/CST-1

Another kinase that may phosphorylate DAF-16 is the *C. elegans Stk4* ortholog, *cst-1*, based on mammalian cell culture. In mammalian cell culture under conditions of oxidative stress, STK4 (serine/threonine kinase 4, formerly known as MST1) can phosphorylate FOXO3A at a site in the forkhead domain that is conserved in *C. elegans* (Lehtinen et al., 2006). Once phosphorylated, nuclear entry of FOXO3A is facilitated (Lehtinen et al., 2006). Interestingly, in vitro, STK4 can robustly phosphorylate *C. elegans* DAF-16 at the conserved STK4 site and overexpression of the *C. elegans Stk4* ortholog, named *cst-1*, promotes longevity in a *daf-16*-dependent manner suggesting a conserved mechanism (Lehtinen et al., 2006). However, once again, there is no direct biochemical data to confirm this interaction.

6. ADDITIONAL INPUTS

6.1 Heat Shock Factor-1

Exposure to heat stress induces a set of specialized molecular chaperones termed the heat shock proteins (Nollen & Morimoto, 2002). Transcriptionally, the heat stress response is modulated by a specialized transcription factor, termed heat shock factor 1 (HSF-1), which binds a heat shock element on promoters (Nollen & Morimoto, 2002). In *C. elegans*, the gene heat shock factor 1, *hsf-1*, was isolated in a genome-wide RNAi screen (Garigan et al., 2002). Thus far, overall data on DAF-16 and HSF-1 have suggested a model where these two proteins may have independent upstream signals but collaborate to promote longevity (Hsu, Murphy, & Kenyon, 2003; Morley & Morimoto, 2004). Most recently, the proteins DDL-1 and DDL-2 have been shown to interact with HSF-1 and formation of this complex is regulated by the IIS (Chiang, Ching, Lee, Mousigian, & Hsu, 2012).

To address DAF-16 specificity, additional cofactors have been identified that positively or negatively regulate DAF-16/FOXO activity. One example is *C. elegans* host-cell factor homolog HCF-1 which has been shown to inhibit nuclear DAF-16 transcriptional activity (Li et al., 2008). RNAi knockdown of *hcf-1* also results in increased life span (Li et al., 2008). A second putative cofactor is beta-catenin, BAR-1, part of the Wnt signaling pathway where DAF-16/FOXO intersects with beta-catenin and its worm homolog BAR-1 in regulation of oxidative stress-related genes (Essers et al., 2005).

7. TARGET GENES

daf-16 is a central regulator for multiple biological processes. How does this one gene receive multiple inputs and then target many downstream genes to regulate these diverse processes? One possible simple model is that DAF-16 directly targets genes for a specific phenotype. For example, under conditions of heat stress, DAF-16 would directly bind certain target genes and this group of target genes would be different under conditions of oxidative stress. However, this is not the case. Moreover, often studies identifying DAF-16 target genes do not specify whether the target genes identified are direct and/or indirect. A direct DAF-16 target gene means *daf-16* binds its promoter. In contrast, an indirect target gene indicates that there is no direct binding of *daf-16* but the expression of the target gene changes in response to *daf-16* expression. Overall, to date, the transcriptional network downstream of *daf-16* is still illusive.

Similar to other members of the FOXO transcription factor family, DAF-16 has been shown to bind DNA in vitro as a monomer to the consensus-binding site (GTAAACA or TGTTTAC), or DAF-16-binding element (DBE) (Furuyama, Nakazawa, Nakano, & Mori, 2000). Microarray analysis in *C. elegans* (Murphy et al., 2003) revealed the DBE to be overrepresented upstream of putative DAF-16 transcriptional targets. This study also identified the existence of a second potential binding site for DAF-16, termed the DAE or DAF-16-associated binding site (Murphy et al., 2003). Further, by focusing on the expression profile differences between the *daf-2* and *daf-16; daf-2* genotypes. Murphy et al. (2003) revealed two different classes of genes referred to as Class 1 and Class II. Genes that follow into Class 1 were induced in animals grown on *daf-2(RNAi)* and reduction of function IIS mutants but were then repressed in animal grown on *daf-16*-(RNAi); *daf-2*(RNAi). Mutation in these genes will result in life span extension. Genes in Class II showed opposite phenotypes where mutations shorten life span. Class II genes were not induced in animals grown on *daf-2(RNAi)* and reduction of function IIS mutants but were induced in animal grown on *daf-16(RNAi); daf-2(RNAi)*. Most recently, Tepper et al. (2013) followed up this analysis and identified a new DAF-16 binding partner protein, PQM-1, a transcriptional activator that binds to the DAE to modulate Class II genes. Following these studies, certain questions still remain, including what are the upstream signals for each of the two transcription factors, DAF-16 and PQM-1? How are the different DAF-16 isoforms involved? And additional questions addressed in Tepper, Murphy, and Bussemaker (2014) including whether there is a functional equivalent in mammals to PQM-1?.

Additional microarrays have focused mainly on the role of the *daf-16* in longevity and the aging process. Golden and Melov (2004) revealed extensive stochastic regulation of the aging process when they performed microarray studies of individual aging worms. McElwee, Bubb, and Thomas (2003) and McElwee, Schuster, Blanc, Thomas, and Gems (2004) revealed similarities between the profiles of *daf-2* mutants and dauers—suggesting the possibility that longevity of the IIS is linked to dauer formation. It has been known for some time that dauers are ageless, i.e., the time an animal spends as a dauer does not affect adult life span. Initial studies (Kenyon, Chang, Gensch, Rudner, & Tabtiang, 1993) suggested this link was not possible since not all *daf-c* mutants exhibited an increase in life span. However, more recent studies suggest that many of the *daf-c* mutants indeed show changes in life span (Narasimhan et al., 2011; Shaw, Luo, Landis, Ashraf, & Murphy, 2007). Together, it is still unclear whether there are distinct target genes for dauer formation and longevity regulation by the IIS and*daf-16*.

Overall, microarray analysis to determine *daf-16* targets have revealed that no single target can recapitulate the full function of the IIS. In fact, there is often little overlap between the putative targets that have been identified due to the fact that often either different alleles, genetic backgrounds, developmental stages, or growth temperatures are used. Additionally, most of the studies use whole worms and there is extensive tissue-specific regulation of *daf-16* which would be missed.

Several studies have looked for direct targets of DAF-16. Oh et al. (2006) performed chromatin immunoprecipitation (ChIP) combined with cloning using a polyclonal anti-DAF-16 antibody that recognized all of the DAF-16 isoforms. This study identified a subset of the 103 direct targets of DAF-16. Phenotypic analysis of these genes found a significant increase in the number of genes that showed both *daf-16*- related phenotypes and expression dependent on *daf-16*. Kumar et al. (2015) performed a genome-wide analysis using the same antibody and reported ~4000 new binding events. The study defined a "core" set of DAF-16 direct targets that were validated phenotypically and then looked for orthologous genes in *Drosophila* and humans. This type of analysis, where direct binding is examined, inclusive of all isoforms will begin to unravel the complex transcriptional network downstream of *C. elegans* FOXO/DAF-16. However, the large number of target genes revealed by this study suggests added complexity at this level of regulation.

Additional genome-wide methods to examine direct targets include two methods that used a *daf-16*a transgene where wild-type worms express exogenous DAF-16::GFP. Niu et al. (2011) (modENCODE project) used ChIP to determine the promoters to which DAF-16a binds directly using a gfp antibody to determine the promoters bound by *daf-16*a. In a separate study, Schuster et al. (2010) used DNA adenine methylation transferase identification (DamID), where DAF-16a was tagged with a DNA adenine methyltransferase. These studies have revealed that DAF-16 functions as both an activator and a repressor and that often DAF-16 binds the promoter of a gene without altering its expression and termed this a parking site (Schuster et al., 2010). Further, Shuster et al., through their findings, suggested a model whereby DAF-16 is part of a gene regulatory network and functions as a central node within this network (Schuster et al., 2010; Tullet et al., 2014).

8. CONCLUSION

The single FOXO homolog DAF-16 has emerged as a central mediator of multiples upstream pathways including the IIS. In response to the signals from multiple pathways, DAF-16 will bind thousands of target genes thereby transducing the upstream signals to regulate distinct biological processes. In this chapter, I have focused on the selected question- how does DAF-16 achieves this specificity? In particular, how does a single transcription factor respond to upstream activator and inhibitory signals, interacting proteins, and cofactors, to then bind direct target genes and result in modulation of multiple biological processes? *C. elegans* provides a powerful system to address these questions by allowing the study of FOXO in the context of a whole organism.

ACKNOWLEDGMENTS

H.A.T. is a William Randolph Hearst Young Investigator. This publication was made possible by an endowment from the William Randolph Hearst Foundation and Grants from the NIA (1R01AG031237) and the American Diabetes Association (1-17-IBS-176). H.A.T. would like to thank Dr. Saghi Ghaffari for support and patience.

REFERENCES

Albert, P. S., Brown, S. J., & Riddle, D. L. (1981). Sensory control of dauer larva formation in *Caenorhabditis elegans*. *The Journal of Comparative Neurology*, *198*, 435–451.

Ankeny, R. A. (2001). The natural history of *Caenorhabditis elegans* research. *Nature Reviews. Genetics*, *2*, 474–479.

Ashrafi, K. (2007). Obesity and the regulation of fat metabolism. In: WormBook (Ed.), *The C. elegans research community*. WormBook (March 9). https://doi.org/10.1895/wormbook.1.130.1, http://www.wormbook.org.

Bansal, A., Kwon, E. S., Conte, D., Jr., Liu, H., Gilchrist, M. J., MacNeil, L. T., et al. (2014). Transcriptional regulation of *Caenorhabditis elegans* FOXO/DAF-16 modulates lifespan. *Longevity & Healthspan, 3*, 5.

Barros, A. G., Liu, J., Lemieux, G. A., Mullaney, B. C., & Ashrafi, K. (2012). Analyses of *C. elegans* fat metabolic pathways. *Methods in Cell Biology, 107*, 383–407.

Berdichevsky, A., Viswanathan, M., Horvitz, H. R., & Guarente, L. (2006). *C. elegans* SIR-2.1 interacts with 14-3-3 proteins to activate DAF-16 and extend life span. *Cell, 125*, 1165–1177.

Birkenkamp, K. U., & Coffer, P. J. (2003). Regulation of cell survival and proliferation by the FOXO (Forkhead box, class O) subfamily of Forkhead transcription factors. *Biochemical Society Transactions, 31*, 292–297.

Brunet, A., Bonni, A., Zigmond, M. J., Lin, M. Z., Juo, P., Hu, L. S., et al. (1999). Akt promotes cell survival by phosphorylating and inhibiting a Forkhead transcription factor. *Cell, 96*, 857–868.

Brunet, A., Datta, S. R., & Greenberg, M. E. (2001). Transcription-dependent and -independent control of neuronal survival by the PI3K-Akt signaling pathway. *Current Opinion in Neurobiology, 11*, 297–305.

Brunet, A., Park, J., Tran, H., Hu, L. S., Hemmings, B. A., & Greenberg, M. E. (2001). Protein kinase SGK mediates survival signals by phosphorylating the forkhead transcription factor FKHRL1 (FOXO3a). *Molecular and Cellular Biology, 21*, 952–965.

Burgering, B. M. (2008). A brief introduction to FOXOlogy. *Oncogene, 27*, 2258–2262.

Burgering, B. M., & Kops, G. J. (2002). Cell cycle and death control: Long live Forkheads. *Trends in Biochemical Sciences, 27*, 352–360.

Burkewitz, K., Weir, H. J., & Mair, W. B. (2016). AMPK as a pro-longevity target. *EXS, 107*, 227–256.

Cahill, C. M., Tzivion, G., Nasrin, N., Ogg, S., Dore, J., Ruvkun, G., et al. (2001). Phosphatidylinositol 3-kinase signaling inhibits DAF-16 DNA binding and function via 14-3-3-dependent and 14-3-3-independent pathways. *The Journal of Biological Chemistry, 276*, 13402–13410.

Chalfie, M., Tu, Y., Euskirchen, G., Ward, W. W., & Prasher, D. C. (1994). Green fluorescent protein as a marker for gene expression. *Science, 263*, 802–805.

Chen, A. T., Guo, C., Itani, O. A., Budaitis, B. G., Williams, T. W., Hopkins, C. E., et al. (2015). Longevity genes revealed by integrative analysis of isoform-specific daf-16/FoxO mutants of *Caenorhabditis elegans*. *Genetics, 201*, 613–629.

Chiang, W. C., Ching, T. T., Lee, H. C., Mousigian, C., & Hsu, A. L. (2012). HSF-1 regulators DDL-1/2 link insulin-like signaling to heat-shock responses and modulation of longevity. *Cell, 148*, 322–334.

Christensen, R., de la Torre-Ubieta, L., Bonni, A., & Colon-Ramos, D. A. (2011). A conserved PTEN/FOXO pathway regulates neuronal morphology during *C. elegans* development. *Development, 138*, 5257–5267.

Cornils, A., Gloeck, M., Chen, Z., Zhang, Y., & Alcedo, J. (2011). Specific insulin-like peptides encode sensory information to regulate distinct developmental processes. *Development, 138*, 1183–1193.

Davis, R. J. (2000). Signal transduction by the JNK group of MAP kinases. *Cell, 103*, 239–252.

Dickinson, D. J., Ward, J. D., Reiner, D. J., & Goldstein, B. (2013). Engineering the *Caenorhabditis elegans* genome using Cas9-triggered homologous recombination. *Nature Methods, 10*(10), 1028–1034. https://doi.org/10.1038/nmeth.2641.

Essers, M. A., de Vries-Smits, L. M., Barker, N., Polderman, P. E., Burgering, B. M., & Korswagen, H. C. (2005). Functional interaction between beta-catenin and FOXO in oxidative stress signaling. *Science, 308*, 1181–1184.

Essers, M. A., Weijzen, S., de Vries-Smits, A. M., Saarloos, I., de Ruiter, N. D., Bos, J. L., et al. (2004). FOXO transcription factor activation by oxidative stress mediated by the small GTPase Ral and JNK. *The EMBO Journal*, *23*, 4802–4812.

Fatt, H. V., & Dougherty, E. C. (1963). Genetic control of differential heat tolerance in two strains of the nematode *Caenorhabditis elegans*. *Science*, *141*, 266–267.

Fernandes de Abreu, D. A., Caballero, A., Fardel, P., Stroustrup, N., Chen, Z., Lee, K., et al. (2014). An insulin-to-insulin regulatory network orchestrates phenotypic specificity in development and physiology. *PLoS Genetics*, *10*, e1004225.

Fielenbach, N., & Antebi, A. (2008). *C. elegans* dauer formation and the molecular basis of plasticity. *Genes & Development*, *22*, 2149–2165.

Furuyama, T., Nakazawa, T., Nakano, I., & Mori, N. (2000). Identification of the differential distribution patterns of mRNAs and consensus binding sequences for mouse DAF-16 homologues. *The Biochemical Journal*, *349*, 629–634.

Garigan, D., Hsu, A. L., Fraser, A. G., Kamath, R. S., Ahringer, J., & Kenyon, C. (2002). Genetic analysis of tissue aging in *Caenorhabditis elegans*: A role for heat-shock factor and bacterial proliferation. *Genetics*, *161*, 1101–1112.

Golden, T. R., & Melov, S. (2004). Microarray analysis of gene expression with age in individual nematodes. *Aging Cell*, *3*, 111–124.

Golden, J. W., & Riddle, D. L. (1982). A pheromone influences larval development in the nematode *Caenorhabditis elegans*. *Science*, *218*, 578–580.

Greer, E. L., Dowlatshahi, D., Banko, M. R., Villen, J., Hoang, K., Blanchard, D., et al. (2007). An AMPK-FOXO pathway mediates longevity induced by a novel method of dietary restriction in *C. elegans*. *Current Biology*, *17*, 1646–1656.

Hertweck, M., Gobel, C., & Baumeister, R. (2004). *C. elegans* SGK-1 is the critical component in the Akt/PKB kinase complex to control stress response and life span. *Developmental Cell*, *6*, 577–588.

Hsu, A. L., Murphy, C. T., & Kenyon, C. (2003). Regulation of aging and age-related disease by DAF-16 and heat-shock factor. *Science*, *300*, 1142–1145.

Hu, P. J., & Dauer (2007). In: WormBook (Ed.), *The C. elegans research community*. WormBook (August 08). https://doi.org/10.1895/wormbook.1.144.1, http://www.wormbook.org.

Jones, K. T., Greer, E. R., Pearce, D., & Ashrafi, K. (2009). Rictor/TORC2 regulates *Caenorhabditis elegans* fat storage, body size, and development through sgk-1. *PLoS Biology*, *7*, e60.

Kaeberlein, M., McVey, M., & Guarente, L. (1999). The SIR2/3/4 complex and SIR2 alone promote longevity in saccharomyces cerevisiae by two different mechanisms. *Genes & Development*, *13*, 2570–2580.

Kenyon, C. (2011). The first long-lived mutants: Discovery of the insulin/IGF-1 pathway for ageing. *Philosophical Transactions of the Royal Society of London. Series B, Biological Sciences*, *366*, 9–16.

Kenyon, C., Chang, J., Gensch, E., Rudner, A., & Tabtiang, R. (1993). A *C. elegans* mutant that lives twice as long as wild type. *Nature*, *366*, 461–464.

Kimura, K. D., Tissenbaum, H. A., Liu, Y., & Ruvkun, G. (1997). daf-2, An insulin receptor-like gene that regulates longevity and diapause in *Caenorhabditis elegans*. *Science*, *277*, 942–946.

Kumar, N., Jain, V., Singh, A., Jagtap, U., Verma, S., & Mukhopadhyay, A. (2015). Genome-wide endogenous DAF-16/FOXO recruitment dynamics during lowered insulin signalling in *C. elegans*. *Oncotarget*, *6*, 41418–41433.

Kwon, E. S., Narasimhan, S. D., Yen, K., & Tissenbaum, H. A. (2010). A new DAF-16 isoform regulates longevity. *Nature*, *466*, 498–502.

Lee, R. Y., Hench, J., & Ruvkun, G. (2001). Regulation of *C. elegans* DAF-16 and its human ortholog FKHRL1 by the daf-2 insulin-like signaling pathway. *Current Biology*, *11*, 1950–1957.

Lehtinen, M. K., Yuan, Z., Boag, P. R., Yang, Y., Villen, J., Becker, E. B., et al. (2006). A conserved MST-FOXO signaling pathway mediates oxidative-stress responses and extends life span. *Cell, 125*, 987–1001.

Li, J., Ebata, A., Dong, Y., Rizki, G., Iwata, T., & Lee, S. S. (2008). *Caenorhabditis elegans* HCF-1 functions in longevity maintenance as a DAF-16 regulator. *PLoS Biology, 6*, e233.

Li, W., Kennedy, S. G., & Ruvkun, G. (2003). daf-28 encodes a *C. elegans* insulin superfamily member that is regulated by environmental cues and acts in the DAF-2 signaling pathway. *Genes & Development, 17*, 844–858.

Li, J., Tewari, M., Vidal, M., & Lee, S. S. (2007). The 14-3-3 protein FTT-2 regulates DAF-16 in *Caenorhabditis elegans. Developmental Biology, 301*, 82–91.

Lin, K., Dorman, J. B., Rodan, A., & Kenyon, C. (1997). daf-16: An HNF-3/forkhead family member that can function to double the life-span of *Caenorhabditis elegans. Science, 278*, 1319–1322.

Lin, K., Hsin, H., Libina, N., & Kenyon, C. (2001). Regulation of the *Caenorhabditis elegans* longevity protein DAF-16 by insulin/IGF-1 and germline signaling. *Nature Genetics, 28*, 139–145.

McElwee, J., Bubb, K., & Thomas, J. H. (2003). Transcriptional outputs of the *Caenorhabditis elegans* forkhead protein DAF-16. *Aging Cell, 2*, 111–121.

McElwee, J. J., Schuster, E., Blanc, E., Thomas, J. H., & Gems, D. (2004). Shared transcriptional signature in *Caenorhabditis elegans* Dauer larvae and long-lived daf-2 mutants implicates detoxification system in longevity assurance. *The Journal of Biological Chemistry, 279*, 44533–44543.

Mendler, M., Riedinger, C., Schlotterer, A., Volk, N., Fleming, T., Herzig, S., et al. (2016). Reduction in ins-7 gene expression in non-neuronal cells of high glucose exposed *Caenorhabditis elegans* protects from reactive metabolites, preserves neuronal structure and head motility, and prolongs lifespan. *Journal of Diabetes and its Complications, 31*, 304–310.

Morley, J. F., & Morimoto, R. I. (2004). Regulation of longevity in *Caenorhabditis elegans* by heat shock factor and molecular chaperones. *Molecular Biology of the Cell, 15*, 657–664.

Morris, J. Z., Tissenbaum, H. A., & Ruvkun, G. (1996). A phosphatidylinositol-3-OH kinase family member regulating longevity and diapause in *Caenorhabditis elegans. Nature, 382*, 536–539.

Murphy, C. T., & Hu, P. J. (2013). Insulin/insulin-like growth factor signaling in *C. elegans*. In: WormBook (Ed.), *The C. elegans research community*. WormBook (December 26) https://doi.org/10.1895/wormbook.1.164.1, http://www.wormbook.org.

Murphy, C. T., McCarroll, S. A., Bargmann, C. I., Fraser, A., Kamath, R. S., Ahringer, J., et al. (2003). Genes that act downstream of DAF-16 to influence the lifespan of *Caenorhabditis elegans. Nature, 424*, 277–283.

Narasimhan, S. D., Yen, K., Bansal, A., Kwon, E. S., Padmanabhan, S., & Tissenbaum, H. A. (2011). PDP-1 links the TGF-beta and IIS pathways to regulate longevity, development, and metabolism. *PLoS Genetics*, 7, e1001377.

Narasimhan, S. D., Yen, K., & Tissenbaum, H. A. (2009). Converging pathways in lifespan regulation. *Current Biology, 19*, R657–666.

Nigon, V., & Dougherty, E. C. (1949). Reproductive patterns and attempts at reciprocal crossing of Rhabditis elegans Maupas, 1900, and Rhabditis briggsae Dougherty and Nigon, 1949 (Nematoda: Rhabditidae). *The Journal of Experimental Zoology, 112*, 485–503.

Niu, W., Lu, Z. J., Zhong, M., Sarov, M., Murray, J. I., Brdlik, C. M., et al. (2011). Diverse transcription factor binding features revealed by genome-wide ChIP-seq in *C. elegans. Genome Research, 21*, 245–254.

Nollen, E. A., & Morimoto, R. I. (2002). Chaperoning signaling pathways: Molecular chaperones as stress-sensing "heat shock" proteins. *Journal of Cell Science, 115*, 2809–2816.

Ogg, S., Paradis, S., Gottlieb, S., Patterson, G. I., Lee, L., Tissenbaum, H. A., et al. (1997). The fork head transcription factor DAF-16 transduces insulin-like metabolic and longevity signals in *C. elegans*. *Nature, 389*, 994–999.

Ogg, S., & Ruvkun, G. (1998). The *C. elegans* PTEN homolog, DAF-18, acts in the insulin receptor-like metabolic signaling pathway. *Molecular Cell, 2*, 887–893.

Oh, S. W., Mukhopadhyay, A., Dixit, B. L., Raha, T., Green, M. R., & Tissenbaum, H. A. (2006). Identification of direct DAF-16 targets controlling longevity, metabolism and diapause by chromatin immunoprecipitation. *Nature Genetics, 38*, 251–257.

Oh, S. W., Mukhopadhyay, A., Svrzikapa, N., Jiang, F., Davis, R. J., & Tissenbaum, H. A. (2005). JNK regulates lifespan in *Caenorhabditis elegans* by modulating nuclear translocation of forkhead transcription factor/DAF-16. *Proceedings of the National Academy of Sciences of the United States of America, 102*, 4494–4499.

Padmanabhan, S., Mukhopadhyay, A., Narasimhan, S., Tesz, G., Czech, M. P., & Tissenbaum, H. A. (2009). A PP2A regulatory subunit regulates *C. elegans* insulin/IGF-1 signaling by modulating AKT-1 phosphorylation. *Cell, 136*, 939–951.

Paradis, S., Ailion, M., Toker, A., Thomas, J. H., & Ruvkun, G. (1999). A PDK1 homolog is necessary and sufficient to transduce AGE-1 PI3 kinase signals that regulate diapause in *Caenorhabditis elegans*. *Genes & Development, 13*, 1438–1452.

Paradis, S., & Ruvkun, G. (1998). *Caenorhabditis elegans* Akt/PKB transduces insulin receptor-like signals from AGE-1 PI3 kinase to the DAF-16 transcription factor. *Genes & Development, 12*, 2488–2498.

Riddle, D. L. (1977). A genetic pathway for dauer larva formation in *Caenorhabditis elegans*. *Stadler Genetics Symposium, 9*, 101–120.

Riddle, D. L., & Albert, P. S. (1997). Genetic and environmental regulation of Dauer Larva Development. In: D. L. Riddle, T. Blumenthal, B. J. Meyer, & J. R. Priess (Eds.), *C. elegans II*. Plainview, NY: Cold Spring Harbor Laboratory Press, pp. 739–768.

Riedel, C. G., Dowen, R. H., Lourenco, G. F., Kirienko, N. V., Heimbucher, T., West, J. A., et al. (2013). DAF-16 employs the chromatin remodeller SWI/SNF to promote stress resistance and longevity. *Nature Cell Biology, 15*, 491–501.

Rine, J., & Herskowitz, I. (1987). Four genes responsible for a position effect on expression from HML and HMR in Saccharomyces cerevisiae. *Genetics, 116*, 9–22.

Ritter, A. D., Shen, Y., Fuxman Bass, J., Jeyaraj, S., Deplancke, B., Mukhopadhyay, A., et al. (2013). Complex expression dynamics and robustness in *C. elegans* insulin networks. *Genome Research, 23*, 954–965.

Robida-Stubbs, S., Glover-Cutter, K., Lamming, D. W., Mizunuma, M., Narasimhan, S. D., Neumann-Haefelin, E., et al. (2012). TOR signaling and rapamycin influence longevity by regulating SKN-1/Nrf and DAF-16/FoxO. *Cell Metabolism, 15*(5), 713–724.

Rogina, B., & Helfand, S. L. (2004). Sir2 mediates longevity in the fly through a pathway related to calorie restriction. *Proceedings of the National Academy of Sciences of the United States of America, 101*, 15998–16003.

Schuster, E., McElwee, J. J., Tullet, J. M., Doonan, R., Matthijssens, F., Reece-Hoyes, J. S., et al. (2010). DamID in *C. elegans* reveals longevity-associated targets of DAF-16/FoxO. *Molecular Systems Biology, 6*, 399.

Shaw, W. M., Luo, S., Landis, J., Ashraf, J., & Murphy, C. T. (2007). The *C. elegans* TGF-beta Dauer pathway regulates longevity via insulin signaling. *Current Biology, 17*, 1635–1645.

Soukas, A. A., Kane, E. A., Carr, C. E., Melo, J. A., & Ruvkun, G. (2009). Rictor/TORC2 regulates fat metabolism, feeding, growth, and life span in *Caenorhabditis elegans*. *Genes & Development, 23*, 496–511.

Stiernagle, T. (2006). Maintenance of *C. elegans*. In: WormBook (Ed.), *The C. elegans research community*. WormBook (February 11). https://doi.org/10.1895/wormbook.1.101.1, http://www.wormbook.org.

Tang, E. D., Nunez, G., Barr, F. G., & Guan, K. L. (1999). Negative regulation of the forkhead transcription factor FKHR by Akt. *The Journal of Biological Chemistry, 274*, 16741–16746.

Tepper, R. G., Ashraf, J., Kaletsky, R., Kleemann, G., Murphy, C. T., & Bussemaker, H. J. (2013). PQM-1 complements DAF-16 as a key transcriptional regulator of DAF-2-mediated development and longevity. *Cell, 154*, 676–690.

Tepper, R. G., Murphy, C. T., & Bussemaker, H. J. (2014). DAF-16 and PQM-1: Partners in longevity. *Aging (Albany NY), 6*, 5–6.

Tissenbaum, H. A., & Guarente, L. (2001). Increased dosage of a sir-2 gene extends lifespan in *Caenorhabditis elegans*. *Nature, 410*, 227–230.

Tran, H., Brunet, A., Griffith, E. C., & Greenberg, M. E. (2003). The many forks in FOXO's road. *Science's STKE, 2003*, RE5.

Troemel, E. R., Chu, S. W., Reinke, V., Lee, S. S., Ausubel, F. M., & Kim, D. H. (2006). p38 MAPK regulates expression of immune response genes and contributes to longevity in *C. elegans*. *PLoS Genetics, 2*, e183.

Tullet, J. M., Araiz, C., Sanders, M. J., Au, C., Benedetto, A., Papatheodorou, I., et al. (2014). DAF-16/FoxO directly regulates an atypical AMP-activated protein kinase gamma isoform to mediate the effects of insulin/IGF-1 signaling on aging in *Caenorhabditis elegans*. *PLoS Genetics, 10*(2), e1004109.

Van Der Heide, L. P., Hoekman, M. F., & Smidt, M. P. (2004). The ins and outs of FoxO shuttling: Mechanisms of FoxO translocation and transcriptional regulation. *The Biochemical Journal, 380*, 297–309.

Vogt, P. K., Jiang, H., & Aoki, M. (2005). Triple layer control: Phosphorylation, acetylation and ubiquitination of FOXO proteins. *Cell Cycle, 4*, 908–913.

Volovik, Y., Moll, L., Marques, F. C., Maman, M., Bejerano-Sagie, M., & Cohen, E. (2014). Differential regulation of the heat shock factor 1 and DAF-16 by neuronal nhl-1 in the nematode *C. elegans*. *Cell Reports, 9*, 2192–2205.

Wang, M. C., Bohmann, D., & Jasper, H. (2005). JNK extends life span and limits growth by antagonizing cellular and organism-wide responses to insulin signaling. *Cell, 121*, 115–125.

Wang, Y., Oh, S. W., Deplancke, B., Luo, J., Walhout, A. J., & Tissenbaum, H. A. (2006). *C. elegans* 14-3-3 proteins regulate life span and interact with SIR-2.1 and DAF-16/FOXO. *Mechanisms of Ageing and Development, 127*, 741–747.

Weigel, D., Jurgens, G., Kuttner, F., Seifert, E., & Jackle, H. (1989). The homeotic gene fork head encodes a nuclear protein and is expressed in the terminal regions of the *Drosophila embryo*. *Cell, 57*(4), 645–658.

Wolff, S., Ma, H., Burch, D., Maciel, G. A., Hunter, T., & Dillin, A. (2006). SMK-1, an essential regulator of DAF-16-mediated longevity. *Cell, 124*, 1039–1053.

Wolkow, C. A., Munoz, M. J., Riddle, D. L., & Ruvkun, G. (2002). Insulin receptor substrate and p55 orthologous adaptor proteins function in the *Caenorhabditis elegans* daf-2/insulin-like signaling pathway. *The Journal of Biological Chemistry, 277*, 49591–49597.

Yen, K., Le, T. T., Bansal, A., Narasimhan, S. D., Cheng, J. X., & Tissenbaum, H. A. (2010). A comparative study of fat storage quantitation in nematode *Caenorhabditis elegans* using label and label-free methods. *PLoS One, 5*(9), e12810. https://doi.org/10.1371/journal.pone.0012810.

Yen, K., Narasimhan, S. D., & Tissenbaum, H. A. (2011). DAF-16/Forkhead box O transcription factor: Many paths to a single fork(head) in the road. *Antioxidants & Redox Signaling, 14*, 623–634.

Zou, C. G., Tu, Q., Niu, J., Ji, X. L., & Zhang, K. Q. (2013). The DAF-16/FOXO transcription factor functions as a regulator of epidermal innate immunity. *PLoS Pathogens, 9*, e1003660.

CHAPTER TWO

Stem Cells Seen Through the FOXO Lens: An Evolving Paradigm

Raymond Liang[*,†], **Saghi Ghaffari**[*,‡,§,¶,1]
[*]Icahn School of Medicine at Mount Sinai, New York, NY, United States
[†]Developmental and Stem Cell Biology, Multidisciplinary Training Area, Icahn School of Medicine at Mount Sinai, New York, NY, United States
[‡]Black Family Stem Cell Institute, Icahn School of Medicine at Mount Sinai, New York, NY, United States
[§]Tisch Cancer Institute, Icahn School of Medicine at Mount Sinai, New York, NY, United States
[¶]Department of Cell, Developmental and Regenerative Biology, Icahn School of Medicine at Mount Sinai, New York, NY, United States
[1]Corresponding author: e-mail address: saghi.ghaffari@mssm.edu

Contents

Abstract

Stem cells self-renew and differentiate to generate all tissues and cells in the body. Stem cell health promotes adaptive responses to tissue damage or loss and is essential for tissue regeneration with age. In the past decade, the evolutionarily conserved transcription factors FOXO with known functions in promoting healthy aging have emerged as key regulators of stem cell homeostasis in various tissues, including blood, neural, and muscle stem cells. Aberrant FOXO functions have been implicated in a variety of disorders including neurodegenerative, blood, cancer, and diabetes some of which are fostered by abnormal stem cell function. As discussed in this chapter, at least in some stem

Current Topics in Developmental Biology, Volume 127
ISSN 0070-2153
https://doi.org/10.1016/bs.ctdb.2017.11.006

cells FOXO regulatory mechanisms and applied functions follow a complex set of rules distinct from that operating in progenitor cell populations and in cultured cell lines. Elucidating the exact nature of FOXO properties in stem cells will be critical for identifying and targeting aberrant FOXO-mediated mechanisms that promote stem cell-derived disease specifically with age.

1. FOXO AND STEM CELLS

The FOXO (FOXO1, FOXO3, FOXO4, and FOXO6 in mammals) (Anderson, Viars, Czekay, Cavenee, & Arden, 1998) subfamily of Forkhead transcription factors is comprised of mammalian orthologs of Dauer Abnormal Formation-16 (DAF-16 in *Caenorhabditis elegans*) and dFoxo proteins (in *Drosophila*), which are targets of insulin and insulin-like growth factor-1 (IGF-1) receptor signaling. These factors are essential for the regulation of stress response and longevity (Giannakou et al., 2004; Hwangbo, Gershman, Tu, Palmer, & Tatar, 2004; Lin, Dorman, Rodan, & Kenyon, 1997; Ogg et al., 1997). While FOXO1, 3, and 4 are widely expressed, FOXO6 is specific to neural tissues and developing brains (Jacobs et al., 2003).

FOXOs are among the very few transcription factors that are essential for stem cell function in both embryos and adults (see Table 1). In adults,

Table 1 Stem Cell Regulation by Transcription Factors FOXO

FOXO1 and FOXO3 in Stem Cells	
Stem Cell Type	**References**
Embryonic stem cells	Zhang et al. (2011) and Ma et al. (2015)
Germ cells	Castrillon, Miao, Kollipara, Horner, and DePinho (2003), Hosaka et al. (2004), Goertz, Wu, Gallardo, Hamra, and Castrillon (2011), and Pelosi et al. (2013)
Neural stem cells	Renault et al. (2009), Paik et al. (2009), and Yeo et al. (2013)
Hematopoietic stem cells	Miyamoto et al. (2007, 2008), Yalcin et al. (2010, 2008), Lee et al. (2010), Li, Du, Maynard, Andreassen, and Pang (2010), Warr et al. (2013), Rimmelé et al. (2015), Mehta et al. (2015), Li et al. (2015), Liang, Rimmele, Bigarella, Yalcin, and Ghaffari (2016), and Bigarella et al. (2017)
Muscle satellite cells	Machida, Spangenburg, and Booth (2003) and Gopinath, Webb, Brunet, and Rando (2014)
Cancer stem cells	Naka et al. (2010), Sykes et al. (2011), Pellicano et al. (2014), and Ahn et al. (2016)

FOXOs are required for stem cell function in multiple tissues. These include FOXO maintenance of hematopoietic, neural, and muscle stem cell pools. Such dependency of stem cells on functional FOXO raises the possibility that FOXO regulations and functions are intertwined with stem cell properties.

2. FOXO IN EMBRYONIC STEM CELLS AND DEVELOPMENT

FOXO1 is required for both homeostasis of spermatogonial stem cells and the initiation of spermatogenesis (Goertz et al., 2011), whereas FOXO3 is critical for the maintenance of oocyte pools (Castrillon et al., 2003; Hosaka et al., 2004; Pelosi et al., 2013). Loss of FOXO3 results in premature depletion of oocytes in mice, while gain of FOXO3 function maintains the ovarian reservoir and enhances ovarian reproductive capacity (Castrillon et al., 2003; Pelosi et al., 2013). In embryonic stem cells (ESCs), FOXO regulates core pluripotency factors (Gonzales & Ng, 2011; Zhang et al., 2011). Specifically in human ESC, FOXO1 is critical for the maintenance of pluripotency state by promoting the expression of pluripotency genes OCT4 and SOX2 but not NANOG. Loss of FOXO1 in human ESCs (hESC) results in hESC differentiation into mesoderm and endoderm but not ectoderm. These findings are in agreement with OCT4 binding to FOXO1 promoter in hESCs (Boyer et al., 2005) and consistent with FOXO1 identified as a target of differentiation-inducing miRNAS in a functional screen of ESC (Ma et al., 2015). In mouse ESC, both FOXO1 and 3 regulate core pluripotency factors and the pluripotency state. FOXO1 regulation of pluripotency in hESC was surprising given that FOXO1-deficient mice do not exhibit an early (pregastrulation) developmental defect (Hosaka et al., 2004). This phenotype is, however, similar to mice generated from deletion of SOX2, TCL1, or TBX3 pluripotency genes (Avilion et al., 2003; Davenport, Jerome-Majewska, & Papaioannou, 2003; Narducci et al., 2002). Lack of early lethality in FOXO1-deficient mice might be due to specie-specific differences or, as is the case for SOX2-deficient mice (Avilion et al., 2003), to the potential presence of long-lived maternal FOXO1 protein at the early stages of FOXO1-deficient embryonic development.

3. FOXO IN ADULT TISSUE-SPECIFIC STEM CELLS

FOXO is implicated in the maintenance of long-lived adult stem cells in various tissues. The nature of FOXO in tissue-specific stem cells and similarities and differences in their regulation and function will be important for our understanding of the contribution of aberrant FOXO function in stem cells to disease.

4. FOXO IN HEMATOPOIETIC STEM CELLS

Adult hematopoietic stem cells (HSCs) are very few in number and constitute a heterogeneous population (Lessard, Faubert, & Sauvageau, 2004). HSCs are identified phenotypically based on the expression of a specific combination of cell surface markers and/or other (e.g., metabolic) properties. Bone marrow transplantation under competitive repopulation constitutes the most stringent assay for identifying HSC with the ability to repopulate all blood cell lineages for a long (4–6 months) period of time (long-term HSC; LT-HSC) in a lethally irradiated mouse in which blood formation is ablated. Using such an approach, HSCs were shown to require FOXO for their maintenance and activity (Miyamoto et al., 2007; Tothova et al., 2007; Yalcin et al., 2008). Conditional ablation of FOXO1/3/4 (Forkhead Box O 1/3/4) transcription factors led to compromised long-term competitive in vivo repopulation capacity of HSC (LT-HSC) in mice (Tothova et al., 2007). Defects of FOXO3-deficient HSC (Miyamoto et al., 2007; Yalcin et al., 2008) recapitulate many if not all defects of FOXO1/3/4-deficient HSC albeit with less severity, suggesting that FOXO3 is the main active FOXO in HSC (Liang et al., 2016; Miyamoto et al., 2007, 2008; Yalcin et al., 2010, 2008).

Although FOXO1, 3, and 4 are all expressed in HSC, their level of expression varies significantly. FOXO1 and FOXO3 transcripts are expressed at relatively similar levels, whereas FOXO4 expression is significantly lower in LT-HSC. In addition, while almost 50% of FOXO1 protein resides in the cytosol, FOXO3 protein is almost entirely nuclear (more than 85%–90%) in LT-HSCs (Liang et al., 2016; Yalcin et al., 2008). These studies suggest that in HSC the regulation of FOXO3 (mostly nuclear) must be distinct from that of FOXO1 (equally distributed in nucleus and cytosol) (see Section 4.6). In agreement with these findings, loss of FOXO3 alone significantly compromises bone marrow HSC frequency, numbers as well as competitive repopulation ability. Furthermore, loss of FOXO3 alone leads to myeloproliferation (Yalcin et al., 2010), anemia (Castrillon et al., 2003; Marinkovic et al., 2007), and immune deficiencies (Dejean et al., 2009; Hinman et al., 2009) (and reviewed in Hedrick, Hess Michelini,

Doedens, Goldrath, & Stone, 2012), similar to the phenotype observed in FOXO1-, 3-, and 4-deleted HSC.

Mechanisms by which FOXO3 regulates HSC quiescence are not fully elucidated. FOXO3 regulates some of the key genes involved in stem cell quiescence such as cyclin-dependent kinase (CDK) inhibitor 1B *Cdkn1b* ($p27^{KIP1}$), *Ccng2* (*cyclin* G2), and p57 (Dijkers et al., 2000; Martinez-Gac, Marques, Garcia, Campanero, & Carrera, 2004), expression of which was profoundly repressed in FOXO3-deficient HSC (Miyamoto et al., 2007; Yalcin et al., 2008). In contrast, expression of the CDK inhibitor *Cdkn1a* ($p21^{CIP1/WAF1/Sdi1}$), which is critical for G1 and G2 checkpoints and known to repress HSC cycling (Cheng et al., 2000), was upregulated in FOXO3-deficient HSC (Yalcin et al., 2008). Upregulation of $p21^{CIP1/WAF1/Sdi1}$ appeared to occur in response to oxidative stress and DNA damage (Bunz et al., 1998; Waldman, Lengauer, Kinzler, & Vogelstein, 1996) as part of a p53-dependent antioxidant response (Yalcin et al., 2008). FOXO3 engages in cross talk with the tumor suppressor p53 (Renault et al., 2011; You & Mak, 2005; You, Yamamoto, & Mak, 2006), which is also required for the maintenance of HSC quiescence (Liu et al., 2009). Loss of FOXO3 activates p53 in HSC (Yalcin et al., 2008); however, the issue of whether the FOXO3/p53 cross talk is implicated in the maintenance of HSC quiescence remains to be determined.

4.1 FOXO3, Reactive Oxygen Species, and Mitochondria in HSC

Reactive oxygen species (ROS), or oxygen radicals, are produced mainly in mitochondria as a result of aberrant, premature oxygen reduction in the electron transfer chain during oxidative phosphorylation. Additional ROS are produced in other cellular compartments, including at the cell membrane.

ROS when deregulated can reach aberrantly high levels that oxidize and damage lipids, membranes, and DNA. Under normal physiological conditions, ROS levels are highly fine-tuned and can act as part of signal transduction pathways (Liang & Ghaffari, 2014). This is especially true in adult stem cells which are localized in hypoxic niches, defined as hypoxic at low O2 levels. Adult stem cells and specifically HSCs are uniquely sensitive to ROS (Jang & Sharkis, 2007; Simon & Keith, 2008) and hypoxic conditions protect against spontaneous generation of ROS. Instead, hypoxia induces a regulated burst of ROS from mitochondria that acts to activate the HIF-mediated hypoxic response (Chandel et al., 1998).

Elevated ROS levels result in stem cell damage, which in turn leads to stem cell exhaustion (Bigarella, Liang, & Ghaffari, 2014). In *Drosophila*, high ROS levels are exhibited by the multipotent hematopoietic progenitors that are analogs of mammalian myeloid progenitors; modulation of ROS directs the differentiation of multipotential hematopoietic progenitor cells, which supports a signaling role for ROS in regulating hematopoietic cell fate (Owusu-Ansah & Banerjee, 2009). Low levels of ROS are essential for the maintenance of adult stem cell quiescence. This may be why stem cells, including HSC, are protected in niches that maintain low ROS levels (Jang & Sharkis, 2007). Increase in ROS levels in HSCs is associated with cell cycle progression, HSC differentiation, and increased production of hematopoietic progenitors (Jang & Sharkis, 2007; Mortensen et al., 2011; Yalcin et al., 2010). Elevated ROS compromise self-renewal potential and engraftment of HSC partly via the activation of p38-MAPK pathway (Ito et al., 2004, 2006; Yahata et al., 2011). Increased ROS levels are also associated with loss of HSC quiescence and activity as measured by competitive repopulation capacity. This has been shown in a number of mutant mice including Bid, DJ1, Meis1 (Kocabas et al., 2012; Maryanovich et al., 2012; Tai-Nagara, Matsuoka, Ariga, & Suda, 2014). The HSC defects in most cases, including in vivo long-term competitive repopulation ability, are entirely reversible through ROS scavenging, which indicates that ROS elevation is the main mediator of HSC defects and specifically long-term competitive repopulation ability.

FOXOs have evolutionarily conserved functions in regulating ROS levels (Lin et al., 1997; Murphy et al., 2003) (see also chapter "DAF-16: FOXO in the context of *C. elegans*" by Tissenbaum). In stem cells, FOXOs are critical for maintaining ROS at low levels (Miyamoto et al., 2007; Paik et al., 2009; Renault et al., 2009; Tothova et al., 2007; Yalcin et al., 2008). FOXO regulates ROS levels by controlling the expression of several antioxidant enzymes—including superoxide dismutase 2, catalase, glutathione peroxidase, and NADPH oxidase and many others—some of which occur through direct transcriptional regulation (Eijkelenboom & Burgering, 2013; Greer & Brunet, 2008). Experiments in *C. elegans* suggest that FOXO maintenance of low ROS levels might contribute to increased longevity (Lee, Kennedy, Tolonen, & Ruvkun, 2003; Murphy et al., 2003). Under harsh living conditions and stress, including oxidative stress, DAF-16—the *C. elegans* ortholog of FOXO—is activated. Activation of DAF-16 promotes the formation of dauer that is in a low metabolic state. The extended dauer state leads to the extension of the worm's lifespan (Lee et al., 2003;

Murphy et al., 2003). The DAF-16 promoted dauer state resembles the quiescent state of adult stem cells and raises the possibility that FOXO maintenance of low ROS levels in stem cells might be involved in FOXO regulation of stem cell longevity (see chapter "FOXO3 and exceptional longevity: Insights from hydra to humans" by Davy et al.).

ROS are also significantly elevated in both triple FOXO1/3/4-deficient and FOXO3-deficient HSC (Miyamoto et al., 2007; Tothova et al., 2007; Yalcin et al., 2008). Elevated ROS mediate enhanced generation of the myeloid progenitors identified as colony-forming unit-spleen (CFU-S) in FOXO1/3/4- as well as FOXO3-deficient mice in vivo (Tothova et al., 2007; Yalcin et al., 2010) and HSC-derived multipotent progenitors known as long-term culture-initiating cells (LTC-IC) in vitro (Miyamoto et al., 2008; Yalcin et al., 2008). In addition, elevated ROS induce oxidative DNA damage (Bigarella et al., 2017) and activate p53/p21 (Yalcin et al., 2008) as well as p38-MAPK pathways (Miyamoto et al., 2007) in FOXO3-deficient HSPC.

FOXO3 also controls ataxia telangiectasia-mutated (ATM) kinase enzymatic activity and expression (Tsai, Chung, Takahashi, Xu, & Hu, 2008; Yalcin et al., 2008). ATM expression is significantly reduced in FOXO3-deficient HSCs in a ROS-independent fashion in contrast to ATM regulation in FOXO1/3/4-deficient HSC. Nonetheless, the extent to which ATM might contribute to the FOXO3 mutant HSC phenotype remains unknown. In agreement with these findings, $ATM^{-/-}$ HSC activity is greatly defective (Ito et al., 2004). The $ATM^{-/-}$ HSC defect is attributed to the ROS-mediated activation of $p16^{Ink4a}$ and retinoblastoma pathway (Ito et al., 2004). It is noteworthy that despite high ROS (Yalcin et al., 2008 #9267), and in contrast to triple FOXO1/3/4-deficient HSC (Tothova et al., 2007), FOXO3-deficient HSCs do not exhibit increased apoptosis (Bigarella et al., 2017; Yalcin et al., 2008) (Fig. 1).

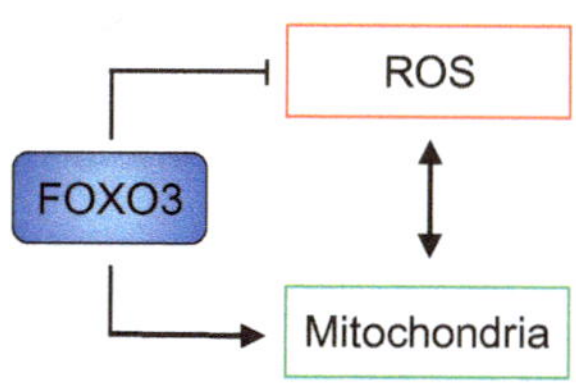

Fig. 1 Independent and interdependent FOXO3 regulation of ROS and mitochondria. FOXO3 maintains low ROS levels through the activation of antioxidant genes. In parallel FOXO3 also regulates mitochondrial function, which in turn influences ROS levels. ROS itself can signal back to mitochondria or cause damage at aberrant levels leading to defective mitochondria.

Thus, it was surprising—given the pivotal ROS regulation of HSPC activity (Bigarella et al., 2014) and multiple ROS-inflicted damages to FOXO3-deficient HSC—that elevated ROS did not mediate the ultimate defects of FOXO3-deficient long-term in vivo competitive repopulation activity (Rimmelé et al., 2015). Furthermore, scavenging ROS in FOXO3-deficient HSC—in contrast to many other mutant HSC with elevated ROS (see earlier; Liang & Ghaffari, 2014)—did not restore their long-term in vivo competitive repopulating activity (Rimmelé et al., 2015). These unexpected findings were associated with defective mitochondrial respiration, morphology, and overall metabolism in FOXO3-deficient HSC. Importantly, FOXO3-deficient HSC mitochondrial defects were not due to the abnormal mTOR activation that was observed in FOXO3-deficient myeloid progenitors (see later; Yalcin et al., 2010), altogether suggesting that the defective long-term in vivo competitive repopulation activity of FOXO3-deficient HSC might be mediated by impaired mitochondrial metabolism rather than elevated ROS. Thus, in FOXO3-deficient HSC, augmented unbalanced ROS levels may be a reflection of mitochondrial abnormalities combined with the reduced expression of antioxidant enzymes (Bigarella et al., 2017; Rimmelé et al., 2015).

Quiescence is a major property of most adult stem cells. Quiescence is also the property that enable cancer stem cells to resist therapy (see also chapter "FOXOs maintaining the equilibrium for better or for worse" by van Doeselaar and Burgering). Quiescence of FOXO3-deficient HSC is lost in a significant subset of mutant cells (Miyamoto et al., 2007; Yalcin et al., 2008). Quiescent HSCs rely upon glycolysis as their main source of energy. Loss of HSC quiescence is associated with a shift to oxidative phosphorylation that becomes the main source of energy in cycling HSCs. One of the unanticipated features of FOXO3-deficient HSC was their increased (rather than decreased) glycolytic flux despite loss of HSC quiescence (Rimmelé et al., 2015). These combined results suggest that loss of FOXO3 function creates a paradoxical HSC state in which mitochondrial respiration and overall metabolism are impaired, raising the question of whether such metabolically compromised HSC might harbor the machinery required for initiating malignancy.

4.2 FOXO and Autophagy in HSC

Autophagy mediates the consumption of damaged or old cellular proteins and components. Autophagy also serves as a source of energy for maintaining HSC homeostasis during cellular stress and starvation (Ho et al., 2017;

Mortensen et al., 2011; Warr et al., 2013; Yang & Klionsky, 2010). FOXOs (FOXO3) are essential for the process of autophagy as they regulate the transcription of many autophagy-related genes (Chiacchiera & Simone, 2009; Liang et al., 2015; Mammucari et al., 2007; Sandri et al., 2004; van der Vos et al., 2012; Warr et al., 2013; Zhao et al., 2010). In agreement with these findings, FOXO3 was identified as the main transcription factor regulating autophagy-related gene expression and process in HSC (Warr et al., 2013). In hematopoietic cells FOXO3 is also a critical regulator of mitophagy, the specific clearance of mitochondria by autophagy (Liang et al., 2015)—a process required for HSC self-renewal (Ito et al., 2016). Accumulation of defective mitochondria as a result of impaired mitophagy (Liang et al., 2015; Liang & Ghaffari, unpublished findings, October 2015) in FOXO3-deficient HSC may compromise HSC long-term stem cell activity in vivo, enhance ROS levels, and result in apoptosis resistance in FOXO3-deficient HSC (Bigarella et al., 2017). Although autophagy is required for the maintenance of HSC activity with age (Warr et al., 2013), the contribution of FOXO3 to autophagy in old HSC remains to be established.

4.3 FOXO and DNA Damage Response in HSC

FOXOs are regulators of DNA damage response (Eijkelenboom, Mokry, Smits, Nieuwenhuis, & Burgering, 2013; Greer & Brunet, 2008). FOXOs induce the expression of DNA damage-inducible gene 45a (Gadd45a). In addition, evidence suggests that XRCC6 (Ku70) protein—a component of DNA double-strand break repair machinery—interacts with FOXO4 in the nucleus (Li et al., 2010, 2015). The Ku70–FOXO4 interaction is dependent on specific levels of ROS. Low levels of ROS promote Ku70–FOXO4 interaction, which in turn inhibits the FOXO4-mediated induction of p27 (kip1) transcription and cell cycle arrest (Li et al., 2010, 2015).

In HSC, the DNA damage response protein Fanconi anemia complementation group D2 (FANCD2) interacts with FOXO3 in a ROS-mediated manner to maintain FOXO3 in the nucleus (Li et al., 2010, 2015). Furthermore, it has recently been shown that FOXO3 protects hematopoietic stem and progenitor cells from oxidative DNA damage under homeostasis and controls specifically base excision repair (BER) that is the main response to oxidative DNA damage (Bigarella et al., 2017). At the steady state, FOXO3-deficient hematopoietic stem and progenitor cells accumulate oxidative DNA damage associated with a G2/M cell cycle delay (Yalcin et al., 2008). In HSPCs, loss of FOXO3 results in the alteration of an oxidative DNA repair machinery gene network related to BER and nucleotide excision repair (Bigarella et al., 2017). ROS elevation in FOXO3-deficient HSC is partially responsible for loss of quiescence and accumulation of damaged DNA in HSCs (Bigarella et al., 2017).

4.4 FOXO, mTOR, and Myeloproliferation

Loss of FOXOs or just FOXO3 leads to leukocytosis, splenomegaly, and mild anemia, associated with increased production of myeloid progenitors (Yalcin et al., 2010). These combined symptoms are reminiscent of human myeloproliferative neoplasms (MPNs) (also known as "myeloproliferative syndromes"). In FOXO3-deficient mice, myeloproliferation is due to the ROS-mediated hyperactivation of the AKT–mTOR signaling pathway. mTOR activation was maintained through a feedback loop mechanism in FOXO3-deficient myeloid progenitors that is not observed in HSC (Rimmelé et al., 2015; Yalcin et al., 2010). Elevated ROS result in reduced LNK (SH2B3) adaptor protein expression in FOXO3-deficient hematopoietic progenitors. LNK negatively regulates cytokine receptor signaling (Takaki, Morita, Tezuka, & Takatsu, 2002; Velazquez et al., 2002) and is a negative regulator of HSC (Bersenev, Wu, Balcerek, & Tong, 2008). Loss of LNK expression results in myeloproliferation in mice (Takaki et al., 2002; Velazquez et al., 2002). Consistent with these results, reduced LNK expression enhanced cytokine receptor signaling specifically AKT–mTOR signaling leading to increased cycling and expansion of FOXO3 mutant hematopoietic progenitor cells (Yalcin et al., 2010). Overexpression of LNK in FOXO3 mutant hematopoietic progenitor cells corrected the myeloproliferation phenotype (Yalcin et al., 2010). In Drosophila, dFOXO represses LNK expression directly and negatively regulates lifespan via control of oxidative stress (Slack et al., 2010). Given that LNK functions as a positive regulator of insulin receptor signalling and PI3-kinase in *Drosophila*, in contrast to its roles in mammals, these findings support an evolutionary conserved function for FOXO regulation of LNK in hematopoietic stem and progenitor cells (HSPC). Overall, $LNK^{-/-}$ HSC become increasingly quiescent over time and exhibit enhanced self-renewal (Bersenev et al., 2008). Loss of LNK may help protect HSC from aging (Bersenev et al., 2012). It is intriguing that this FOXO–mTOR–LNK evolutionarily conserved network of proteins connected by ROS and implicated in aging—and HSC aging—is also implicated in blood malignancies derived from HSC (Ahn et al., 2016; Oh et al., 2010).

4.5 FOXO and Leukemic Stem Cells

FOXOs were originally identified in chromosomal translocation breakpoints in human tumors. Specifically, the mixed lineage leukemia (MLL)-FOXO3 and MLL-FOXO4 fusion genes in acute myeloid leukemias which suggested that FOXOs may be involved in malignancy (Anderson et al., 1998; Borkhardt et al., 1997; Davis, D'Cruz, Lovell, Biegel, & Barr, 1994; Galili et al., 1993; Hillion, Le Coniat, Jonveaux, Berger, & Bernard, 1997). In addition, FOXO exert many tumor-suppressive functions, including inhibition of cell cycle progression and activation of apoptosis (Eijkelenboom et al., 2013; Greer & Brunet, 2008). However, evidence suggests that FOXO induction of apoptosis is cell context dependent (Brunet et al., 2004). Initial studies in cancer cell lines showed that FOXO activity is suppressed in transformed cells (Ghaffari, Jagani, Kitidis, Lodish, & Khosravi-Far, 2003; Hu et al., 2004; Kikuchi et al., 2007; Modur, Nagarajan, Evers, & Milbrandt, 2002; Sakoe, Sakoe, Kirito, Ozawa, & Komatsu, 2010). In addition, several studies showed that activation of FOXOs results in apoptotic death of malignant cells (Ghaffari et al., 2003; Gu et al., 2004; Hu et al., 2004; Kikuchi et al., 2007; Modur et al., 2002; Sakoe et al., 2010; Scheijen, Ngo, Kang, & Griffin, 2004).

FOXOs also interact with known tumor suppressors. For instance, there is a significant overlap between both the regulation and downstream targets of FOXO and the tumor suppressor p53. Furthermore, as mentioned earlier FOXO3 and p53 are in cross talk (Renault et al., 2011; You & Mak, 2005; You et al., 2006). FOXO3 also regulates the expression and activity of tumor-suppressant protein ATM (Tsai et al., 2008; Yalcin et al., 2008), further substantiating the tumor-suppressant claim of FOXO (Burgering & Medema, 2003; You et al., 2004; You & Mak, 2005). In line with FOXO tumor suppression ability, loss of FOXO1, 3, 4 or FOXO3 alone results in a myeloproliferative-like syndrome, a condition that in humans may precede leukemia (Yalcin et al., 2010). However, single deletion of other FOXOs did not result in spontaneous tumors (Castrillon et al., 2003; Hosaka et al., 2004). In contrast, the triple deletion of FOXO (FOXO1, 3, and 4) also resulted in hemangiomas and thymic lymphomas in mice (Paik et al., 2007). Nonetheless, the limited scope of these tumors, despite the widespread expression and functions of FOXO proteins, is surprising and further supports the notion that the tumor-suppressing function of FOXO may be context dependent (see also chapter "FOXOs maintaining the equilibrium for better or for worse" by van Doeselaar and Burgering).

Leukemic stem cells (LSCs) are a subset of leukemia cells that are highly resistant to therapy and able to reconstitute the malignancy up to years after therapy. LSCs are highly sensitive to increased ROS levels (Diehn et al., 2009), a property that is exploited specifically to target LSC (Diehn et al., 2009; Lagadinou et al., 2013). Several reports suggest that FOXOs—FOXO3 and 1—are essential for the maintenance of mouse and human LSCs, respectively (see also chapter "FOXOs maintaining the equilibrium for better or for worse" by van Doeselaar and Burgering) (Naka et al., 2010; Pellicano et al., 2014). This notion is based at least partly on data showing that loss of FOXO prevents leukemia development in mice. Moreover, loss of FOXO increases the sensitivity to tyrosine kinase inhibitors in chronic myeloid leukemias (Pellicano et al., 2014). FOXO3 has also been identified as a main regulator of LSC in a MPN mouse model as well as patient samples (Ahn et al., 2016). These findings are intriguing as loss of FOXO (or FOXO3), as mentioned earlier leads to a myeloproliferative-like syndrome which interestingly does not lead to full-blast leukemia (Tothova et al., 2007; Yalcin et al., 2010, 2008). In addition, the FOXO requirement for LSC contradicts the tumor-suppressive brand of FOXO. Lack of leukemia development in the absence of FOXO3 may be due to impaired mitochondrial metabolism in HSPC (Rimmelé et al., 2015). Greater clarity can be brought to this debate by further investigations into properties shared by normal and leukemic stem cells such as mechanisms by which FOXO maintains stem cell quiescence, the heterogeneity of the stem cell population, and the varying dosage of FOXO activity in stem and progenitor cells subsets involved in normal homeostasis vs leukemic development/progression.

4.6 FOXO, HSC Longevity, and the Role of Sirtuins and miRNAs

FOXO is implicated in promoting healthy longevity from hydra (Boehm et al., 2012) to humans (Willcox et al., 2008) (see chapter "FOXO3 and exceptional longevity: Insights from hydra to humans" by Davy et al.). Healthy longevity may depend at least in part on the maintenance of long-lived tissue-specific stem cells with age. Healthy stem cells preserve their ability to respond to acute damage with age, regenerate tissues, and overall maintain tissue homeostasis that would otherwise be compromised with age (Signer & Morrison, 2013) (Fig. 2). Although the role of FOXO in preserving stem cell longevity has yet to be determined, evidence discussed later suggests that FOXO may protect stem cells from age-associated damage.

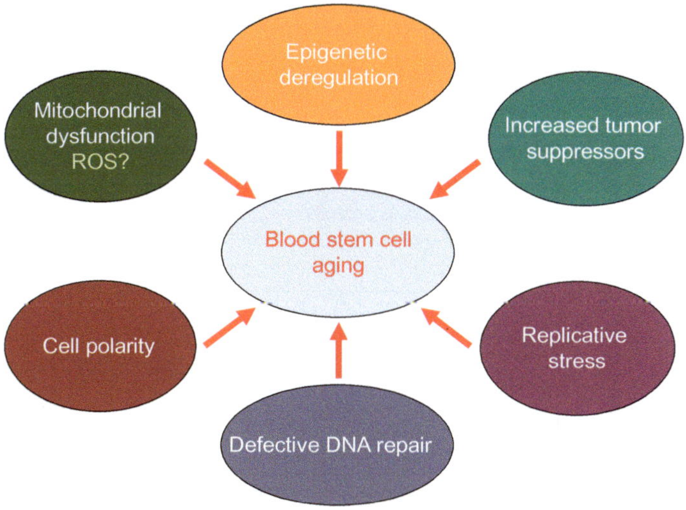

Fig. 2 Age-associated defective processes implicated in HSC aging. The decline of various cellular processes with age (*outer circles*) known to contribute to HSC aging results in the overall decline of blood stem cell function (*center circle*) with age.

The sirtuin family of NAD-dependent deacetylases (including seven members in mammals) is implicated in regulating healthy aging, oxidative stress, mitochondrial biogenesis, and metabolism (Haigis & Sinclair, 2010). Recent evidence shows that SIRT1 is key for maintaining the HSC pool in young adult mice both at the steady state and under stress (Ou et al., 2011; Rimmelé et al., 2014; Singh et al., 2013). Loss of SIRT1 results in an aging-like HSC phenotype associated with defective HSC lineage specification—myeloid-biased production of hematopoietic progenitors—and many other hallmarks of stem cell aging, such as accumulation of ROS and DNA damage in HSC in young adult mice (Rimmelé et al., 2014). The striking similarity of Sirt1$^{-/-}$ HSC to the phenotype observed in old HSC has suggested that SIRT1 may be required for protecting HSC from age-associated damage (Fig. 3) (Beerman & Rossi, 2015; Bigarella et al., 2014).

FOXOs are deacetylated by sirtuins but also regulated by a number of additional posttranslational modifications—including phosphorylation, acetylation, ubiquitination, redox modulation, and methylation—that together delineate FOXO functional output (see chapter "Regulation of FOXO factors in mammalian cells" by Brown and Webb). Phosphorylation by AKT serine threonine protein kinase inhibits FOXO activity by promoting their cytosolic localization and constitutes the main regulation of FOXO in response to growth factors and cytokine signaling. In HSC,

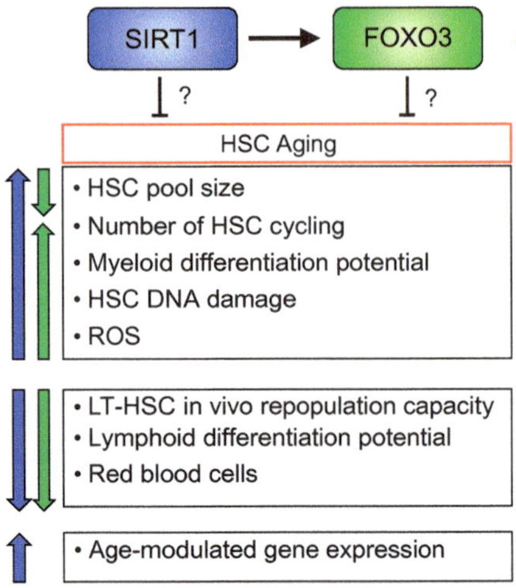

Fig. 3 Model for FOXO3-regulated network in protecting HSC (stem cells) from age-related damage. SIRT1 maintains FOXO3 in an active form in HSCs. The *upper box* lists aberrant activities of HSCs that increase with aging, while the *lower box* lists normal activities of HSCs that decrease with aging. Defects observed with loss of SIRT1 (*blue arrows*) or loss of FOXO3 (*green arrows*) in young HSC are compared to defects of old HSC. Note the degree of overlap of SIRT1-deficient HSC and FOXO3-deficient HSC to old HSC (difference observed might be due to the conditional deletion of SIRT1 but not FOXO3 in the adult mice).

FOXO3 is the principal active FOXO and exists mainly in an active nuclear form. Interestingly, the constitutive activation of AKT does not promote cytosolic localization of FOXO3 in HSC (Lee et al., 2010). Similar to normal HSC, FOXO3 in MPN LSC (Ahn et al., 2016) supports the notion that FOXO activity is modulated differently in stem cells than in more differentiated cells (Ghaffari, 2008; Yalcin et al., 2008; Zhang et al., 2011). These unexpected results suggest that in primary HSC and LSC, endogenous FOXO3 is regulated by a mechanism independent from the canonical pathway.

FOXO3 (FOXO) is a known substrate of the NAD-dependent SIRT1 deacetylase (Brunet et al., 2004; Motta et al., 2004). SIRT1 deacetylation activates or inhibits FOXO in a context-dependent manner (Brunet et al., 2004; Motta et al., 2004). In HSC, FOXO3 is mainly controlled by SIRT1, which—presumably through deacetylation—maintains FOXO3 in an active nuclear form in these cells (Rimmelé et al., 2014) (Fig. 4). FOXO3 localization becomes sensitized to AKT inhibition only in the absence of SIRT1, providing evidence that SIRT1 regulation may override AKT

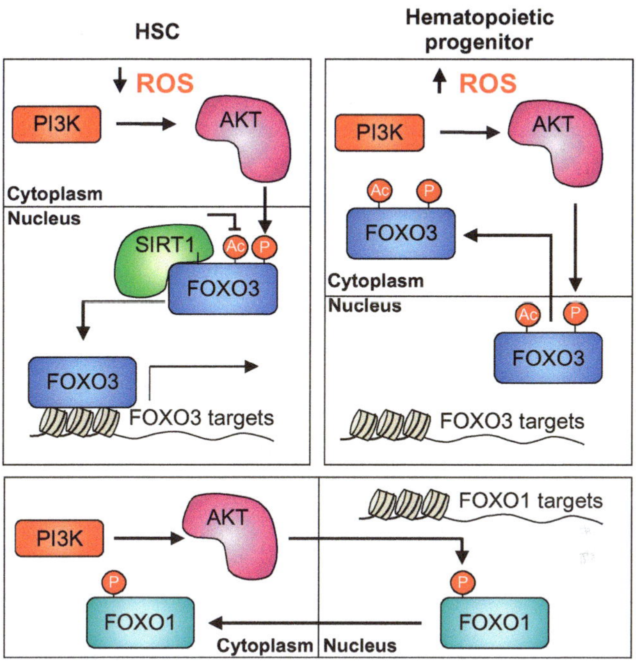

Fig. 4 Model for distinct regulation of FOXO1 and FOXO3 in stem cells. In HSCs, ROS levels are kept low and FOXO3 is localized in the nucleus despite the inhibitory phosphorylation by PI3K/AKT. Phosphorylation (P) and acetylation (Ac) are only designating the reaction and not the site or number of residues. SIRT1 activation of FOXO3, likely through deacetylation is dominant and leads to FOXO3 to be retained within the nucleus. However, in hematopoietic progenitor cells where ROS levels are higher as compared to HSC, FOXO3 is distributed between the nucleus and cytoplasm and also more responsive to PI3K/AKT mediated inhibition and cytoplasmic translocation. In contrast, FOXO1 activity is much lower in both HSCs and hematopoietic progenitor cells, which is seen with greater cytoplasmic localization. FOXO1 is also more sensitive to PI3K/AKT phosphorylation and subsequent cytoplasmic translocation.

phosphorylation of FOXO3 in HSPC (Liang et al., 2016). These combined findings support the notion that, under homeostatic conditions, AKT is not the dominant determinant of endogenous FOXO3 activity in HSC. Similar observations have also been made regarding regulation of FOXO1 function in human ESCs (Zhang et al., 2011).

Some of the Sirt1$^{-/-}$ HSC defects may be attributed to the relative loss of FOXO3 activity and deregulated expression of FOXO3 targets (Rimmelé et al., 2014). Many of these age-associated HSC defects are also observed in FOXO3-deficient HSC from young adult mice (Hinman et al., 2009; Kerdiles et al., 2010; Miyamoto et al., 2007; Yalcin et al., 2010, 2008) together, raising the interesting possibility that modulation of FOXO3 activity in HSC with age might be implicated in the aging of HSC (Fig. 3). Although direct evidence of FOXO3 protecting age-associated damage in old HSC is still missing, some available reports already support this notion (Mehta et al., 2015; Miyamoto et al., 2008; Warr et al., 2013). Therefore, the SIRT1 regulation of FOXO3 might be part of a protein network that promotes healthy aging and protects HSC from age-associated damage. FOXO regulation of a metabolic/mitochondrial program in stem cells is in agreement with this model (Rimmelé et al., 2015; Yeo et al., 2013) (see Fig. 1). This model requires to be tested in old HSC. SIRT3 and SIRT7 members of the sirtuin family have also been implicated in promoting healthy HSC aging (Brown et al., 2013; Mohrin et al., 2015). SIRT3 in particular—known to regulate FOXO (and specifically FOXO3)—is critical for the maintenance of the blood-forming stem cell pool in mice that are old or under stress (Brown et al., 2013). Whether and under which conditions FOXO (specifically FOXO3) might function downstream of SIRT3 and 7 in HSC requires further investigation.

miRNAs are increasingly implicated in regulating FOXO (Brett, Renault, Rafalski, Webb, & Brunet, 2011; Kim, Hwang, Muller, & Paik, 2015; Mehta et al., 2015; Yu et al., 2010). Specifically, microRNA-212/132 cluster (Mirc19) is enriched in HSC and upregulated with age. Altered expression of this miRNA cluster leads to HSC abnormalities with age. FOXO3 is the main target of Mirc19 in HSC. The data suggest that Mirc19 may buffer FOXO3 expression in HSC with age (Mehta et al., 2015). In neural stem and progenitor cells, miR-9 regulates FOXO1 expression (Kim et al., 2015). Similarly, miR-106b∼25 (miR-106b, miR-93, and miR-25) regulate neural stem/progenitor cells function as part of a network involving the insulin/IGF-FOXO pathway (Brett et al., 2011).

5. FOXO AND NEURAL STEM CELLS

FOXO (FOXO3)-mediated redox balance and transcriptional control of metabolic genes is also implicated in the maintenance of neural stem cells (NSCs) (Paik et al., 2009; Renault et al., 2009; Yeo et al., 2013) (see also chapter "FOXO in neural cells and diseases of the nervous system" by Santo and Paik). FOXO-deficient mice exhibit initially increased brain size and proliferation of neural progenitor cells during early postnatal life, followed by premature decline in the NSC pool and overall neurogenesis in adult brains. Similarly, NSC isolated from adult FOXO3-deficient mice display decreased self-renewal and impaired ability to generate different neural lineages. The FOXO (and FOXO3)-dependent gene expression profile in NSC suggests that FOXO (FOXO3) regulates the NSC pool by inducing a program of genes that preserves quiescence, prevents premature differentiation, and controls oxygen metabolism (Paik et al., 2009; Renault et al., 2009).

It is noteworthy that in neural progenitor cells, FOXO3-bound sites are enriched for bHLH transcription factors motifs. FOXO3 shares common targets with the proneuronal bHLH transcription factor ASCL1/MASH1 in neural progenitor cells (Webb et al., 2013). FOXO3 inhibits ASCL1-dependent neurogenesis in neural progenitor cells; thus, restraining ASCL1-dependent neurogenesis may help FOXO3 to preserve the NSC pool. FOXO (FOXO3) is also shown to regulate a metabolic program that includes glycolysis, glutaminolysis, and pentose phosphate pathway, which together support the redox balance and neurogenic potential of neural stem and progenitor cells (Yeo et al., 2013).

6. FOXO3 AND MUSCLE STEM CELLS

Recent evidence suggests that FOXO3 is key for the maintenance of muscle stem cell quiescence (Gopinath et al., 2014). In skeletal muscle, satellite cells constitute the myogenic stem cells that are quiescent and in low metabolic state (Cheung & Rando, 2013). In response to muscle fiber injury, satellite cells are activated, proliferate and differentiate into multinucleated myofibers, or self-renew. Upon completion of self-renewal, a subset of satellite cells returns to quiescence. FOXO3 was recently shown to be required for the self-renewal of satellite cells during muscle regeneration. The FOXO3 target in this process was identified as Notch signaling (Gopinath et al., 2014). Interestingly, FOXO3 is also the mediator of satellite cells' apoptosis when these cells—while harboring an imbalance in thyroid hormone signaling—are exposed to acute skeletal muscle injury (Dentice et al., 2014).

7. CONCLUDING REMARKS

In many species ranging from hydra to humans, FOXOs are key regulators of stem cells in both embryonic and adult tissues. Future research is needed to elucidate the mechanisms whereby FOXO regulates stem cell activity and in particular quiescence in adult stem cells is maintained by FOXO. Another fertile direction for future research is to elucidate the modulations of FOXO with age in tissue-specific stem cells. Lastly, the mechanisms of regulation of FOXO in stem cells remain to be uncovered. A major outstanding question is how modulations of FOXO in stem cells might contribute to disease, including to malignancy in particular.

ACKNOWLEDGMENTS

The work in Ghaffari lab is supported by National Institutes of Health (NIH) Grants R01HL136255 and R01CA205975 to S.G. R.L. is partially supported by an American Heart Association predoctoral award.

REFERENCES

Ahn, J. S., Li, J., Chen, E., Kent, D. G., Park, H. J., & Green, A. R. (2016). JAK2V617F mediates resistance to DNA damage-induced apoptosis by modulating FOXO3A localization and Bcl-xL deamidation. *Oncogene*, *35*, 2235–2246.

Anderson, M. J., Viars, C. S., Czekay, S., Cavenee, W. K., & Arden, K. C. (1998). Cloning and characterization of three human forkhead genes that comprise an FKHR-like gene subfamily. *Genomics*, *47*, 187–199.

Avilion, A. A., Nicolis, S. K., Pevny, L. H., Perez, L., Vivian, N., & Lovell-Badge, R. (2003). Multipotent cell lineages in early mouse development depend on SOX2 function. *Genes & Development*, *17*, 126–140.

Beerman, I., & Rossi, D. J. (2015). Epigenetic control of stem cell potential during homeostasis, aging, and disease. *Cell Stem Cell*, *16*, 613–625.

Bersenev, A., Rozenova, K., Balcerek, J., Jiang, J., Wu, C., & Tong, W. (2012). Lnk deficiency partially mitigates hematopoietic stem cell aging. *Aging Cell*, *11*, 949–959.

Bersenev, A., Wu, C., Balcerek, J., & Tong, W. (2008). Lnk controls mouse hematopoietic stem cell self-renewal and quiescence through direct interactions with JAK2. *The Journal of Clinical Investigation*, *118*, 2832–2844.

Bigarella, C. L., Li, J., Rimmele, P., Liang, R., Sobol, R. W., & Ghaffari, S. (2017). FOXO3 is essential for protecting hematopoietic stem and progenitor cells from oxidative DNA damage. *The Journal of Biological Chemistry*, *292*, 3005–3015.

Bigarella, C. L., Liang, R., & Ghaffari, S. (2014). Stem cells and the impact of ROS signaling. *Development*, *141*, 4206–4218.

Boehm, A. M., Khalturin, K., Erxleben, F. A., Hemmrich, G., Klostermeier, U. C., Lopez-Quintero, J. A., et al. (2012). FoxO is a critical regulator of stem cell maintenance in immortal Hydra. *Proceedings of the National Academy of Sciences of the United States of America*, *109*, 19697–19702.

Borkhardt, A., Repp, R., Haas, O. A., Leis, T., Harbott, J., Kreuder, J., et al. (1997). Cloning and characterization of AFX, the gene that fuses to MLL in acute leukemias with a t(X;11)(q13;q23). *Oncogene*, *14*, 195–202.

Boyer, L. A., Lee, T. I., Cole, M. F., Johnstone, S. E., Levine, S. S., Zucker, J. P., et al. (2005). Core transcriptional regulatory circuitry in human embryonic stem cells. *Cell*, *122*, 947–956.

Brett, J. O., Renault, V. M., Rafalski, V. A., Webb, A. E., & Brunet, A. (2011). The microRNA cluster miR-106b~25 regulates adult neural stem/progenitor cell proliferation and neuronal differentiation. *Aging*, *3*, 108–124.

Brown, K., Xie, S., Qiu, X., Mohrin, M., Shin, J., Liu, Y., et al. (2013). SIRT3 reverses aging-associated degeneration. *Cell Reports*, *3*, 319–327.

Brunet, A., Sweeney, L. B., Sturgill, J. F., Chua, K. F., Greer, P. L., Lin, Y., et al. (2004). Stress-dependent regulation of FOXO transcription factors by the SIRT1 deacetylase. *Science*, *303*, 2011–2015.

Bunz, F., Dutriaux, A., Lengauer, C., Waldman, T., Zhou, S., Brown, J. P., et al. (1998). Requirement for p53 and p21 to sustain G2 arrest after DNA damage. *Science*, *282*, 1497–1501.

Burgering, B. M., & Medema, R. H. (2003). Decisions on life and death: FOXO forkhead transcription factors are in command when PKB/Akt is off duty. *Journal of Leukocyte Biology*, *73*, 689–701.

Castrillon, D. H., Miao, L., Kollipara, R., Horner, J. W., & DePinho, R. A. (2003). Suppression of ovarian follicle activation in mice by the transcription factor Foxo3a. *Science*, *301*, 215–218.

Chandel, N. S., Maltepe, E., Goldwasser, E., Mathieu, C. E., Simon, M. C., & Schumacker, P. T. (1998). Mitochondrial reactive oxygen species trigger hypoxia-induced transcription. *Proceedings of the National Academy of Sciences of the United States of America*, *95*, 11715–11720.

Cheng, T., Rodrigues, N., Shen, H., Yang, Y., Dombkowski, D., Sykes, M., et al. (2000). Hematopoietic stem cell quiescence maintained by p21cip1/waf1. *Science*, *287*, 1804–1808.

Cheung, T. H., & Rando, T. A. (2013). Molecular regulation of stem cell quiescence. *Nature Reviews. Molecular Cell Biology*, *14*, 329–340.

Chiacchiera, F., & Simone, C. (2009). Inhibition of p38alpha unveils an AMPK-FoxO3A axis linking autophagy to cancer-specific metabolism. *Autophagy*, *5*, 1030–1033.

Davenport, T. G., Jerome-Majewska, L. A., & Papaioannou, V. E. (2003). Mammary gland, limb and yolk sac defects in mice lacking Tbx3, the gene mutated in human ulnar mammary syndrome. *Development*, *130*, 2263–2273.

Davis, R. J., D'Cruz, C. M., Lovell, M. A., Biegel, J. A., & Barr, F. G. (1994). Fusion of PAX7 to FKHR by the variant t(1;13)(p36;q14) translocation in alveolar rhabdomyosarcoma. *Cancer Research*, *54*, 2869–2872.

Dejean, A. S., Beisner, D. R., Ch'en, I. L., Kerdiles, Y. M., Babour, A., Arden, K. C., et al. (2009). Transcription factor Foxo3 controls the magnitude of T cell immune responses by modulating the function of dendritic cells. *Nature Immunology*, *10*, 504–513.

Dentice, M., Ambrosio, R., Damiano, V., Sibilio, A., Luongo, C., Guardiola, O., et al. (2014). Intracellular inactivation of thyroid hormone is a survival mechanism for muscle stem cell proliferation and lineage progression. *Cell Metabolism*, *20*, 1038–1048.

Diehn, M., Cho, R. W., Lobo, N. A., Kalisky, T., Dorie, M. J., Kulp, A. N., et al. (2009). Association of reactive oxygen species levels and radioresistance in cancer stem cells. *Nature*, *458*, 780–783.

Dijkers, P. F., Medema, R. H., Pals, C., Banerji, L., Thomas, N. S., Lam, E. W., et al. (2000). Forkhead transcription factor FKHR-L1 modulates cytokine-dependent transcriptional regulation of p27(KIP1). *Molecular and Cellular Biology*, *20*, 9138–9148.

Eijkelenboom, A., & Burgering, B. M. (2013). FOXOs: Signalling integrators for homeostasis maintenance. *Nature Reviews. Molecular Cell Biology*, *14*, 83–97.

Eijkelenboom, A., Mokry, M., Smits, L. M., Nieuwenhuis, E. E., & Burgering, B. M. (2013). FOXO3 selectively amplifies enhancer activity to establish target gene regulation. *Cell Reports*, *5*, 1664–1678.

Galili, N., Davis, R. J., Fredericks, W. J., Mukhopadhyay, S., Rauscher, F. J., 3rd, Emanuel, B. S., et al. (1993). Fusion of a fork head domain gene to PAX3 in the solid tumour alveolar rhabdomyosarcoma. *Nature Genetics*, *5*, 230–235.

Ghaffari, S. (2008). Oxidative stress in the regulation of normal and neoplastic hematopoiesis. *Antioxidants & Redox Signaling*, *10*, 1923–1940.

Ghaffari, S., Jagani, Z., Kitidis, C., Lodish, H. F., & Khosravi-Far, R. (2003). Cytokines and BCR-ABL mediate suppression of TRAIL-induced apoptosis through inhibition of forkhead FOXO3a transcription factor. *Proceedings of the National Academy of Sciences of the United States of America*, *100*, 6523–6528.

Giannakou, M. E., Goss, M., Junger, M. A., Hafen, E., Leevers, S. J., & Partridge, L. (2004). Long-lived drosophila with overexpressed dFOXO in adult fat body. *Science*, *305*, 361.

Goertz, M. J., Wu, Z., Gallardo, T. D., Hamra, F. K., & Castrillon, D. H. (2011). Foxo1 is required in mouse spermatogonial stem cells for their maintenance and the initiation of spermatogenesis. *The Journal of Clinical Investigation*, *121*, 3456–3466.

Gonzales, K. A., & Ng, H. H. (2011). FoxO: A new addition to the ESC cartel. *Cell Stem Cell*, *9*, 181–183.

Gopinath, S. D., Webb, A. E., Brunet, A., & Rando, T. A. (2014). FOXO3 promotes quiescence in adult muscle stem cells during the process of self-renewal. *Stem Cell Reports*, *2*, 414–426.

Greer, E. L., & Brunet, A. (2008). FOXO transcription factors in ageing and cancer. *Acta Physiologica (Oxford, England)*, *192*, 19–28.

Gu, T. L., Tothova, Z., Scheijen, B., Griffin, J. D., Gilliland, D. G., & Sternberg, D. W. (2004). NPM-ALK fusion kinase of anaplastic large-cell lymphoma regulates survival and proliferative signaling through modulation of FOXO3a. *Blood*, *103*, 4622–4629.

Haigis, M. C., & Sinclair, D. A. (2010). Mammalian sirtuins: Biological insights and disease relevance. *Annual Review of Pathology*, *5*, 253–295.

Hedrick, S. M., Hess Michelini, R., Doedens, A. L., Goldrath, A. W., & Stone, E. L. (2012). FOXO transcription factors throughout T cell biology. *Nature Reviews. Immunology*, *12*, 649–661.

Hillion, J., Le Coniat, M., Jonveaux, P., Berger, R., & Bernard, O. A. (1997). AF6q21, a novel partner of the MLL gene in t(6;11)(q21;q23), defines a forkhead transcriptional factor subfamily. *Blood*, *90*, 3714–3719.

Hinman, R. M., Nichols, W. A., Diaz, T. M., Gallardo, T. D., Castrillon, D. H., & Satterthwaite, A. B. (2009). Foxo3 −/− mice demonstrate reduced numbers of pre-B and recirculating B cells but normal splenic B cell sub-population distribution. *International Immunology*, *21*, 831–842.

Ho, T. T., Warr, M. R., Adelman, E. R., Lansinger, O. M., Flach, J., Verovskaya, E. V., et al. (2017). Autophagy maintains the metabolism and function of young and old stem cells. *Nature*, *543*, 205–210.

Hosaka, T., Biggs, W. H., 3rd, Tieu, D., Boyer, A. D., Varki, N. M., Cavenee, W. K., et al. (2004). Disruption of forkhead transcription factor (FOXO) family members in mice reveals their functional diversification. *Proceedings of the National Academy of Sciences of the United States of America*, *101*, 2975–2980.

Hu, M. C., Lee, D. F., Xia, W., Golfman, L. S., Ou-Yang, F., Yang, J. Y., et al. (2004). IkappaB kinase promotes tumorigenesis through inhibition of forkhead FOXO3a. *Cell*, *117*, 225–237.

Hwangbo, D. S., Gershman, B., Tu, M. P., Palmer, M., & Tatar, M. (2004). Drosophila dFOXO controls lifespan and regulates insulin signalling in brain and fat body. *Nature*, *429*, 562–566.

Ito, K., Hirao, A., Arai, F., Matsuoka, S., Takubo, K., Hamaguchi, I., et al. (2004). Regulation of oxidative stress by ATM is required for self-renewal of haematopoietic stem cells. *Nature*, *431*, 997–1002.

Ito, K., Hirao, A., Arai, F., Takubo, K., Matsuoka, S., Miyamoto, K., et al. (2006). Reactive oxygen species act through p38 MAPK to limit the lifespan of hematopoietic stem cells. *Nature Medicine*, *12*(4), 446–451.

Ito, K., Turcotte, R., Cui, J., Zimmerman, S. E., Pinho, S., Mizoguchi, T., et al. (2016). Self-renewal of a purified Tie2+ hematopoietic stem cell population relies on mitochondrial clearance. *Science*, *354*, 1156–1160.

Jacobs, F. M., van der Heide, L. P., Wijchers, P. J., Burbach, J. P., Hoekman, M. F., & Smidt, M. P. (2003). FoxO6, a novel member of the FoxO class of transcription factors with distinct shuttling dynamics. *The Journal of Biological Chemistry*, *278*, 35959–35967.

Jang, Y. Y., & Sharkis, S. J. (2007). A low level of reactive oxygen species selects for primitive hematopoietic stem cells that may reside in the low-oxygenic niche. *Blood*, *110*, 3056–3063.

Kerdiles, Y. M., Stone, E. L., Beisner, D. R., McGargill, M. A., Ch'en, I. L., Stockmann, C., et al. (2010). Foxo transcription factors control regulatory T cell development and function. *Immunity*, *33*, 890–904.

Kikuchi, S., Nagai, T., Kunitama, M., Kirito, K., Ozawa, K., & Komatsu, N. (2007). Active FKHRL1 overcomes imatinib resistance in chronic myelogenous leukemia-derived cell lines via the production of tumor necrosis factor-related apoptosis-inducing ligand. *Cancer Science*, *98*, 1949–1958.

Kim, D. Y., Hwang, I., Muller, F. L., & Paik, J. H. (2015). Functional regulation of FoxO1 in neural stem cell differentiation. *Cell Death and Differentiation*, *22*, 2034–2045.

Kocabas, F., Zheng, J., Thet, S., Copeland, N. G., Jenkins, N. A., DeBerardinis, R. J., et al. (2012). Meis1 regulates the metabolic phenotype and oxidant defense of hematopoietic stem cells. *Blood*, *120*, 4963–4972.

Lagadinou, E. D., Sach, A., Callahan, K., Rossi, R. M., Neering, S. J., Minhajuddin, M., et al. (2013). BCL-2 inhibition targets oxidative phosphorylation and selectively eradicates quiescent human leukemia stem cells. *Cell Stem Cell*, *12*, 329–341.

Lee, S. S., Kennedy, S., Tolonen, A. C., & Ruvkun, G. (2003). DAF-16 target genes that control *C. elegans* life-span and metabolism. *Science*, *300*, 644–647.

Lee, J. Y., Nakada, D., Yilmaz, O. H., Tothova, Z., Joseph, N. M., Lim, M. S., et al. (2010). mTOR activation induces tumor suppressors that inhibit leukemogenesis and deplete hematopoietic stem cells after Pten deletion. *Cell Stem Cell*, *7*, 593–605.

Lessard, J., Faubert, A., & Sauvageau, G. (2004). Genetic programs regulating HSC specification, maintenance and expansion. *Oncogene*, *23*, 7199–7209.

Li, J., Du, W., Maynard, S., Andreassen, P. R., & Pang, Q. (2010). Oxidative stress-specific interaction between FANCD2 and FOXO3a. *Blood*, *115*, 1545–1548.

Li, X., Li, J., Wilson, A., Sipple, J., Schick, J., & Pang, Q. (2015). Fancd2 is required for nuclear retention of foxo3a in hematopoietic stem cell maintenance. *The Journal of Biological Chemistry*, *290*, 2715–2727.

Liang, R., Camprecios, G., Kou, Y., McGrath, K., Nowak, R., Catherman, S., et al. (2015). A systems approach identifies essential FOXO3 functions at key steps of terminal erythropoiesis. *PLoS Genetics*, *11*, e1005526.

Liang, R., & Ghaffari, S. (2014). Stem cells, redox signaling, and stem cell aging. *Antioxidants & Redox Signaling*, *20*, 1902–1916.

Liang, R., Rimmele, P., Bigarella, C. L., Yalcin, S., & Ghaffari, S. (2016). Evidence for AKT-independent regulation of FOXO1 and FOXO3 in haematopoietic stem and progenitor cells. *Cell Cycle*, *15*, 861–867.

Lin, K., Dorman, J. B., Rodan, A., & Kenyon, C. (1997). daf-16: An HNF-3/forkhead family member that can function to double the life-span of *Caenorhabditis elegans*. *Science*, *278*, 1319–1322.

Liu, Y., Elf, S. E., Miyata, Y., Sashida, G., Huang, G., Di Giandomenico, S., et al. (2009). p53 regulates hematopoietic stem cell quiescence. *Cell Stem Cell*, *4*, 37–48.

Ma, Y., Yao, N., Liu, G., Dong, L., Liu, Y., Zhang, M., et al. (2015). Functional screen reveals essential roles of miR-27a/24 in differentiation of embryonic stem cells. *The EMBO Journal, 34*, 361–378.

Machida, S., Spangenburg, E. E., & Booth, F. W. (2003). Forkhead transcription factor FoxO1 transduces insulin-like growth factor's signal to p27Kip1 in primary skeletal muscle satellite cells. *Journal of Cellular Physiology, 196*, 523–531.

Mammucari, C., Milan, G., Romanello, V., Masiero, E., Rudolf, R., Del Piccolo, P., et al. (2007). FoxO3 controls autophagy in skeletal muscle in vivo. *Cell Metabolism, 6*, 458–471.

Marinkovic, D., Zhang, X., Yalcin, S., Luciano, J. P., Brugnara, C., Huber, T., et al. (2007). Foxo3 is required for the regulation of oxidative stress in erythropoiesis. *The Journal of Clinical Investigation, 117*, 2133–2144.

Martinez-Gac, L., Marques, M., Garcia, Z., Campanero, M. R., & Carrera, A. C. (2004). Control of cyclin G2 mRNA expression by forkhead transcription factors: Novel mechanism for cell cycle control by phosphoinositide 3-kinase and forkhead. *Molecular and Cellular Biology, 24*, 2181–2189.

Maryanovich, M., Oberkovitz, G., Niv, H., Vorobiyov, L., Zaltsman, Y., Brenner, O., et al. (2012). The ATM-BID pathway regulates quiescence and survival of haematopoietic stem cells. *Nature Cell Biology, 14*, 535–541.

Mehta, A., Zhao, J. L., Sinha, N., Marinov, G. K., Mann, M., Kowalczyk, M. S., et al. (2015). The microRNA-132 and microRNA-212 cluster regulates hematopoietic stem cell maintenance and survival with age by buffering FOXO3 expression. *Immunity, 42*, 1021–1032.

Miyamoto, K., Araki, K. Y., Naka, K., Arai, F., Takubo, K., Yamazaki, S., et al. (2007). Foxo3a is essential for maintenance of the hematopoietic stem cell pool. *Cell Stem Cell, 1*, 101–112.

Miyamoto, K., Miyamoto, T., Kato, R., Yoshimura, A., Motoyama, N., & Suda, T. (2008). FoxO3a regulates hematopoietic homeostasis through a negative feedback pathway in conditions of stress or aging. *Blood, 112*, 4485–4493.

Modur, V., Nagarajan, R., Evers, B. M., & Milbrandt, J. (2002). FOXO proteins regulate tumor necrosis factor-related apoptosis inducing ligand expression. Implications for PTEN mutation in prostate cancer. *The Journal of Biological Chemistry, 277*, 47928–47937.

Mohrin, M., Shin, J. Y., Liu, Y. F., Brown, K., Luo, H. Z., Xi, Y. N., et al. (2015). A mitochondrial UPR-mediated metabolic checkpoint regulates hematopoietic stem cell aging. *Science, 347*, 1374–1377.

Mortensen, M., Soilleux, E. J., Djordjevic, G., Tripp, R., Lutteropp, M., Sadighi-Akha, E., et al. (2011). The autophagy protein Atg7 is essential for hematopoietic stem cell maintenance. *The Journal of Experimental Medicine, 208*, 455–467.

Motta, M. C., Divecha, N., Lemieux, M., Kamel, C., Chen, D., Gu, W., et al. (2004). Mammalian SIRT1 represses forkhead transcription factors. *Cell, 116*, 551–563.

Murphy, C. T., McCarroll, S. A., Bargmann, C. I., Fraser, A., Kamath, R. S., Ahringer, J., et al. (2003). Genes that act downstream of DAF-16 to influence the lifespan of *Caenorhabditis elegans*. *Nature, 424*, 277–283.

Naka, K., Hoshii, T., Muraguchi, T., Tadokoro, Y., Ooshio, T., Kondo, Y., et al. (2010). TGF-beta-FOXO signalling maintains leukaemia-initiating cells in chronic myeloid leukaemia. *Nature, 463*, 676–680.

Narducci, M. G., Fiorenza, M. T., Kang, S. M., Bevilacqua, A., Di Giacomo, M., Remotti, D., et al. (2002). TCL1 participates in early embryonic development and is overexpressed in human seminomas. *Proceedings of the National Academy of Sciences of the United States of America, 99*, 11712–11717.

Ogg, S., Paradis, S., Gottlieb, S., Patterson, G. I., Lee, L., Tissenbaum, H. A., et al. (1997). The fork head transcription factor DAF-16 transduces insulin-like metabolic and longevity signals in *C. elegans*. *Nature*, *389*, 994–999.

Oh, S. T., Simonds, E. F., Jones, C., Hale, M. B., Goltsev, Y., Gibbs, K. D., Jr., et al. (2010). Novel mutations in the inhibitory adaptor protein LNK drive JAK-STAT signaling in patients with myeloproliferative neoplasms. *Blood*, *116*, 988–992.

Ou, X., Chae, H. D., Wang, R. H., Shelley, W. C., Cooper, S., Taylor, T., et al. (2011). SIRT1 deficiency compromises mouse embryonic stem cell hematopoietic differentiation, and embryonic and adult hematopoiesis in the mouse. *Blood*, *117*, 440–450.

Owusu-Ansah, E., & Banerjee, U. (2009). Reactive oxygen species prime Drosophila haematopoietic progenitors for differentiation. *Nature*, *461*, 537–541.

Paik, J. H., Ding, Z., Narurkar, R., Ramkissoon, S., Muller, F., Kamoun, W. S., et al. (2009). FoxOs cooperatively regulate diverse pathways governing neural stem cell homeostasis. *Cell Stem Cell*, *5*, 540–553.

Paik, J. H., Kollipara, R., Chu, G., Ji, H., Xiao, Y., Ding, Z., et al. (2007). FoxOs are lineage-restricted redundant tumor suppressors and regulate endothelial cell homeostasis. *Cell*, *128*, 309–323.

Pellicano, F., Scott, M. T., Helgason, G. V., Hopcroft, L. E., Allan, E. K., Aspinall-O'Dea, M., et al. (2014). The antiproliferative activity of kinase inhibitors in chronic myeloid leukemia cells is mediated by FOXO transcription factors. *Stem Cells*, *32*, 2324–2337.

Pelosi, E., Omari, S., Michel, M., Ding, J., Amano, T., Forabosco, A., et al. (2013). Constitutively active Foxo3 in oocytes preserves ovarian reserve in mice. *Nature Communications*, *4*, 1843.

Renault, V. M., Rafalski, V. A., Morgan, A. A., Salih, D. A., Brett, J. O., Webb, A. E., et al. (2009). FoxO3 regulates neural stem cell homeostasis. *Cell Stem Cell*, *5*, 527–539.

Renault, V. M., Thekkat, P. U., Hoang, K. L., White, J. L., Brady, C. A., Kenzelmann Broz, D., et al. (2011). The pro-longevity gene FoxO3 is a direct target of the p53 tumor suppressor. *Oncogene*, *30*, 3207–3221.

Rimmelé, P., Bigarella, C. L., Liang, R., Izac, B., Dieguez-Gonzalez, R., Barbet, G., et al. (2014). Aging-like phenotype and defective lineage specification in SIRT1-deleted hematopoietic stem and progenitor cells. *Stem Cell Reports*, *3*, 44–59.

Rimmelé, P., Liang, R., Bigarella, C. L., Kocabas, F., Xie, J., Serasinghe, M. N., et al. (2015). Mitochondrial metabolism in hematopoietic stem cells requires functional FOXO3. *EMBO Reports*, *16*, 1164–1176.

Sakoe, Y., Sakoe, K., Kirito, K., Ozawa, K., & Komatsu, N. (2010). FOXO3A as a key molecule for all-trans retinoic acid-induced granulocytic differentiation and apoptosis in acute promyelocytic leukemia. *Blood*, *115*, 3787–3795.

Sandri, M., Sandri, C., Gilbert, A., Skurk, C., Calabria, E., Picard, A., et al. (2004). Foxo transcription factors induce the atrophy-related ubiquitin ligase atrogin-1 and cause skeletal muscle atrophy. *Cell*, *117*, 399–412.

Scheijen, B., Ngo, H. T., Kang, H., & Griffin, J. D. (2004). FLT3 receptors with internal tandem duplications promote cell viability and proliferation by signaling through Foxo proteins. *Oncogene*, *23*, 3338–3349.

Signer, R. A., & Morrison, S. J. (2013). Mechanisms that regulate stem cell aging and life span. *Cell Stem Cell*, *12*, 152–165.

Simon, M. C., & Keith, B. (2008). The role of oxygen availability in embryonic development and stem cell function. *Nature Reviews. Molecular Cell Biology*, *9*, 285–296.

Singh, S. K., Williams, C. A., Klarmann, K., Burkett, S. S., Keller, J. R., & Oberdoerffer, P. (2013). Sirt1 ablation promotes stress-induced loss of epigenetic and genomic hematopoietic stem and progenitor cell maintenance. *The Journal of Experimental Medicine*, *210*, 987–1001.

Slack, C., Werz, C., Wieser, D., Alic, N., Foley, A., Stocker, H., et al. (2010). Regulation of lifespan, metabolism, and stress responses by the Drosophila SH2B protein, Lnk. *PLoS Genetics*, *6*, e1000881.

Sykes, S. M., Lane, S. W., Bullinger, L., Kalaitzidis, D., Yusuf, R., Saez, B., et al. (2011). AKT/FOXO signaling enforces reversible differentiation blockade in myeloid leukemias. *Cell*, *146*(5), 697–708.

Tai-Nagara, I., Matsuoka, S., Ariga, H., & Suda, T. (2014). Mortalin and DJ-1 coordinately regulate hematopoietic stem cell function through the control of oxidative stress. *Blood*, *123*, 41–50.

Takaki, S., Morita, H., Tezuka, Y., & Takatsu, K. (2002). Enhanced hematopoiesis by hematopoietic progenitor cells lacking intracellular adaptor protein, Lnk. *The Journal of Experimental Medicine*, *195*, 151–160.

Tothova, Z., Kollipara, R., Huntly, B. J., Lee, B. H., Castrillon, D. H., Cullen, D. E., et al. (2007). FoxOs are critical mediators of hematopoietic stem cell resistance to physiologic oxidative stress. *Cell*, *128*, 325–339.

Tsai, W. B., Chung, Y. M., Takahashi, Y., Xu, Z., & Hu, M. C. (2008). Functional interaction between FOXO3a and ATM regulates DNA damage response. *Nature Cell Biology*, *10*, 460–467.

van der Vos, K. E., Eliasson, P., Proikas-Cezanne, T., Vervoort, S. J., van Boxtel, R., Putker, M., et al. (2012). Modulation of glutamine metabolism by the PI(3)K-PKB-FOXO network regulates autophagy. *Nature Cell Biology*, *14*, 829–837.

Velazquez, L., Cheng, A. M., Fleming, H. E., Furlonger, C., Vesely, S., Bernstein, A., et al. (2002). Cytokine signaling and hematopoietic homeostasis are disrupted in Lnk-deficient mice. *The Journal of Experimental Medicine*, *195*, 1599–1611.

Waldman, T., Lengauer, C., Kinzler, K. W., & Vogelstein, B. (1996). Uncoupling of S phase and mitosis induced by anticancer agents in cells lacking p21. *Nature*, *381*, 713–716.

Warr, M. R., Binnewies, M., Flach, J., Reynaud, D., Garg, T., Malhotra, R., et al. (2013). FOXO3A directs a protective autophagy program in haematopoietic stem cells. *Nature*, *494*, 323–327.

Webb, A. E., Pollina, E. A., Vierbuchen, T., Urban, N., Ucar, D., Leeman, D. S., et al. (2013). FOXO3 shares common targets with ASCL1 genome-wide and inhibits ASCL1-dependent neurogenesis. *Cell Reports*, *4*, 477–491.

Willcox, B. J., Donlon, T. A., He, Q., Chen, R., Grove, J. S., Yano, K., et al. (2008). FOXO3A genotype is strongly associated with human longevity. *Proceedings of the National Academy of Sciences of the United States of America*, *105*, 13987–13992.

Yahata, T., Takanashi, T., Muguruma, Y., Ibrahim, A. A., Matsuzawa, H., Uno, T., et al. (2011). Accumulation of oxidative DNA damage restricts the self-renewal capacity of human hematopoietic stem cells. *Blood*, *118*(11), 2941–2950.

Yalcin, S., Marinkovic, D., Mungamuri, S. K., Zhang, X., Tong, W., Sellers, R., et al. (2010). ROS-mediated amplification of AKT/mTOR signalling pathway leads to myeloproliferative syndrome in Foxo3(−/−) mice. *The EMBO Journal*, *29*, 4118–4131.

Yalcin, S., Zhang, X., Luciano, J. P., Mungamuri, S. K., Marinkovic, D., Vercherat, C., et al. (2008). Foxo3 is essential for the regulation of ataxia telangiectasia mutated and oxidative stress-mediated homeostasis of hematopoietic stem cells. *The Journal of Biological Chemistry*, *283*(37), 25692–25705.

Yang, Z., & Klionsky, D. J. (2010). Mammalian autophagy: Core molecular machinery and signaling regulation. *Current Opinion in Cell Biology*, *22*, 124–131.

Yeo, H., Lyssiotis, C. A., Zhang, Y., Ying, H., Asara, J. M., Cantley, L. C., et al. (2013). FoxO3 coordinates metabolic pathways to maintain redox balance in neural stem cells. *The EMBO Journal*, *32*, 2589–2602.

You, H., Jang, Y., You-Ten, A. I., Okada, H., Liepa, J., Wakeham, A., et al. (2004). p53-dependent inhibition of FKHRL1 in response to DNA damage through protein kinase SGK1. *Proceedings of the National Academy of Sciences of the United States of America, 101*, 14057–14062.
You, H., & Mak, T. W. (2005). Crosstalk between p53 and FOXO transcription factors. *Cell Cycle, 4*, 37–38.
You, H., Yamamoto, K., & Mak, T. W. (2006). Regulation of transactivation-independent proapoptotic activity of p53 by FOXO3a. *Proceedings of the National Academy of Sciences of the United States of America, 103*, 9051–9056.
Yu, D., dos Santos, C. O., Zhao, G., Jiang, J., Amigo, J. D., Khandros, E., et al. (2010). miR-451 protects against erythroid oxidant stress by repressing 14-3-3zeta. *Genes & Development, 24*, 1620–1633.
Zhang, X., Yalcin, S., Lee, D. F., Yeh, T. Y., Lee, S. M., Su, J., et al. (2011). FOXO1 is an essential regulator of pluripotency in human embryonic stem cells. *Nature Cell Biology, 13*, 1092–1099.
Zhao, Y., Yang, J., Liao, W., Liu, X., Zhang, H., Wang, S., et al. (2010). Cytosolic FoxO1 is essential for the induction of autophagy and tumour suppressor activity. *Nature Cell Biology, 12*, 665–675.

CHAPTER THREE

FOXOs Maintaining the Equilibrium for Better or for Worse

Sabina van Doeselaar, Boudewijn M.T. Burgering[1]
Molecular Cancer Research, Center Molecular Medicine, Oncode Institute, University Medical Center Utrecht, Utrecht University, Utrecht, The Netherlands
[1]Corresponding author: e-mail address: b.m.t.burgering@umcutrecht.nl

Contents

Abstract

A paradigm shift is emerging within the FOXO field and accumulating evidence indicates that we need to reappreciate the role of FOXOs, at least in cancer development. Here, we discuss the possibility that FOXOs are both tumor suppressors as well as promoters of tumor progression. This is mostly dependent on the biological context. Critical to this dichotomous role is the notion that FOXOs are central in preserving cellular homeostasis in redox control, genomic stability, and protein turnover. From this perspective, a paradoxical role in both suppressing and enhancing tumor progression

Current Topics in Developmental Biology, Volume 127
ISSN 0070-2153
https://doi.org/10.1016/bs.ctdb.2017.10.003

can be reconciled. As many small molecules targeting the PI3K pathway are developed by big pharmaceutical companies and/or are in clinical trial, we will discuss what the consequences may be for the context-dependent role of FOXOs in tumor development in treatment options based on active PI3K signaling in tumors.

1. INTRODUCTION: FOXO REGULATION GETTING UP TO PAR

Following the discovery of insulin in 1922 by Banting and Best, and the establishment that insulin is the causative link in diabetes, it became essential to understand the cellular consequences of insulin action (Banting, Best, Collip, Campbell, & Fletcher, 1922). By now, much has been learned as to how insulin regulates glucose metabolism through activation of signal transduction pathways, but also how insulin's ability to regulate glucose metabolism relates to the other functions of insulin, i.e., regulation of cell survival, cell proliferation, and cell growth. Insulin-induced activation of the phosphoinositide 3-kinase (PI3K)/protein kinase B (PKB) signaling pathway is considered to be a key regulatory event in these actions of insulin (reviewed in Hemmings & Restuccia, 2012). This is illustrated by the identification of a mutation in the PKBβ gene within a family showing autosomal dominant inheritance of severe insulin resistance and diabetes mellitus. Expression of the resulting mutant kinase in cultured cells disrupts insulin signaling (George et al., 2004).

The PI3K/PKB pathway is not only causative to type II diabetes, but deregulated PI3K/PKB signaling is also strongly implicated in the development of cancer. Mutations that result in the hyperactivation of PI3K signaling occur at high frequency in many human tumors (Samuels et al., 2004).

The PI3K/PKB pathway is not only regulated through insulin action. By now many extracellular triggers have been shown to activate this pathway. Also PKB, being a serine/threonine kinase, has been shown to phosphorylate many different substrates. The relevance of all these PKB-mediated phosphorylation events is not understood in full. However, a general picture emerges and genetic interaction studies have convincingly demonstrated that several of the PKB substrates are in vivo relevant substrates (Manning & Toker, 2017).

The Forkhead box O (FOXO) family of transcription factors consists of four members, FOXO1 (FKHR), FOXO3 (FKHRL), FOXO4 (AFX), and FOXO6. It is a subclass of the larger Forkhead family of transcription factors.

The FOXO family is a conserved family of transcription factors that binds to the consensus DNA sequence 5′-TTGTTTAC-3′. Others and we have shown that FOXO transcription factors act as essential downstream nuclear components of the PI3K/PKB pathway. The insulin signaling pathway through PI3K and PKB/AKT negatively regulates FOXOs. Phosphorylation by PKB leads to nuclear exclusion of FOXO, and as a result, its function as a transcriptional activator is inhibited. Consequently, many studies have shown that in line with the critical role for PI3K/PKB in insulin action, also FOXOs are critically involved in mediating the cellular and organismal responses to insulin (reviewed in Burgering, 2008).

Studies in model organisms, notably the nematode *Caenorhabditis elegans*, have shown that the molecular details of insulin/PI3K/PKB/FOXO signaling are conserved through evolution. Importantly, this pathway affects life span. Namely, high FOXO (DAF-16) activity increases adult life span (Kenyon, Chang, Gensch, Rudner, & Tabtiang, 1993). Thus, whereas normal signaling through this pathway affects organismal life span, deregulation of this pathway significantly contributes to two of the most common age-related diseases in humans, cancer and diabetes.

Increased levels of reactive oxygen species (ROS) are implicated in aging and the onset and/or progression of numerous diseases, including cancer. In agreement with the proposed role of ROS in aging and the life span-extending effect of increased DAF-16/FOXO activity, others and we have shown that FOXO activity can reduce cellular oxidative stress. This happens via gene regulation of MnSOD, catalase, and other genes involved in maintaining the cellular redox balance (Furukawa-Hibi, Yoshida-Araki, Ohta, Ikeda, & Motoyama, 2002; Kops, Dansen, et al., 2002; Nemoto & Finkel, 2002; Tran et al., 2002). Importantly, FOXO activity is also in return controlled by changes in cellular redox (for reviews: van der Horst & Burgering, 2007). From the aging perspective, it is interesting to note that redox regulation of FOXOs is also evolutionary conserved (*Hydra vulgaris*, *Drosophila melanogaster*, *C. elegans*). Cellular oxidative stress regulates FOXO in a complex manner, involving a plethora of posttranslational modifications and regulatory proteins (reviewed in Eijkelenboom & Burgering, 2013; van der Horst & Burgering, 2007).

Thus, two evolutionary conserved pathways regulate FOXO activity. In the presence of growth factors, FOXOs are negatively regulated by the canonical insulin signaling pathway through PI3K/PKB. On the other hand, FOXOs are activated in the presence of oxidative stress through JNK signaling and additional mechanisms. In addition, several other signaling pathways have been shown to modulate FOXO activity. Downstream gene programs induced by FOXOs are also diverse and affect many cellular processes, including cell cycle regulation, cell survival, and metabolism (reviewed in Eijkelenboom & Burgering, 2013; van der Horst & Burgering, 2007).

2. THE PI3K/PKB(AKT)/FOXO SIGNALING PATHWAY AND CANCER

In cancer, FOXOs were initially identified as partners of chromosomal translocations resulting in FOXO1-PAX3/7, FOXO3-MLL, and FOXO4-MLL fusion proteins in alveolar rhabdomyosarcoma and acute leukemia (Galili et al., 1993; Hillion, Le Coniat, Jonveaux, Berger, & Bernard, 1997; Mercado & Barr, 2007). In all these translocations, the transcriptional transactivation domain of FOXOs is fused to the DNA binding moiety of either MLL or PAX3/7. Interestingly, it has been shown that PAX/MLL-FOXO fusions potently activate PI3K signaling, which results in the repression of the remaining normal FOXO alleles. Thus, the PAX/MLL-FOXO fusions resulting from a translocation event of a single allele of FOXO appear to result in a reduced activity of all FOXO isoforms (Schmitt-Ney & Camussi, 2015; So & Cleary, 2002).

As described before, hyperactivation of PI3K signaling is observed in many human cancers (Samuels et al., 2004). This is mostly due to gain-of-function mutations in PIK3CA or loss-of-function mutations in PTEN. However, also overexpression or otherwise hyperactivation of receptor tyrosine kinases (e.g., HER2) results in aberrant PI3K activation. Furthermore, whole-genome sequencing of tumors has also revealed gain-of-function mutations in PKB/AKT to be present in a small percentage of human cancers (reviewed in Yuan & Cantley, 2008).

Also single-nucleotide mutations of FOXO1 can be observed in some tumor types, i.e., non-Hodgkin lymphomas, follicular lymphoma, and diffuse B-cell lymphoma. These mutations are predominantly located in the N-terminus of the FOXO1 protein. They result either in expression of a truncated form of FOXO1 or, in case of R19, R21, and T24 mutation, they result in reduced PKB/AKT-mediated phosphorylation of the first PKB/AKT site of FOXO1 (Morin et al., 2011; Trinh et al., 2013). This impairs PKB/AKT-mediated FOXO inhibition and indeed these mutations increase FOXO1 nuclear localization, implicating enhanced FOXO1 activity in these tumors. This would be contradictory to what one would expect based on all other means of PI3K deregulation in tumors that will inactivate

FOXOs. These mutations did also correlate to poor prognosis and already point toward a more complex role for FOXOs in cancer.

3. FOXOs AND THE HALLMARKS OF CANCER

The hallmarks of cancer as originally defined by Hanahan and Weinberg in 2000 and revised in 2011 propose six biological capabilities acquired during the multistep development of human tumors. They include sustaining proliferative signaling, resisting cell death, enabling replicative immortality, inducing angiogenesis, evading growth suppressors, and activating invasion and metastasis. The hallmarks of cancer as defined by Hanahan and Weinberg provide a framework of prerequisites for tumor formation and progression. PI3K signaling toward FOXOs impacts on several of these hallmarks (Hanahan & Weinberg, 2000, 2011).

3.1 FOXOs and the Hallmarks: Sustaining Proliferative Signaling

Initially it was shown that activation of FOXO, either by pharmacological inhibition of PI3K-PKB or by the ectopic overexpression of FOXO mutants that are insensitive to PKB/AKT-mediated inhibition, resulted in a robust cell cycle arrest in the G1 phase through the transcriptional regulation of $p27^{kip1}$ (CDKN1B) expression status (Brunet et al., 1999; Kops et al., 1999; Kops, Medema, et al., 2002; Medema, Kops, Bos, & Burgering, 2000). Following this observation, FOXOs have also been described as regulators of $p21^{cip1}$ (CDKN1A), $p57^{cip2}$ (CDKN1C), and the INK4 family of CKIs, $p15^{INK4b}$ (CDKN2B), $p16^{INK4a}$ (CDKN2A), $p18^{INK4c}$ (CDKN2C), and $p19^{INK4d}$ (CDKN2D) (Bouchard et al., 2007; de Keizer et al., 2010; Katayama, Nakamura, Sugimoto, Tsuruo, & Fujita, 2008; Pellicano et al., 2014). FOXO-mediated induction of cyclin-dependent kinase inhibitor (CKI) expression leads to inhibition of the cyclin/CDK complexes responsible for progression through the different phases of the cell cycle. This results in a robust cell cycle arrest in G_0/G_1, G_2/M, or even senescence (Alvarez, Martínez-A, Burgering, & Carrera, 2001; Courtois-Cox et al., 2006; de Keizer et al., 2010; Medema et al., 2000). In addition, FOXO has also been shown to inhibit G_1–S progression via the transcriptional downregulation of cyclin D (Schmidt et al., 2002). The inhibition of cell cycle progression by FOXOs complies with the ability of active PI3K/PKB/AKT to inhibit FOXO and to stimulate cell cycle progression. Interestingly, some targets that are increased by FOXO, such as p27^{kip1}, are inhibited by PKB/AKT through direct phosphorylation and consequent degradation (Liang et al., 2002; Shin et al., 2002; Viglietto et al., 2002). This bilayered control of targets appears a common theme for the PKB-FOXO signaling module.

3.2 FOXOs and the Hallmarks: Resisting Cell Death

Following initial reports that ectopic expression of FOXO alleles insensitive to PKB inhibition caused G1 cell cycle arrest in a variety of cancer cell lines, it was noted that in some cell lines (e.g., pre-B cells, T cells) expression of the same FOXO alleles induces apoptosis (Dijkers, Medema, Lammers, Koenderman, & Coffer, 2000). Apoptosis in these cells is preceded by a cell cycle arrest, whereas in DLD-1 cells a G1 cell cycle arrest precedes a G0 arrest (Kops, Medema, et al., 2002).

Apoptosis occurs via two main pathways: an intrinsic pathway mediated by mitochondrial release of cytochrome *c* and caspase 9 activation, and an extrinsic pathway, in which activation of death receptors by ligands like TNFα, Trail, and Fas ligand activates caspase 8. Both pathways converge on caspase 3 and caspase 7 activation, which are the final and irreversible executioners. Mitochondrial membrane integrity is regulated by BAX and BAK protein located at the outer mitochondrial membrane. Activity of BAX/BAK is regulated through a competitive interaction with either proapoptotic BH3-only proteins (e.g., BIM, BAD, BID, PUMA, NOXA, BOK, and BMF) or antiapoptotic factors BCL2, BCL-XL, or MCL1. Antiapoptotic factors render BAK/BAX inactive. However, upon binding to proapoptotic proteins, BAK/BAX becomes activated, which leads to mitochondrial outer membrane permeabilization, cytochrome *c* release, and subsequently apoptosis (Strasser, Cory, & Adams, 2011). FOXOs can transcriptionally regulate factors involved in both the intrinsic and extrinsic apoptotic machinery. FOXOs regulate expression of proapoptotic BH3-only proteins BIM, PUMA, Bnip3, and BMF, but also expression of antiapoptotic BCL-XL (Chaanine et al., 2016; Dijkers et al., 2000; Hornsveld et al., 2016; Tang et al., 2002; You et al., 2006). This suggests that FOXOs do not induce apoptosis as a consequence of inducing proapoptotic proteins, but are more likely to regulate the homeostasis between pro- and antiapoptotic factors. Within the extrinsic pathway, FOXOs are shown to regulate FasL (Brunet et al., 1999; Wang & Li, 2010), TRAIL (Ghaffari, Jagani, Kitidis, Lodish, & Khosravi-Far, 2003; Modur, Nagarajan, Evers, & Milbrandt, 2002), and TNFR (Ding, Kirkiles-Smith, & Pober, 2009).

3.3 FOXOs and the Hallmarks: Enabling Replicative Immortality

Senescence is a state of permanent cell cycle arrest in the response to sustained stress over longer periods of times to prevent the accumulation of damaged cells. Several triggers can induce senescence, among which telomere attrition, DNA damage, and oncogene activation. These triggers promote the activation of p53/p21 or p16/RB, resulting in cell cycle arrest. Indeed, senescent cells usually have sustained high levels of the cell cycle inhibitors p21 and/or p16. Senescent cells remain nonproliferative; however, they are still metabolically active. Senescent cells differ from proliferating cells, as they are resistant to apoptosis and oncogenic transformation. Also, senescent cells often have a senescence-associated secretory phenotype (SASP), whereby they secrete a heterogeneous mixture of proinflammatory molecules (reviewed in de Keizer, 2017; Loaiza & Demaria, 2016).

Senescence has an important role in both aging and cancer. In aging, senescence is mostly detrimental for organismal functioning. The amount of senescent cells increases with age and with that the SASP secretion increases. Chronic SASP secretion contributes to inflammation and impairs tissue functioning (reviewed in de Keizer, 2017). However in cancer, senescence is considered as a tumor-suppressive mechanism. Senescence stops the proliferation of premalignant cells and thereby lowers the risk for cancer initiation. Additionally, the secreted SASP is also important in the recruitment of immune cells to remove malignant cells. However, SASP can also be oncogenic, as it can form a microenvironment that is supportive for tumor cell invasion, tumor vascularization, and immunosuppression (reviewed in Lecot, Alimirah, Desprez, Campisi, & Wiley, 2016).

FOXO has been shown to induce senescence in response to an oncogene-induced increase in ROS levels and subsequent JNK signaling. FOXO then upregulates p21, without concomitant changes in p16 or $p27^{kip1}$. In this way, FOXO protects the organism by removing possible premalignant cells via the induction of senescence. However, the tumor-suppressive effect of senescence goes at the expense of FOXO's positive effect on aging. On the contrary, in the case when FOXO is activated in the response to growth factor suppression, this usually results in a temporary G1 arrest or quiescence, thereby positively affecting both tumor suppression and aging (de Keizer et al., 2010; Medema et al., 2000).

Since p53 mediates senescence (Chen et al., 2005; Moiseeva, Mallette, Mukhopadhyay, Moores, & Ferbeyre, 2006) and FOXO and p53 directly interact with each other (Wang et al., 2008), it seemed likely that there would be a connection between the involvement of FOXO in senescence and FOXOs binding with p53. Indeed, it was shown that a peptide (FOXO4-DRI) that inhibits the binding of p53 and FOXO clears senescent cells by apoptosis (Baar et al., 2017). This puts FOXO at the interplay between p53-mediated senescence and apoptosis.

3.4 FOXOs and the Hallmarks: Inducing Angiogenesis

The formation of new blood vessels requires concerted action of proliferation and migration. In contrast to most cells, endothelial cells will proliferate and migrate under proangiogenic conditions, such as low nutrient and oxygen conditions, in order to increase oxygen and nutrient conditions at distant sites. Limiting conditions, such as low oxygen, are known to induce nuclear translocation and increased transcriptional activity of FOXOs (Bakker, Harris, & Mak, 2007). FOXO1 −/− mice die of vascular defects during embryonic development. This is probably due to an improper vascular endothelial growth factor A response of endothelial cells (ECs) (Furuyama et al., 2004). In return, VEGF binds to the VEGFR2 receptor, which activates PI3K signaling and inactivates FOXOs. In agreement with the role of VEGF in EC proliferation, EC-specific deletion of FOXO1 in adult mice resulted in hyperproliferation and a disorganized and dilated vascular network in the developing mouse retina. In contrast, introduction of a FOXO1 allele insensitive to PKB inhibition restricted proliferation and resulted in a sparse retinal vasculature composed of fewer than normal endothelial cells (Wilhelm et al., 2016). These results show that FOXO involvement is primarily to control cell proliferation as a prerequisite for migration in case of angiogenesis. In agreement herewith, conditional deletion of FOXO1/3 and 4 results in hemangiomas, indicating a tumor-suppressive role for FOXOs in ECs (Paik et al., 2007).

3.5 FOXOs and the Hallmarks: Activating Invasion and Metastasis

PI3K signaling is critical to many aspects of cellular migration and hence tumor invasiveness and metastasis. For example, PI3K signaling is an important denominator in the control of the actin polymerization status mostly through controlling the RAC small GTPase (reviewed in Reif, Nobes, Thomas, Hall, & Cantrell, 1996). However, relatively little is known with respect to the role of FOXOs in migration. Various studies suggest a role for FOXO in controlling cellular migration of prostate cancer cells.

Astrocyte-elevated gene-1 (AEG1) appears to reduce FOXO3 expression and it increases invasiveness of prostate cancer. Upon knockdown of AEG1, FOXO3 nuclear localization and activity is increased (Kikuno et al., 2007). FOXO3 represses expression of TWIST1, an important transcription factor for EMT. In agreement, reduced FOXO3 expression results in increased TWIST1 expression (Shiota et al., 2010). Also, FOXO1 and FOXO4 can limit migration of prostate cancer cells by antagonizing RUNX2 transcriptional activity (Zhang et al., 2011).

One of the prerequisites for metastasis is that tumor cells need to counteract apoptosis due to matrix detachment (anoikis). PI3K signaling regulates anoikis at various levels (reviewed in Buchheit, Weigel, & Schafer, 2014) and FOXOs have recently been shown to play a role in mediating anoikis as well. FOXOs mediate the expression of proapoptotic proteins BMF and BIM upon loss of contact with the extracellular matrix, thereby FOXO prevents metastasis (Gan et al., 2009; Hornsveld et al., 2016).

In T cells, FOXOs partake in the transcriptional control of the membrane expression of receptors that function to home T cells to secondary lymphoid tissue (Fabre et al., 2008; Sinclair et al., 2008). Here, FOXOs regulate transcription of KLF2, which in turn controls CD62L and other key lymph node homing receptors, including CCR7 and the sphingosine 1 phosphate receptor (S1P1) (Bai, Hu, Yeung, & Chen, 2007; Carlson et al., 2006; Kerdiles et al., 2009). The relevance of FOXO-mediated control of T-cell homing for cancer is not yet clear.

3.6 FOXOs Genuine or "Bona Fide" Tumor Suppressor

Based on the roles of FOXO in the hallmarks of cancer as discussed earlier, FOXOs can best be considered a tumor suppressor gene. Indeed, forced FOXO activation acts cytostatic in cancer cell lines expressing oncogenes that activate PI3K signaling (Kops, Medema, et al., 2002). Furthermore, in the classical two-step c-Myc-RAS model of oncogene transformation, inhibition of FOXO activity was shown to be sufficient to replace c-Myc and to cause cell transformation of primary MEFs together with RAS (Bouchard, Marquardt, Bras, Medema, & Eilers, 2004). Further studies extended this observation and showed that FOXOs are potent suppressors of the oncogenic function of c-Myc in mouse models of c-Myc-driven lymphomagenesis (Bouchard et al., 2007). Expression of either a dominant-negative version of FOXO or genetic deletion of FOXO3 enhances lymphomagenesis in the Eμ-myc mice model. Additional loss of p53 is commonly observed during tumor progression in this model and interestingly expression of dominant-negative FOXO alleviated the selective pressure to lose p53 during lymphoma development, whereas this was not observed after genetic deletion of FOXO3 (Bouchard et al., 2007). Whether in the latter model this is related to redundant activity of FOXO1 and/or FOXO4 is unknown, but in either case these observations point at a redundant role for p53 and FOXO in tumorigenesis. To circumvent the possibility of redundancy between the three major FOXO alleles, Paik et al. generated an inducible triple floxed FOXO1, FOXO3, and FOXO4 mouse model which upon conditional deletion showed increased frequency of spontaneous development of lymphoma and hemangiomas. This led to the conclusion that indeed FOXOs act as "bona fide" tumor suppressors (Paik et al., 2007).

3.7 FOXOs: Tumor Suppressors or Promoters?

Despite the large body of evidence that FOXOs act as tumor suppressor, evidence for the opposite is also emerging, namely that FOXO activity is beneficial for tumor progression.

Tumor cells encounter various forms of stress, predominantly through nutrient and oxygen deprivation. Both may act through increased ROS to induce nuclear retention of FOXO3 and this was shown to increase breast cancer cell invasion. Mechanistically, this involved FOXO3-mediated transcriptional activation of metalloproteases (MMP-9 and MMP-13; Storz, Doppler, Copland, Simpson, & Toker, 2009). Also in colon cancer cells, FOXO3 was shown to promote cell migration and metastasis and this involved the concerted action of β-catenin and FOXO3 (Tenbaum et al., 2012). As discussed earlier, in prostate cancer most evidence indicates that FOXO3 represses tumor cell migration and WNT/β-catenin deregulation is uncommon in prostate cancer (Liu et al., 2015). Thus, the status of WNT signaling and the level of nuclear β-catenin may provide the context that switches FOXO from a repressor to a promoter of metastasis. Similarly, depending on the type of movement (reviewed by Friedl & Wolf, 2003) and the type of tumor environment, migration and metastasis may be more or less dependent on extracellular matrix breakdown by metalloproteases. This then also provides the context, which determines whether FOXO can promote migration and metastasis.

Clearly, a transcription factor receives signals that certain genes require transcription activation, but a transcription factor has no knowledge of the consequence of this activation, i.e., whether it supports function of a normal cell or a cancer cell. A similar finding was described for the transcriptional control of IDH1 by FOXOs. IDH1 is found mutated mostly in AML and glioblastoma. Normally IDH1 converts isocitrate to α-ketoglutaric acid (a-KG), yet mutant IDH1 further converts a-KG into 2-hydroxyglutarate (2-HG). FOXOs equally regulate both wild-type and mutant alleles of IDH1. As such, FOXO can contribute to 2-HG formation in the case of cells carrying a mutant IDH1 allele. 2-HG is coined an "onco-metabolite." It acts as a competitive inhibitor of a-KG in the regulation of a large group dioxygenases, enzymes that require O_2 for their catalytic reaction. These include histone and DNA-modifying enzymes such as TET enzymes involved in DNA demethylation. This example illustrates how the same function, transcriptional regulation of IDH1, can turn FOXO into a tumor-promoting factor (Charitou et al., 2015).

Preservation of stem cell function is key for life span of multicellular organisms. The ability to replace damaged cells requires stem cell function to be tightly regulated and depletion of the stem cell reservoir is considered to be a cause of aging. The involvement of PI3K signaling toward FOXO has been mostly studied within the hematopoietic system. Conditional deletion of all FOXOs results in hematopoietic stem cell exhaustion probably due to increased proliferation (Tothova et al., 2007). Loss of only FOXO3 also results in stem cell depletion, showing that FOXO3 is the dominant allele in this compartment (Liang, Rimmelé, Bigarella, Yalcin, & Ghaffari, 2016; Miyamoto et al., 2007; Yalcin et al., 2008). Also neuronal stem cells require FOXO function, as loss of FOXO3 impairs neuronal stem cell function (Renault et al., 2009).

With respect to cancer, several studies also suggest involvement for FOXOs in regulating the cell fate of cancer stem cells (CSCs, also named stem cell-like cancer cells or cancer-initiating cells). Knockdown of FOXO3 results in increased CSC number, increased self-renewal, and increased expression of stem cell features in prostate, glioblastoma, ovarian, breast, liver, and colorectal cancer (Dubrovska et al., 2009; Ning, Luo, Ren, Quan, & Cao, 2014; Prabhu, Allen, Dicker, & El-Deiry, 2015; Smit et al., 2016; Sunayama et al., 2011; Zhou et al., 2015). However, in CML the opposite was observed, where loss of FOXO3 impaired leukemia-initiating cell function (Naka et al., 2010; Pellicano et al., 2014). High FOXO3 expression was also shown to correlate with poor prognosis in AML (Sykes et al., 2011). Clearly, the role of FOXOs from a stem cell perspective is not yet clear. It still needs to be defined in what context (i.e., microenvironment, additional mutations in pathways interacting with PI3K signaling, tumor type, tumor stage, etc.) FOXO activity will enhance or inhibit (cancer) stem cell function.

As we will discuss later, FOXO-dependent regulation of transcription serves to preserve homeostasis of the cell. Especially for long-living stem cells preserving homeostasis in time is essential, yet for cancer stem cells a similar scenario applies. Thus, the notion that the function of FOXO is similar within a cancer stem cell to that in a normal stem cell, namely to maintain homeostasis, generates the paradoxical outcome that within an existing tumor FOXO function may actually aid tumorigenesis. However, leukemia-initiating cells seemed to need FOXO3 for stem cell maintenance in CML (Naka et al., 2010; Pellicano et al., 2014). Indeed, active FOXO was observed in ~40% of human AML and inhibition of FOXO resulted in myeloid maturation and subsequent AML cell death (Sykes et al., 2011). Interestingly, FOXO activity was linked to active JNK in these AML samples suggesting a balance between PKB and JNK signaling in determining the final outcome of FOXO activity, as was also proposed previously by us (Essers et al., 2004). This again points toward the importance of stress conditions emerging during tumor development in activating FOXO programs that then become beneficial for cancer cells. Thus, knowledge of context is essential to understand FOXO regulation in detail and the role of PI3K signaling in tumorigenesis and this is not simply a linear readout. Consequently, reports on FOXO nuclear localization as prognostic factor are conflicting (Chen, Gomes, et al., 2010; Habashy et al., 2011).

Besides the homeostatic role of FOXO that may turn its function from a tumor suppressor in healthy cells to a tumor promoter in cancer cells, there are also other mechanisms by which FOXO can promote cancer growth directly. In anaplastic thyroid carcinoma, FOXO3 promotes cell cycle progression and proliferation by upregulating cyclin A1 transcription. This is caused by a decrease in Akt phosphorylation at Ser-473, due to low levels of the mTORC2-complex, in anaplastic thyroid carcinoma cell lines and patient samples, resulting in FOXO3 nuclear localization (Marlow et al., 2012). Besides this, FOXO1 has a role in the relapse of t(8;21) AML by increasing the self-renewal of preleukemic cells (Lin et al., 2017). This shows nonhomeostatic functions whereby FOXO may also increase tumorigenesis. This review will continue to focus on the homeostatic functions of FOXO in relation to cancer.

4. FOXOs AS RHEOSTATS IN HOMEOSTASIS

Because of the apparent major role of FOXOs in providing the ability to adapt to changing external conditions, this may equally account for cancer cells. Also cancer cells suffer from harsh conditions, such as lack of nutrients and hypoxia, but also treatment with anticancer drugs. Thus, in cancer cells, FOXOs may also act to preserve the cancer cell state and hence aid to tumor survival. This appears counterintuitive given that many cancers display (hyper)active PI3K/PKB(AKT) signaling, which would consequently inhibit FOXO transcriptional activity. However, some studies have suggested an exclusive role for cytosolic FOXO, independent of its ability to function as a transcription factor, including regulation of autophagy (Zhao et al., 2010). The nontranscriptional role of FOXO is relatively less well studied, and in case of autophagy, regulation concurs with FOXOs' role as a tumor suppressor. Therefore, it remains to be shown for FOXO that a nontranscriptional cytosolic function contributes to tumor progression. Alternatively, many cancer types sustain high levels of ROS and this is for cancer cells nondeleterious because of a concomitant increase in antioxidant levels (Liou & Storz, 2010). However, an imbalance between ROS production and antioxidant systems will cause so-called cellular oxidative stress. We have shown that activation of FOXO by cellular oxidative stress signaling may override inhibition by PI3K signaling in a dose-dependent manner (Essers et al., 2004). Thus, FOXO will protect cells during a change from the equilibrium between ROS and antioxidant capacity, be it in a normal cell or in a cancer cell even driven by deranged PI3K signaling. This will provide cancer cells with the plasticity and ability to respond to a changing environment, especially if this environment is signaling through redox changes.

4.1 Redox

Regulation of FOXOs by cellular oxidative stress and the ability of FOXOs to transcriptionally regulate antioxidant genes have revealed a rheostat role for FOXOs in maintaining redox homeostasis (Fig. 1). ROS are produced by oxygen metabolism at various cellular locations. Most ROS are extremely short lived (<ms) and react with molecules, proteins, lipids, DNA, etc. in the immediate proximity. These oxidation reactions will in most cases result in what is coined "oxidative damage," i.e., molecules are beyond repair inhibited in their normal function. This requires

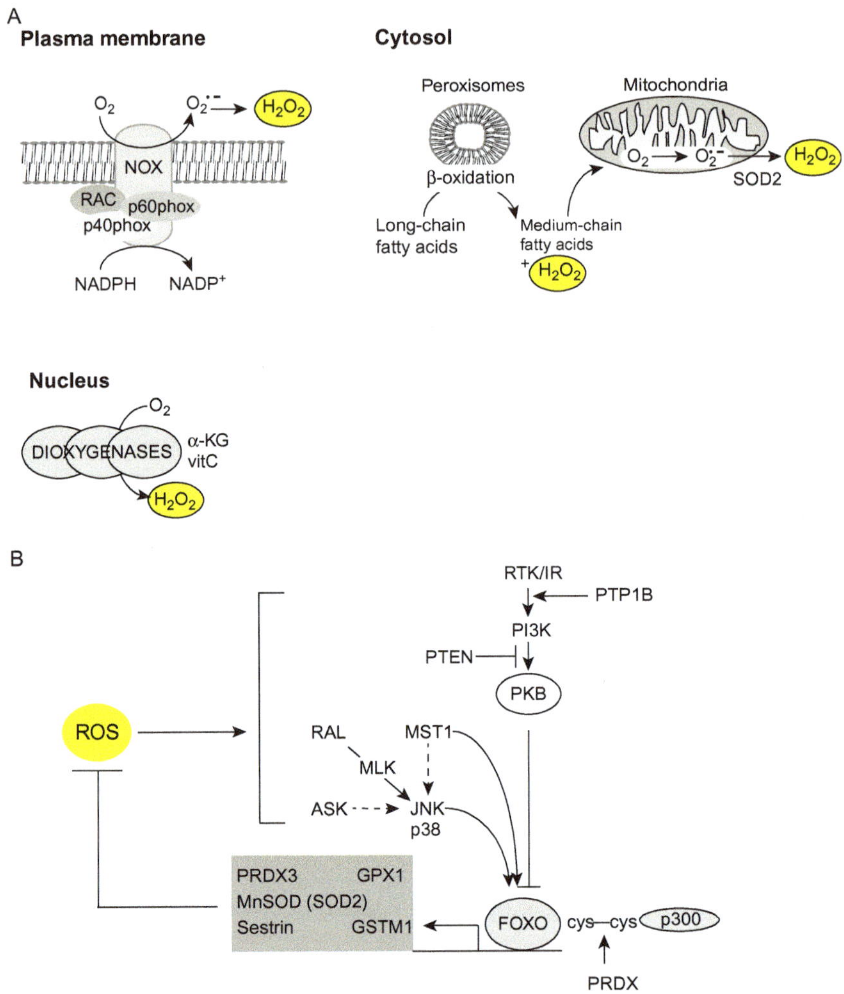

Fig. 1 FOXOs and homeostasis of cellular redox. (A) Hydrogen peroxide can be produced at several cellular locations via the conversion of superoxide by NOX, SOD, or dioxygenases, all of which can be regulated by FOXO. (B) ROS can activate FOXO in two ways: via activation of JNK by RAL and ASK1 or via oxidation of the cysteines of FOXO. Both of them result in nuclear accumulation of FOXO. In response, FOXO can regulate transcription of the antioxidant genes *PRDX3*, *SOD2*, *sestrin*, *GPX1*, and *GSTM1*.

degradation and de novo synthesis to restore function. However, superoxide can be converted by superoxide dismutases in hydrogen peroxide (H_2O_2), which is paradoxically also a ROS and has a considerably longer half-life (10–20 s). In principal, this reaction would thus enhance or spread oxidative damage. However, research over the years has provided compelling evidence that H_2O_2 has a role in normal signal transduction and other cellular processes (reviewed in Veal, Day, & Morgan, 2007). For example, following activation of the insulin receptor, NOX4 activation results in membrane-localized production of H_2O_2 (Mahadev et al., 2004). Inhibition of the negative regulator of insulin signaling (PTP1b) by H_2O_2 is required for increased and prolonged insulin signaling (Salmeen et al., 2003). Insulin-induced H_2O_2 production is thus required for temporal inhibition of negative feedback on insulin signaling. In addition, it is noted for a long time that high levels of antioxidants inhibit cell cycle progression, although the mechanism here is unclear (Menon et al., 2003). The notion that ROS production also serves beneficial processes indicates that ROS production and clearance requires tight control. This is illustrated among others by the fact that peroxiredoxin, an important family of redox regulators, makes up almost 1% of all soluble protein in mammalian cells (Chae, Kim, Kang, & Rhee, 1999).

When high levels of ROS are produced in the cell, FOXOs move to the nucleus and become active in two different ways. The first one is JNK mediated. JNK is a member of the mitogen-activated protein kinase (MAPK) family and is activated by a variety of cellular stresses including genotoxic stress. JNK activation occurs by the redox-sensitive kinase ASK1 and the small GTPase Ral. Activated JNK inactivates RTK signaling by phosphorylation of insulin receptor substrate adaptor proteins IRS1/2 and FOXOs directly. JNK phosphorylates FOXO4 at Thr-447 and Thr-451 in the response to H_2O_2. This overrides GF signaling-dependent inactivation of FOXO, resulting in nuclear translocation and increased transcriptional activity of FOXO (Essers et al., 2004; van den Berg & Burgering, 2011).

Second, cysteines in FOXOs can be oxidized by H_2O_2. Oxidized FOXOs form disulfide bridges with nuclear importers TNPO1, IPO7, and IPO8, resulting in strong binding and nuclear translocation independently of other signals (Putker et al., 2013, 2015). In order to counteract elevated ROS production in the cell, FOXO mediates the transcription of antioxidant genes like *Catalase*, *SESN1/2/3*, *SOD2*, *PRDX3*, *GPX1*, *GSTM1*, and genes involved in the metabolic generation of the glutathione antioxidant system and reductive entities like NADPH (Klotz et al., 2015; Yeo et al., 2013).

4.2 DNA Damage Response

It is important for cells to maintain genomic stability in order to prevent the accumulation of DNA mutations and genomic rearrangements with a subsequent risk of cancer initiation. To maintain genomic stability, cells need to properly sense and repair DNA lesions. These lesions can result from everyday exogenous stressors such as ultraviolet irradiation and chemical exposure, as well as from endogenous processes such as replication errors. There are a multitude of DNA lesions that can occur in a cell, such as covalent modifications or breakage of the phosphate backbone. Cells have evolved a plethora of tightly regulated mechanisms for surveillance, detection, and repair of DNA lesions. These mechanisms are collectively referred to as the DNA damage response (DDR) (Ciccia & Elledge, 2010; Hoeijmakers, 2009).

Following detection of DNA damage, the initial response includes temporary cell cycle arrest to prevent lesions to be processed into fixed mutations, which requires DNA replication. During cell cycle arrest, the cell repairs the DNA lesions if possible. If repair was successful, the cell restarts cell cycle progression without consequent mutations or compromised genomic stability. Cells normally detect unsuccessful repair and this can evoke two major responses to prevent propagation of cells with DNA damage. Apoptosis results in permanent deletion of these damaged cells, whereas senescence results in a permanent withdrawal from the cell cycle by which the damaged cell can still contribute to organ integrity and mass, without the possibility of proceeding into a hyperplastic or cancerous lesion. Senescent cells in principal do not cycle but remain metabolically active. Secretion of interleukines by senescent cells is shown to be potentially harmful to the neighboring nonsenescent cells in as much that senescent cells can induce tissue damage and even promote cancer progression. Unsuccessful repair is not always sensed by the cell, which can lead to an accumulation of cells with unrepaired or not properly repaired DNA. This is associated with cancer and aging and has resulted in the "DNA damage theory of aging" (Freitas & de Magalhaes, 2011).

Critical mediators of the DDR are ATM/ATR kinases. They phosphorylate many substrates that are directly involved in the actual repair process. They also phosphorylate effector kinases such as Chk1 and Chk2, which are involved in regulating cell cycle machinery. An important key downstream mediator of DDR signaling is p53, which is directly and indirectly stabilized upon DNA damage. This results in an increase in $p21^{cip1}$ transcription and protein expression. $p21^{cip1}$ acts as a direct inhibitor of cyclin/cdk complexes, which are the main drivers of cell cycle phase transitions. Indeed, cells lacking p53 or $p21^{cip1}$ display an impaired DDR response. Proteins involved in DDR are often implicated as being tumor suppressors. Examples are sensors of DNA damage (ATM/ATR), regulators of cell cycle arrest (Chk1/Chk2), and inducers of apoptosis if the DNA is not correctly repaired (p53) (Motoyama & Naka, 2004). Also, human genetic disorders with disruptions in DNA repair proteins are associated with increased cancer risk. Known examples are BRCA1/2, ATM, and NBS1. Mutations in these genes usually result in syndromes with multiple symptoms, including an increased risk for certain types of cancer (reviewed in Romero-Laorden & Castro, 2017). Importantly, the loss of activity of some DNA repair proteins is not associated with increased cancer risk, but with premature aging (reviewed in Hoeijmakers, 2009; Schumacher, 2009). This illustrates the tight connection between cancer and aging and a pivotal role for DNA repair processes. Of note, whereas loss of DNA repair may result in premature aging the opposite, increased life span as a result of increased DNA repair has not yet been reported. This is in contrast to FOXO/DAF-16 for which in model organisms, loss of activity decreases life span and increased activity extends life span.

p53 and FOXO share an extended similarity, which has been reviewed by us and others (van der Horst & Burgering, 2007; You & Mak, 2005). FOXO and p53 are similar in their modes of regulation as well as their transcriptional output. Together this suggests similarity in function, in which both p53 and FOXO act as rheostat regulators for homeostasis maintenance. The action of both p53 and FOXO is required during periods of stress, when the normal healthy balance ensuring cell viability is temporarily or definitely disturbed. Given the pivotal role of both the DDR and FOXOs in aging and cancer, a link between the DDR and FOXOs is to be expected. This hypothesis is further strengthened by the similarity between p53 and FOXO, and the role of p53 in the DDR.

FOXOs are phosphorylated and regulated by several kinases involved in the response to genotoxic stress. JNK activates FOXO in response to oxidative stress, resulting in nuclear accumulation (Essers et al., 2004). Also other MAPK family proteins are able to activate FOXO. MK5/PRAK promotes nuclear accumulation of FOXO3 in response to DNA damage by the phosphorylation of Ser-215 (Kress et al., 2011). Similarly, MK5 was shown to also phosphorylate FOXO1 and activate its nuclear localization (Chow, Timblin, McWhirter, & Schlissel, 2013). In addition, p38 phosphorylates FOXO3 at Ser-7 in response to DNA damage mediated by doxorubicin, which results in nuclear accumulation and activation of FOXO3 (Ho et al., 2012).

Cyclin-dependent kinases (CDKs) are key regulators of cell cycle progression and through phosphorylation control many essential cell cycle components. CDK1 and CDK2 activity is inhibited upon DNA damage, thereby linking DNA damage to regulating cell cycle arrest. CDK1 and 2 can phosphorylate FOXO1 at Ser-249. This phosphorylation appears to have an inhibitory effect on FOXO1, as it mediates FOXOs nuclear exclusion (Huang, Regan, Lou, Chen, & Tindall, 2006; Liu, Kao, & Huang, 2008). In agreement, the use of a FOXO peptide that blocks the phosphorylation of FOXO by CDK2 resulted in increased nuclear accumulation and transcriptional activity by FOXO (Lu, Liu, Pan, & Huang, 2011). Contrary to these findings, it was reported that in neuronal cells CDK1 also mediates the phosphorylation of FOXO1 at Ser-249, yet phosphorylation mediates in this case the nuclear accumulation and thereby activation of FOXO via the inhibition of the binding to the 14-3-3 proteins (Yuan et al., 2008). Clearly, FOXO activity is regulated by CDK1 and CDK2 phosphorylation, the activity of which is in turn regulated by DNA damage. However, it is not yet clear how the context, e.g., cell type and other stresses, determines whether FOXOs activity is downregulated or upregulated upon CDK1 phosphorylation.

FOXOs have been shown to indirectly function in DNA damage repair. This function is directly related to FOXOs function as a transcription factor. In this way, growth arrest and DNA damage-inducible A (gadd45a) and damage-specific DNA binding protein 1 (DDB1) have been shown to be increased in the response to DNA damage as a direct result of FOXO binding to their promoter. In addition, FOXO has been shown to promote developmental growth in response to UV irradiation via a change in its transcriptional profile.

The role of Gadd45a in DDR is not entirely clear, but it is reported to be involved in nucleotide and base excision repair. It is also involved in the DNA damage response upon UV irradiation. Part of its response is mediated via binding to PCNA. The *GADD45A* gene has a FOXO3 consensus binding site within its promoter region. In agreement, overexpression of FOXO3 or inhibition of PI3K with LY294002, which induces FOXO activity, increases the expression of Gadd45a, indicating that FOXO3 directly regulates the expression of Gadd45a. The regulation is DNA damage dependent, since treatment with UV induces a FOXO3-dependent increase in expression of Gadd45a. Furthermore, treatment with the DNA replication stress inducer aphidicolin causes a G2/M delay important for DNA repair. This delay is mediated by FOXO3 and could be directly attributed to the FOXO3-dependent increase in Gadd45a expression (Tran et al., 2002). Similar results were also reported for FOXO4 in response to oxidative stress. Treatment of cells with H_2O_2 results in the induction of Gadd45a, which is regulated by FOXO4. This similarly induces a G2 cell cycle arrest regulated by Gadd45a (Furukawa-Hibi et al., 2002). This shows that FOXO in response to DNA damage increases the expression of Gadd45a to promote a proper DNA damage response.

The FOXO-dependent expression of Gadd45a is further regulated by myc, Sirt1, and APRIL/BCMA. In response to serum, nuclear FOXO goes down via PI3K-PKB/AKT signaling and myc occupancy of the Gadd45a promoter goes up. Combined this results in a reduction of Gadd45a expression (Amente et al., 2011). Besides that, Sirt1 regulates the deacetylation of FOXO3. Sirt1 is dependent on NAD+ to deacetylate its target proteins. Inhibition of NAMPT, the enzyme that regulates the conversion from NADH to NAD+, results in a reduction of cellular NAD+. So the reduction of NAD+ by inhibition of NAMPT results in a decreased functioning of Sirt1 and an increase in FOXO3 acetylation. This decreases DNA binding by FOXO and consequently decreases Gadd45a expression (Thakur et al., 2012). Furthermore, APRIL/BCMA are part of the TNF family. They are mainly involved in B cell maturation, but seem to have also functions in other cell types. It was shown that in hepatocytes, APRIL/BCMA seem to stimulate the phosphorylation of JNK and thereby the activity of FOXO. The increased activity of FOXO increases Gadd45a expression (Notas et al., 2012). This shows that the FOXO-dependent Gadd45a expression can be further tuned by several stress-related stimuli such as growth factor deprivation or energy deprivation.

FOXO also regulates the expression of DNA damage-related protein DDB1. DDB1 is involved in NER and regulates the expression of UV-induced genes. It also regulates the cell cycle arrest following DNA damage by influencing the speed of degradation of $p27^{kip1}$ (Iovine, Iannella, & Bevilacqua, 2011). The expression levels of DDB1 were also found to be increased by FOXO1, again indicating a positive effect of FOXO on DNA repair (Ramaswamy, Nakamura, Sansal, Bergeron, & Sellers, 2002). FOXO-dependent DDB1 expression can be regulated by LMP1, an oncogene involved in among others DNA repair. In response to UV damage, LMP1 decreases DNA repair via the activation of AKT/PKB. This subsequently inactivates FOXO and decreases DDB1 expression, thereby hampering DNA repair following UV irradiation (Chen et al., 2008). So besides Gadd45a, also DDB1 expression is regulated by FOXO and in this way FOXO indirectly promotes DNA repair.

In addition, FOXO regulates developmental growth in the response to UV irradiation. A study in *C. elegans* shows that UV irradiation of worms results in activation of DAF16 (the *C. elegans* homologue of FOXO). Furthermore, they show that DAF16 activation overrides the developmental arrest normally initiated in response to DNA damage, thereby DAF16 promotes developmental progression. This UV-induced regulation of gene expression resulting in growth seems counteractive to DAF16's known induction of growth arrest in response to starvation (Mueller et al., 2014). This system could potentially also be exploited by cancer cells, where FOXO could induce growth upon treatment with DNA-damaging agents.

FOXO also has a direct function in the DNA damage response independent of its transcription factor function. This has been shown by its direct interaction with ATM upon DNA damage. FOXO has also been shown to be important for DNA repair in response to UV irradiation in a manner independent of its transcription factor function via binding RPA1. Additionally, FOXO interacts with Fanconi anemia factor FANCD2 in response to oxidative stress.

ATM is a phosphatidylinositol 3-kinase-related kinase (PIKK) that is important in the initial steps of the DNA damage response. FOXO3 indeed interacts with ATM, but surprisingly this interaction appears to regulate ATM activity. ATM binds to the CR3 domain of FOXO3 (amino acid 616–623), whereas FOXO binds to the FRAP-ATM-TRRAP (FAT; amino acid 1802–2571) or the FAT-C-terminal (FATC; amino acid 2842–3056) domain of ATM. The binding of FOXO to ATM also mediates the interaction of Tip60 (KAT5) and ATM (Tsai, Chung, Takahashi, Xu, & Hu, 2008). In turn, Tip60-mediated acetylation of ATM is important in the activation of ATM by autophosphorylation, one of the most upstream events in the DNA damage response (Adamowicz, Vermezovic, & d'Adda di Fagagna, 2016; Tsai et al., 2008). Initially, a proteomic screen identified FOXO1 also as a potential substrate for ATM/ATR kinases, phosphorylated at a consensus ATM site (S/T-Q) upon irradiation, namely, at Ser-509 (Matsuoka et al., 2007). In addition, our lab shows phosphorylation of FOXO4 at an SQ site in response to DNA damage, which is absent upon ATM inhibition (manuscript in preparation). Furthermore, in hematopoietic stem cells, it has been shown that FOXO regulates the expression of ATM (Yalcin et al., 2008). Overexpression or knockdown of FOXO results in a respective increase or decrease of ATM RNA and protein levels. In agreement with the involvement of FOXO in ATM autophosphorylation, FOXO-knockout cells also show a decrease in pATM at Ser-1981 (Yalcin et al., 2008). These studies indicate a role for FOXO in regulating ATM and vice versa, but the exact process still needs to be elucidated.

In the DNA damage response upon UV irradiation, it was shown that knockdown of FOXO1, but not FOXO3, induces sensitivity and thereby increases apoptosis. Normally, UV lesions are repaired by NER. However, if repair does not occur before S-phase of the cell cycle, the lesion can be repaired by translesion synthesis in order to prevent replication fork collapses. During translesion synthesis, specialized polymerases can incorporate new bases at the opposite strand of the UV lesion. The recruitment of these polymerases is promoted via the monoubiquitination of PCNA by Rad18. Rad18 is activated by ssDNA coated with RPA1. It is shown that FOXO1 binds RPA1 and decreases the monoubiquitination of PCNA, indicating the involvement of FOXO with translesion synthesis (Daitoku et al., 2016).

Fanconi anemia (FA) is a syndrome that arises from mutations in the FA pathway, which results in bone marrow depletion, hypersensitivity to interstrand cross-link agents, and increased cancer risk. The FA pathway consists of 16 genes that function together in DNA damage repair. FANCD2 is a component of the FA pathway that is monoubiquitinated by the FA core complex upon DNA damage and during S-phase of the cell. In response to monoubiquitination it localizes to the damage site and effects downstream FA components that function directly in DNA repair (Longerich, Li, Xiong, Sung, & Kupfer, 2014). FOXO3 has been shown to interact with FANCD2 in the response to oxidative stress, but only when the FA core complex is intact and FANCD2 is monoubiquitinated. FOXO3–FANCD2 binding results in less ROS accumulation and an increase in antioxidant gene expression (Li, Du, Maynard, Andreassen, & Pang, 2010). This suggests an oxidative stress pathway via the interaction of FANCD2 and FOXO3 that results in an improved cellular antioxidant defense.

The above indicates both direct and indirect involvement of FOXO in the DNA damage response (Fig. 2), although the exact mechanism by which FOXO influences DNA repair is not yet fully established. Proper repair of DNA lesions is important for a cell to maintain its genomic integrity. So FOXO is needed in healthy cells for proper DNA repair and in this way, FOXO maintains genomic stability and decreases the risk of tumor initiation or progression. On the other hand, cancer cells are often treated with chemotherapeutics with a mode of action that increases DNA damage and it thereby targets rapidly dividing cancer cells. In cancer cells that are treated with such chemotherapeutics, the presence of FOXO would theoretically

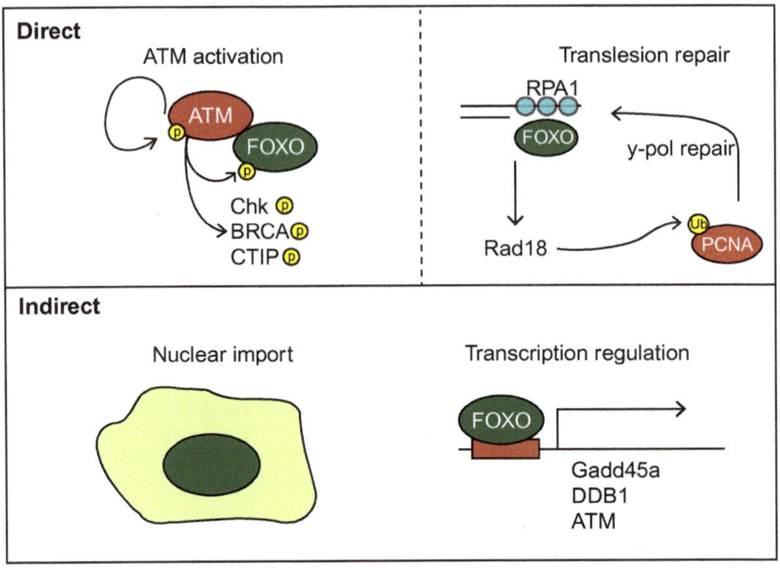

Fig. 2 FOXOs and homeostasis of genomic stability. FOXO affects the DNA damage response both directly and indirectly. FOXO directly binds to ATM and thereby promotes the autophosphorylation of ATM. In response, ATM becomes activated to phosphorylate downstream proteins, like checkpoint protein (Chk), repair protein (BRCA/CTIP), and also FOXO itself. It is also directly involved in translesion repair. FOXOs bind to ssDNA coating protein RPA1. This activates Rad18 and results in the monoubiquitination of PCNA, which in turn attracts specialized polymerases for repair. During indirect repair, the induction of damage promotes nuclear localization of FOXO and thereby FOXO is able to change the expression of target genes that are directly involved in the DNA damage response, such as Gadd45a, DDB1, and ATM.

provide a resistance mechanism since the cancer cells expressing FOXO will cope better with DNA damage. So far, no studies have been reported that focus on this potential mode of chemotherapeutic resistance caused by FOXO presence in cancer cells.

4.3 Protein Homeostasis

Protein homeostasis involves both breakdown and renewed synthesis of proteins. PI3K signaling toward FOXO is involved in the two major pathways of protein breakdown, autophagy and the ubiquitin–proteasome-mediated protein breakdown.

Autophagy is a complex process in which substrates destined for degradation are sequestered in autophagosomes, small membrane vesicles that ultimately fuse with lysosomes, in which the content is degraded by lysosomal hydrolases. This process requires the concerted action of Atg (autophagy-specific gene) proteins and a multitude of related proteins (Beclin-1, ULK1, Vsp34, etc.) (Gallagher, Williamson, & Chan, 2016).

PKB regulates Tsc2 through direct phosphorylation (Potter, Pedraza, & Xu, 2002), which in complex with Tsc1 acts to regulate mTORC1 complex (reviewed in Huang & Manning, 2008). This complex is key in integrating autophagy as a cellular response toward adverse conditions such as nutrient, energy, and growth factor deprivation. PKB activity results in increased mTORC1 activity, thereby preventing induction of autophagy. Growth factor deprivation results in decreased PKB and hence mTORC1 activity, resulting in autophagy induction (reviewed in Jung, Ro, Cao, Otto, & Kim, 2010).

The Beclin1/vacuole protein sorting 34 (Vps34)/(PI3K) complex is important for the initiation of the autophagic process (Kihara, Kabeya, Ohsumi, & Yoshimori, 2001; Matsuura, Tsukada, Wada, & Ohsumi, 1997). Interestingly, Beclin-1 has been reported as PKB substrate where PKB-mediated phosphorylation of Beclin-1 results in 14-3-3 binding and sequestering of Beclin-1 to vimentin filaments and consequent Beclin-1 inhibition. This indicates that next to mTORC1 regulation PKB also through Beclin-1 regulation inhibits autophagy and this was shown to be involved in PI3K-driven tumor formation.

FOXOs have been reported to regulate gene expression of a large number of genes encoding proteins involved in autophagy (ATG5, ATG8, ATG12, ATG14, beclin1, ULK1, LC3, Gabarapl1, and Bnip3) and as such can regulate in a transcriptionally dependent manner autophagy (reviewed by van der Vos, Gomez-Puerto, & Coffer, 2012). In addition, cytosolic FOXO1 has also been shown to directly interact with ATG7 forming a complex that directly may initiate autophagy (Zhao et al., 2010). FOXOs also regulate transcription of glutamine synthase (GS) and IDH1, two key enzymes in glutamine metabolism. Glutamine metabolism has been linked to mTORC1 activity and it has been proposed that L-glutamine regulates mTORC1 activity through SLC7A5/SLC3A2, a bidirectional transporter that regulates L-glutamine influx followed by a rapid simultaneous efflux of L-glutamine out of cells and transport of L-leucine/EAA into cells (Nicklin et al., 2009). However, this mechanism is debated. Alternatively, it has been shown that glutaminolysis, the conversion of glutamine to a-KG, through two subsequent deamination steps catalyzed by glutaminase and glutamate dehydrogenase, regulates mTORC1 activity (Duran et al., 2012). FOXOs also control glutamine metabolism by regulating GS which would result in a shift in the net balance of a-KG and glutamine to a FOXO-induced increase of a-KG. However, a-KG can also be produced from isocitrate through the action of IDH1. FOXOs also control IDH1 expression and when directly measured, loss of FOXO reduces a-KG levels, suggesting the latter reaction to be determining a-KG levels rather than indirectly by increased GS. Clearly, metabolic flux measurements need to be performed to understand the details of the regulation of glutamine metabolism by FOXOs and its impact on mTORC1 regulation, but these have yet to be published. FOXOs also regulate mTORC1 via the expression of sestrins. Sestrins represent a small conserved gene family, *sesn1*, *sesn2*, and *sesn3*, expression of which is induced under conditions of stress including DNA damage and oxidative stress through p53 and FOXOs. Sestrins inhibit mTORC1 via the activation of AMPK (Budanov & Karin, 2008).

All the above could indicate an involvement of FOXOs in tumor suppression via activation of autophagy. This would be in line with PKB activity repressing autophagy. However, PKB activation will also result in cytosolic accumulation of FOXOs and it has been shown that when FOXOs are transcriptionally silenced through PKB-mediated phosphorylation and nuclear exclusion, the cytosolic FOXO binds ATG7 in an acetylation-dependent manner. Binding of FOXO to ATG7 would activate autophagy (Zhao et al., 2010).

Besides autophagy, FOXOs also regulate ubiquitin–proteasome-dependent protein turnover. FOXO4 directly regulates the transcription of PSMD11, a 19S proteasome subunit. Increased expression of FOXO4 results in increased proteasome activity via the upregulation of PSMD11 (Vilchez et al., 2012). Furthermore, several E3 ubiquitin ligases are transcriptionally regulated by FOXOs and this has been best studied in the context of muscle atrophy (Sandri, 2013). The expression of the E3 ligases MAFbx/atrogin-1 and MuRF1 is controlled by FOXO1 and FOXO3 (Waddell et al., 2008; Zheng, Ohkawa, Li, Roberts-Wilson, & Price, 2010). These E3 ligases are overexpressed in various atrophy models like fasting and muscle denervation (Li et al., 2011; Tang et al., 2014). MAFbx/atrogin-1 activity leads to the ubiquitination of several proteins involved in skeletal muscle maintenance, including the myogenic transcription factors MyoD and myogenin (Jogo, Shiraishi, & Tamura, 2009; Tintignac et al., 2005).

5. FOXOs AND RESISTANCE TO THERAPY

Resistance to therapy is a major drawback of all cancer treatments, including novel targeted therapies using small molecules. One consequence of the central role FOXOs play in a number of homeostatic processes is that FOXOs will likely counteract drug treatments as these treatments disturb the existing homeostasis equilibrium in cells. Recent studies have shown that the PI3K signaling pathway indeed is a major determinant of therapy resistance (Berns et al., 2007; Eichhorn et al., 2008). PI3K signaling directly promotes survival and thereby resistance, but in addition PI3K signaling regulates feedback signaling that may indirectly result in a signaling bypass that renders therapy no longer efficient. Intriguingly, transcriptional targets of FOXO have been reported in several cases to be the consequence of feedback that leads to resistance (e.g., ERBB3; Sun et al., 2014). Here, we discuss three possible general mechanisms whereby FOXOs can contribute to therapy resistance, namely, induction of quiescence, regulation of growth factor feedback signaling, and cross talk with other pathways.

To gain more insight into the direct involvement of FOXO in sensitivity or resistance, we determined the effect of FOXO to some often-used chemotherapeutics, namely, cisplatin, doxorubicin, etoposide, and 5-fluorouracil (Table 1). For these compounds, FOXO often seemed to increase drug sensitivity, mostly via the upregulation of apoptosis. Besides this, FOXO could also induce chemoresistance for these compounds via the upregulation of drug efflux pumps (Aldonza et al., 2016; Han et al., 2008; Hui, Francis, et al., 2008).

5.1 Quiescence as a Mechanism of Resistance

In mammalian systems, quiescence is considered a specific type of prolonged G1 cell cycle arrest named G0. Quiescence displays features, which may discriminate this G0 state from a state of extreme slow cell cycling. It has been shown that various experimental conditions of inducing quiescence (i.e., growth factor withdrawal, contact inhibition, etc.) result in differential changes in gene expression of which some sort of common quiescence gene "signature" can be extracted (Coller, Sang, & Roberts, 2006). Importantly, this gene signature differs from prolonged cell cycle arrest imposed by artificial activation of cell cycle inhibitors, such as $p21^{cip1}$ (Sang, Coller, & Roberts, 2008).

Previously we have shown that acute FOXO activation results in a cell cycle arrest in G1 through regulation of $p27^{kip1}$ and reduced CDK-dependent Rb phosphorylation (Medema et al., 2000). Following prolonged arrest cells can proceed either toward apoptosis (Birkenkamp & Coffer, 2003), to quiescence (Kops, Medema, et al., 2002) or senescence (de Keizer et al., 2010). Importantly, these FOXO-dependent choices of cell fate are highly context dependent, for example, under conditions of

Table 1 FOXO Can Promote Resistance or Sensitivity to Chemotherapeutic Agents

CPT	Pro/Con	FOXO	ind/dec	By	Tissue/Cell Line	Cancer Type	Timing	Concentration	Mechanism
Park et al. (2017)	Pro	FOXO1	ind	OE FOXO1	Cell line	Gastric cancer	72h	10 μg/mL	Reduced Her2 and MET
Zhang et al. (2017)	Con	FOXO1	ind	CAFS	Cell line	Esophageal carcinoma	24h	20 μg/mL	Induced TGFβ
Wei et al. (2016)	Pro	FOXO3	ind	miR-i	Primary cell lines/ xenografts	Ovarian cancer	72h	8 μg/mL max	ND
Chi et al. (2016)	Pro	FOXO1	dec	TH/TR	Cell line/ xenograft	Hepatoma cells	1 month	10 mg/kg/ week	Decreased apoptosis
Guan et al. (2016)	Pro	FOXO3	dec	shFOXO3	Cell line/ xenograft	Cholangiocarcinoma cell	48h/ 1 month	5 mg/kg	NRF2 redox response
Altan et al. (2016)	Con	FOXO1	ind	siPRMT1	Cell line	Gastric cancer	48h	16 μg/mL	ND
Germani et al. (2014)	Pro	FOXO3	dec	siFOXO3	Cell line	Colorectal carcinoma	36h	30 μM	Reduced apoptosis
Zhang et al. (2013)	Pro	FOXO3	dec	miR-153	Cell line	Colon carcinoma	48h	50 μg/mL	Reduced apoptosis
Tezil, Bodur, Kutuk, and Basaga (2012)	Pro	FOXO3	Both	siFOXO3/OE FOXO3	Cell line	Breast cancer	48h	30 μM	Autophagy

Chock, Allison, Shimizu, and ElShamy (2010)	Pro	FOXO1/3	ind	BRCA-iris OE	Cell lines	Ovarian cancer	48 h	100 μM	Induced apoptosis
Shiota et al. (2010)	Pro	FOXO3	dec	siFOXO3	Cell line	Bladder carcinoma	48 h	10 μM	ND
Lei and Quelle (2009)	Con	ND	dec	FOXO-kd	Cell line	Bone marrow	24 h	5 μM	Dysregulated cell cycle arrest
Arimoto-Ishida et al. (2004)	Pro	FOXO3	dec	OE TM.FOXO3	Cell line	Ovarian carcinoma	120 h	100 μM	ND

DXR	**Pro/ Con**	**FOXO**	**ind/ dec**	**By**	**Tissue/Cell Line**	**Cancer Type**	**Timing**	**Concentration**	**Mechanism**
Shen et al. (2017)	Pro	FOXO1	dec	miR-222	Cell line	Breast cancer	48 h	3.2 μM	ND
Chi et al. (2016)	Pro	FOXO1	dec	Thyroid hormone	Cell line	Hepatoma	24 h	2.5 μM	Reduced apoptosis
Tubi et al. (2017)	Pro	FOXO1/3	ind	CK2 block	Cell line	AML	18 h	Suppl	Induced apoptosis
Aldonza, Hong, Alinsug, Song, and Lee (2016)	Con	FOXO3	Both	OE FOXO3/ siFOXO3	Cell line	Adenocarcinoma/ prostate cancer	8 h	2 μM	Upregulation efflux pumps
Ryu et al. (2017)	Con	FOXO4	Both	OE FOXO4/ siFOXO4	Cell line	B-cell lymphoma	48 h	1 μM	Increased stem cell features
Shimizu et al. (2016)	Con	FOXO3	dec	Pin1	Cell line	MEF	16 h	0.4 μg/mL	Downregulation efflux pumps

Continued

Table 1 FOXO Can Promote Resistance or Sensitivity to Chemotherapeutic Agents—cont'd

DXR	Pro/ Con	FOXO	ind/ dec	By	Tissue/Cell Line	Cancer Type	Timing	Concentration	Mechanism
Zhou et al. (2015)	Pro	FOXO3	Both	OE FOXO3/ siFOXO3	Cell line	Hepatocellular carcinoma	48 h	0.25 μg/mL	Prevent EMT
Go et al. (2015)	Pro	FOXO1	dec	FOXO1i	Cell line	Lymphoma	ND	ND	ND
Yang et al. (2014)	Pro	FOXO3	ind	mdm2i	Xenograft	Adenocarcinoma	25 days	1 mg/kg	Reduced DNMT3B
Simioni et al. (2013)	Pro	FOXO3	ind	MK-2206	Cell line	Hepatocellular carcinoma	72 h	1.5 μM	ND
Choi et al. (2013)	Con	FOXO1	dec	Nicotinamide	Cell line	Breast cancer	36 h	30 μM	Downregulation efflux pumps
Obrador-Hevia, Serra-Sitjar, Rodriguez, Villalonga, and Fernandez de Mattos (2012)	Pro	FOXO3	ind	Psammaplysene A	Cell line	Lymphoma	48 h	1 μM	ND
Hagenbuchner et al. (2012)	Pro	FOXO3	Both	FOXO3.A3/ shFOXO3	Cell line	Neuroblastoma	4 h	0.5 μg/mL	ROS accumulation
Dieudonne, Marion, Marie, and Modrowski (2012)	Pro	FOXO3	ind	OE FOXO3	Cell line	Osteosarcoma	24 h	50 ng/mL	Syndecan increase

Obexer et al. (2009)	Pro	FOXO3	ind	OE FOXO3	Cell line	Neuroblastoma	72 h	0.25 μg/mL	Induced apoptosis
Wang and Li (2010)	Pro	FOXO3	Both	OE FOXO3/ siFOXO3	Cell line	Lung cancer/ neuroblastoma	12 h	2 μM	Induced apoptosis
Chen et al. (2010)	Con	FOXO3	ind	OE FOXO3.A3	Cell line	Breast cancer	48 h	5 μM	ND
Hui, Francis, et al. (2008) and Hui, Gomes, et al. (2008)	Con	FOXO3	ind	OE FOXO3.ER	Cell line	CML	24 h	1 μM	Upregulation efflux pumps
Lupertz et al. (2008)	Pro	FOXO4	ind	OE FOXO4/ siFOXO4	Cell line	Colorectal carcinoma	3 h	10 μM	Induced apoptosis
Kojima et al. (2010)	Pro	FOXO1	dec	siFOXO1	Cell line	Renal cell carcinoma	48 h	1 μM	Induced apoptosis
Hui, Francis, et al. (2008) and Hui, Gomes, et al. (2008)	Pro	FOXO3	dec	siFOXO3	Cell line	CML	48 h	1 mM	Feedback PI3K up
Han, Cho, Choi, Han, and Kang (2008)	Con	FOXO1	dec	siFOXO1	Cell line	Breast cancer	224 h	30 μM	Upregulation efflux pumps

Continued

Table 1 FOXO Can Promote Resistance or Sensitivity to Chemotherapeutic Agents—cont'd

ETP	Pro/Con	FOXO	ind/dec	By	Tissue/Cell Line	Cancer Type	Timing	Concentration	Mechanism
Zhang et al. (2016)	Pro	FOXO1	ind	sirt1-kd	Cell line	CML	4h	20 μM	Reduced proliferation
Li et al. (2015)	Pro	FOXO1	ind	sirt1-kd	Cell line	ATL	4h	20 μM	Reduced proliferation
Hagenbuchner et al. (2012)	Pro	FOXO3	Both	FOXO3.A3/shFOXO3	Cell line	Neuroblastoma	2h	20 μg/mL	ROS accumulation
Morfouace et al. (2012)	Pro	FOXO3	ind	DCA	Primary cell lines	Glioma	12h	50 μM	Induced apoptosis
Obexer et al. (2009)	Pro	FOXO3	ind	OE FOXO3	Cell line	Neuroblastoma	48h	2.5 μg/mL	Induced apoptosis
Qiong et al. (2010)	Pro	FOXO1	ind	Wortmannin	Cell line	ATL	48h	5 μM	Induced apoptosis
Nogueira et al. (2008)	Pro	FOXO3	dec	Akt	Xenograft	Ovarian cancer	50 days	10mg/kg	ROS accumulation
Nyakern, Cappellini, Mantovani, and Martelli (2006)	Pro	FOXO1	ind	Akt/PI3Ki	Cell line	ATL	48h	10 μM	Induced apoptosis
Liu et al. (2005)	Both	FOXO3	ind	ROS	Cell line	Prostate cancer	24h	25 μM	ROS protection/induced apoptosis

5FU	Pro/ Con	FOXO	ind/ dec	By	Tissue/Cell Line	Cancer Type	Timing	Concentration	Mechanism
Zhao et al. (2016)	Pro	FOXO1	Both	siFOXO1/OE FOXO1	Cell lines	Nasopharyngeal carcinoma	48 h	100 μM	Increased miR-3188
Song et al. (2017)	Pro	FOXO3	dec	REP1	Cell line	Colon carcinoma	24 h	20 μM	Reduced apoptosis
Sun et al. (2014)	Pro	FOXO1/ 3/4	ind	PTEN	Cell lines	Colon carcinoma	48 h	10 μM	Increased apoptosis
Wang et al. (2015)	Pro	FOXO3	ind	Akt-i	Cell line	Colon carcinoma	24 h	400 μM	Increased apoptosis
Xie et al. (2012)	Pro	FOXO4	ind	Tcf4-kd	Cell line	Colon carcinoma	72 h	8 μM	ND

Indicated are studies on the role of FOXO in resistance to the chemotherapeutics cisplatin (CPT), doxorubicin (DXR), etoposide (ETP), and 5-fluorouracil (5FU). Per study, the following is summarized: whether FOXO is pro tumor suppressor (pro) or con tumor suppressor (con); which FOXO was studied; whether the activity of FOXO was induced (ind) or decreased (dec); how the activity was manipulated (OE, over expression; kd, knockdown); whether the studies were performed in cell lines or xenograft models; what cancer type was used (CML, chronic myeloid leukemia; AML, acute myeloid leukemia; MEF, mouse embryonic fibroblasts; ATL, adult T-cell leukemia-lymphoma); how long the treatment lasted; what the used concentration was; and what the mechanism was whereby FOXO affected resistance (ND, not described).

oncogene (BRAF^{V600E})-induced oxidative stress FOXOs can switch from a p27^{kip1} G0/G1 arrest toward a p21^{cip1}-dependent senescence arrest (de Keizer et al., 2010). Alternatively, induction of apoptosis appears confined mostly to immune cells deprived of essential growth factors (Birkenkamp & Coffer, 2003).

We have shown that FOXO-dependent entry into quiescence involves regulation of p130Rb2 (Kops, Medema, et al., 2002). Interestingly, a recent study highlighted the importance of CDK2 activity regulation to determine entry into quiescence or prolonged proliferation (Spencer et al., 2013). In agreement, FOXOs regulate CDK2 activity through transcriptional regulation of the cell cycle inhibitors p27^{kip1} and p21^{cip1}. In line with the importance of cell cycle inhibitors and Rb/CDK signaling, it has been shown that p16, a negative regulator of the Rb pathway, critically determines whether muscle satellite cells enter GO or senescence (Sousa-Victor et al., 2014). Whereas entry into quiescence apparently depends on cell cycle regulation, maintenance of quiescence will likely involve additional gene regulation such as those involved in stress regulation, e.g., MnSOD (Sarsour, Venkataraman, Kalen, Oberley, & Goswami, 2008). Finally, cross-examination of aforementioned quiescence gene expression signature shows further overlap with potential FOXO-regulated gene expression (e.g., CTD-SP2 (unpublished) and cyclin G2 (Martinez-Gac, Marques, Garcia, Campanero, & Carrera, 2004)), suggesting further a role for FOXO in regulating quiescence.

The notion that FOXOs through induction of a timely arrest promote survival is compatible with the hypothesis that quiescence in general represents a cell fate that allows survival. Consequently, ample evidence show that increased cell survival at least correlates with quiescence (reviewed in Wells, Griffith, Wells, & Taylor, 2013). However, the survival of quiescent cancer cells in response to stress induced by cancer treatments increases tumor resistance and risk for relapse.

5.2 The Role of Feedback Signaling in Acquiring Resistance

Feedback mechanisms are a common feature of biological systems and are crucial for dynamic control of biological processes. PI3K/PKB signaling is no exception to this and two important feedback mechanisms have been described to date (Fig. 3). The first example of feedback to be described was the negative effect of mTORC1 and its direct target S6 kinase (S6K) on IRS1/2 stability (reviewed in Carracedo & Pandolfi, 2008). More dispersed over the last decade, several papers have also reported FOXO targets that

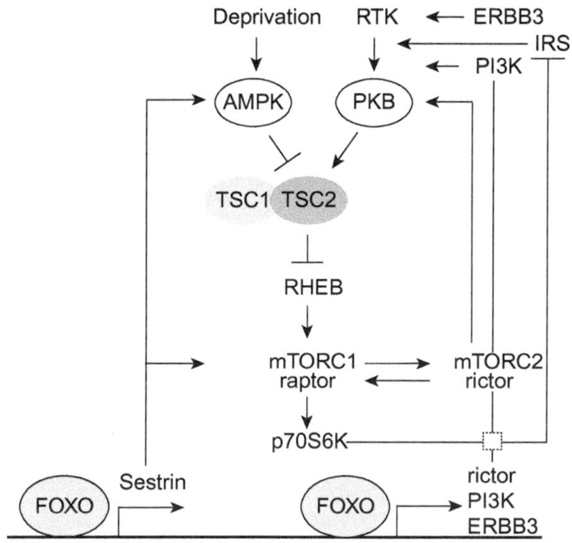

Fig. 3 Common feedback signaling in the PI3K pathway toward FOXO. PI3K/PKB signaling is partly controlled via feedback mechanisms. FOXO regulates the expression of sestrins, which leads to activation of AMPK and the inhibition of mTORC1. Sestrins also inhibit mTORC1 by binding Rag GTPase/GATOR. Further, FOXO regulates the expression of genes that increase PKB activation (IRS/PI3K/mTORC2/RICTOR/ERBB3), thereby reducing its own activity.

potentially result in increased PKB activation, including IRS2 (Zhang et al., 2001), the insulin receptor (Chandarlapaty et al., 2011; Puig, Marr, Ruhf, & Tjian, 2003), PI3KCA (Hui, Gomes, et al., 2008), mTORC2 component RICTOR (Chen, Jeon, et al., 2010; Lin et al., 2014), human EGF receptor 3 (HER3)/v-erb-b2 avian erythroblastic leukemia viral oncogene homologue 3 (ERBB3) (Chakrabarty, Sanchez, Kuba, Rinehart, & Arteaga, 2012; Chandarlapaty et al., 2011; Garrett et al., 2011), and IGF receptor 1 (IGF-IR) (Chandarlapaty et al., 2011; Huo et al., 2014). Indeed, PKB activation in response to FOXO overexpression has been documented previously (Chen, Jeon, et al., 2010; Hui, Gomes, et al., 2008; Skurk et al., 2005). Recently this has received extra attention because mTORC1/FOXO feedback regulation of PKB has major implications for targeting PI3K/mTORC signaling in cancer (reviewed in Rodon, Dienstmann, Serra, & Tabernero, 2013). In brief, while inhibition of PI3K or mTORC1/2 by small-molecule inhibitors initially results in PKB inactivation, it was shown that PKB is rapidly reactivated. Two feedback signaling cascades have been shown to be important for this: relief of the negative regulation of PI3K by mTORC1/S6K (O'Reilly et al., 2006) and

FOXO-dependent regulation of several RTKs, including ERBB3 and the insulin receptor (Chakrabarty et al., 2012; Chandarlapaty et al., 2011; Garrett et al., 2011) or RICTOR (Lin et al., 2014).

Importantly, reactivation of PKB through FOXO-mediated transcription occurs simultaneously with the inhibition of mTORC1 through FOXO-induced sestrin expression (described before). Sestrins activate AMPK and thereby inhibit mTORC1 (Budanov & Karin, 2008). AMPK in turn phosphorylates and activates FOXO by a yet to be identified mechanism (Greer et al., 2007). It is not known whether this FOXO-sestrin-AMPK positive feedback loop is required for prolonged mTORC1 inhibition. Also, how mTORC1 activity is restored via inhibition of FOXO by PI3K/PKB instead of activation by AMPK is unclear. Besides via AMPK, sestrins can also inhibit mTORC1 activity through binding Rag GTPase/GATOR. The Rag GTPase/GATOR complex is required for mTORC1 regulation through amino acid availability (Parmigiani et al., 2014). At present it is not known whether amino acid deprivation regulates FOXOs directly, yet this indicates that FOXO/sestrin activation also short circuits this input into mTORC1 regulation.

The ability to reenter the cell cycle discriminates quiescence from senescence. Interestingly, FOXOs are known to induce feedback signaling and transcriptionally increase many signaling components required for proliferation (e.g., IRS2, PI3K, TORC2, and ERBB2/3; for review, see Eijkelenboom & Burgering, 2013). A suggested rationale is that there is competition for available resources following resolution of stress and that therefore cells need to be prepared for better times (see for discussion Kloet & Burgering, 2011). In cancer there is an obvious downside for this feedback signaling. FOXO activation in cancer will therefore not only be instrumental in requiring therapy resistance through its ability to impose a quiescent state, but also through enhancing the ability of cancer cells to respond to limiting resources of growth factors by increased expression of growth factor signaling components. Indeed, several studies describe the importance of PI3K signaling and regulation of ERBB2/3, in particular in therapy resistance (Berns et al., 2007; Chakrabarty et al., 2012; Eichhorn et al., 2008; Prahallad et al., 2012; Sun et al., 2014).

5.3 Cross Talk With Other Pathways

Most cancers are heterogeneous by nature, meaning that cancer cells genetically differ within one tumor. The common driver mutations are present in the large majority of cancer cells within a tumor, but as a consequence of genetic instability cancer cells will differ in a variety of additional genetic lesions. These heterogeneic genetic makeups will impact on the interplay between the various signaling pathways that are affected within a cancer cell and their response to treatment. Therefore, it is of importance to understand how PI3K signaling in the context of active WNT signaling differs from PI3K signaling in the context of activated K-RAS. These could be determinants for response to targeted cancer therapies.

GSK3 activity is a pivot in regulating the stability of β-catenin, the key factor in WNT signaling. In resting cells, GSK3 activity is high and a substantial fraction of GSK3 is present in the so-called destruction complex consisting of Axin, APC, casein kinase 2, and GSK3. β-Catenin binds to the destruction complex and in this complex GSK3 phosphorylation targets β-catenin for proteasomal degradation. Following WNT binding to the Frizzled receptor, Dishevelled is recruited and GSK3/destruction complex is inactivated in a manner not yet completely understood. As a consequence, GSK3-mediated β-catenin phosphorylation is reduced and β-catenin is stabilized. Following nuclear translocation, β-catenin will act as a transcriptional coactivator for members of the TCF/LEF-1 transcription factor family (reviewed in MacDonald, Tamai, & He, 2009). GSK3 is the first identified substrate protein for PKB and following stimulation of resting cells with ligands that activate PI3K, PKB-mediated phosphorylation of GSK3 results in inactivation of GSK3 (Cross, Alessi, Cohen, Andjelkovich, & Hemmings, 1995; Stambolic & Woodgett, 1994). This suggested direct involvement of PI3K/PKB in regulating β-catenin stability and a synergism between WNT and PI3K signaling in activating β-catenin/TCF transcriptional output. However, this appeared not straightforward and numerous conflicting studies have either disclaimed (Deming et al., 2014; Ng et al., 2009) or supported a role for PI3K/PKB signaling in regulating β-catenin (Lee et al., 2010). In addition, others and we have shown a functional interaction between FOXO and β-catenin, especially under conditions of increased cellular oxidative stress. FOXOs can compete with TCF for binding β-catenin and inhibition of FOXO by PI3K/PKB signaling indirectly stimulates TCF activity by increasing the availability of β-catenin for binding to TCF (Essers et al., 2005; Hoogeboom et al., 2008). This mechanism of competition between FOXO and TCF has been shown to occur during osteoblast differentiation (Almeida, Han, Martin-Millan, O'Brien, & Manolagas, 2007) and during nutrient deprivation (Liu et al., 2011). Therefore, it can be suggested that in tumors harboring gene mutations that activate both PI3K and WNT signaling, e.g., PTEN and APC deletion, the redox balance becomes an important denominator for the outcome TCF/β-catenin activity. Cancer treatments often cause genotoxic or oxidative stress, thereby challenging the redox balance of cells and changing the interplay between FOXO and β-catenin.

In general oxidative stress has an important role in the relay between different signaling pathways, as many proteins within pathways are sensitive to oxidation. Oxidation in most cases will impair function, but in the case of ASK1 or ATM, it will also result in activation of signaling proteins. Therefore, it is hard to predict a priori what the impact of a redox imbalance in tumor cells will be on signaling activity of a given pathway, let alone its interaction with another pathway. For example, in insulin signaling toward PI3K, both PTEN and PTP1B are inactivated by cysteine oxidation (Barrett et al., 1999; Lee et al., 2002). This will steer insulin signaling clearly to PKB activation and hence FOXO inactivation. On the other hand, we have shown that FOXOs are also subject to oxidation and this results in activation of FOXOs (Putker et al., 2013, 2015). At present it is unclear how this paradox can be reconciled, but one possibility maybe that different levels of oxidative stress discriminate between FOXO inactivation and activation and whether a change in cellular redox occurs in a resting cell or a cell stimulated with insulin.

An important downstream gene target of classical WNT signaling is c-Myc. By now there is ample evidence that a reciprocal antagonism between FOXO and c-Myc is an important mechanism to enable either FOXO or c-Myc to function. Initially, it was shown that ectopic expression of c-Myc counteracts FOXO-induced transcriptional regulation of $p27^{kip1}$ (Chandramohan, Jeay, Pianetti, & Sonenshein, 2004), possibly through a direct interaction between FOXO and c-Myc (Chandramohan et al., 2008). In a reciprocal manner, FOXO inhibits c-Myc, but does so through multiple mechanisms. First, FOXO induces transcription of Mxi1, an antagonist of c-Myc that competes with c-Myc for the cofactor MAX (Delpuech et al., 2007). Second, FOXO induces expression of Fwbx7, an E3 ligase that targets c-Myc for proteasomal degradation (Mei et al., 2015; Sandri et al., 2004). FOXO has been shown to reduce c-Myc transcription (possibly) and to reduce mRNA expression through induction of miR-145 (Sachdeva et al., 2009). This extensive negative regulation of c-Myc by FOXO has been shown to be of biological importance and mediates several processes, including angiogenesis (Wilhelm et al., 2016). This shows that the interplay between FOXO and c-Myc is important in cancer initiation and cancer progression. Cancer treatments can influence FOXO and c-Myc levels and could thereby tip the balance toward cancer progression or cancer cell death.

Besides the WNT pathway, there is extensive cross talk between FOXO and the TGFβ pathway. Initial genetic experiments in *C. elegans* already indicated that DAF-7, encoding a *C. elegans* SMAD orthologue, cooperates with DAF-16 in regulating the transcription of key metabolic and developmental control genes (Ogg et al., 1997). Also in higher eukaryotes it was shown that Smad proteins activated by TGFβ form a physical complex with FOXO proteins (Seoane, Le, Shen, Anderson, & Massague, 2004). This complex was later shown to regulate a subset of all TGFβ-inducible genes, mostly genes involved in stress response and adaptive responses that accompany the cytostatic response of keratinocytes to TGFβ (Gomis et al., 2006). Interestingly, similar to the picture emerging for FOXOs, TGFβ is already known for a long time to exhibit dual activities in cancer. It stimulates metastasis and angiogenesis, on the one hand, while it exhibits also cytostatic tumor-suppressive effects. This dual effect of TGFβ is explained by the interaction of TGFβ with stromal components and fibroblast cells within the tumor microenvironment (reviewed in Calon, Tauriello, & Batlle, 2014).

Signaling pathways are already complex by themselves; however, these signaling pathways also often cross talk within one cell. In this way, when treating a cell with a targeted inhibitor for a certain pathway, other pathways can be affected as well. This may result in the acquisition of resistance or increased invasion or metastasis risk. FOXO is a component in one of the major growth factor signaling cascades and it shows extensive cross talk with WNT signaling and TGFβ signaling. So, if a chemotherapeutic targets any component of the PKB, WNT, or TGFβ pathway, either directly or indirectly, other pathways may be affected as well. The response in the pathway that is stimulated via cross talk may change the sensitivity of the cell to the chemotherapeutic. In this way, cross talk between pathways can result in the occurrence of resistance via FOXO.

6. CONCLUSION

In this review, we summarized the role of FOXO and the mechanisms whereby it exerts its role in tumorigenesis. Initially, most studies complied with a role of FOXO as a tumor suppressor, leading to the qualification of FOXO as "bona fide" tumor suppressor. However, evidence is accumulating that FOXOs can also promote tumor progression in various ways. These results altogether indicate context as a discriminatory factor whether FOXOs act to inhibit or promote tumor progression. As indicated by the hallmarks, cancer progression can be divided into several discrete steps, namely, tumor initiation, primary tumor growth, invasive growth, and metastasis. At these various stages, the requirements for tumor progression differ. This provides one type of context where potentially FOXO function may be irrelevant for tumorigenesis for one stage, but highly relevant for a different stage of progression. Relevance here applies equally to tumor suppression or promotion. In any case, a "bona fide" tumor suppressor would prevent tumor initiation as a first line of defense against tumor formation. Reducing tumor mass in later stages of tumor progression by inhibiting tumor growth, although repressive does not eliminate tumor cells and maybe qualifies as second best in term of tumor suppression.

Here, we highlight homeostasis as an additional context. A normal cell can be considered a system in equilibrium. In thermodynamics, for a system in equilibrium, a change in a system's parameter will be compensated for by changing other system's parameter to return to equilibrium, or to transit to a novel equilibrium. In the case of cellular stress, the main cause for changing a system's parameters originates from endogenous or exogenous sources. For cells, the main function of FOXOs, but also for p53, is to enable return to equilibrium. Consequently, if FOXO and p53 fail to perform this task, cells will enter a new equilibrium that is the equilibrium of a cancer cell. However, the role of FOXO in equilibrium maintenance has not changed and its

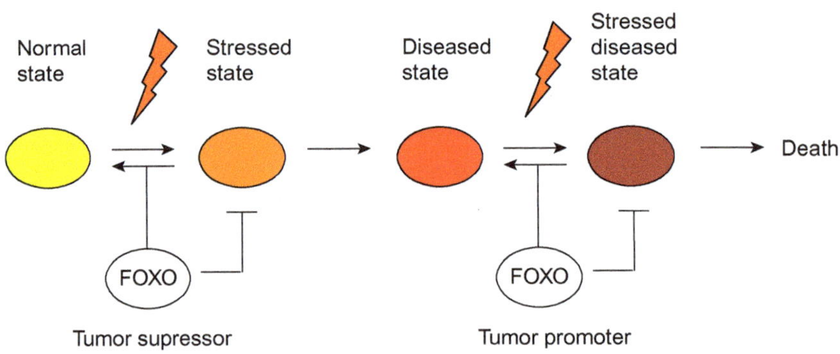

Fig. 4 Model on the dual function of FOXO in cancer. In normal cells, FOXOs act as tumor suppressors by trying to maintain equilibrium. FOXOs will try to convert a stress cell to its original state. If FOXO is unable to bring the equilibrium back to its normal state, the stressed or diseased state may become the new equilibrium (i.e., cancer develops). In the diseased state, FOXO will still try to maintain equilibrium. In this case, FOXO might be tumor promoting, since it will protect stressed cancer cells from dying.

role as rheostat in essential parameters, such as cellular redox, predicts that a change in a system's parameter of a cancer cell will still be counteracted as to return to equilibrium. For a cancer cell, a transition to a novel equilibrium can be either a transition to a more aggressive state or in the case of chemotherapy as an inducer of parameter change this can be ultimately cell death. As illustrated in Fig. 4, this line of reasoning may rationalize FOXOs to act both as tumor suppressor and tumor promoter.

Given the efforts to develop targeted therapy toward signaling pathways such as the PI3K pathway, it is of importance to understand at what stage of tumor progression therapy may be targeted effectively. More specifically, it will be relevant to determine whether FOXO is active and whether the activity of FOXO occurs in the presence of active PI3K signaling or other signaling pathway, such as WNT signaling. Furthermore, effective cancer therapy may require better understanding of how equilibrium is maintained and what the most effective approach will be to disturb this equilibrium in a manner that this will result in cancer cell death.

REFERENCES

Adamowicz, M., Vermezovic, J., & d'Adda di Fagagna, F. (2016). NOTCH1 inhibits activation of ATM by impairing the formation of an ATM-FOXO3a-KAT5/Tip60 complex. *Cell Reports*, *16*(8), 2068–2076. https://doi.org/10.1016/j.celrep.2016.07.038.

Aldonza, M. B., Hong, J. Y., Alinsug, M. V., Song, J., & Lee, S. K. (2016). Multiplicity of acquired cross-resistance in paclitaxel-resistant cancer cells is associated with feedback control of TUBB3 via FOXO3a-mediated ABCB1 regulation. *Oncotarget*, 7(23), 34395–34419. https://doi.org/10.18632/oncotarget.9118.

Almeida, M., Han, L., Martin-Millan, M., O'Brien, C. A., & Manolagas, S. C. (2007). Oxidative stress antagonizes Wnt signaling in osteoblast precursors by diverting beta-catenin from T cell factor- to forkhead box O-mediated transcription. *The Journal of Biological Chemistry*, *282*(37), 27298–27305. https://doi.org/10.1074/jbc.M702811200.

Altan, B., Yokobori, T., Ide, M., Mochiki, E., Toyomasu, Y., Kogure, N., et al. (2016). Nuclear PRMT1 expression is associated with poor prognosis and chemosensitivity in gastric cancer patients. *Gastric Cancer*, *19*(3), 789–797. https://doi.org/10.1007/s10120-015-0551-7.

Alvarez, B., Martínez-A, C., Burgering, B. M., & Carrera, A. C. (2001). Forkhead transcription factors contribute to execution of the mitotic programme in mammals. *Nature*, *413*(6857), 744–747. https://doi.org/10.1038/35099574.

Amente, S., Zhang, J., Lavadera, M. L., Lania, L., Avvedimento, E. V., & Majello, B. (2011). Myc and PI3K/AKT signaling cooperatively repress FOXO3a-dependent PUMA and GADD45a gene expression. *Nucleic Acids Research*, *39*(22), 9498–9507. https://doi.org/10.1093/nar/gkr638.

Arimoto-Ishida, E., Ohmichi, M., Mabuchi, S., Takahashi, T., Ohshima, C., Hayakawa, J., et al. (2004). Inhibition of phosphorylation of a forkhead transcription factor sensitizes human ovarian cancer cells to cisplatin. *Endocrinology*, *145*(4), 2014–2022. https://doi.org/10.1210/en.2003-1199.

Baar, M. P., Brandt, R. M., Putavet, D. A., Klein, J. D., Derks, K. W., Bourgeois, B. R., et al. (2017). Targeted apoptosis of senescent cells restores tissue homeostasis in response to chemotoxicity and aging. *Cell, 169*(1), 132–147.e116. https://doi.org/10.1016/j.cell.2017.02.031.

Bai, A., Hu, H., Yeung, M., & Chen, J. (2007). Kruppel-like factor 2 controls T cell trafficking by activating L-selectin (CD62L) and sphingosine-1-phosphate receptor 1 transcription. *The Journal of Immunology, 178*(12), 7632–7639.

Bakker, W. J., Harris, I. S., & Mak, T. W. (2007). FOXO3a is activated in response to hypoxic stress and inhibits HIF1-induced apoptosis via regulation of CITED2. *Molecular Cell, 28*(6), 941–953. https://doi.org/10.1016/j.molcel.2007.10.035.

Banting, F. G., Best, C. H., Collip, J. B., Campbell, W. R., & Fletcher, A. A. (1922). Pancreatic extracts in the treatment of diabetes mellitus. *Canadian Medical Association Journal, 12*(3), 141–146.

Barrett, W. C., DeGnore, J. P., Konig, S., Fales, H. M., Keng, Y. F., Zhang, Z. Y., et al. (1999). Regulation of PTP1B via glutathionylation of the active site cysteine 215. *Biochemistry, 38*(20), 6699–6705. https://doi.org/10.1021/bi990240v.

Berns, K., Horlings, H. M., Hennessy, B. T., Madiredjo, M., Hijmans, E. M., Beelen, K., et al. (2007). A functional genetic approach identifies the PI3K pathway as a major determinant of trastuzumab resistance in breast cancer. *Cancer Cell, 12*(4), 395–402. https://doi.org/10.1016/j.ccr.2007.08.030.

Birkenkamp, K. U., & Coffer, P. J. (2003). Regulation of cell survival and proliferation by the FOXO (Forkhead box, class O) subfamily of Forkhead transcription factors. *Biochemical Society Transactions, 31*(Pt. 1), 292–297. https://doi.org/10.1042/.

Bouchard, C., Lee, S., Paulus-Hock, V., Loddenkemper, C., Eilers, M., & Schmitt, C. A. (2007). FoxO transcription factors suppress Myc-driven lymphomagenesis via direct activation of Arf. *Genes & Development, 21*(21), 2775–2787. https://doi.org/10.1101/gad.453107.

Bouchard, C., Marquardt, J., Bras, A., Medema, R. H., & Eilers, M. (2004). Myc-induced proliferation and transformation require Akt-mediated phosphorylation of FoxO proteins. *The EMBO Journal, 23*(14), 2830–2840. https://doi.org/10.1038/sj.emboj.7600279.

Brunet, A., Bonni, A., Zigmond, M. J., Lin, M. Z., Juo, P., Hu, L. S., et al. (1999). Akt promotes cell survival by phosphorylating and inhibiting a Forkhead transcription factor. *Cell, 96*(6), 857–868.

Buchheit, C. L., Weigel, K. J., & Schafer, Z. T. (2014). Cancer cell survival during detachment from the ECM: Multiple barriers to tumour progression. *Nature Reviews. Cancer, 14*(9), 632–641. https://doi.org/10.1038/nrc3789.

Budanov, A. V., & Karin, M. (2008). The p53-regulated Sestrin gene products inhibit mTOR signaling. *Cell, 134*(3), 451–460.

Burgering, B. M. (2008). A brief introduction to FOXOlogy. *Oncogene, 27*(16), 2258–2262. https://doi.org/10.1038/onc.2008.29.

Calon, A., Tauriello, D. V., & Batlle, E. (2014). TGF-beta in CAF-mediated tumor growth and metastasis. *Seminars in Cancer Biology, 25*, 15–22. https://doi.org/10.1016/j.semcancer.2013.12.008.

Carlson, C. M., Endrizzi, B. T., Wu, J., Ding, X., Weinreich, M. A., Walsh, E. R., et al. (2006). Kruppel-like factor 2 regulates thymocyte and T-cell migration. *Nature, 442*(7100), 299–302. https://doi.org/10.1038/nature04882.

Carracedo, A., & Pandolfi, P. P. (2008). The PTEN-PI3K pathway: Of feedbacks and cross-talks. *Oncogene, 27*(41), 5527–5541. https://doi.org/10.1038/onc.2008.247.

Chaanine, A. H., Kohlbrenner, E., Gamb, S. I., Guenzel, A. J., Klaus, K., Fayyaz, A. U., et al. (2016). FOXO3a regulates BNIP3 and modulates mitochondrial calcium, dynamics, and function in cardiac stress. *American Journal of Physiology. Heart and Circulatory Physiology, 311*(6), H1540–H1559. https://doi.org/10.1152/ajpheart.00549.2016.

Chae, H. Z., Kim, H. J., Kang, S. W., & Rhee, S. G. (1999). Characterization of three isoforms of mammalian peroxiredoxin that reduce peroxides in the presence of thioredoxin. *Diabetes Research and Clinical Practice*, *45*(2–3), 101–112.

Chakrabarty, A., Sanchez, V., Kuba, M. G., Rinehart, C., & Arteaga, C. L. (2012). Feedback upregulation of HER3 (ErbB3) expression and activity attenuates antitumor effect of PI3K inhibitors. *Proceedings of the National Academy of Sciences of the United States of America*, *109*(8), 2718–2723. https://doi.org/10.1073/pnas.1018001108.

Chandarlapaty, S., Sawai, A., Scaltriti, M., Rodrik-Outmezguine, V., Grbovic-Huezo, O., Serra, V., et al. (2011). AKT inhibition relieves feedback suppression of receptor tyrosine kinase expression and activity. *Cancer Cell*, *19*(1), 58–71. https://doi.org/10.1016/j.ccr.2010.10.031.

Chandramohan, V., Jeay, S., Pianetti, S., & Sonenshein, G. E. (2004). Reciprocal control of Forkhead box O 3a and c-Myc via the phosphatidylinositol 3-kinase pathway coordinately regulates p27Kip1 levels. *The Journal of Immunology*, *172*(9), 5522–5527.

Chandramohan, V., Mineva, N. D., Burke, B., Jeay, S., Wu, M., Shen, J., et al. (2008). c-Myc represses FOXO3a-mediated transcription of the gene encoding the p27 (Kip1) cyclin dependent kinase inhibitor. *Journal of Cellular Biochemistry*, *104*(6), 2091–2106. https://doi.org/10.1002/jcb.21765.

Charitou, P., Rodriguez-Colman, M., Gerrits, J., van Triest, M., Groot Koerkamp, M., Hornsveld, M., et al. (2015). FOXOs support the metabolic requirements of normal and tumor cells by promoting IDH1 expression. *EMBO Reports*, *16*(4), 456–466. https://doi.org/10.15252/embr.201439096.

Chen, J., Gomes, A. R., Monteiro, L. J., Wong, S. Y., Wu, L. H., Ng, T. T., et al. (2010). Constitutively nuclear FOXO3a localization predicts poor survival and promotes Akt phosphorylation in breast cancer. *PLoS One*, *5*(8), e12293. https://doi.org/10.1371/journal.pone.0012293.

Chen, C. C., Jeon, S. M., Bhaskar, P. T., Nogueira, V., Sundararajan, D., Tonic, I., et al. (2010). FoxOs inhibit mTORC1 and activate Akt by inducing the expression of Sestrin3 and Rictor. *Developmental Cell*, *18*(4), 592–604. https://doi.org/10.1016/j.devcel.2010.03.008.

Chen, Y. R., Liu, M. T., Chang, Y. T., Wu, C. C., Hu, C. Y., & Chen, J. Y. (2008). Epstein-Barr virus latent membrane protein 1 represses DNA repair through the PI3K/Akt/FOXO3a pathway in human epithelial cells. *Journal of Virology*, *82*(16), 8124–8137. https://doi.org/10.1128/JVI.00430-08.

Chen, Z., Trotman, L. C., Shaffer, D., Lin, H. K., Dotan, Z. A., Niki, M., et al. (2005). Crucial role of p53-dependent cellular senescence in suppression of Pten-deficient tumorigenesis. *Nature*, *436*(7051), 725–730. https://doi.org/10.1038/nature03918.

Chi, H. C., Chen, S. L., Cheng, Y. H., Lin, T. K., Tsai, C. Y., Tsai, M. M., et al. (2016). Chemotherapy resistance and metastasis-promoting effects of thyroid hormone in hepatocarcinoma cells are mediated by suppression of FoxO1 and Bim pathway. *Cell Death & Disease*, 7(8), e2324. https://doi.org/10.1038/cddis.2016.227.

Chock, K. L., Allison, J. M., Shimizu, Y., & ElShamy, W. M. (2010). BRCA1-IRIS overexpression promotes cisplatin resistance in ovarian cancer cells. *Cancer Research*, *70*(21), 8782–8791. https://doi.org/10.1158/0008-5472.CAN-10-1352.

Choi, H. K., Cho, K. B., Phuong, N. T., Han, C. Y., Han, H. K., Hien, T. T., et al. (2013). SIRT1-mediated FoxO1 deacetylation is essential for multidrug resistance-associated protein 2 expression in tamoxifen-resistant breast cancer cells. *Molecular Pharmaceutics*, *10*(7), 2517–2527. https://doi.org/10.1021/mp400287p.

Chow, K. T., Timblin, G. A., McWhirter, S. M., & Schlissel, M. S. (2013). MK5 activates Rag transcription via Foxo1 in developing B cells. *The Journal of Experimental Medicine*, *210*(8), 1621–1634. https://doi.org/10.1084/jem.20130498.

Ciccia, A., & Elledge, S. J. (2010). The DNA damage response: Making it safe to play with knives. *Molecular Cell*, *40*(2), 179–204. https://doi.org/10.1016/j.molcel.2010.09.019.
Coller, H. A., Sang, L., & Roberts, J. M. (2006). A new description of cellular quiescence. *PLoS Biology*, *4*(3), e83. https://doi.org/10.1371/journal.pbio.0040083.
Courtois-Cox, S., Genther Williams, S. M., Reczek, E. E., Johnson, B. W., McGillicuddy, L. T., Johannessen, C. M., et al. (2006). A negative feedback signaling network underlies oncogene-induced senescence. *Cancer Cell*, *10*(6), 459–472. https://doi.org/10.1016/j.ccr.2006.10.003.
Cross, D. A., Alessi, D. R., Cohen, P., Andjelkovich, M., & Hemmings, B. A. (1995). Inhibition of glycogen synthase kinase-3 by insulin mediated by protein kinase B. *Nature*, *378*(6559), 785–789. https://doi.org/10.1038/378785a0.
Daitoku, H., Kaneko, Y., Yoshimochi, K., Matsumoto, K., Araoi, S., Sakamaki, J. I., et al. (2016). Non-transcriptional function of FOXO1/DAF-16 contributes to translesion DNA synthesis. *Molecular and Cellular Biology*, *36*(21), 2755–2766. https://doi.org/10.1128/MCB.00265-16.
de Keizer, P. L. (2017). The fountain of youth by targeting senescent cells? *Trends in Molecular Medicine*, *23*(1), 6–17. https://doi.org/10.1016/j.molmed.2016.11.006.
de Keizer, P. L., Packer, L. M., Szypowska, A. A., Riedl-Polderman, P. E., van den Broek, N. J., de Bruin, A., et al. (2010). Activation of forkhead box O transcription factors by oncogenic BRAF promotes p21cip1-dependent senescence. *Cancer Research*, *70*(21), 8526–8536. https://doi.org/10.1158/0008-5472.CAN-10-1563.
Delpuech, O., Griffiths, B., East, P., Essafi, A., Lam, E. W., Burgering, B., et al. (2007). Induction of Mxi1-SR alpha by FOXO3a contributes to repression of Myc-dependent gene expression. *Molecular and Cellular Biology*, *27*(13), 4917–4930. https://doi.org/10.1128/MCB.01789-06.
Deming, D. A., Leystra, A. A., Nettekoven, L., Sievers, C., Miller, D., Middlebrooks, M., et al. (2014). PIK3CA and APC mutations are synergistic in the development of intestinal cancers. *Oncogene*, *33*(17), 2245–2254. https://doi.org/10.1038/onc.2013.167.
Dieudonne, F. X., Marion, A., Marie, P. J., & Modrowski, D. (2012). Targeted inhibition of T-cell factor activity promotes syndecan-2 expression and sensitization to doxorubicin in osteosarcoma cells and bone tumors in mice. *Journal of Bone and Mineral Research*, *27*(10), 2118–2129. https://doi.org/10.1002/jbmr.1650.
Dijkers, P. F., Medema, R. H., Lammers, J. W., Koenderman, L., & Coffer, P. J. (2000). Expression of the pro-apoptotic Bcl-2 family member Bim is regulated by the forkhead transcription factor FKHR-L1. *Current Biology*, *10*(19), 1201–1204.
Ding, B., Kirkiles-Smith, N. C., & Pober, J. S. (2009). FOXO3a regulates oxygen-responsive expression of tumor necrosis factor receptor 2 in human dermal microvascular endothelial cells. *The Journal of Biological Chemistry*, *284*(29), 19331–19339. https://doi.org/10.1074/jbc.M109.006536.
Dubrovska, A., Kim, S., Salamone, R. J., Walker, J. R., Maira, S. M., Garcia-Echeverria, C., et al. (2009). The role of PTEN/Akt/PI3K signaling in the maintenance and viability of prostate cancer stem-like cell populations. *Proceedings of the National Academy of Sciences of the United States of America*, *106*(1), 268–273. https://doi.org/10.1073/pnas.0810956106.
Duran, R. V., Oppliger, W., Robitaille, A. M., Heiserich, L., Skendaj, R., Gottlieb, E., et al. (2012). Glutaminolysis activates Rag-mTORC1 signaling. *Molecular Cell*, *47*(3), 349–358. https://doi.org/10.1016/j.molcel.2012.05.043.
Eichhorn, P. J., Gili, M., Scaltriti, M., Serra, V., Guzman, M., Nijkamp, W., et al. (2008). Phosphatidylinositol 3-kinase hyperactivation results in lapatinib resistance that is reversed by the mTOR/phosphatidylinositol 3-kinase inhibitor NVP-BEZ235. *Cancer Research*, *68*(22), 9221–9230. https://doi.org/10.1158/0008-5472.CAN-08-1740.
Eijkelenboom, A., & Burgering, B. M. (2013). FOXOs: Signalling integrators for homeostasis maintenance. *Nature Reviews. Molecular Cell Biology*, *14*(2), 83–97. https://doi.org/10.1038/nrm3507.

Essers, M. A., de Vries-Smits, L. M., Barker, N., Polderman, P. E., Burgering, B. M., & Korswagen, H. C. (2005). Functional interaction between beta-catenin and FOXO in oxidative stress signaling. *Science*, *308*(5725), 1181–1184. https://doi.org/10.1126/science.1109083.

Essers, M. A., Weijzen, S., de Vries-Smits, A. M., Saarloos, I., de Ruiter, N. D., Bos, J. L., et al. (2004). FOXO transcription factor activation by oxidative stress mediated by the small GTPase Ral and JNK. *The EMBO Journal*, *23*(24), 4802–4812. https://doi.org/10.1038/sj.emboj.7600476.

Fabre, S., Carrette, F., Chen, J., Lang, V., Semichon, M., Denoyelle, C., et al. (2008). FOXO1 regulates L-selectin and a network of human T cell homing molecules downstream of phosphatidylinositol 3-kinase. *Journal of Immunology*, *181*(5), 2980–2989.

Freitas, A. A., & de Magalhaes, J. P. (2011). A review and appraisal of the DNA damage theory of ageing. *Mutation Research*, *728*(1–2), 12–22. https://doi.org/10.1016/j.mrrev.2011.05.001.

Friedl, P., & Wolf, K. (2003). Proteolytic and non-proteolytic migration of tumour cells and leucocytes. *Biochemical Society Symposium*, (70), 277–285. https://www.ncbi.nlm.nih.gov/pubmed/14587300.

Furukawa-Hibi, Y., Yoshida-Araki, K., Ohta, T., Ikeda, K., & Motoyama, N. (2002). FOXO forkhead transcription factors induce G(2)-M checkpoint in response to oxidative stress. *The Journal of Biological Chemistry*, *277*(30), 26729–26732. https://doi.org/10.1074/jbc.C200256200.

Furuyama, T., Kitayama, K., Shimoda, Y., Ogawa, M., Sone, K., Yoshida-Araki, K., et al. (2004). Abnormal angiogenesis in Foxo1 (Fkhr)-deficient mice. *The Journal of Biological Chemistry*, *279*(33), 34741–34749. https://doi.org/10.1074/jbc.M314214200.

Galili, N., Davis, R. J., Fredericks, W. J., Mukhopadhyay, S., Rauscher, F. J., 3rd, Emanuel, B. S., et al. (1993). Fusion of a fork head domain gene to PAX3 in the solid tumour alveolar rhabdomyosarcoma. *Nature Genetics*, *5*(3), 230–235. https://doi.org/10.1038/ng1193-230.

Gallagher, L. E., Williamson, L. E., & Chan, E. Y. (2016). Advances in autophagy regulatory mechanisms. *Cell*, *5*(2), 24. https://doi.org/10.3390/cells5020024.

Gan, L., Liu, P., Lu, H., Chen, S., Yang, J., McCarthy, J. B., et al. (2009). Cyclin D1 promotes anchorage-independent cell survival by inhibiting FOXO-mediated anoikis. *Cell Death and Differentiation*, *16*(10), 1408–1417. https://doi.org/10.1038/cdd.2009.86.

Garrett, J. T., Olivares, M. G., Rinehart, C., Granja-Ingram, N. D., Sanchez, V., Chakrabarty, A., et al. (2011). Transcriptional and posttranslational up-regulation of HER3 (ErbB3) compensates for inhibition of the HER2 tyrosine kinase. *Proceedings of the National Academy of Sciences of the United States of America*, *108*(12), 5021–5026. https://doi.org/10.1073/pnas.1016140108.

George, S., Rochford, J. J., Wolfrum, C., Gray, S. L., Schinner, S., Wilson, J. C., et al. (2004). A family with severe insulin resistance and diabetes due to a mutation in AKT2. *Science*, *304*(5675), 1325–1328. https://doi.org/10.1126/science.1096706.

Germani, A., Matrone, A., Grossi, V., Peserico, A., Sanese, P., Liuzzi, M., et al. (2014). Targeted therapy against chemoresistant colorectal cancers: Inhibition of p38alpha modulates the effect of cisplatin in vitro and in vivo through the tumor suppressor FoxO3A. *Cancer Letters*, *344*(1), 110–118. https://doi.org/10.1016/j.canlet.2013.10.035.

Ghaffari, S., Jagani, Z., Kitidis, C., Lodish, H. F., & Khosravi-Far, R. (2003). Cytokines and BCR-ABL mediate suppression of TRAIL-induced apoptosis through inhibition of forkhead FOXO3a transcription factor. *Proceedings of the National Academy of Sciences of the United States of America*, *100*(11), 6523–6528.

Go, H., Jang, J. Y., Kim, P. J., Kim, Y. G., Nam, S. J., Paik, J. H., et al. (2015). MicroRNA-21 plays an oncogenic role by targeting FOXO1 and activating the PI3K/AKT pathway in diffuse large B-cell lymphoma. *Oncotarget*, *6*(17), 15035–15049. https://doi.org/10.18632/oncotarget.3729.

Gomis, R. R., Alarcon, C., He, W., Wang, Q., Seoane, J., Lash, A., et al. (2006). A FoxO-Smad synexpression group in human keratinocytes. *Proceedings of the National Academy of Sciences of the United States of America*, *103*(34), 12747–12752. https://doi.org/10.1073/pnas.0605333103.

Greer, E. L., Oskoui, P. R., Banko, M. R., Maniar, J. M., Gygi, M. P., Gygi, S. P., et al. (2007). The energy sensor AMP-activated protein kinase directly regulates the mammalian FOXO3 transcription factor. *The Journal of Biological Chemistry*, *282*(41), 30107–30119. https://doi.org/10.1074/jbc.M705325200.

Guan, L., Zhang, L., Gong, Z., Hou, X., Xu, Y., Feng, X., et al. (2016). FoxO3 inactivation promotes human cholangiocarcinoma tumorigenesis and chemoresistance through Keap1-Nrf2 signaling. *Hepatology*, *63*(6), 1914–1927. https://doi.org/10.1002/hep.28496.

Habashy, H. O., Rakha, E. A., Aleskandarany, M., Ahmed, M. A., Green, A. R., Ellis, I. O., et al. (2011). FOXO3a nuclear localisation is associated with good prognosis in luminal-like breast cancer. *Breast Cancer Research and Treatment*, *129*(1), 11–21. https://doi.org/10.1007/s10549-010-1161-z.

Hagenbuchner, J., Kuznetsov, A., Hermann, M., Hausott, B., Obexer, P., & Ausserlechner, M. J. (2012). FOXO3-induced reactive oxygen species are regulated by BCL2L11 (Bim) and SESN3. *Journal of Cell Science*, *125*(Pt. 5), 1191–1203. https://doi.org/10.1242/jcs.092098.

Han, C. Y., Cho, K. B., Choi, H. S., Han, H. K., & Kang, K. W. (2008). Role of FoxO1 activation in MDR1 expression in adriamycin-resistant breast cancer cells. *Carcinogenesis*, *29*(9), 1837–1844. https://doi.org/10.1093/carcin/bgn092.

Hanahan, D., & Weinberg, R. A. (2000). The hallmarks of cancer. *Cell*, *100*(1), 57–70.

Hanahan, D., & Weinberg, R. A. (2011). Hallmarks of cancer: The next generation. *Cell*, *144*(5), 646–674. https://doi.org/10.1016/j.cell.2011.02.013.

Hemmings, B. A., & Restuccia, D. F. (2012). PI3K-PKB/Akt pathway. *Cold Spring Harbor Perspectives in Biology*, *4*(9), a011189. https://doi.org/10.1101/cshperspect.a011189.

Hillion, J., Le Coniat, M., Jonveaux, P., Berger, R., & Bernard, O. A. (1997). AF6q21, a novel partner of the MLL gene in t(6;11)(q21;q23), defines a forkhead transcriptional factor subfamily. *Blood*, *90*(9), 3714–3719.

Ho, K. K., McGuire, V. A., Koo, C. Y., Muir, K. W., de Olano, N., Maifoshie, E., et al. (2012). Phosphorylation of FOXO3a on Ser-7 by p38 promotes its nuclear localization in response to doxorubicin. *The Journal of Biological Chemistry*, *287*(2), 1545–1555. https://doi.org/10.1074/jbc.M111.284224.

Hoeijmakers, J. H. (2009). DNA damage, aging, and cancer. *The New England Journal of Medicine*, *361*(15), 1475–1485. https://doi.org/10.1056/NEJMra0804615.

Hoogeboom, D., Essers, M. A., Polderman, P. E., Voets, E., Smits, L. M., & Burgering, B. M. (2008). Interaction of FOXO with beta-catenin inhibits beta-catenin/T cell factor activity. *The Journal of Biological Chemistry*, *283*(14), 9224–9230. https://doi.org/10.1074/jbc.M706638200.

Hornsveld, M., Tenhagen, M., van de Ven, R. A., Smits, A. M., van Triest, M. H., van Amersfoort, M., et al. (2016). Restraining FOXO3-dependent transcriptional BMF activation underpins tumour growth and metastasis of E-cadherin-negative breast cancer. *Cell Death and Differentiation*, *23*(9), 1483–1492. https://doi.org/10.1038/cdd.2016.33.

Huang, J., & Manning, B. D. (2008). The TSC1-TSC2 complex: A molecular switchboard controlling cell growth. *The Biochemical Journal*, *412*(2), 179–190. https://doi.org/10.1042/BJ20080281.

Huang, H., Regan, K. M., Lou, Z., Chen, J., & Tindall, D. J. (2006). CDK2-dependent phosphorylation of FOXO1 as an apoptotic response to DNA damage. *Science*, *314*(5797), 294–297. https://doi.org/10.1126/science.1130512.

Hui, R. C., Francis, R. E., Guest, S. K., Costa, J. R., Gomes, A. R., Myatt, S. S., et al. (2008). Doxorubicin activates FOXO3a to induce the expression of multidrug resistance gene ABCB1 (MDR1) in K562 leukemic cells. *Molecular Cancer Therapeutics*, *7*(3), 670–678. https://doi.org/10.1158/1535-7163.MCT-07-0397.

Hui, R. C., Gomes, A. R., Constantinidou, D., Costa, J. R., Karadedou, C. T., Fernandez de Mattos, S., et al. (2008). The forkhead transcription factor FOXO3a increases phosphoinositide-3 kinase/Akt activity in drug-resistant leukemic cells through induction of PIK3CA expression. *Molecular and Cellular Biology*, *28*(19), 5886–5898. https://doi.org/10.1128/MCB.01265-07.

Huo, X., Liu, S., Shao, T., Hua, H., Kong, Q., Wang, J., et al. (2014). GSK3 protein positively regulates type I insulin-like growth factor receptor through forkhead transcription factors FOXO1/3/4. *The Journal of Biological Chemistry*, *289*(36), 24759–24770. https://doi.org/10.1074/jbc.M114.580738.

Iovine, B., Iannella, M. L., & Bevilacqua, M. A. (2011). Damage-specific DNA binding protein 1 (DDB1) is involved in ubiquitin-mediated proteolysis of p27Kip1 in response to UV irradiation. *Biochimie*, *93*(5), 867–875. https://doi.org/10.1016/j.biochi.2010.12.017.

Jogo, M., Shiraishi, S., & Tamura, T. A. (2009). Identification of MAFbx as a myogenin-engaged F-box protein in SCF ubiquitin ligase. *FEBS Letters*, *583*(17), 2715–2719. https://doi.org/10.1016/j.febslet.2009.07.033.

Jung, C. H., Ro, S. H., Cao, J., Otto, N. M., & Kim, D. H. (2010). mTOR regulation of autophagy. *FEBS Letters*, *584*(7), 1287–1295. https://doi.org/10.1016/j.febslet.2010.01.017.

Katayama, K., Nakamura, A., Sugimoto, Y., Tsuruo, T., & Fujita, N. (2008). FOXO transcription factor-dependent p15(INK4b) and p19(INK4d) expression. *Oncogene*, *27*(12), 1677–1686. https://doi.org/10.1038/sj.onc.1210813.

Kenyon, C., Chang, J., Gensch, E., Rudner, A., & Tabtiang, R. (1993). A C. elegans mutant that lives twice as long as wild type. *Nature*, *366*(6454), 461–464. https://doi.org/10.1038/366461a0.

Kerdiles, Y. M., Beisner, D. R., Tinoco, R., Dejean, A. S., Castrillon, D. H., DePinho, R. A., et al. (2009). Foxo1 links homing and survival of naive T cells by regulating L-selectin, CCR7 and interleukin 7 receptor. *Nature Immunology*, *10*(2), 176–184. https://doi.org/10.1038/ni.1689.

Kihara, A., Kabeya, Y., Ohsumi, Y., & Yoshimori, T. (2001). Beclin-phosphatidylinositol 3-kinase complex functions at the trans-Golgi network. *EMBO Reports*, *2*(4), 330–335. https://doi.org/10.1093/embo-reports/kve061.

Kikuno, N., Shiina, H., Urakami, S., Kawamoto, K., Hirata, H., Tanaka, Y., et al. (2007). Knockdown of astrocyte-elevated gene-1 inhibits prostate cancer progression through upregulation of FOXO3a activity. *Oncogene*, *26*(55), 7647–7655. https://doi.org/10.1038/sj.onc.1210572.

Kloet, D. E., & Burgering, B. M. (2011). The PKB/FOXO switch in aging and cancer. *Biochimica et Biophysica Acta*, *1813*(11), 1926–1937. https://doi.org/10.1016/j.bbamcr.2011.04.003.

Klotz, L. O., Sanchez-Ramos, C., Prieto-Arroyo, I., Urbanek, P., Steinbrenner, H., & Monsalve, M. (2015). Redox regulation of FoxO transcription factors. *Redox Biology*, *6*, 51–72. https://doi.org/10.1016/j.redox.2015.06.019.

Kojima, T., Shimazui, T., Horie, R., Hinotsu, S., Oikawa, T., Kawai, K., et al. (2010). FOXO1 and TCF7L2 genes involved in metastasis and poor prognosis in clear cell renal cell carcinoma. *Genes, Chromosomes & Cancer*, *49*(4), 379–389. https://doi.org/10.1002/gcc.20750.

Kops, G. J., Dansen, T. B., Polderman, P. E., Saarloos, I., Wirtz, K. W., Coffer, P. J., et al. (2002). Forkhead transcription factor FOXO3a protects quiescent cells from oxidative stress. *Nature*, *419*(6904), 316–321. https://doi.org/10.1038/nature01036.

Kops, G. J., de Ruiter, N. D., De Vries-Smits, A. M., Powell, D. R., Bos, J. L., & Burgering, B. M. (1999). Direct control of the Forkhead transcription factor AFX by protein kinase B. *Nature*, *398*(6728), 630–634. https://doi.org/10.1038/19328.

Kops, G. J., Medema, R. H., Glassford, J., Essers, M. A., Dijkers, P. F., Coffer, P. J., et al. (2002). Control of cell cycle exit and entry by protein kinase B-regulated forkhead transcription factors. *Molecular and Cellular Biology*, *22*(7), 2025–2036.

Kress, T. R., Cannell, I. G., Brenkman, A. B., Samans, B., Gaestel, M., Roepman, P., et al. (2011). The MK5/PRAK kinase and Myc form a negative feedback loop that is disrupted during colorectal tumorigenesis. *Molecular Cell*, *41*(4), 445–457. https://doi.org/10.1016/j.molcel.2011.01.023.

Lecot, P., Alimirah, F., Desprez, P. Y., Campisi, J., & Wiley, C. (2016). Context-dependent effects of cellular senescence in cancer development. *British Journal of Cancer*, *114*(11), 1180–1184. https://doi.org/10.1038/bjc.2016.115.

Lee, G., Goretsky, T., Managlia, E., Dirisina, R., Singh, A. P., Brown, J. B., et al. (2010). Phosphoinositide 3-kinase signaling mediates beta-catenin activation in intestinal epithelial stem and progenitor cells in colitis. *Gastroenterology*, *139*(3), 869–881. 881. e1–9. https://doi.org/10.1053/j.gastro.2010.05.037.

Lee, S. R., Yang, K. S., Kwon, J., Lee, C., Jeong, W., & Rhee, S. G. (2002). Reversible inactivation of the tumor suppressor PTEN by H2O2. *The Journal of Biological Chemistry*, *277*(23), 20336–20342. https://doi.org/10.1074/jbc.M111899200.

Lei, H., & Quelle, F. W. (2009). FOXO transcription factors enforce cell cycle checkpoints and promote survival of hematopoietic cells after DNA damage. *Molecular Cancer Research*, 7(8), 1294–1303. https://doi.org/10.1158/1541-7786.MCR-08-0531.

Li, J., Du, W., Maynard, S., Andreassen, P. R., & Pang, Q. (2010). Oxidative stress-specific interaction between FANCD2 and FOXO3a. *Blood*, *115*(8), 1545–1548. https://doi.org/10.1182/blood-2009-07-234385.

Li, Q., Li, J., Lan, H., Wang, N., Hu, X., Chen, L., et al. (2011). Effects of fasting and refeeding on expression of MAFbx and MuRF1 in chick skeletal muscle. *Science China. Life Sciences*, *54*(10), 904–907. https://doi.org/10.1007/s11427-011-4226-2.

Li, L., Ye, S., Yang, M., Yu, W., Fan, Z., Zhang, H., et al. (2015). SIRT1 downregulation enhances chemosensitivity and survival of adult T-cell leukemia-lymphoma cells by reducing DNA double-strand repair. *Oncology Reports*, *34*(6), 2935–2942. https://doi.org/10.3892/or.2015.4287.

Liang, R., Rimmelé, P., Bigarella, C. L., Yalcin, S., & Ghaffari, S. (2016). Evidence for AKT-independent regulation of FOXO1 and FOXO3 in haematopoietic stem and progenitor cells. *Cell Cycle*, *15*(6), 861–867.

Liang, J., Zubovitz, J., Petrocelli, T., Kotchetkov, R., Connor, M. K., Han, K., et al. (2002). PKB/Akt phosphorylates p27, impairs nuclear import of p27 and opposes p27-mediated G1 arrest. *Nature Medicine*, *8*(10), 1153–1160. https://doi.org/10.1038/nm761.

Lin, A., Piao, H. L., Zhuang, L., Sarbassov dos, D., Ma, L., & Gan, B. (2014). FoxO transcription factors promote AKT Ser473 phosphorylation and renal tumor growth in response to pharmacologic inhibition of the PI3K-AKT pathway. *Cancer Research*, *74*(6), 1682–1693. https://doi.org/10.1158/0008-5472.CAN-13-1729.

Lin, S., Ptasinska, A., Chen, X., Shrestha, M., Assi, S. A., Chin, P. S., et al. (2017). A FOXO1-induced oncogenic network defines the AML1-ETO pre-leukemic program. *Blood*, *130*(10), 1213–1222. https://doi.org/10.1182/blood-2016-11-750976.

Liou, G. Y., & Storz, P. (2010). Reactive oxygen species in cancer. *Free Radical Research*, *44*(5), 479–496. https://doi.org/10.3109/10715761003667554.

Liu, J. W., Chandra, D., Rudd, M. D., Butler, A. P., Pallotta, V., Brown, D., et al. (2005). Induction of prosurvival molecules by apoptotic stimuli: Involvement of FOXO3a and ROS. *Oncogene*, *24*(12), 2020–2031. https://doi.org/10.1038/sj.onc.1208385.

Liu, H., Fergusson, M. M., Wu, J. J., Rovira, I. I., Liu, J., Gavrilova, O., et al. (2011). Wnt signaling regulates hepatic metabolism. *Science Signaling*, *4*(158), ra6. https://doi.org/10.1126/scisignal.2001249.

Liu, P., Kao, T. P., & Huang, H. (2008). CDK1 promotes cell proliferation and survival via phosphorylation and inhibition of FOXO1 transcription factor. *Oncogene*, *27*(34), 4733–4744. https://doi.org/10.1038/onc.2008.104.

Liu, H., Yin, J., Wang, H., Jiang, G., Deng, M., Zhang, G., et al. (2015). FOXO3a modulates WNT/beta-catenin signaling and suppresses epithelial-to-mesenchymal transition in prostate cancer cells. *Cellular Signalling*, *27*(3), 510–518. https://doi.org/10.1016/j.cellsig.2015.01.001.

Loaiza, N., & Demaria, M. (2016). Cellular senescence and tumor promotion: Is aging the key? *Biochimica et Biophysica Acta*, *1865*(2), 155–167. https://doi.org/10.1016/j.bbcan.2016.01.007.

Longerich, S., Li, J., Xiong, Y., Sung, P., & Kupfer, G. M. (2014). Stress and DNA repair biology of the Fanconi anemia pathway. *Blood*, *124*(18), 2812–2819. https://doi.org/10.1182/blood-2014-04-526293.

Lu, H., Liu, P., Pan, Y., & Huang, H. (2011). Inhibition of cyclin-dependent kinase phosphorylation of FOXO1 and prostate cancer cell growth by a peptide derived from FOXO1. *Neoplasia*, *13*(9), 854–863.

Lupertz, R., Chovolou, Y., Unfried, K., Kampkotter, A., Watjen, W., & Kahl, R. (2008). The forkhead transcription factor FOXO4 sensitizes cancer cells to doxorubicin-mediated cytotoxicity. *Carcinogenesis*, *29*(11), 2045–2052. https://doi.org/10.1093/carcin/bgn184.

MacDonald, B. T., Tamai, K., & He, X. (2009). Wnt/beta-catenin signaling: Components, mechanisms, and diseases. *Developmental Cell*, *17*(1), 9–26. https://doi.org/10.1016/j.devcel.2009.06.016.

Mahadev, K., Motoshima, H., Wu, X., Ruddy, J. M., Arnold, R. S., Cheng, G., et al. (2004). The NAD(P)H oxidase homolog Nox4 modulates insulin-stimulated generation of H2O2 and plays an integral role in insulin signal transduction. *Molecular and Cellular Biology*, *24*(5), 1844–1854.

Manning, B. D., & Toker, A. (2017). AKT/PKB signaling: Navigating the network. *Cell*, *169*(3), 381–405. https://doi.org/10.1016/j.cell.2017.04.001.

Marlow, L. A., von Roemeling, C. A., Cooper, S. J., Zhang, Y., Rohl, S. D., Arora, S., et al. (2012). Foxo3a drives proliferation in anaplastic thyroid carcinoma through transcriptional regulation of cyclin A1: A paradigm shift that impacts current therapeutic strategies. *Journal of Cell Science*, *125*(Pt. 18), 4253–4263. https://doi.org/10.1242/jcs.097428.

Martinez-Gac, L., Marques, M., Garcia, Z., Campanero, M. R., & Carrera, A. C. (2004). Control of cyclin G2 mRNA expression by forkhead transcription factors: Novel mechanism for cell cycle control by phosphoinositide 3-kinase and forkhead. *Molecular and Cellular Biology*, *24*(5), 2181–2189.

Matsuoka, S., Ballif, B. A., Smogorzewska, A., McDonald, E. R., 3rd., Hurov, K. E., Luo, J., et al. (2007). ATM and ATR substrate analysis reveals extensive protein networks responsive to DNA damage. *Science*, *316*(5828), 1160–1166. https://doi.org/10.1126/science.1140321.

Matsuura, A., Tsukada, M., Wada, Y., & Ohsumi, Y. (1997). Apg1p, a novel protein kinase required for the autophagic process in Saccharomyces cerevisiae. *Gene*, *192*(2), 245–250.

Medema, R. H., Kops, G. J., Bos, J. L., & Burgering, B. M. (2000). AFX-like Forkhead transcription factors mediate cell-cycle regulation by Ras and PKB through p27kip1. *Nature*, *404*(6779), 782–787. https://doi.org/10.1038/35008115.

Mei, Z., Zhang, D., Hu, B., Wang, J., Shen, X., & Xiao, W. (2015). FBXO32 targets c-Myc for proteasomal degradation and inhibits c-Myc activity. *The Journal of Biological Chemistry*, *290*(26), 16202–16214. https://doi.org/10.1074/jbc.M115.645978.

Menon, S. G., Sarsour, E. H., Spitz, D. R., Higashikubo, R., Sturm, M., Zhang, H., et al. (2003). Redox regulation of the G1 to S phase transition in the mouse embryo fibroblast cell cycle. *Cancer Research*, *63*(9), 2109–2117.

Mercado, G. E., & Barr, F. G. (2007). Fusions involving PAX and FOX genes in the molecular pathogenesis of alveolar rhabdomyosarcoma: Recent advances. *Current Molecular Medicine*, 7(1), 47–61.

Miyamoto, K., Araki, K. Y., Naka, K., Arai, F., Takubo, K., Yamazaki, S., et al. (2007). Foxo3a is essential for maintenance of the hematopoietic stem cell pool. *Cell Stem Cell*, *1*(1), 101–112. https://doi.org/10.1016/j.stem.2007.02.001.

Modur, V., Nagarajan, R., Evers, B. M., & Milbrandt, J. (2002). FOXO proteins regulate tumor necrosis factor-related apoptosis inducing ligand expression. Implications for PTEN mutation in prostate cancer. *The Journal of Biological Chemistry*, *277*(49), 47928–47937. https://doi.org/10.1074/jbc.M207509200.

Moiseeva, O., Mallette, F. A., Mukhopadhyay, U. K., Moores, A., & Ferbeyre, G. (2006). DNA damage signaling and p53-dependent senescence after prolonged beta-interferon stimulation. *Molecular Biology of the Cell*, *17*(4), 1583–1592. https://doi.org/10.1091/mbc.E05-09-0858.

Morfouace, M., Lalier, L., Bahut, M., Bonnamain, V., Naveilhan, P., Guette, C., et al. (2012). Comparison of spheroids formed by rat glioma stem cells and neural stem cells reveals differences in glucose metabolism and promising therapeutic applications. *The Journal of Biological Chemistry*, *287*(40), 33664–33674. https://doi.org/10.1074/jbc.M111.320028.

Morin, R. D., Mendez-Lago, M., Mungall, A. J., Goya, R., Mungall, K. L., Corbett, R. D., et al. (2011). Frequent mutation of histone-modifying genes in non-Hodgkin lymphoma. *Nature*, *476*(7360), 298–303. https://doi.org/10.1038/nature10351.

Motoyama, N., & Naka, K. (2004). DNA damage tumor suppressor genes and genomic instability. *Current Opinion in Genetics & Development*, *14*(1), 11–16. https://doi.org/10.1016/j.gde.2003.12.003.

Mueller, M. M., Castells-Roca, L., Babu, V., Ermolaeva, M. A., Muller, R. U., Frommolt, P., et al. (2014). DAF-16/FOXO and EGL-27/GATA promote developmental growth in response to persistent somatic DNA damage. *Nature Cell Biology*, *16*(12), 1168–1179. https://doi.org/10.1038/ncb3071.

Naka, K., Hoshii, T., Muraguchi, T., Tadokoro, Y., Ooshio, T., Kondo, Y., et al. (2010). TGF-beta-FOXO signalling maintains leukaemia-initiating cells in chronic myeloid leukaemia. *Nature*, *463*(7281), 676–680. https://doi.org/10.1038/nature08734.

Nemoto, S., & Finkel, T. (2002). Redox regulation of forkhead proteins through a p66shc-dependent signaling pathway. *Science*, *295*(5564), 2450–2452. https://doi.org/10.1126/science.1069004.

Ng, S. S., Mahmoudi, T., Danenberg, E., Bejaoui, I., de Lau, W., Korswagen, H. C., et al. (2009). Phosphatidylinositol 3-kinase signaling does not activate the wnt cascade. *The Journal of Biological Chemistry*, *284*(51), 35308–35313. https://doi.org/10.1074/jbc.M109.078261.

Nicklin, P., Bergman, P., Zhang, B., Triantafellow, E., Wang, H., Nyfeler, B., et al. (2009). Bidirectional transport of amino acids regulates mTOR and autophagy. *Cell*, *136*(3), 521–534. https://doi.org/10.1016/j.cell.2008.11.044.

Ning, Y., Luo, C., Ren, K., Quan, M., & Cao, J. (2014). FOXO3a-mediated suppression of the self-renewal capacity of sphere-forming cells derived from the ovarian cancer SKOV3 cell line by 7-difluoromethoxyl-5,4′-di-n-octyl genistein. *Molecular Medicine Reports*, *9*(5), 1982–1988. https://doi.org/10.3892/mmr.2014.2012.

Nogueira, V., Park, Y., Chen, C. C., Xu, P. Z., Chen, M. L., Tonic, I., et al. (2008). Akt determines replicative senescence and oxidative or oncogenic premature senescence and sensitizes cells to oxidative apoptosis. *Cancer Cell*, *14*(6), 458–470. https://doi.org/10.1016/j.ccr.2008.11.003.

Notas, G., Alexaki, V. I., Kampa, M., Pelekanou, V., Charalampopoulos, I., Sabour-Alaoui, S., et al. (2012). APRIL binding to BCMA activates a JNK2-FOXO3-GADD45 pathway and induces a G2/M cell growth arrest in liver cells. *Journal of Immunology, 189*(10), 4748–4758. https://doi.org/10.4049/jimmunol.1102891.

Nyakern, M., Cappellini, A., Mantovani, I., & Martelli, A. M. (2006). Synergistic induction of apoptosis in human leukemia T cells by the Akt inhibitor perifosine and etoposide through activation of intrinsic and Fas-mediated extrinsic cell death pathways. *Molecular Cancer Therapeutics, 5*(6), 1559–1570. https://doi.org/10.1158/1535-7163.MCT-06-0076.

Obexer, P., Hagenbuchner, J., Unterkircher, T., Sachsenmaier, N., Seifarth, C., Bock, G., et al. (2009). Repression of BIRC5/survivin by FOXO3/FKHRL1 sensitizes human neuroblastoma cells to DNA damage-induced apoptosis. *Molecular Biology of the Cell, 20*(7), 2041–2048. https://doi.org/10.1091/mbc.E08-07-0699.

Obrador-Hevia, A., Serra-Sitjar, M., Rodriguez, J., Villalonga, P., & Fernandez de Mattos, S. (2012). The tumour suppressor FOXO3 is a key regulator of mantle cell lymphoma proliferation and survival. *British Journal of Haematology, 156*(3), 334–345. https://doi.org/10.1111/j.1365-2141.2011.08951.x.

Ogg, S., Paradis, S., Gottlieb, S., Patterson, G. I., Lee, L., Tissenbaum, H. A., et al. (1997). The Fork head transcription factor DAF-16 transduces insulin-like metabolic and longevity signals in C. elegans. *Nature, 389*(6654), 994–999. https://doi.org/10.1038/40194.

O'Reilly, K. E., Rojo, F., She, Q. B., Solit, D., Mills, G. B., Smith, D., et al. (2006). mTOR inhibition induces upstream receptor tyrosine kinase signaling and activates Akt. *Cancer Research, 66*(3), 1500–1508. https://doi.org/10.1158/0008-5472.CAN-05-2925.

Paik, J. H., Kollipara, R., Chu, G., Ji, H., Xiao, Y., Ding, Z., et al. (2007). FoxOs are lineage-restricted redundant tumor suppressors and regulate endothelial cell homeostasis. *Cell, 128*(2), 309–323. https://doi.org/10.1016/j.cell.2006.12.029.

Park, J., Choi, Y., Ko, Y. S., Kim, Y., Pyo, J. S., Jang, B. G., et al. (2017). FOXO1 suppression is a determinant of acquired lapatinib-resistance in HER2-positive gastric cancer cells through MET upregulation. *Cancer Research and Treatment.* https://doi.org/10.4143/crt.2016.580. (Epub ahead of print).

Parmigiani, A., Nourbakhsh, A., Ding, B., Wang, W., Kim, Y. C., Akopiants, K., et al. (2014). Sestrins inhibit mTORC1 kinase activation through the GATOR complex. *Cell Reports, 9*(4), 1281–1291. https://doi.org/10.1016/j.celrep.2014.10.019.

Pellicano, F., Scott, M. T., Helgason, G. V., Hopcroft, L. E., Allan, E. K., Aspinall-O'Dea, M., et al. (2014). The antiproliferative activity of kinase inhibitors in chronic myeloid leukemia cells is mediated by FOXO transcription factors. *Stem Cells, 32*(9), 2324–2337. https://doi.org/10.1002/stem.1748.

Potter, C. J., Pedraza, L. G., & Xu, T. (2002). Akt regulates growth by directly phosphorylating Tsc2. *Nature Cell Biology, 4*(9), 658–665. https://doi.org/10.1038/ncb840.

Prabhu, V. V., Allen, J. E., Dicker, D. T., & El-Deiry, W. S. (2015). Small-molecule ONC201/TIC10 targets chemotherapy-resistant colorectal cancer stem-like cells in an Akt/Foxo3a/TRAIL-dependent manner. *Cancer Research, 75*(7), 1423–1432. https://doi.org/10.1158/0008-5472.CAN-13-3451.

Prahallad, A., Sun, C., Huang, S., Di Nicolantonio, F., Salazar, R., Zecchin, D., et al. (2012). Unresponsiveness of colon cancer to BRAF(V600E) inhibition through feedback activation of EGFR. *Nature, 483*(7387), 100–103. https://doi.org/10.1038/nature10868.

Puig, O., Marr, M. T., Ruhf, M. L., & Tjian, R. (2003). Control of cell number by Drosophila FOXO: Downstream and feedback regulation of the insulin receptor pathway. *Genes & Development, 17*(16), 2006–2020. https://doi.org/10.1101/gad.1098703.

Putker, M., Madl, T., Vos, H. R., de Ruiter, H., Visscher, M., van den Berg, M. C., et al. (2013). Redox-dependent control of FOXO/DAF-16 by transportin-1. *Molecular Cell, 49*(4), 730–742. https://doi.org/10.1016/j.molcel.2012.12.014.

Putker, M., Vos, H. R., van Dorenmalen, K., de Ruiter, H., Duran, A. G., Snel, B., et al. (2015). Evolutionary acquisition of cysteines determines FOXO paralog-specific redox signaling. *Antioxidants & Redox Signaling*, *22*(1), 15–28. https://doi.org/10.1089/ars.2014.6056.

Qiong, Z., Ruofan, H., Xiaohua, L., Xinli, Z., Jingwei, J., & Zhaohui, C. (2010). Role of dephosphorylation of FOXO1 on apoptosis induced by wortmannin for non-Hodgkin's lymphoma cells. *Molecular Biology Reports*, *37*(5), 2397–2402. https://doi.org/10.1007/s11033-009-9748-3.

Ramaswamy, S., Nakamura, N., Sansal, I., Bergeron, L., & Sellers, W. R. (2002). A novel mechanism of gene regulation and tumor suppression by the transcription factor FKHR. *Cancer Cell*, *2*(1), 81–91.

Reif, K., Nobes, C. D., Thomas, G., Hall, A., & Cantrell, D. A. (1996). Phosphatidylinositol 3-kinase signals activate a selective subset of Rac/Rho-dependent effector pathways. *Current Biology*, *6*(11), 1445–1455.

Renault, V. M., Rafalski, V. A., Morgan, A. A., Salih, D. A., Brett, J. O., Webb, A. E., et al. (2009). FoxO3 regulates neural stem cell homeostasis. *Cell Stem Cell*, *5*(5), 527–539. https://doi.org/10.1016/j.stem.2009.09.014.

Rodon, J., Dienstmann, R., Serra, V., & Tabernero, J. (2013). Development of PI3K inhibitors: Lessons learned from early clinical trials. *Nature Reviews. Clinical Oncology*, *10*(3), 143–153. https://doi.org/10.1038/nrclinonc.2013.10.

Romero-Laorden, N., & Castro, E. (2017). Inherited mutations in DNA repair genes and cancer risk. *Current Problems in Cancer*, *41*, 251–264. https://doi.org/10.1016/j.currproblcancer.2017.02.009.

Ryu, K. J., Park, C., Hong, M., Ko, Y. H., Kim, W. S., & Kim, S. J. (2017). FOXO4 expression is related to stem cell-like properties and resistance to treatment in diffuse large B-cell lymphoma. *Oncotarget*, *8*(2), 2466–2476. https://doi.org/10.18632/oncotarget.13690.

Sachdeva, M., Zhu, S., Wu, F., Wu, H., Walia, V., Kumar, S., et al. (2009). p53 represses c-Myc through induction of the tumor suppressor miR-145. *Proceedings of the National Academy of Sciences of the United States of America*, *106*(9), 3207–3212. https://doi.org/10.1073/pnas.0808042106.

Salmeen, A., Andersen, J. N., Myers, M. P., Meng, T. C., Hinks, J. A., Tonks, N. K., et al. (2003). Redox regulation of protein tyrosine phosphatase 1B involves a sulphenyl-amide intermediate. *Nature*, *423*(6941), 769–773. https://doi.org/10.1038/nature01680.

Samuels, Y., Wang, Z., Bardelli, A., Silliman, N., Ptak, J., Szabo, S., et al. (2004). High frequency of mutations of the PIK3CA gene in human cancers. *Science*, *304*(5670), 554. https://doi.org/10.1126/science.1096502.

Sandri, M. (2013). Protein breakdown in muscle wasting: Role of autophagy-lysosome and ubiquitin-proteasome. *The International Journal of Biochemistry & Cell Biology*, *45*(10), 2121–2129. https://doi.org/10.1016/j.biocel.2013.04.023.

Sandri, M., Sandri, C., Gilbert, A., Skurk, C., Calabria, E., Picard, A., et al. (2004). Foxo transcription factors induce the atrophy-related ubiquitin ligase atrogin-1 and cause skeletal muscle atrophy. *Cell*, *117*(3), 399–412.

Sang, L., Coller, H. A., & Roberts, J. M. (2008). Control of the reversibility of cellular quiescence by the transcriptional repressor HES1. *Science*, *321*(5892), 1095–1100. https://doi.org/10.1126/science.1155998.

Sarsour, E. H., Venkataraman, S., Kalen, A. L., Oberley, L. W., & Goswami, P. C. (2008). Manganese superoxide dismutase activity regulates transitions between quiescent and proliferative growth. *Aging Cell*, 7(3), 405–417. https://doi.org/10.1111/j.1474-9726.2008.00384.x.

Schmidt, M., Fernandez de Mattos, S., van der Horst, A., Klompmaker, R., Kops, G. J., Lam, E. W., et al. (2002). Cell cycle inhibition by FoxO forkhead transcription factors involves downregulation of cyclin D. *Molecular and Cellular Biology*, *22*(22), 7842–7852.

Schmitt-Ney, M., & Camussi, G. (2015). The PAX3-FOXO1 fusion protein present in rhabdomyosarcoma interferes with normal FOXO activity and the TGF-beta pathway. *PLoS One*, *10*(3), e0121474. https://doi.org/10.1371/journal.pone.0121474.

Schumacher, B. (2009). Transcription-blocking DNA damage in aging: A mechanism for hormesis. *Bioessays*, *31*(12), 1347–1356. https://doi.org/10.1002/bies.200900107.

Seoane, J., Le, H. V., Shen, L., Anderson, S. A., & Massague, J. (2004). Integration of Smad and forkhead pathways in the control of neuroepithelial and glioblastoma cell proliferation. *Cell*, *117*(2), 211–223.

Shen, H., Wang, D., Li, L., Yang, S., Chen, X., Zhou, S., et al. (2017). MiR-222 promotes drug-resistance of breast cancer cells to adriamycin via modulation of PTEN/Akt/FOXO1 pathway. *Gene*, *596*, 110–118. https://doi.org/10.1016/j.gene.2016.10.016.

Shimizu, T., Bamba, Y., Kawabe, Y., Fukuda, T., Fujimori, F., Takahashi, K., et al. (2016). Prolyl isomerase Pin1 regulates doxorubicin-inducible P-glycoprotein level by reducing Foxo3 stability. *Biochemical and Biophysical Research Communications*, *471*(2), 328–333. https://doi.org/10.1016/j.bbrc.2016.02.014.

Shin, I., Yakes, F. M., Rojo, F., Shin, N. Y., Bakin, A. V., Baselga, J., et al. (2002). PKB/Akt mediates cell-cycle progression by phosphorylation of p27(Kip1) at threonine 157 and modulation of its cellular localization. *Nature Medicine*, *8*(10), 1145–1152. https://doi.org/10.1038/nm759.

Shiota, M., Song, Y., Yokomizo, A., Kiyoshima, K., Tada, Y., Uchino, H., et al. (2010). Foxo3a suppression of urothelial cancer invasiveness through Twist1, Y-box-binding protein 1, and E-cadherin regulation. *Clinical Cancer Research*, *16*(23), 5654–5663. https://doi.org/10.1158/1078-0432.CCR-10-0376.

Simioni, C., Martelli, A. M., Cani, A., Cetin-Atalay, R., McCubrey, J. A., Capitani, S., et al. (2013). The AKT inhibitor MK-2206 is cytotoxic in hepatocarcinoma cells displaying hyperphosphorylated AKT-1 and synergizes with conventional chemotherapy. *Oncotarget*, *4*(9), 1496–1506. https://doi.org/10.18632/oncotarget.1236.

Sinclair, L. V., Finlay, D., Feijoo, C., Cornish, G. H., Gray, A., Ager, A., et al. (2008). Phosphatidylinositol-3-OH kinase and nutrient-sensing mTOR pathways control T lymphocyte trafficking. *Nature Immunology*, *9*(5), 513–521. https://doi.org/10.1038/ni.1603.

Skurk, C., Izumiya, Y., Maatz, H., Razeghi, P., Shiojima, I., Sandri, M., et al. (2005). The FOXO3a transcription factor regulates cardiac myocyte size downstream of AKT signaling. *The Journal of Biological Chemistry*, *280*(21), 20814–20823. https://doi.org/10.1074/jbc.M500528200.

Smit, L., Berns, K., Spence, K., Ryder, W. D., Zeps, N., Madiredjo, M., et al. (2016). An integrated genomic approach identifies that the PI3K/AKT/FOXO pathway is involved in breast cancer tumor initiation. *Oncotarget*, 7(3), 2596–2610. https://doi.org/10.18632/oncotarget.6354.

So, C. W., & Cleary, M. L. (2002). MLL-AFX requires the transcriptional effector domains of AFX to transform myeloid progenitors and transdominantly interfere with forkhead protein function. *Molecular and Cellular Biology*, *22*(18), 6542–6552.

Song, K. H., Woo, S. R., Chung, J. Y., Lee, H. J., Oh, S. J., Hong, S. O., et al. (2017). REP1 inhibits FOXO3-mediated apoptosis to promote cancer cell survival. *Cell Death & Disease*, *8*(1), e2536. https://doi.org/10.1038/cddis.2016.462.

Sousa-Victor, P., Gutarra, S., Garcia-Prat, L., Rodriguez-Ubreva, J., Ortet, L., Ruiz-Bonilla, V., et al. (2014). Geriatric muscle stem cells switch reversible quiescence into senescence. *Nature*, *506*(7488), 316–321. https://doi.org/10.1038/nature13013.

Spencer, S. L., Cappell, S. D., Tsai, F. C., Overton, K. W., Wang, C. L., & Meyer, T. (2013). The proliferation-quiescence decision is controlled by a bifurcation in CDK2 activity at mitotic exit. *Cell*, *155*(2), 369–383. https://doi.org/10.1016/j.cell.2013.08.062.

Stambolic, V., & Woodgett, J. R. (1994). Mitogen inactivation of glycogen synthase kinase-3 beta in intact cells via serine 9 phosphorylation. *The Biochemical Journal, 303*(Pt. 3), 701–704.

Storz, P., Doppler, H., Copland, J. A., Simpson, K. J., & Toker, A. (2009). FOXO3a promotes tumor cell invasion through the induction of matrix metalloproteinases. *Molecular and Cellular Biology, 29*(18), 4906–4917. https://doi.org/10.1128/MCB.00077-09.

Strasser, A., Cory, S., & Adams, J. M. (2011). Deciphering the rules of programmed cell death to improve therapy of cancer and other diseases. *The EMBO Journal, 30*(18), 3667–3683. https://doi.org/10.1038/emboj.2011.307.

Sun, C., Hobor, S., Bertotti, A., Zecchin, D., Huang, S., Galimi, F., et al. (2014). Intrinsic resistance to MEK inhibition in KRAS mutant lung and colon cancer through transcriptional induction of ERBB3. *Cell Reports*, 7(1), 86–93. https://doi.org/10.1016/j.celrep.2014.02.045.

Sunayama, J., Sato, A., Matsuda, K., Tachibana, K., Watanabe, E., Seino, S., et al. (2011). FoxO3a functions as a key integrator of cellular signals that control glioblastoma stem-like cell differentiation and tumorigenicity. *Stem Cells, 29*(9), 1327–1337. https://doi.org/10.1002/stem.696.

Sykes, S. M., Lane, S. W., Bullinger, L., Kalaitzidis, D., Yusuf, R., Saez, B., et al. (2011). AKT/FOXO signaling enforces reversible differentiation blockade in myeloid leukemias. *Cell, 146*(5), 697–708. https://doi.org/10.1016/j.cell.2011.07.032.

Tang, T. T., Dowbenko, D., Jackson, A., Toney, L., Lewin, D. A., Dent, A. L., et al. (2002). The forkhead transcription factor AFX activates apoptosis by induction of the BCL-6 transcriptional repressor. *The Journal of Biological Chemistry*, 277(16), 14255–14265. https://doi.org/10.1074/jbc.M110901200.

Tang, H., Inoki, K., Lee, M., Wright, E., Khuong, A., Khuong, A., et al. (2014). mTORC1 promotes denervation-induced muscle atrophy through a mechanism involving the activation of FoxO and E3 ubiquitin ligases. *Science Signaling*, 7(314), ra18. https://doi.org/10.1126/scisignal.2004809.

Tenbaum, S. P., Ordonez-Moran, P., Puig, I., Chicote, I., Arques, O., Landolfi, S., et al. (2012). Beta-catenin confers resistance to PI3K and AKT inhibitors and subverts FOXO3a to promote metastasis in colon cancer. *Nature Medicine, 18*(6), 892–901. https://doi.org/10.1038/nm.2772.

Tezil, T., Bodur, C., Kutuk, O., & Basaga, H. (2012). IKK-beta mediates chemoresistance by sequestering FOXO3; a critical factor for cell survival and death. *Cellular Signalling, 24*(6), 1361–1368. https://doi.org/10.1016/j.cellsig.2012.01.012.

Thakur, B. K., Lippka, Y., Dittrich, T., Chandra, P., Skokowa, J., & Welte, K. (2012). NAMPT pathway is involved in the FOXO3a-mediated regulation of GADD45A expression. *Biochemical and Biophysical Research Communications, 420*(4), 714–720. https://doi.org/10.1016/j.bbrc.2012.03.017.

Tintignac, L. A., Lagirand, J., Batonnet, S., Sirri, V., Leibovitch, M. P., & Leibovitch, S. A. (2005). Degradation of MyoD mediated by the SCF (MAFbx) ubiquitin ligase. *The Journal of Biological Chemistry, 280*(4), 2847–2856. https://doi.org/10.1074/jbc.M411346200.

Tothova, Z., Kollipara, R., Huntly, B. J., Lee, B. H., Castrillon, D. H., Cullen, D. E., et al. (2007). FoxOs are critical mediators of hematopoietic stem cell resistance to physiologic oxidative stress. *Cell, 128*(2), 325–339. https://doi.org/10.1016/j.cell.2007.01.003.

Tran, H., Brunet, A., Grenier, J. M., Datta, S. R., Fornace, A. J., Jr., DiStefano, P. S., et al. (2002). DNA repair pathway stimulated by the forkhead transcription factor FOXO3a through the Gadd45 protein. *Science, 296*(5567), 530–534. https://doi.org/10.1126/science.1068712.

Trinh, D. L., Scott, D. W., Morin, R. D., Mendez-Lago, M., An, J., Jones, S. J., et al. (2013). Analysis of FOXO1 mutations in diffuse large B-cell lymphoma. *Blood, 121*(18), 3666–3674. https://doi.org/10.1182/blood-2013-01-479865.

Tsai, W. B., Chung, Y. M., Takahashi, Y., Xu, Z., & Hu, M. C. (2008). Functional interaction between FOXO3a and ATM regulates DNA damage response. *Nature Cell Biology, 10*(4), 460–467. https://doi.org/10.1038/ncb1709.

Tubi, L., Nunes, S. C., Brancalion, A., Breatta, E. D., Manni, S., Mandato, E., et al. (2017). Protein kinase CK2 regulates AKT, NF-κB and STAT3 activation, stem cell viability and proliferation in acute myeloid leukemia. *Leukemia, 31*, 292–300. https://doi.org/10.1038/leu.2016.209.

van den Berg, M. C. W., & Burgering, B. M. T. (2011). Integrating opposing signals toward forkhead box O. *Antioxidants & Redox Signaling, 14*(4), 607–621. https://doi.org/10.1089/ars.2010.3415.

van der Horst, A., & Burgering, B. M. (2007). Stressing the role of FoxO proteins in lifespan and disease. *Nature Reviews. Molecular Cell Biology, 8*(6), 440–450. https://doi.org/10.1038/nrm2190.

van der Vos, K. E., Gomez-Puerto, C., & Coffer, P. J. (2012). Regulation of autophagy by Forkhead box (FOX) O transcription factors. *Advances in Biological Regulation, 52*(1), 122–136. https://doi.org/10.1016/j.advenzreg.2011.10.002.

Veal, E. A., Day, A. M., & Morgan, B. A. (2007). Hydrogen peroxide sensing and signaling. *Molecular Cell, 26*(1), 1–14. https://doi.org/10.1016/j.molcel.2007.03.016.

Viglietto, G., Motti, M. L., Bruni, P., Melillo, R. M., D'Alessio, A., Califano, D., et al. (2002). Cytoplasmic relocalization and inhibition of the cyclin-dependent kinase inhibitor p27(Kip1) by PKB/Akt-mediated phosphorylation in breast cancer. *Nature Medicine, 8*(10), 1136–1144. https://doi.org/10.1038/nm762.

Vilchez, D., Boyer, L., Morantte, I., Lutz, M., Merkwirth, C., Joyce, D., et al. (2012). Increased proteasome activity in human embryonic stem cells is regulated by PSMD11. *Nature, 489*(7415), 304–308. https://doi.org/10.1038/nature11468.

Waddell, D. S., Baehr, L. M., van den Brandt, J., Johnsen, S. A., Reichardt, H. M., Furlow, J. D., et al. (2008). The glucocorticoid receptor and FOXO1 synergistically activate the skeletal muscle atrophy-associated MuRF1 gene. *American Journal of Physiology. Endocrinology and Metabolism, 295*(4), E785–797. https://doi.org/10.1152/ajpendo.00646.2007.

Wang, K., & Li, P. F. (2010). Foxo3a regulates apoptosis by negatively targeting miR-21. *The Journal of Biological Chemistry, 285*(22), 16958–16966. https://doi.org/10.1074/jbc.M109.093005.

Wang, F., Marshall, C. B., Yamamoto, K., Li, G. Y., Plevin, M. J., You, H., et al. (2008). Biochemical and structural characterization of an intramolecular interaction in FOXO3a and its binding with p53. *Journal of Molecular Biology, 384*(3), 590–603. https://doi.org/10.1016/j.jmb.2008.09.025.

Wang, H., Zhang, L., Yang, X., Jin, Y., Pei, S., Zhang, D., et al. (2015). PUMA mediates the combinational therapy of 5-FU and NVP-BEZ235 in colon cancer. *Oncotarget, 6*(16), 14385–14398. https://doi.org/10.18632/oncotarget.3775.

Wei, Z., Liu, Y., Wang, Y., Zhang, Y., Luo, Q., Man, X., et al. (2016). Downregulation of Foxo3 and TRIM31 by miR-551b in side population promotes cell proliferation, invasion, and drug resistance of ovarian cancer. *Medical Oncology, 33*(11), 126. https://doi.org/10.1007/s12032-016-0842-9.

Wells, A., Griffith, L., Wells, J. Z., & Taylor, D. P. (2013). The dormancy dilemma: Quiescence versus balanced proliferation. *Cancer Research, 73*(13), 3811–3816. https://doi.org/10.1158/0008-5472.CAN-13-0356.

Wilhelm, K., Happel, K., Eelen, G., Schoors, S., Oellerich, M. F., Lim, R., et al. (2016). FOXO1 couples metabolic activity and growth state in the vascular endothelium. *Nature, 529*(7585), 216–220. https://doi.org/10.1038/nature16498.

Xie, J., Xiang, D. B., Wang, H., Zhao, C., Chen, J., Xiong, F., et al. (2012). Inhibition of Tcf-4 induces apoptosis and enhances chemosensitivity of colon cancer cells. *PLoS One, 7*(9), e45617. https://doi.org/10.1371/journal.pone.0045617.

Yalcin, S., Zhang, X., Luciano, J. P., Mungamuri, S. K., Marinkovic, D., Vercherat, C., et al. (2008). Foxo3 is essential for the regulation of ataxia telangiectasia mutated and oxidative stress-mediated homeostasis of hematopoietic stem cells. *The Journal of Biological Chemistry*, *283*(37), 25692–25705. https://doi.org/10.1074/jbc.M800517200.

Yang, Y. C., Tang, Y. A., Shieh, J. M., Lin, R. K., Hsu, H. S., & Wang, Y. C. (2014). DNMT3B overexpression by deregulation of FOXO3a-mediated transcription repression and MDM2 overexpression in lung cancer. *Journal of Thoracic Oncology*, *9*(9), 1305–1315. https://doi.org/10.1097/JTO.0000000000000240.

Yeo, H., Lyssiotis, C. A., Zhang, Y., Ying, H., Asara, J. M., Cantley, L. C., et al. (2013). FoxO3 coordinates metabolic pathways to maintain redox balance in neural stem cells. *The EMBO Journal*, *32*(19), 2589–2602. https://doi.org/10.1038/emboj.2013.186.

You, H., & Mak, T. W. (2005). Crosstalk between p53 and FOXO transcription factors. *Cell Cycle*, *4*(1), 37–38. https://doi.org/10.4161/cc.4.1.1401.

You, H., Pellegrini, M., Tsuchihara, K., Yamamoto, K., Hacker, G., Erlacher, M., et al. (2006). FOXO3a-dependent regulation of Puma in response to cytokine/growth factor withdrawal. *The Journal of Experimental Medicine*, *203*(7), 1657–1663. https://doi.org/10.1084/jem.20060353.

Yuan, Z., Becker, E. B., Merlo, P., Yamada, T., DiBacco, S., Konishi, Y., et al. (2008). Activation of FOXO1 by Cdk1 in cycling cells and postmitotic neurons. *Science*, *319*(5870), 1665–1668. https://doi.org/10.1126/science.1152337.

Yuan, T. L., & Cantley, L. C. (2008). PI3K pathway alterations in cancer: Variations on a theme. *Oncogene*, *27*(41), 5497–5510. https://doi.org/10.1038/onc.2008.245.

Zhang, J., Ou, J., Bashmakov, Y., Horton, J. D., Brown, M. S., & Goldstein, J. L. (2001). Insulin inhibits transcription of IRS-2 gene in rat liver through an insulin response element (IRE) that resembles IREs of other insulin-repressed genes. *Proceedings of the National Academy of Sciences of the United States of America*, *98*(7), 3756–3761. https://doi.org/10.1073/pnas.071054598.

Zhang, H., Pan, Y., Zheng, L., Choe, C., Lindgren, B., Jensen, E. D., et al. (2011). FOXO1 inhibits Runx2 transcriptional activity and prostate cancer cell migration and invasion. *Cancer Research*, *71*(9), 3257–3267. https://doi.org/10.1158/0008-5472.CAN-10-2603.

Zhang, L., Pickard, K., Jenei, V., Bullock, M. D., Bruce, A., Mitter, R., et al. (2013). miR-153 supports colorectal cancer progression via pleiotropic effects that enhance invasion and chemotherapeutic resistance. *Cancer Research*, *73*(21), 6435–6447. https://doi.org/10.1158/0008-5472.CAN-12-3308.

Zhang, W., Wu, H., Yang, M., Ye, S., Li, L., Zhang, H., et al. (2016). SIRT1 inhibition impairs non-homologous end joining DNA damage repair by increasing Ku70 acetylation in chronic myeloid leukemia cells. *Oncotarget*, 7(12), 13538–13550. https://doi.org/10.18632/oncotarget.6455.

Zhang, H., Xie, C., Yue, J., Jiang, Z., Zhou, R., Xie, R., et al. (2017). Cancer-associated fibroblasts mediated chemoresistance by a FOXO1/TGFbeta1 signaling loop in esophageal squamous cell carcinoma. *Molecular Carcinogenesis*, *56*(3), 1150–1163. https://doi.org/10.1002/mc.22581.

Zhao, M., Luo, R., Liu, Y., Gao, L., Fu, Z., Fu, Q., et al. (2016). miR-3188 regulates nasopharyngeal carcinoma proliferation and chemosensitivity through a FOXO1-modulated positive feedback loop with mTOR-p-PI3K/AKT-c-JUN. *Nature Communications*, 7, 11309. https://doi.org/10.1038/ncomms11309.

Zhao, Y., Yang, J., Liao, W., Liu, X., Zhang, H., Wang, S., et al. (2010). Cytosolic FoxO1 is essential for the induction of autophagy and tumour suppressor activity. *Nature Cell Biology*, *12*(7), 665–675. https://doi.org/10.1038/ncb2069.

Zheng, B., Ohkawa, S., Li, H., Roberts-Wilson, T. K., & Price, S. R. (2010). FOXO3a mediates signaling crosstalk that coordinates ubiquitin and atrogin-1/MAFbx expression during glucocorticoid-induced skeletal muscle atrophy. *The FASEB Journal*, *24*(8), 2660–2669. https://doi.org/10.1096/fj.09-151480.

Zhou, Y., Liang, C., Xue, F., Chen, W., Zhi, X., Feng, X., et al. (2015). Salinomycin decreases doxorubicin resistance in hepatocellular carcinoma cells by inhibiting the beta-catenin/TCF complex association via FOXO3a activation. *Oncotarget*, *6*(12), 10350–10365. https://doi.org/10.18632/oncotarget.3585.

FURTHER READING

van der Vos, K. E., & Coffer, P. J. (2012). Glutamine metabolism links growth factor signaling to the regulation of autophagy. *Autophagy*, *8*(12), 1862–1864. https://doi.org/10.4161/auto.22152.

CHAPTER FOUR

FOXO in Neural Cells and Diseases of the Nervous System

Evan E. Santo, Jihye Paik[1]
Weill Cornell Medicine, New York, NY, United States
[1]Corresponding author: e-mail address: jep2025@med.cornell.edu

Contents

Abstract

The evolutionarily conserved FOXO family of transcription factors has emerged as a significant arbiter of neural cell fate and function in mammals. From the neural stem cell (NSC) state through mature neurons under both physiological and pathological conditions, they have been found to modulate neural cell survival, stress responses, lineage commitment, and neuronal signaling. Lineage-specific FOXO knockout mice have provided an invaluable tool for the dissection of FOXO biology in the nervous system. Within the NSC compartments of the brain, FOXOs are required for the maintenance of NSC quiescence and for the clearance of reactive oxygen species. Within mature neurons, FOXO transcriptional activity is essential for the prevention of age-dependent axonal degeneration. Acutely, FOXO3 has been found to cause axonal degeneration upon withdrawal of neurotrophic factors. In more active neural signaling, FOXO6 promotes increased dendritic spine density of hippocampal neurons and is required for the consolidation of memories. In addition to the central nervous system (CNS), FOXOs also influence the functionality of the peripheral nervous system (PNS). FOXO1 knockout within the PNS results in a reduction of sympathetic tone and decreased levels of brain-derived norepinephrine and lower energy expenditure. FOXO3 knockout mice have impaired hearing which may be due to defects in synapse localization within the ear. Given the scope of FOXO activities in both the CNS and PNS, it will be of interest to study FOXOs within the context of neurodegenerative diseases such as Alzheimer's, Parkinson's, Huntington's, and amyotrophic lateral sclerosis. From within the nervous system, FOXOs may also regulate important parameters such as whole-body metabolism, motor function, and catecholamine production, making FOXOs key players in physiologic homeostasis.

Current Topics in Developmental Biology, Volume 127
ISSN 0070-2153
https://doi.org/10.1016/bs.ctdb.2017.10.002

1. INTRODUCTION

In recent years, mammalian FOXO transcription factors have arisen as crucial regulators of cell fate and function in the nervous system. Of the four mammalian homologs, FOXO1, 3, and 6 are the ones most implicated in neuronal function (Hoekman, Jacobs, Smidt, & Burbach, 2006). FOXOs have roles ranging from neural stem cell (NSC) maintenance, reactive oxygen species (ROS) suppression, induction of apoptosis, promotion of survival by engagement of autophagy, and regulation of catecholamine biosynthesis (Doan et al., 2016; Gilley, Coffer, & Ham, 2003; Kajimura, Paone, Mann, & Karsenty, 2014; Paik et al., 2009; Renault et al., 2009; Xu, Das, Reilly, & Davis, 2011). The response of FOXOs to a given input (i.e., growth factor withdrawal or oxidative stress) is typically governed by posttranslational modifications of the proteins which in-turn control FOXO subcellular localization and/or transcriptional activity. Due to their diversity of function, gain or loss of FOXO activity in the mammalian nervous system can have a number of consequences under both physiologic and pathophysiologic conditions. In this review, we will explore these context-dependent roles of FOXO in the mammalian nervous system and the phenotypic consequences they elicit.

2. NEURAL DEVELOPMENT AND STEM CELL MAINTENANCE

The NSCs of the subventricular zone and the subgranular zone are a pool of brain stem cells capable of giving rise to new neurons throughout the life span of a mammal (Zhao, Deng, & Gage, 2008). In order to accomplish this task without premature depletion (they have a limited replicative life span), their cell cycle and lineage commitment must be tightly regulated. Knockout of FOXO1/3/4 or even FOXO3 alone is sufficient to reduce the quiescence of NSCs causing them to hyperproliferate and therefore strongly deplete this stem cell pool early in life (Paik et al., 2009; Renault et al., 2009). In addition to hyperproliferation, FOXO knockout NSCs also exhibit decreased self-renewal, increased apoptosis, decreased glycolytic flux, decreased glutathione, increased oxidative metabolism, and high ROS levels (Paik et al., 2009; Yeo et al., 2013). Increased oxidative metabolism in conjunction with the decrease in glutathione is the likely cause of the increased ROS. Treatment with the ROS scavenging antioxidant

N-acetylcysteine can rescue the apoptosis and self-renewal defects but it cannot rescue the hyperproliferation. FOXO3 knockout NSCs from adult mice also have a bias in lineage commitment toward the astrocyte and away from the oligodendrocyte and neuronal lineages (Renault et al., 2009). In neural progenitors, FOXO3 inhibits ASCL1-dependent neuronal differentiation and also restrains neurogenesis in vivo (Webb et al., 2013). Similarly, constitutively active FOXO1 inhibits while its knockdown increases neuronal differentiation of neural progenitors (Kim, Hwang, Muller, & Paik, 2015). These findings together raise the possibility that FOXOs suppress lineage commitment and such activity may help maintaining the NSC reserves throughout the lifetime.

A phenotypic consequence of the above-described NSC dysregulation is that FOXO1/3/4 and FOXO3 knockout mice have enlarged brains due to increased cell number and a limited ability to generate new neurons later in life (Paik et al., 2009; Renault et al., 2009). Although FOXO1/3/4 knockout mice have enlarged brains, they are developmentally and histopathologically normal otherwise (Paik et al., 2009). The same is mainly true of the FOXO6 knockout mouse, although FOXO6 knockout neurons do have a somewhat decreased dendritic spine density (Salih et al., 2012). They also have defects in the radial migration of cortical neurons (Paap et al., 2016). Interestingly, combined knockdown of FOXO1/3/6 by shRNA in vitro and in vivo results in Pak1 loss and severe defects in the establishment of neural polarity (de la Torre-Ubieta et al., 2010). This phenotype was only observed upon targeting FOXO1/3/6 simultaneously and not caused by knockdown of any FOXO alone; consistent with the normal brain architecture and development seen in other brain FOXO knockout mouse models. However, the polarity defect was best rescued by overexpression of FOXO6 rather than FOXO1 or FOXO3 which could not rescue as well. It may be that FOXO6 has a specialized role in the establishment of neuronal polarity, although its loss is compensated by FOXO1 or FOXO3 to a certain extent.

3. NEURONAL SURVIVAL AND STRESS RESPONSES

The FOXO family is well known for its responsiveness to extracellular cues and execution of stress resistance or apoptotic programs; in the nervous system this is no exception. One of the prime regulators of FOXO activity is the PI3K/AKT pathway (Brunet et al., 1999; Kops et al., 1999; Nakae, Park, & Accili, 1999). In response to a wide swath of growth factors that include insulin and neurotrophic factors (i.e., NGF and BDNF), AKT is activated which then phosphorylates FOXOs at three different sites (Fig. 1A and B). This phosphorylation results in their nuclear exclusion and cytoplasmic sequestration; rendering them unable to regulate gene expression. The relevance of this signaling pathway to neural cells was first shown in PC12-derived and primary cultured sympathetic spinal cord neurons where overexpression of a dominant-negative FOXO3 construct could rescue BIM-induced cell death triggered by NGF withdrawal (Gilley et al., 2003).

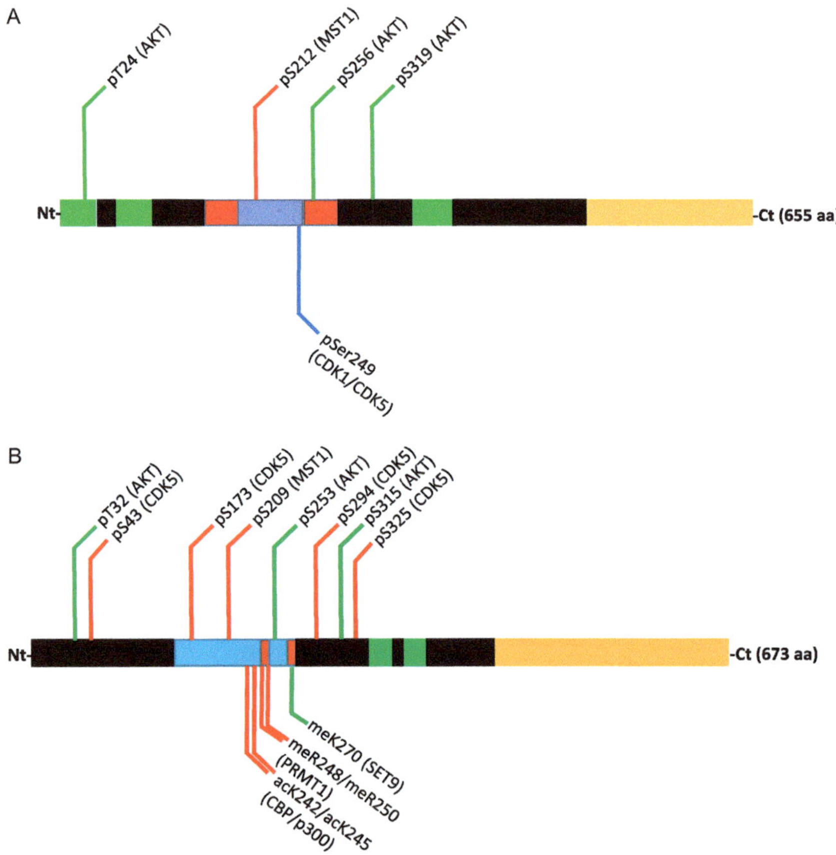

Fig. 1 Domain map and posttranslational modifications that influence FOXO1 and FOXO3 apoptotic activity within neurons. (A) Annotated FOXO1 protein numbered vs the human sequence. Within the domain structure, the forkhead domain is colored *blue*, nuclear localization signals are *red*, nuclear export signals are *green*, and the transactivation domain is *yellow*. Modifications that promote cell death are marked with *red lines*, *green lines* mark modifications that promote survival, and *blue lines* mark modifications with conflicting reports on survival. (B) Annotated FOXO3 protein numbered vs the human sequence. Same color scheme as (A).

Shortly thereafter this pathway was also described in motor neurons and it was suggested that FasL and JNK signaling might be partially responsible for FOXO3-induced death (Barthelemy, Henderson, & Pettmann, 2004). Most recently, AKT/FOXO3 signaling has been shown to be of central importance in neuronal cell death triggered from the axon that signals back to the cell body to commence apoptosis following NGF withdrawal (Simon et al., 2016). FOXO1 has also been shown to regulate apoptosis downstream of AKT in neurons. Synapse activity-dependent NMDA receptor signaling activates PI3K/AKT which was shown to be neuroprotective under conditions of ROS accumulation (Papadia et al., 2008) by modulating in part TXNIP (Thioredoxin-interacting protein). TXNIP enhances oxidative stress by binding thioredoxin and inhibiting its activity. Mechanistically, AKT-mediated inhibition of FOXO1 reduced TXNIP levels which resulted in increased thioredoxin activity and enhanced ROS detoxification. FOXO1 is also shown to repress TXNIP transcription directly in other cell types (De Candia, Blekhman, Chabot, Oshlack, & Gilad, 2008; Kibbe, Chen, Xu, Jing, & Shalev, 2013).

FOXOs also regulate apoptosis in neurons in response to activity deprivation and ROS via the activation of alternative signaling routes. FOXO3 and FOXO1 have been found to be phosphorylated by the MST1 kinase in response to either ROS or neuronal activity deprivation (depolarization), respectively (Lehtinen et al., 2006; Yuan et al., 2009; Fig. 1A and B). MST1/MST2 are core components of the so-called Hippo pathway which regulates organ size by coordinating differentiation, proliferation, and apoptosis (Qin, Tian, Zhou, & Chen, 2013). In cerebellar granular neurons (CGNs), ROS triggers MST1 to phosphorylate FOXO3 on Ser209 (mislabeled as Ser207 in Lehtinen et al., 2006; Fig. 1B) which causes it to translocate to the nucleus and induce cell death (Lehtinen et al., 2006). In a related study, FOXO1 was found to be phosphorylated by MST1 on Ser212 (equivalent of FOXO3 Ser209) in response to depolarization or growth factor withdrawal which also caused apoptosis (Yuan et al., 2009; Fig. 1A). Surprisingly, the cell cycle regulatory kinase CDK1 is activated in postmitotic CGNs by depolarization and phosphorylates FOXO1 on Ser249 causing its nuclear translocation and apoptosis (Yuan et al., 2008; Fig. 1A). In cortical neurons under oxidative stress, CDK5 was also found to phosphorylate Ser249 on FOXO1 (Zhou et al., 2015). In contrast to the CDK1 report this event resulted in the cytosolic sequestration of FOXO1 and attenuation of apoptosis. Interestingly, it was also found that Ser249 phosphorylation was dominant to AKT-mediated phosphorylation in that

Ser249 phosphorylation could localize FOXO1 to the cytoplasm even when all three AKT sites were mutated to alanine. Increased CDK activity and Ser249 phosphorylation of FOXO1 have also been associated with the promotion of autophagy and neuronal survival. JNK triple knockout CGNs exhibited increased autophagic flux that is CDK and FOXO1 dependent (Xu et al., 2011). In this JNK-deficient background knockdown of FOXO1 by shRNA or pharmacological CDK inhibition resulted in inhibition of autophagy and increased CGN death. Additional studies will be needed to resolve the discrepant actions of FOXO1 in these distinct contexts.

In addition to phosphorylation other modifications were identified as modulating FOXO's proapoptotic activity in neurons. In response to ROS the methyltransferase SET9 was found to methylate FOXO3 on Lys270 which impaired its transcriptional activity resulting in reduced apoptosis in CGNs (Xie et al., 2012; Fig. 1B). It has also been shown that deacetylation of FOXO3 at Lys242 and Lys245 by the class III histone deacetylase SIRT1 (this deacetylation also occurs on FOXO1 at the same sites) can attenuate FOXO3's proapoptotic activity in CGNs (Brunet et al., 2004; Daitoku et al., 2004; Fig. 1B). SIRT1 is known to be activated by ROS as well as other stress stimuli such as DNA damage (Houtkooper, Pirinen, & Auwerx, 2012). It will be of interest to further explore the interplay of these modifications and how they undoubtedly collaborate to regulate FOXO function and neuronal survival.

4. PHENOTYPIC CONSEQUENCES OF FOXO LOSS

Although it is clear that FOXOs can induce death in neurons in response to a variety of stressors, they are also essential for the long-term maintenance of established neurons under physiologic conditions. Deletion of FOXO in neurons results in an age-dependent degeneration (Hwang et al., 2017).

Whole-body FOXO3 null mice have adult onset auditory neuropathy rendering them partially deaf (Gilels, Paquette, Zhang, Rahman, & White, 2013). This was traced to the synapses being not correctly positioned within the cochlea. In this study, it was also found that FOXO3 may function cell autonomously within spiral ganglion neurons where auditory stimulation could actually induce its nuclear localization. This activity-dependent response has also been seen with FOXO6 in the process of memory consolidation (Salih et al., 2012). FOXO6 knockout mice are capable of learning but have a reduced ability to form long-term contextual and object recognition memories. Furthermore, specific inhibition of FOXO6 in the hippocampus of wild-type mice by a dominant-negative construct is sufficient to recapitulate this phenotype. Molecularly, learning tasks actually activate FOXO6 which then binds and activates the promoters of genes involved in synapse formation.

More restricted deletions of FOXOs have also provided insight into their nervous system function. Brain-specific FOXO1 deletion increased the anxiety of mice in a forced swim test, while deletion of FOXO3 decreased their anxiety (Polter et al., 2009). Strikingly, FOXO3 knockout mice showed decreased anxiety similarly to the level detected in mice treated with the tricyclic antidepressant imipramine. The combination of FOXO3 knockout and imipramine gave the least anxiety, suggesting that imipramine may actually function in part through FOXO3. In agreement, imipramine treatment was found to inactivate FOXO3 in the cerebral cortex by cytosolic sequestration. Together with the FOXO1 results, this study suggests that FOXOs may have distinct effects on depressive behavior.

FOXO1 has also been investigated by specific deletion in the sympathetic nervous system. Deletion of FOXO1 in the sympathetic nervous system by crossing $FOXO1^{Flox}$ mice with dopamine-β-hydroxylase (DBH) Cre mice resulted in reduced catecholamine (adrenaline) biosynthesis, reduced sympathetic nervous system activity (sympathetic tone), reduced energy expenditure, improved glucose clearance, and increased bone mass (Kajimura et al., 2014). Notably, catecholamine production was reduced in the brain but was normal in the adrenaline-producing chromaffin cells of the adrenal medulla (adrenal gland); in these cells, FOXO1 is not expressed so deletion had no direct effect. This effect on catecholamine biosynthesis was attributed to reduced levels of DBH expression due to loss of FOXO1 binding to the DBH promoter. When challenged with glucose FOXO1 sympathetic knockout mice exhibited increased insulin secretion which may explain their enhanced glucose clearance. FOXO1 has also been specifically knocked out in dopaminergic (DA) neurons using DAT-IRES-Cre (Doan et al., 2016). These mice demonstrated resistance to weight gain, improved glucose clearance, increased serum levels of norepinephrine, increased midbrain levels of dopamine, and increased interscapular brown adipose tissue thermogenesis. The effect on dopamine biosynthesis was attributed to direct suppression of the tyrosine hydroxylase promoter by FOXO1 binding. These results are additionally supported by the finding that FOXO1 deletion using synapsin Cre has a similar metabolic profile including resistance to weight gain on a high-fat diet (Ren et al., 2013).

From these studies, it is clear that FOXO1 is capable of regulating whole-body metabolism from its actions within the nervous system. However, it remains to be explained what role if any catecholamine metabolism actually plays in the common increased glucose clearance phenotype given the point that catecholamine biosynthesis is decreased in the DBH-Cre model while it is increased in the DAT-IRES-Cre model.

5. NEURODEGENERATIVE DISEASE

Huntington's, Parkinson's, and Alzheimer's diseases are all caused by the accumulation of insoluble protein aggregates (mHtt, α-synuclein/Lewy Bodies, and amyloid beta (Aβ) plaques, respectively) that generate considerable oxidative and proteotoxic stress within neurons. This chronic stress eventually leads to axonal degeneration and/or apoptosis of the affected neurons (Montie & Durcan, 2013). Given the pivotal role, FOXOs can play in triggering either neuronal apoptosis or survival in response to stress they are potentially very relevant players in the pathologies of these afflictions.

In Huntington's disease (HD) mutant huntingtin protein (mHtt) was found to specifically activate the XBP1 transcription factor (Vidal et al., 2012). XBP1 knockout in the background of the mHtt mouse model reduced mHtt levels, enhanced neuronal survival, and improved motor performance. These improvements were attributed to an increase in autophagy. Loss of XBP1 caused FOXO1 to accumulate on the protein level in the nucleus and coexpression of wild-type FOXO1 with mHtt in a cell line model was also able to increase autophagy and clearance of mHtt. Increased expression and nuclear localization have also been observed for FOXO3 in HD but the functional meaning of it has not yet been determined (Kannike, Sepp, Zuccato, Cattaneo, & Timmusk, 2014).

FOXO3's proapoptotic and survival activities have been evaluated in a Parkinson's disease (PD) model by overexpression of wild-type, constitutively active, and dominant-negative constructs in DA neurons of the substantia nigra (Pino et al., 2014). In agreement with many findings in other neuronal types, overexpression of constitutively active FOXO3 induced acute apoptosis in DA neurons. However, high-level expression of dominant-negative FOXO3 was also toxic in DA neurons, and this was accompanied by signs of oxidative damage, indicating the necessity of FOXO3 for basic neuronal survival and resistance to oxidative stress. When coinjected with α-synuclein constitutively active FOXO3 was still acutely toxic, while dominant-negative FOXO3 was completely protective of DA neurons. There was no explanation offered for this striking level of protection provided by dominant-negative FOXO3 in light of the fact that α-synuclein levels and autophagic vacuole formation were largely unchanged by dominant-negative FOXO3 in DA neurons. FOXO1 and putative FOXO1 target genes (i.e., SMOX) are increased in the prefrontal cortex of PD patients which is a region of the brain also pathologically affected in PD but not as degenerated as the substantia nigra by the time of autopsy (Dumitriu et al., 2012). Whether this is also true in the DA neurons of the substantia nigra and the implications of this observation remain unclear.

So far only FOXO3 has been studied in Alzheimer's disease (AD) models and seems to largely be an inducer of cell death in response to Aβ plaques. Injection of $A\beta_{1-42}$ into the hippocampus and cortex of adult rats resulted in a series of posttranslational modifications to FOXO3 followed by its nuclear translocation and induction of cell death by BIM induction which could be rescued by FOXO3-directed shRNA (Sanphui & Biswas, 2013). $A\beta_{1-42}$ induced a decrease in AKT-mediated phosphorylation, increased MST1 phosphorylation at Ser209, and increased methylation by PRMT1, the methylation at Arg248 and Arg250 being known to promote the nuclear retention and activation of FOXOs (Yamagata et al., 2008; Fig. 1B). Additional modifications of FOXO3 have been uncovered that are particularly relevant in the pathogenesis of Alzheimer's. CDK5 can phosphorylate FOXO3 at Ser43, Ser173, Ser294, and Ser325 in response to glutamate-induced excitotoxicity in immortalized mouse hippocampal (HT22) cells or primary cortical neurons (Shi, Viccaro, Lee, & Shah, 2016; Fig. 1B). These phosphorylations can induce FOXO3 nuclear translocation and initially promote stress resistance through upregulation of MnSOD. At later time points, FOXO3 drives BIM expression and cell death and also promotes neurotoxic cleavage and excretion of $A\beta_{1-42}$. Knockdown of CDK5 or FOXO3 is sufficient to prevent these events. In a p25-induced mouse, AD model early activation of both CDK5 and FOXO3 were observed, preceding Aβ plaque formation and cell death. Further support for FOXO3 as a pivotal player in early AD pathogenesis comes from the discovery that the miR-132/212 microRNAs are strongly suppressed in early and late stage AD-affected neurons and that this corresponds with upregulation of PTEN and FOXO3 (Wong et al., 2013). These microRNAs were found to directly bind to the 3pUTRs of PTEN and FOXO3 inhibiting their expression. Experimentally, inhibition of miR-132/212 increased baseline apoptosis in hippocampal neurons and also enhances apoptosis in response to oxidative stress. These effects could be rescued by knockdown of PTEN and FOXO3.

6. FUTURE DIRECTIONS

Although much has been revealed about the role of FOXOs in the mammalian nervous system much remains to be understood. Physiologically, FOXO1 and FOXO3 seem to be necessary for the maintenance of neuronal integrity from mid-life on. They may work to counteract and/or prevent the cumulative damage caused by proteotoxic and oxidative stress. It will be of interest to study the neuron-specific FOXO knockout model further to identify additional signaling and metabolic pathways FOXO may regulate that offer neuronal protection. This will increase our understanding of exactly how neurons age and what may be done to maintain their functionality. A mammalian genetic model that could enhance FOXO activity throughout the nervous system would be most welcome in order to assess whether or not FOXO can slow or prevent the onset of age-related neuropathies and accompanying motor and cognitive decline. From a developmental perspective further exploration of FOXO6 loss in conjunction with the deletion of other FOXO isoforms will be necessary to establish the full meaning of FOXO in neural development. This may lead to new understandings about the pathways and signaling cues that drive mammalian neurogenesis and neural circuit construction. The extensive posttranslational modifications that regulate FOXO activity indicate that FOXOs act as hubs for the integration of signaling outputs from diverse pathways. A deeper understanding of the interplay of these modifications and how they influence FOXO-mediated gene regulation will be a necessary step toward understanding the full range of their contextual functions.

Further study of the roles of FOXOs within the full spectrum of neurodegenerative diseases is warranted. It may prove fruitful to delete FOXOs within the backgrounds of established PD, HD, and AD models to clarify their functions in these pathologies. Studies should also be extended to the role of FOXOs in amyotrophic lateral sclerosis given the progressive degeneration observed in the neuronal FOXO knockout model. Finally, efforts should be made to evaluate pharmacological interventions to restore or inhibit FOXO functionality as appropriate. This could be accomplished by targeting known or novel FOXO regulatory enzymes as this may lead to potent treatments for these destructive conditions.

REFERENCES

Barthelemy, C., Henderson, C. E., & Pettmann, B. (2004). Foxo3a induces motoneuron death through the Fas pathway in cooperation with JNK. *BMC Neuroscience*, *5*(48), 1–9.

Brunet, A., Bonni, A., Zigmond, M. J., Lin, M. Z., Juo, P., Hu, L. S., et al. (1999). Akt promotes cell survival by phosphorylating and inhibiting a Forkhead transcription factor. *Cell, 96*(6), 857–868.

Brunet, A., Sweeney, L. B., Sturgill, J. F., Chua, K. F., Greer, P. L., Lin, Y., et al. (2004). Stress-dependent regulation of FOXO transcription factors by the SIRT1 deacetylase. *Science, 303*(5666), 2011–2015.

Daitoku, H., Hatta, M., Matsuzaki, H., Aratani, S., Ohshima, T., Miyagishi, M., et al. (2004). Silent information regulator 2 potentiates Foxo1-mediated transcription through its deacetylase activity. *Proceedings of the National Academy of Sciences of the United States of America, 101*(27), 10042–10047.

De Candia, P., Blekhman, R., Chabot, A. E., Oshlack, A., & Gilad, Y. (2008). A combination of genomic approaches reveals the role of FOXO1a in regulating an oxidative stress response pathway. *PLoS One, 3*(2), e1670.

de la Torre-Ubieta, L., Gaudilliere, B., Yang, Y., Ikeuchi, Y., Yamada, T., DiBacco, S., et al. (2010). A FOXO-Pak1 transcriptional pathway controls neuronal polarity. *Genes & Development, 24*(8), 799–813.

Doan, K. V., Kinyua, A. W., Yang, D. J., Ko, C. M., Moh, S. H., Shong, K. E., et al. (2016). FoxO1 in dopaminergic neurons regulates energy homeostasis and targets tyrosine hydroxylase. *Nature Communications*, 7, 12733.

Dumitriu, A., Latourelle, J. C., Hadzi, T. C., Pankratz, N., Garza, D., Miller, J. P., et al. (2012). Gene expression profiles in Parkinson disease prefrontal cortex implicate FOXO1 and genes under its transcriptional regulation. *PLoS Genetics, 8*(6), e1002794.

Gilels, F., Paquette, S. T., Zhang, J., Rahman, I., & White, P. M. (2013). Mutation of Foxo3 causes adult onset auditory neuropathy and alters cochlear synapse architecture in mice. *The Journal of Neuroscience, 33*(47), 18409–18424.

Gilley, J., Coffer, P. J., & Ham, J. (2003). FOXO transcription factors directly activate *bim* gene expression and promote apoptosis in sympathetic neurons. *The Journal of Cell Biology, 162*(4), 613–622.

Hoekman, M. F., Jacobs, F. M., Smidt, M. P., & Burbach, J. P. (2006). Spatial and temporal expression of FoxO transcription factors in the developing and adult murine brain. *Gene Expression Patterns, 6*(2), 134–140.

Houtkooper, R. H., Pirinen, E., & Auwerx, J. (2012). Sirtuins as regulators of metabolism and healthspan. *Nature Reviews Molecular Cell Biology, 13*(4), 225–238.

Hwang, I., Oh, H., Santo, E., Kim, D., Chen, J., Bronson, R., et al. (2017). FOXO protects against age-progressive axonal degeneration. *Aging Cell,* e12701, in press. https://doi.org/10.1111/acel.12701.

Kajimura, D., Paone, R., Mann, J. J., & Karsenty, G. (2014). Foxo1 regulates Dbh expression and the activity of the sympathetic nervous system in vivo. *Molecular Metabolism, 3*(7), 770–777.

Kannike, K., Sepp, M., Zuccato, C., Cattaneo, E., & Timmusk, T. (2014). Forkhead transcription factor FOXO3a levels are increased in Huntington disease because of overactivated positive autofeedback loop. *The Journal of Biological Chemistry, 289*(47), 32845–32857.

Kibbe, C., Chen, J., Xu, G., Jing, G., & Shalev, A. (2013). FOXO1 competes with carbohydrate response element-binding protein (ChREBP) and inhibits thioredoxin-interacting protein (TXNIP) transcription in pancreatic beta cells. *The Journal of Biological Chemistry, 288*(32), 23194–23202.

Kim, D. Y., Hwang, I., Muller, F. L., & Paik, J. H. (2015). Functional regulation of FoxO1 in neural stem cell differentiation. *Cell Death and Differentiation, 22*(12), 2034–2045.

Kops, G. J., de Ruiter, N. D., De Vries-Smits, A. M., Powell, D. R., Bos, J. L., & Burgering, B. M. (1999). Direct control of the Forkhead transcription factor AFX by protein kinase B. *Nature, 398*(6728), 630–634.

Lehtinen, M. K., Yuan, Z., Boag, P. R., Yang, Y., Villen, J., Becker, E. B., et al. (2006). A conserved MST-FOXO signaling pathway mediates oxidative-stress responses and extends life span. *Cell, 125*(5), 987–1001.

Montie, H. L., & Durcan, T. M. (2013). The cell and molecular biology of neurodegenerative diseases: An overview. *Frontiers in Neurology, 4,* 194.

Nakae, J., Park, B. C., & Accili, D. (1999). Insulin stimulates phosphorylation of the forkhead transcription factor FKHR on serine 253 through a Wortmannin-sensitive pathway. *The Journal of Biological Chemistry, 274*(23), 15982–15985.

Paap, R. H., Oosterbroek, S., Wagemans, C. M., von, O. L., Schellevis, R. D., Vastenhouw-van der Linden, A. J., et al. (2016). FoxO6 affects Plxna4-mediated neuronal migration during mouse cortical development. *Proceedings of the National Academy of Sciences of the United States of America, 113*(45), E7087–E7096.

Paik, J. H., Ding, Z., Narurkar, R., Ramkissoon, S., Muller, F., Kamoun, W. S., et al. (2009). FoxOs cooperatively regulate diverse pathways governing neural stem cell homeostasis. *Cell Stem Cell, 5*(5), 540–553.

Papadia, S., Soriano, F. X., Leveille, F., Martel, M. A., Dakin, K. A., Hansen, H. H., et al. (2008). Synaptic NMDA receptor activity boosts intrinsic antioxidant defenses. *Nature Neuroscience, 11*(4), 476–487.

Pino, E., Amamoto, R., Zheng, L., Cacquevel, M., Sarria, J. C., Knott, G. W., et al. (2014). FOXO3 determines the accumulation of alpha-synuclein and controls the fate of dopaminergic neurons in the substantia nigra. *Human Molecular Genetics, 23*(6), 1435–1452.

Polter, A., Yang, S., Zmijewska, A. A., van, G. T., Paik, J. H., DePinho, R. A., et al. (2009). Forkhead box, class O transcription factors in brain: Regulation and behavioral manifestation. *Biological Psychiatry, 65*(2), 150–159.

Qin, F., Tian, J., Zhou, D., & Chen, L. (2013). Mst1 and Mst2 kinases: Regulations and diseases. *Cell & Bioscience, 3*(1), 31.

Ren, H., Plum-Morschel, L., Gutierrez-Juarez, R., Lu, T. Y., Kim-Muller, J. Y., Heinrich, G., et al. (2013). Blunted refeeding response and increased locomotor activity in mice lacking FoxO1 in synapsin-Cre-expressing neurons. *Diabetes, 62*(10), 3373–3383.

Renault, V. M., Rafalski, V. A., Morgan, A. A., Salih, D. A., Brett, J. O., Webb, A. E., et al. (2009). FoxO3 regulates neural stem cell homeostasis. *Cell Stem Cell, 5*(5), 527–539.

Salih, D. A., Rashid, A. J., Colas, D., de la Torre-Ubieta, L., Zhu, R. P., Morgan, A. A., et al. (2012). FoxO6 regulates memory consolidation and synaptic function. *Genes & Development, 26*(24), 2780–2801.

Sanphui, P., & Biswas, S. C. (2013). FoxO3a is activated and executes neuron death via Bim in response to beta-amyloid. *Cell Death & Disease, 4,* e625.

Shi, C., Viccaro, K., Lee, H. G., & Shah, K. (2016). Cdk5-Foxo3 axis: Initially neuroprotective, eventually neurodegenerative in Alzheimer's disease models. *Journal of Cell Science, 129*(9), 1815–1830.

Simon, D. J., Pitts, J., Hertz, N. T., Yang, J., Yamagishi, Y., Olsen, O., et al. (2016). Axon degeneration gated by retrograde activation of somatic pro-apoptotic signaling. *Cell, 164*(5), 1031–1045.

Vidal, R. L., Figueroa, A., Court, F. A., Thielen, P., Molina, C., Wirth, C., et al. (2012). Targeting the UPR transcription factor XBP1 protects against Huntington's disease through the regulation of FoxO1 and autophagy. *Human Molecular Genetics, 21*(10), 2245–2262.

Webb, A. E., Pollina, E. A., Vierbuchen, T., Urban, N., Ucar, D., Leeman, D. S., et al. (2013). FOXO3 shares common targets with ASCL1 genome-wide and inhibits ASCL1-dependent neurogenesis. *Cell Reports, 4*(3), 477–491.

Wong, H. K., Veremeyko, T., Patel, N., Lemere, C. A., Walsh, D. M., Esau, C., et al. (2013). De-repression of FOXO3a death axis by microRNA-132 and -212 causes neuronal apoptosis in Alzheimer's disease. *Human Molecular Genetics*, *22*(15), 3077–3092.

Xie, Q., Hao, Y., Tao, L., Peng, S., Rao, C., Chen, H., et al. (2012). Lysine methylation of FOXO3 regulates oxidative stress-induced neuronal cell death. *EMBO Reports*, *13*(4), 371–377.

Xu, P., Das, M., Reilly, J., & Davis, R. J. (2011). JNK regulates FoxO-dependent autophagy in neurons. *Genes & Development*, *25*(4), 310–322.

Yamagata, K., Daitoku, H., Takahashi, Y., Namiki, K., Hisatake, K., Kako, K., et al. (2008). Arginine methylation of FOXO transcription factors inhibits their phosphorylation by Akt. *Molecular Cell*, *32*(2), 221–231.

Yeo, H., Lyssiotis, C. A., Zhang, Y., Ying, H., Asara, J. M., Cantley, L. C., et al. (2013). FoxO3 coordinates metabolic pathways to maintain redox balance in neural stem cells. *EMBO Journal*, *32*(19), 2589–2602.

Yuan, Z., Becker, E. B., Merlo, P., Yamada, T., DiBacco, S., Konishi, Y., et al. (2008). Activation of FOXO1 by Cdk1 in cycling cells and postmitotic neurons. *Science*, *319*(5870), 1665–1668.

Yuan, Z., Lehtinen, M. K., Merlo, P., Villen, J., Gygi, S., & Bonni, A. (2009). Regulation of neuronal cell death by MST1-FOXO1 signaling. *The Journal of Biological Chemistry*, *284*(17), 11285–11292.

Zhao, C., Deng, W., & Gage, F. H. (2008). Mechanisms and functional implications of adult neurogenesis. *Cell*, *132*(4), 645–660.

Zhou, J., Li, H., Li, X., Zhang, G., Niu, Y., Yuan, Z., et al. (2015). The roles of Cdk5-mediated subcellular localization of FOXO1 in neuronal death. *Journal of Neuroscience*, *35*(6), 2624–2635.

FURTHER READING

Biggs, W. H., III, Meisenhelder, J., Hunter, T., Cavenee, W. K., & Arden, K. C. (1999). Protein kinase B/Akt-mediated phosphorylation promotes nuclear exclusion of the winged helix transcription factor FKHR1. *Proceedings of the National Academy of Sciences of the United States of America*, *96*(13), 7421–7426.

Brownawell, A. M., Kops, G. J., Macara, I. G., & Burgering, B. M. (2001). Inhibition of nuclear import by protein kinase B (Akt) regulates the subcellular distribution and activity of the forkhead transcription factor AFX. *Molecular and Cellular Biology*, *21*(10), 3534–3546.

Brunet, A., Kanai, F., Stehn, J., Xu, J., Sarbassova, D., Frangioni, J. V., et al. (2002). 14-3-3 transits to the nucleus and participates in dynamic nucleocytoplasmic transport. *The Journal of Cell Biology*, *156*(5), 817–828.

Jacobs, F. M., van der Heide, L. P., Wijchers, P. J., Burbach, J. P., Hoekman, M. F., & Smidt, M. P. (2003). FoxO6, a novel member of the FoxO class of transcription factors with distinct shuttling dynamics. *The Journal of Biological Chemistry*, *278*(38), 35959–35967.

Nasrin, N., Ogg, S., Cahill, C. M., Biggs, W., Nui, S., Dore, J., et al. (2000). DAF-16 recruits the CREB-binding protein coactivator complex to the insulin-like growth factor binding protein 1 promoter in HepG2 cells. *Proceedings of the National Academy of Sciences of the United States of America*, *97*(19), 10412–10417.

Rena, G., Prescott, A. R., Guo, S., Cohen, P., & Unterman, T. G. (2001). Roles of the forkhead in rhabdomyosarcoma (FKHR) phosphorylation sites in regulating 14-3-3 binding, transactivation and nuclear targetting. *Biochemical Journal*, *354*(Pt 3), 605–612.

So, C. W., & Cleary, M. L. (2002). MLL-AFX requires the transcriptional effector domains of AFX to transform myeloid progenitors and transdominantly interfere with forkhead protein function. *Molecular and Cellular Biology*, *22*(18), 6542–6552.

Wang, F., Marshall, C. B., Yamamoto, K., Li, G. Y., Gasmi-Seabrook, G. M., Okada, H., et al. (2012). Structures of KIX domain of CBP in complex with two FOXO3a transactivation domains reveal promiscuity and plasticity in coactivator recruitment. *Proceedings of the National Academy of Sciences of the United States of America*, *109*(16), 6078–6083.

Zhao, X., Gan, L., Pan, H., Kan, D., Majeski, M., Adam, S. A., et al. (2004). Multiple elements regulate nuclear/cytoplasmic shuttling of FOXO1: Characterization of phosphorylation- and 14-3-3-dependent and -independent mechanisms. *Biochemical Journal*, *378*(Pt 3), 839–849.

CHAPTER FIVE

Regulation of Hepatic Glucose Metabolism by FoxO Proteins, an Integrated Approach

Terry G. Unterman[1]
University of Illinois at Chicago College of Medicine, Chicago, IL, United States
Medical Service, Jesse Brown VA Medical Center, Chicago, IL, United States
[1]Corresponding author: e-mail address: unterman@uic.edu

Contents

Abstract

FoxO proteins are ancient targets of insulin action and play an important role in mediating effects of insulin on gene expression and metabolism. Regulation of FoxO function in the liver is critical for the ability of insulin to maintain glucose homeostasis and suppress hepatic glucose production (HGP), and dysregulation of FoxO function is thought to contribute to the pathogenesis of diabetes mellitus. Signaling by the insulin/PI3 kinase/Akt pathway suppresses FoxO function, and FoxO proteins also are regulated by counterregulatory factors and sirtuin deacetylases which increase their activity. FoxO proteins promote gluconeogenic gene expression; however, effects of FoxO proteins on glycolytic, lipogenic, and lipid catabolic pathways also are important in mediating the effects of FoxOs on HGP. Recent studies indicate that hepatic FoxO proteins also exert important effects on extrahepatic tissues, including white and brown fat, that contribute to the regulation of HGP and systemic glucose utilization. Together, these observations indicate that FoxO proteins contribute to the regulation of systemic and hepatic glucose metabolism as part of a larger complex system engaged in the maintenance of metabolic homeostasis and the adaptation to changes in nutrient availability.

Current Topics in Developmental Biology, Volume 127
ISSN 0070-2153
https://doi.org/10.1016/bs.ctdb.2017.10.005

1. OVERVIEW

FoxO Forkhead transcription factors are major targets of insulin action (Barthel, Schmoll, & Unterman, 2005; Gross, Wan, & Birnbaum, 2009), and dysregulation of FoxO function is thought to promote the development of diabetes mellitus (Pajvani & Accili, 2015). FoxO proteins are expressed in many tissues where they contribute to the adaptation to changes in nutrient availability, including adipose tissue (Armoni et al., 2006; Jing, Gesta, & Kahn, 2007; Nakae et al., 2008), pancreatic islets (Kim-Muller et al., 2014; Talchai, Xuan, Lin, Sussel, & Accili, 2012; Xiao et al., 2017), gut (Talchai, Xuan, Kitamura, DePinho, & Accili, 2012), skeletal muscle (Bastie et al., 2005; Kamei et al., 2004; Sandri et al., 2004), brain (Jacobs et al., 2003; Plum et al., 2009; Wang et al., 2014), and liver (Matsumoto, Pocai, Rossetti, Depinho, & Accili, 2007; O-Sullivan et al., 2015; Zhang et al., 2012). The regulation of hepatic glucose production (HGP) is an important determinant of glucose homeostasis in both fasting and fed states (Campbell, Mandarino, & Gerich, 1988; Consoli, Nurjhan, Capani, & Gerich, 1989; Defronzo, 2009), and studies in this laboratory have focused on understanding the role of FoxO proteins in mediating effects of insulin on gene expression and metabolism in the liver. Based on the recognition that insulin regulates gene expression through the phosphatidylinositol-3′ kinase/Akt pathway (Cichy et al., 1998) and that FoxO proteins are ancient targets of the insulin/PI3K/Akt pathway (Lin, Dorman, Rodan, & Kenyon, 1997; Ogg et al., 1997), early studies focused on the role of FoxO proteins in mediating effects of insulin on the expression of gluconeogenic genes, including phosphoenolpyruvate carboxykinase, which mediates a rate-limiting step in gluconeogenesis, and glucose-6 phosphatase (G6Pase), which dephosphorylates glucose-6 phosphate and thereby promotes secretion of glucose by the liver (Nakae, Kitamura, Silver, & Accili, 2001; Schmoll et al., 2000; Yeagley, Guo, Unterman, & Quinn, 2001). However, subsequent studies have shown that FoxO proteins also contribute to the regulation of glycolytic, lipogenic, and lipid catabolic pathways in the liver, and that the regulation of these pathways occurs in a highly integrated fashion that contribute to the regulation of HGP (Haeusler et al., 2014; Zhang et al., 2016, 2006). Recent studies in liver-specific knockout mice where insulin signaling is disrupted in the liver indicate that FoxO1, the major FoxO protein expressed in the liver (Zhang et al., 2016), plays a critical role in the ability of insulin to maintain glucose homeostasis and regulate hepatic glucose production through both intrahepatic and extrahepatic mechanisms (O-Sullivan et al., 2015; Titchenell et al., 2016). As will be discussed, these observations indicate that hepatic FoxO proteins contribute the regulation of glucose metabolism as part of a complex homeostatic system engaged in the regulation of energy metabolism and adaptation to changes in nutrient status.

2. EARLY STUDIES AND EARLY INSIGHTS

Early studies provided important insight into the mechanisms by which insulin regulates the function of FoxO proteins, and the role that FoxO proteins play in mediating the adaptation to nutritional restriction at the level of the whole organism. As described below, these findings pointed toward two key characteristics of FoxO proteins that are important for understanding their role in the regulation of glucose metabolism: in addition to being direct targets of insulin action which can mediate effects on gene expression in specific target tissues, FoxO proteins also function within the context of a larger system that serves to orchestrate a highly integrated response to changes in nutrient availability that impacts diverse functions that are critical for the survival of the organism and perpetuation of the species, including energy storage, reproductive activity, and life span (Lin et al., 1997; Ogg et al., 1997). With this as a framework, it is not surprising that FoxO proteins contribute to the regulation of glucose metabolism as part of a complex system involving multiple metabolic pathways and tissues.

Early studies indicated that hepatic production of growth promoting and inhibitory factors contribute to the regulation of growth and anabolism at the whole body level in response to changes in nutritional state (Phillips & Unterman, 1984), including hepatic production of insulin-like growth factor binding protein-1 (IGFBP-1) (Unterman, Oehler, Murphy, & Lacson, 1991; Unterman et al., 1990). The identification of insulin-regulated insulin-like growth factor binding protein-1 (IGFBP-1), which is expressed in the liver and rapidly suppressed by insulin at the level of gene transcription (Ooi et al., 1990; Unterman et al., 1991, 1990; Yeoh & Baxter, 1988), provided an important target for understanding the role of FoxO proteins in insulin action. IGFBP-1 gene expression and circulating levels are increased in fasting and insulin-deficient states and suppressed by refeeding and/or insulin treatment (Ooi et al., 1990; Unterman et al., 1990; Yeoh & Baxter, 1988). Cell culture studies in liver-derived cells showed that the expression of IGFBP-1 is suppressed by insulin in a cell autonomous fashion (Powell, Lee, & Suwanichkul, 1993; Unterman et al., 1991), and analysis of the IGFBP-1 promoter revealed that insulin suppresses IGFBP-1 gene expression through two insulin responsive elements located in the proximal 5′-upstream IGFBP-1 promoter, each of which is able to function independently, and is highly related to the insulin response sequence (IRS) previously shown to mediate negative effects on insulin on the promoter of the PEPCK gene (Goswami, Lacson, Yang, Sam, & Unterman, 1994; O'Brien, Lucas, Forest, Magnuson, & Granner, 1990; Suwanichkul, Allander, Morris, & Powell, 1994). These findings suggested that the same *trans*-acting factors might mediate negative effects of insulin on the expression of IGFBP-1 and PEPCK, and possibly other genes in the liver.

In contrast to the PEPCK gene, which required activation of promoter activity by cyclic AMP agonists and/or glucocorticoids PEPCK promoter for studies of insulin action (O'Brien et al., 1990; Schmoll et al., 2000; Yeagley et al., 2001), insulin is able to suppress basal IGFBP-1 promoter activity (Goswami et al., 1994) so that the mechanisms mediating the effects of insulin on gene expression could be studied in the absence of potentially confounding effects of counterregulatory factors. Studies using the IGFBP-1 promoter as a model system showed that the phosphatidylinositol-3′ kinase/Akt pathway mediates the ability of insulin to suppress IGFBP-1 promoter activity through IRSs (Cichy et al., 1998). DNA sequence analysis predicted that this region of the IGFBP-1 promoter binds hepatocyte nuclear factor-3 transcription factor (HNF-3) proteins (Unterman et al., 1994), which now are classified as members of the FoxA subgroup of the Forkhead family of transcription factors (Kaestner, Knochel, & Martinez, 2000). However, methylation interference studies (Rausa et al., 2000) and functional analysis of constructs containing point mutations within this region of the IGFBP-1 promoter (Guo et al., 1999) revealed that FoxA proteins were unlikely to mediate effects of insulin on IGFBP-1 promoter activity, and studies in the PEPCK promoter yielded similar results (O'Brien et al., 1995). Also, replacing the IGFBP-1 insulin response element with a consensus FoxA binding site failed to recapitulate the insulin effect (Guo et al., 1999). These results indicated that HNF-3/FoxA Forkhead proteins were not likely to mediate Akt-dependent sequence-specific effects of insulin through the IRSs, but did not exclude that possibility that other Forkhead family members might be involved.

Genetic studies in *C. elegans* identified DAF-16, a member of the FoxO subfamily of Forkhead proteins, as a major target of the insulin receptor homologue, DAF-2, and the downstream PI3K/Akt signaling pathway (Lin et al., 1997; Ogg et al., 1997). Disruption of the insulin receptor homologue and its downstream effectors was shown to arrest reproductive development, reduce locomotor and metabolic activity, and prolong life span, mimicking the effects of nutrient restriction, and these effects were reversed by disruption of the DAF-16 gene (Ogg et al., 1997). Although direct evidence demonstrating effects of insulin/PI3K/Akt signaling on DAF-16 function were not provided by this genetic analysis, these results strongly supported the concept that DAF-16 promotes these adaptations to nutritional restriction and that, when environmental conditions are favorable, insulin/PI3K/Akt signaling disrupts these effects of DAF-16 and allows metabolic activity and reproductive development to resume.

Previous studies of FoxO proteins had focused largely on their role in the pathogenesis of soft tissue tumors due to translocations involving the transactivation domain of FoxO1 and the DNA binding domain PAX3 (Xia & Barr, 2004). In contrast, DAF-16 was identified as a member of the FoxO subfamily of Forkhead proteins based on the homology of the amino acid sequence of its DNA binding domain with mammalian FoxO proteins (Kaestner et al., 2000; Ogg et al., 1997). The structure of the DNA binding motif of Forkhead proteins, known as the Forkhead box (or "Fox box"), provides the basis for naming these proteins as Fox proteins, and Forkhead proteins are classified based on homologies within the DNA binding domain (Kaestner et al., 2000). The amino acid sequence of the DNA binding domain of the FoxO subfamily, including the sequence of alpha helix 3, which interacts directly with DNA target sequences (Obsil & Obsilova, 2008; Tsai et al., 2007) (Fig. 1) and is therefore important in determining binding specificity, is the most divergent among the family of Forkhead proteins, which resulted in their being identified as Fox "Other," or FoxO, proteins (Kaestner et al., 2000). Thus, the identification of DAF-16 as a member of the FoxO subfamily of Forkhead proteins was consistent with the possibility that FoxO proteins may interact with a distinct set of DNA target sites within the genome, including, perhaps, IRSs like those identified in the IGFBP-1 and PEPCK genes, where they might function as targets of insulin action downstream from PI3K and Akt. As will be discussed, other features of the DNA binding domain that also are unique to FoxO proteins also contribute to the ability of insulin/PI3K/Akt and nutrient-sensing pathways to regulate their function.

Subsequent studies confirmed that mammalian FoxO proteins interact with and stimulate gene expression through IRSs from the IGFBP-1 promoter and that their function could be inhibited by insulin and other growth factors in a PI3K/Akt-dependent fashion (Brunet et al., 1999; Guo et al., 1999; Kops et al., 1999; Nakae, Park, & Accili, 1999; Tang, Nunez, Barr, & Guan, 1999). Using a series of constructs containing point mutations

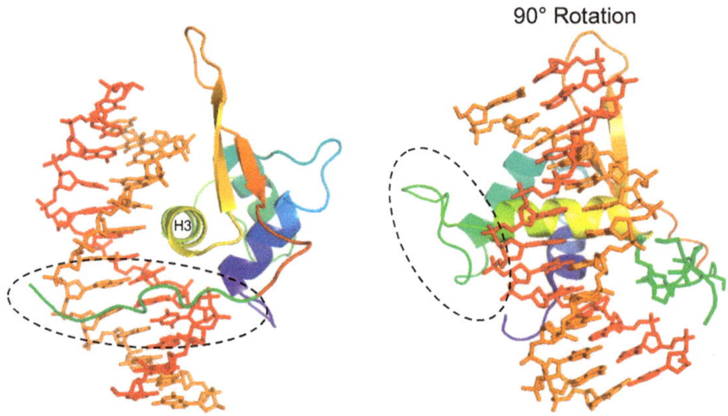

Fig. 1 FoxO DNA binding domain. The FoxO DNA binding domain is a winged-helix structure, which contains three alpha helices, with helix 1 and 2 presenting helix 3 (H3) to the DNA target site in the major groove of the double helix, providing sequence-specific recognition of target sites. The basic region located at the C-terminal end of the DNA binding domain (enclosed in *dashed oval*) forms a random coil structure so that basic (positively charged) amino acid residues within this region can interact with negatively charged phosphate residues along the backbone of the DNA (shown on the *right*). These sequence-nonspecific, electrostatic interactions are required to stabilize the binding FoxO proteins to target sites. Basic residues located at the N-terminal end of the DNA binding domain (not shown) also provide additional binding stability. The diagram was created using PyMol based on structural coordinates (2UZK) provided in the PDB protein data bank.

within an IRS from the IGFBP-1 promoter, we found that FoxO proteins interact with IRSs in a highly sequence-specific fashion that mirrors the ability of insulin to regulate promoter activity (Furuyama, Nakazawa, Nakano, & Mori, 2000; Guo et al., 1999), and, using an unbiased site-selected amplification procedure, Mori et al. derived a consensus sequence for FoxO proteins that was consistent with these results (Furuyama et al., 2000). Further, the expression of an antisense FoxO1 mRNA was sufficient to disrupt IRS-mediated effects of insulin on promoter function in liver-derived cells (Guo et al., 1999), supporting the concept that FoxO proteins, and especially FoxO1, play a major role in mediating squence-specific IRS-dependent effects of insulin on gene expression in the liver.

More recent studies in liver-specific knockout mice indicate that multiple FoxO proteins contribute to the regulation of hepatic glucose and lipid metabolism, including FoxO1, which is thought to exert its predominant effect on glucose-related gene expression, and FoxO3, which also contributes to glucoregulation while preferentially affecting lipid metabolism (Haeusler, Kaestner, & Accili, 2010; Zhang et al., 2012), while FoxO4 is expressed at very low levels in the liver (Zhang et al., 2016). Recent studies suggest that FoxO6, which was originally identified in the central nervous system and lacks the N-terminal Akt phosphorylation site corresponding to threonine 24 in human FoxO1 (Jacobs et al., 2003), also may contribute to the regulation of glucose and lipid metabolism in the liver (Sun et al., 2016). Thus, while specific FoxO proteins may play somewhat different roles in the liver, increasing evidence supports the concept that they function together in a complementary and overlapping fashion, reflecting the close homology of their DNA binding domains, to mediate effects of insulin on gene expression and glucose metabolism in the liver.

3. REGULATION OF FoxO FUNCTION BY INSULIN/PI3K/Akt SIGNALING AND DEACETYLASES

To better understand the role of FoxO proteins in regulating hepatic glucose metabolism, it also is important to understand the ways in which their function is regulated by insulin, counterregulatory hormones, and nutrient-sensing pathways. Based on a consensus sequence identified for phosphorylation by Akt (Alessi, Caudwell, Andjelkovic, Hemmings, & Cohen, 1996), we and others identified three highly conserved Akt phosphorylation sites in FoxO proteins, corresponding to threonine 24, serine 256, and serine 319 in human FoxO1, and determined that phosphorylation at these sites is sufficient to inhibit the function of FoxO proteins, and that phosphorylation of serine 256, located within the basic region at the C-terminal end of the FoxO DNA binding domain, plays a critical role in this process (Nakae, Barr, & Accili, 2000; Rena, Prescott, Guo, Cohen, & Unterman, 2001). In addition to contributing to the stability of DNA/protein interactions (Fig. 1) (Obsil & Obsilova, 2008; Tsai et al., 2007), this relatively unstructured region of the DNA binding domain region also contains nuclear localization signal(s) responsible for promoting the translocation of FoxO proteins from the cytoplasmic compartment into the nucleus (Obsilova et al., 2005; Zhao et al., 2004).

Phosphorylation of serine 256 (or the substitution of serine with aspartate, a negatively charged residue) and the introduction of a negative charge at this site can limit the function of FoxO proteins by reducing the stability of protein/DNA interactions (Zhang et al., 2002). In addition, phosphorylation of serine 256 also is required for the phosphorylation of threonine 24 and serine 319 by Akt and/or related kinases (Brunet et al., 2001; Nakae et al., 2000; Nakae, Kitamura, Ogawa, Kasuga, & Accili, 2001; Rena et al., 2001). Phosphorylation of serine 319 initiates a cascade of phosphorylations involving nearby residues (Rena et al., 2002; Woods et al., 2001), resulting in the activation of a nuclear export signal (Biggs, Meisenhelder, Hunter, Cavenee, & Arden, 1999) which promotes the translocation of FoxO proteins to the cytoplasmic compartment. At the same time, phosphorylation of threonine 24 results in the formation of a high-affinity 14-3-3 binding site, which together with phosphoserine 256 binds a dimer of 14-3-3 proteins (Rena et al., 2001). The binding of 14-3-3 proteins to serine 256 interferes with DNA binding and the function of the nuclear localization signal located at the C-terminal end of the DNA binding domain (Obsilova et al., 2005; Silhan et al., 2009), resulting in the sequestration of FoxO proteins in the cytoplasmic compartment following their transport from the nucleus. Thus, phosphorylation of FoxO proteins by Akt disrupts the function of FoxO proteins by multiple mechanisms, by interfering with DNA binding, and by promoting both their translocation from the nucleus and their sequestration in the cytoplasmic compartment.

Other posttranslational modifications also play an important role in regulating the function of FoxO proteins in response to changes in nutrient status, including the acetylation of positively charged lysine residues which are located near serine 256 within the C-terminal basic region of the DNA binding domain (Obsil & Obsilova, 2008). Upon binding to DNA target sites, FoxO transcription factors recruit coactivator proteins, including P300/cyclic AMP response element binding protein (CBP) and P300/CBP-associated factor (PCAF) proteins, which possess acetyltransferase activity and promote gene expression by acetylating nearby histones. At the same time, these same acetyltransferases also can acetylate specific positively charged lysine residues located near serine 256 within the basic region at the C-terminal region of the FoxO DNA binding domain (Fukuoka et al., 2003; Yoshimochi, Daitoku, & Fukamizu, 2010). Acetylation at these sites reduces the positive charge in this region and the stability of DNA binding, and thereby renders serine 256 more accessible to phosphorylation by Akt (Matsuzaki et al., 2005). Thus, the recruitment of P300/CBP and PCAF by FoxO proteins can both mediate and, at the same time, limit the effects of FoxO proteins on gene expression due to effects of acetylation.

Conversely, deacetylation of FoxO proteins can enhance their retention in the nucleus and function (Frescas, Valenti, & Accili, 2005; Qiang, Banks, & Accili, 2010). Sirtuin 1 (Sirt1), an NAD+-dependent histone deacetylase (HDAC), deacetylates FoxO proteins, as well as peroxisomal proliferator gamma coactivator-1α (PGC-1α) (Brunet et al., 2004; Frescas et al., 2005; Puigserver et al., 2003; Rodgers et al., 2005), which functions as a coactivator with FoxO1 to promote HGP

(Puigserver et al., 2003). In fasting, when NAD+ levels are high, sirtuins can associate with hepatic FoxO proteins and help to maintain them in a deacetylated, active state, which can function together with P300/CBP, PCAF, and PGC-1α to drive gluconeogenic gene expression. Glucagon also may promote the deacetylation of FoxO proteins due to the activation of Sirt1 downstream from protein kinase A (Gerhart-Hines et al., 2011) and the mobilization of class IIa HDACs, which function together with HDAC3 to deacetylate FoxO1 (Mihaylova et al., 2011). Thus, multiple factors combine to promote the deacetylation of FoxO proteins and promote gluconeogenic gene expression in the liver in fasting. Conversely, in the fed state, reduced levels of glucagon and NAD+, and, decreased sirtuin and class IIa HDAC/HDAC3 activity result in increased acetylation and reduced DNA binding by FoxO proteins, and, together with increased insulin levels and Akt activity, promotes increased phosphorylation and reduced function of FoxO proteins in the liver (Fig. 2).

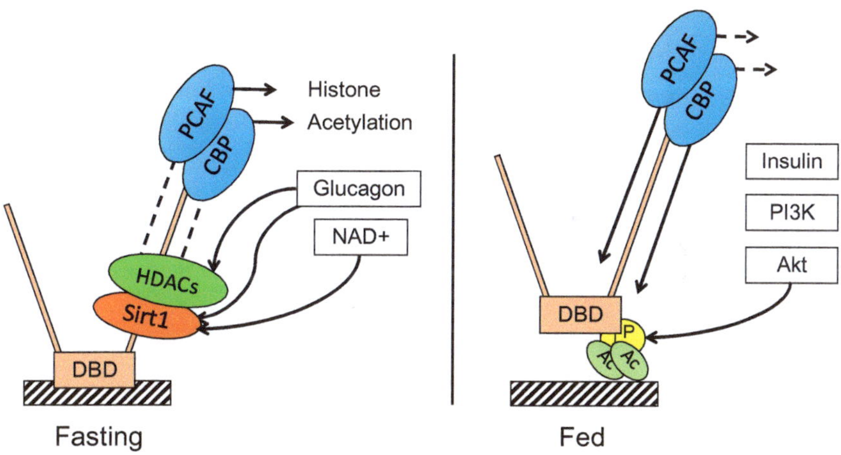

Fig. 2 Acetylation and phosphorylation of FoxO DNA binding domain. Acetylation and phosphorylation of the C-terminal basic domain contribute to the regulation of FoxO function. *Left panel*: In fasting, insulin levels and Akt activity are low, phosphorylation of FoxO proteins is reduced, and FoxO proteins are available to bind to DNA target sites. FoxO proteins recruit P300/CBP and PCAF, which possess acetyltransferase activity and promote acetylation of histones, and thereby enhances gene expression. P300/CBP and PCAF also can acetylate multiple lysine residues located within in the C-terminal region of the DNA binding domain. However, recruitment and activation of sirtuin 1 and other HDACs (by glucagon and NAD+) maintain FoxO proteins in a deacetylated, active state. *Right panel*: In the fed state, decreased sirtuin and HDAC activity result in increased acetylation of FoxOs, which reduces DNA binding activity and thereby promotes Akt phosphorylation of serine 256 within the basic region of the DNA binding domain. This further reduces DNA binding, and promotes translocation from the nucleus and sequestration in the cytoplasmic compartment in combination with 14-3-3 proteins (not shown).

Thus, hepatic FoxO proteins and their effects on gene expression are coordinately regulated by multiple factors in response to changes in nutrient status.

4. EFFECTS ON GLUCONEOGENIC GENE EXPRESSION

Based on the recognition that FoxO proteins bind to and mediate effects of insulin gene expression, and the ability of insulin to suppress gluconeogenic gene expression and hepatic glucose production, early studies regarding the effect of FoxO proteins on glucose metabolism largely focused on understanding their role the regulation of gluconeogenic genes in the liver, including PEPCK and the catalytic subunit of G6Pase, both of which contain insulin response sequences that can function as FoxO response elements (FREs) in their promoters (Barthel et al., 2005; Hall et al., 2000; Nakae, Kitamura, Silver, et al., 2001; Yeagley et al., 2001). Studies in transgenic mouse models expressing constitutively active FoxO1 in the liver confirmed that FoxO proteins can drive the expression of both PEPCK and G6Pase (Zhang et al., 2006), and studies in knockout mice also support the concept that endogenous FoxO proteins contribute to the regulation of gluconeogenic gene expression in the liver (Haeusler et al., 2010; Zhang et al., 2006). Cell culture studies with reporter gene constructs confirm that FoxO proteins can stimulate PEPCK and G6Pase promoter activity by direct effects due to their binding to specific IRS/FREs and mediate effects of insulin on promoter activity when they are overexpressed in cell culture models (Hall et al., 2000; Schmoll et al., 2000; Yeagley et al., 2001). In addition, FoxO proteins also function cooperatively with glucocorticoids to stimulate PEPCK and G6Pase promoter activity, and the ability of insulin to suppress glucocorticoid-stimulated promoter function is enhanced in the presence of FoxO proteins (Schmoll et al., 2000; Yeagley et al., 2001). However, mutation of the IRS/FRE in these promoters does not completely disrupt the ability of insulin to suppress glucocorticoid-stimulated promoter activity (Schmoll et al., 2000; Vander Kooi et al., 2003; Yeagley et al., 2001). Together, these findings indicate that FoxO proteins contribute to the ability of insulin to regulate the expression of gluconeogenic gene expression through both direct effects, and due to their ability to function cooperatively with other *trans*-acting factors, including glucocorticoids. In this way, suppressing the function of FoxO proteins provides a mechanism by which insulin also can reduce the function of counterregulatory factors on gluconeogenesis, including glucocorticoids.

5. EFFECTS ON GLYCOLYSIS AND GLUCOSE UTILIZATION

Studies in this and other laboratories have shown that FoxO proteins also have significant effects on glucose utilization in the liver, which can impact net HGP in several ways. Studies in transgenic mice expressing a constitutively active form of FoxO1 in the liver revealed that, in addition to promoting gluconeogenic gene expression, FoxO1 also suppresses the expression of glucokinase, which phosphorylates glucose in the liver, and thereby promotes the utilization of glucose by glycolysis, the pentose phosphate shunt, and utilization of glucose-derived pyruvate for oxidative metabolism and de novo lipogenesis (Zhang et al., 2006). FoxO1 also suppresses the activity of the beta cell-specific glucokinase promoter (Yoshida et al., 2007) and glucokinase expression in pancreatic beta cells (Buteau, Shlien, Foisy, & Accili, 2007) indicating that FoxO proteins also may suppress glucose utilization in other tissues where glucokinase is expressed. Activation of FoxO1 also suppresses glucose utilization in skeletal muscle cells (Bastie et al., 2005).

Studies in isolated hepatocytes confirm that FoxO proteins suppress the expression of glucokinase in a cell autonomous fashion (Zhang et al., 2006), and reporter gene studies suggest several mechanisms may be involved. Multiple factors contribute to the expression of glucokinase in the liver, including hepatocyte nuclear factor-4 (HNF-4), hypoxia-inducible factor-1α (HIF-1α), peroxisome proliferator receptor gamma (PPARγ), and others (Kim et al., 2009; Roth, Curth, Unterman, & Kietzmann, 2004). FoxO1 interacts directly with HNF-4 and inhibits transactivation by HNF-4 (Ganjam, Dimova, Unterman, & Kietzmann, 2009; Hirota et al., 2003), and a putative FRE has been identified in the glucokinase promoter which may promote increased local interaction between FoxO proteins with HNF-4 within the context of the glucokinase promoter (Ganjam et al., 2009). Studies with resveratrol, an activator of Sirt1, indicate that the interaction of FoxO1 and HNF-4, and the expression of glucokinase are determined, at least in part, by changes in the acetylation of FoxO proteins and/or other proteins, including HNF-4 (Ganjam et al., 2009). FoxO proteins also interact with and inhibit the function of PPARγ in an acetylation-dependent fashion (Armoni et al., 2006; Dowell, Otto, Adi, & Lane, 2003; Fan et al., 2009; Jing et al., 2007). Thus, in addition to enhancing effects of FoxO proteins on gluconeogenic gene expression, deacetylation by sirtuins also enhances the ability of FoxO to suppress the expression of glucokinase due to interaction with other nuclear transcription factors. Recent studies indicate that FoxO proteins may promote the degradation of HIF-1α (Brucker, Maurer, Harter, Rieger, & Steinbach, 2016). In addition, a recent paper published while this review was in revision indicates that Sin3, a corepressor protein, plays in important role in mediating effects of FoxO proteins on the expression of glucokinase and hepatic glucose handling (Langlet et al., 2017), providing an additional mechanism by which FoxO proteins might inhibit the expression of glucokinase in the liver.

Studies in liver-specific FoxO knockout mice confirm that endogenous FoxO proteins suppress the expression of glucokinase in the liver (Haeusler et al., 2014; Xiong, Tao, DePinho, & Dong, 2013; Zhang et al., 2012). However, although glucokinase mRNA and protein levels are increased in fasting FoxO knockout mice compared to floxed controls (Haeusler et al., 2014), they do not reach the level achieved after refeeding in wild-type mice (Zhang et al., 2006). This discrepancy indicates that, in addition to suppressing FoxO function, other factors also are required for the marked induction of glucokinase expression that occurs in response to feeding. A recent study in mice where Akt-1 and -2 are disrupted in the liver indicates that Akt-dependent expression of glucokinase requires both the suppression of FoxO1 function and the activation of signaling by the mammalian target of rapamycin complex-1 (mTORC1) (Titchenell et al., 2016). Similarly, Akt-dependent induction of lipogenic gene expression also required both the activation of mTORC1 signaling as well as the disruption of the inhibitory effects of FoxO1 on lipogenic gene expression (Titchenell et al., 2016). Thus, while suppression of FoxO function contributes to the ability of Akt to suppress gluconeogenic gene expression, it appears that both positive (activation of mTORC1) and negative (suppression of FoxO function) effects are involved in mediating effects of Akt on glycolytic and lipogenic gene expression and metabolism in the liver.

Studies in liver-specific FoxO knockout mice also provide insight into the role that changes in glucokinase expression play in mediating effects of FoxO proteins on glucose homeostasis. Pyruvate tolerance, which is thought to reflect HGP, is enhanced liver-specific FoxO knockout mice and knocking down glucokinase disrupts this effect (Xiong et al., 2013), suggesting that the effects of FoxO proteins on glucose homeostasis are dependent on their impact on glucokinase. The recognition that FoxO proteins exert opposing effects on the expression of glucokinase (which promotes the formation of glucose-6 phosphate and its metabolism), and G6Pase (which dephosphorylates glucose-6 phosphate and promotes glucose secretion) (Haeusler et al., 2014; Zhang et al., 2006), has suggested that FoxO proteins promote a net increase in HGP, at least in part, by tipping the balance of these functions in the direction of net HGP (Haeusler et al., 2014; Zhang et al., 2006).

Suppressing the expression of glucokinase also may contribute to other effects of FoxO proteins, including lipogenesis and HGP. Under fed conditions, glucokinase is required for carbohydrate flux through the pentose phosphate shunt, a major source of NADPH for lipogenesis. Glycolytic metabolism also activates the carbohydrate response element binding protein (ChREBP) which promotes the expression of other glycolytic enzymes, including liver pyruvate kinase, and lipogenic genes (Zhang et al., 2006). ChREBP also can suppress the expression of Sirt1 (Noriega et al., 2011), which may limit the effects of FoxO proteins on gluconeogenic gene expression. Also, glycolysis results in the production of pyruvate and ATP, both of which can be used for the synthesis of lipids, including malonyl CoA, which inhibits the function of carnitine palmitoyltransferase 1 (McGarry, Leatherman, & Foster, 1978) and thereby suppresses fatty acid oxidation (FAO), an important determinant of gluconeogenesis (see below). Recent studies indicate that increasing the flux of pyruvate to mitochondria can by itself suppress FAO (Danial et al., 2003; Diaz-Ruiz, Rigoulet, & Devin, 2011; Giménez-Cassina et al., 2014), which also may limit gluconeogenesis. Conversely, the suppression of glucokinase expression by FoxO proteins may promote HGP both by limiting glucose utilization, and by promoting FAO and gluconeogenesis in the liver.

In addition, suppressing glycolysis reduces the production of cytosolic NADH from NAD+, which increases the ratio of cytosolic NAD+/NADH and promotes the formation of pyruvate from lactate by lactate dehydrogenase, and, conversely, limits the formation of lactate from pyruvate. This is important for maintaining the availability of pyruvate in fasting, when pyruvate, derived from amino acids and lactate, is a major substrate for gluconeogenesis. Increasing the NAD+/NADH ratio also may enhance sirtuin activity (Imai, Armstrong, Kaeberlein, & Guarente, 2000) and further promote glucose production through other mechanisms (Rodgers et al., 2005).

Taken together, these observations suggest that the suppression of glucokinase expression and glycolysis by FoxO proteins may promote glucose production in the liver by multiple mechanisms.

6. EFFECTS OF FATTY ACID OXIDATION ON GLYCOLYSIS AND GLUCONEOGENESIS

FAO plays an important role in the regulation of hepatic glucose production and utilization (Delisle & Fritz, 1967; Perry et al., 2015; Rebrin, Steil, Getty, & Bergman, 1995; Ruderman, Toews, & Shafrir, 1969). Recent studies indicate that FoxO proteins promote lipid catabolism and

FAO in the liver (Tao et al., 2011; Zhang et al., 2016) and that these effects are important in mediating the effects of FoxO proteins on glucose metabolism (Zhang et al., 2016). While gene profiling studies in transgenic mice expressing constitutively active FoxO1 in the liver do not demonstrate marked effects of FoxO proteins on the expression of genes directly involved in promoting FAO (Zhang et al., 2006), recent studies indicate that FoxO proteins promote lipid catabolism in the liver by other mechanisms (Tao et al., 2011; Zhang et al., 2016).

FAO impacts glucose metabolism in several ways. Beta oxidation results in the production of mitochondrial acetyl CoA which combines with oxaloacetate to form citrate, an allosteric inhibitor of phosphofructokinase (PFK) (Garland, Randle, & Newsholme, 1963), the rate-limiting step of glycolysis. Inhibition of PFK limits the utilization of glucose for the production of pyruvate and allows fructose-6 phosphate to be converted back to glucose-6 phosphate, which can be converted to glucose by glucose-6 phosphatase and released into the circulation.

The accumulation of mitochondrial acetyl CoA also inhibits the activity of pyruvate dehydrogenase (PDH) (Hue & Taegtmeyer, 2009) and thereby limits the utilization of pyruvate (derived from lactate and amino acids under fasting conditions) for oxidative and lipogenic metabolism. At the same time, acetyl CoA activates pyruvate carboxylase (PCO) which promotes the conversion of pyruvate to oxaloacetate, which can be used for glucose production (Freedman & Kohn, 1964; Williamson, Kreisberg, & Felts, 1966).

Metabolism of acetyl CoA by the TCA cycle results in the production of mitochondrial NADH, which also limits PDH activity, sparing pyruvate for conversion into oxaloacetate. Mitochondrial NADH also promotes the conversion of oxaloacetate to malate by mitochondrial malate dehydrogenase (MDH), and malate can exit the mitochondria and enter the cytoplasmic compartment where gluconeogenesis occurs. Metabolism of acetyl CoA by the TCA cycle also promotes the production of GTP, which is used by PEPCK in the formation of phosphoenolpyruvate, and ATP, which is required at other energy-requiring steps in the process of glucose production.

In the cytoplasm, the suppression of glycolysis results in a relatively high NAD+/NADH ratio, which promotes the conversion of malate to oxaloacetate, the substrate for PEPCK, by cytosolic MDH. The generation of NADH from this reaction can be used later in the process of gluconeogenesis to generate glyceraldehyde-3 phosphate. Thus, FAO suppresses glycolysis and promotes the utilization of pyruvate for glucose production in multiple ways.

7. EFFECTS ON INTRA- AND EXTRAHEPATIC LIPID CATABOLISM

FoxO proteins promote FAO in the liver through several mechanisms. Tao et al. noted that the accumulation of triacylglycerol (TAG) is increased in liver-specific knockout mice where the alleles for FoxO1, FoxO3, and FoxO4 have been disrupted in the liver and they also found that FAO is decreased in hepatocytes from these mice (Tao et al., 2011). This reduction in FAO was associated with a decrease in the expression of nicotinamide phosphoribosyltransferase (NAMPT), which is important for maintaining levels of NAD+, a necessary substrate for sirtuin deacetylase activity, which also was decreased in hepatocytes from FoxO knockout mice (Tao et al., 2011). Conversely, restoration of FoxO or NAMPT expression rescued NAD+ levels and sirtuin activity, as well as FAO in hepatocytes derived from liver-specific FoxO knockout mice (Tao et al., 2011). These results indicate that FoxO proteins promote FAO, at least in part, by maintaining NAMPT expression and the abundance of NAD+ required to support sirtuin-dependent effects on FAO under fasting conditions.

Recently, we reported that FoxO proteins also promote FAO in the liver by regulating the hydrolysis of TAG stores in lipid droplets, a subcellular organelle that specializes in the storage of TAG (Mashek, Khan, Sathyanarayan, Ploeger, & Franklin, 2015). Adipose triacylglycerol lipase (ATGL) mediates the first step in hydrolysis of TAG in lipid droplets (Zimmermann et al., 2004), and FoxO1 promotes the expression of ATGL in adipocytes (Chakrabarti & Kandror, 2009). FoxO1 promotes ATGL expression and suppresses the expression of the G0/S1 switch gene 2 (G0S2) (Zhang et al., 2016), which binds and inhibits the function of ATGL (Yang et al., 2010) in the liver and isolated hepatocytes, and endogenous FoxO proteins are required for the effects of insulin on ATGL and G0S2 expression in isolated hepatocytes (Zhang et al., 2016). FoxO1 promotes TAG turnover and FAO oxidation in the liver and hepatocytes in an ATGL-dependent fashion, and ATGL knockdown and G0S2 rescue studies indicate that ATGL-dependent lipolysis also is required for effects of FoxO1 on glucose homeostasis (Zhang et al., 2016). Further, in addition to being regulated by FoxO proteins, knockdown indicate that ATGL-dependent lipolysis also is required for effects of FoxO1 on glucokinase and gluconeogenic gene expression (Zhang et al., 2016), possibly due to effects of ATGL on Sirt1 activity in the liver (Khan et al., 2015). These studies indicate that FoxO proteins promote intrahepatic lipolysis in an ATGL-dependent fashion, and that intrahepatic lipolysis also plays an important role in mediating or facilitating the effects of FoxO proteins on glycolytic and gluconeogenic gene expression and metabolism in the liver.

Recent studies indicate that hepatic FoxO proteins also may promote FAO and glucose production by the liver due to indirect, extra-hepatic effects on lipolysis in white adipose tissue (WAT) (Titchenell et al., 2016). Early studies showed that liver-specific insulin receptor knockout (LIRKO) mice are severely glucose intolerant and resistant to the ability of insulin to suppress HGP (Michael et al., 2000), indicating that insulin receptor signaling in the liver is essential for the ability of insulin to suppress HGP and maintain glucose homeostasis. Subsequent studies in this and other laboratories have shown that knocking out FoxO1 in the liver is sufficient to restore glucose tolerance and the ability of insulin to suppress HGP in mice where insulin signaling has been disrupted in the liver (Dong et al., 2008; Lu et al., 2012; O-Sullivan et al., 2015; Titchenell, Chu, Monks, & Birnbaum, 2015). These results indicate that extrahepatic effects of insulin (e.g., on adipose tissue) are sufficient to maintain glucose homeostasis and suppress HGP when FoxO1 is disrupted in the liver. And studies using S691, a specific insulin receptor blocking peptide, showed that extrahepatic effects of insulin also are necessary to maintain glucose homeostasis when both insulin signaling and FoxO1 are disrupted in the liver (O-Sullivan et al., 2015).

What are the mechanisms by which disrupting FoxO1 in the liver restores the ability of insulin to regulate HGP in the absence of hepatic insulin receptor signaling? Perry et al. reported that the inhibition of lipolysis in WAT (measured as glycerol turnover) and the subsequent lowering of circulating levels of nonesterified fatty acids (NEFAs) and intrahepatic levels of acetyl CoA is required for the ability of insulin to suppress HGP, both in wild-type mice and in mice where insulin/Akt signaling and FoxO1 have been disrupted in the liver (Perry et al., 2015). Subsequently, Titchenell et al. reported that the ability of insulin to suppress NEFA levels and HGP is impaired in liver-specific Akt-1 and -2 knockout mice, and is restored when FoxO1 also is knocked out in the liver (Titchenell et al., 2016). These results suggest that insulin is unable to suppress HGP when insulin signaling is disrupted in the liver, at least in part, because the ability of insulin to inhibit lipolysis in WAT and the flux of NEFAs to the liver is impaired, due to FoxO1-dependent effects in the liver. However, previous studies indicate that, although the ability of insulin to suppress HGP is disrupted, the ability of insulin to lower NEFA levels may not be entirely disrupted in LIRKO mice (Fisher & Kahn, 2003), suggesting that other mechanisms also may be involved in limiting the ability of insulin to suppress HGP when insulin signaling is disrupted in the liver.

Gluconeogenic gene expression is increased when insulin signaling is disrupted in the liver, and knocking out FoxO1 also largely disrupts this effect of hepatic insulin resistance (Dong et al., 2008; Lu et al., 2012; O-Sullivan et al., 2015). Also, the expression of glucokinase is suppressed when insulin signaling is disrupted in the liver, and knocking out FoxO1 can promote partial (but not complete) recovery of glucokinase expression in the absence of the hepatic insulin receptor (O-Sullivan et al., 2015). FoxO1-dependent effects on mitochondrial structure and function also have been reported in liver-specific IRS-1/2 knockout mice (Cheng et al., 2009), and might contribute to effects on glucose homeostasis. These and other FoxO1-dependent effects may impact the response of the liver to extra-hepatic effects of insulin, including effects on WAT lipolysis and NEFA levels, when insulin signaling is disrupted in the liver (Fig. 3).

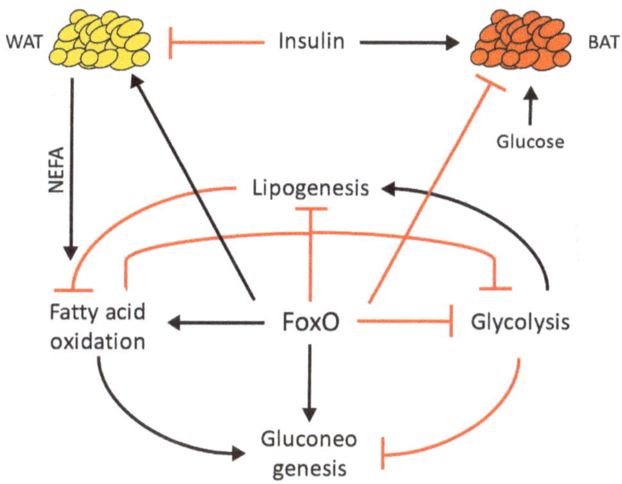

Fig. 3 Integrated effects of FoxO proteins impact hepatic glucose production through multiple mecahnisms. FoxO proteins promote hepatic glucose production directly, by enhancing gluconeogenic gene expression, and indirectly, by promoting intrahepatic lipolysis and fatty acid oxidation, due to ATGL- and sirtuin-dependent effects, and by suppressing glycolysis, which promotes lipogenesis and thereby suppresses fatty acid oxidation. In addition, when insulin signaling is disrupted in the liver, hepatic FoxO1-dependent effects impair the ability of insulin to suppress lipolysis in WAT, thereby limiting the ability of insulin to suppress the flux of fatty acids to the liver, where fatty acid oxidation promotes gluconeogenesis. Hepatic insulin resistance also may impair the ability of insulin to stimulate glucose utilization in BAT, and further impair glucose homeostasis in a FoxO1-dependent fashion.

8. EFFECTS ON GLUCOSE UTILIZATION

In addition to restricting the ability of insulin to suppress HGP, increasing evidence suggests that hepatic insulin resistance also promotes glucose intolerance, at least in part, by limiting glucose utilization, and that hepatic FoxO proteins contribute to this effect on glucose utilization. Insulin-stimulated whole-body glucose utilization is reduced in LIRKO mice compared to floxed controls (Michael et al., 2000), and glucose utilization is improved when FoxO1 is knocked out in the liver of LIRKO mice (O-Sullivan et al., 2015). Studies utilizing uniformly labeled ^{13}C-glucose indicate that glycolytic metabolism (producing ^{13}C-lactate from ^{13}C-glucose) is reduced in LIRKO mice, and restored when FoxO1 is knocked out in the liver (O-Sullivan et al., 2015). While changes in glucokinase expression and hepatic glucose utilization may contribute to these effects, glucose utilization also is reported to be decreased in brown adipose tissue (BAT) when insulin signaling is disrupted in the liver (Emanuelli et al., 2014; Stoehr, 2017), and preliminary studies suggest that glucose uptake is restored in BAT when FoxO1 is knocked out in the liver of LIRKO mice (I. O-Sullivan & T. Unterman, unpublished observations). Together, these findings are consistent with the concept that, in addition to effects on HGP, hepatic FoxO1 also impacts glucose utilization, due, at least in part, to effects on glucokinase expression in the liver and effects on glucose utilization in extrahepatic tissues, including BAT (Fig. 3).

9. HEPATOKINES

The mechanism(s) by which hepatic insulin resistance impacts glucose metabolism and insulin action in extrahepatic tissues is not yet understood. LIRKO mice are severely hyperinsulinemic (Michael et al., 2000), and hyperinsulinemia has been suggested to promote insulin resistance (Shanik et al., 2008). However, reducing insulin levels by treatment with streptozotocin failed to restore the ability of insulin to suppress HGP or significantly improve glucose utilization in LIRKO mice (Fisher & Kahn, 2003), suggesting that other factors are involved in promoting peripheral insulin resistance in the setting of hepatic insulin resistance.

Increasing evidence suggests that circulating factors produced by the liver (hepatokines) have important effects on glucose and lipid metabolism in other tissues (Meex & Watt, 2017). Several laboratories have observed that hepatic expression and circulating levels of FGF21, which promotes glucose utilization and enhances insulin sensitivity in BAT (BonDurant et al., 2017), are suppressed when insulin signaling is disrupted in the liver (Emanuelli et al., 2014; Stoehr, 2017), and restoring circulating levels of FGF21 levels, by infusing FGF21 with subcutaneous osmotic pumps or targeting its expression in the liver using an adeno-associated virus, restores insulin-stimulated glucose metabolism in BAT and glucose tolerance in these mice (Emanuelli et al., 2014; Stoehr, 2017). Preliminary studies indicate that the expression of FGF21 is recovered when FoxO1 is knocked out in the liver of LIRKO mice (I. O-Sullivan & T. Unterman, unpublished observations), suggesting that, in addition to its effects on hepatic glucose metabolism and WAT, hepatic FoxO1 also may impact insulin sensitivity and glucose utilization in BAT and other tissues due, at least in part, to changes in hepatic production of FGF21.

Changes in the production of other FoxO1-regulated factors by the liver also are likely to contribute to the metabolic impact of hepatic insulin resistance on peripheral tissues (Meex & Watt, 2017). Hepatic expression and production of insulin-like growth factor I are reduced when insulin signaling is disrupted in the liver (Dong et al., 2006; Michael et al., 2000) and restored when FoxO1 also is knocked out in the liver (Dong et al., 2006) (I. O-Sullivan & T. Unterman, unpublished observations). Hepatic production of IGF-I suppresses the secretion of growth hormone by the pituitary, and reduced production of IGF-I by the liver has been shown to result in peripheral insulin resistance due to increased growth hormone levels (Yakar et al., 2004).

FoxO proteins also have been shown to regulate the production and quality of bile acids in the liver (Biddinger et al., 2008; Haeusler, Pratt-Hyatt, Welch, Klaassen, & Accili, 2012). In addition to their effects on the absorption of cholesterol and other lipids, bile acids also promote GLP-1 and FGF15/19 production and secretion by the gut (Katsuma, Hirasawa, & Tsujimoto, 2005; Kliewer & Mangelsdorf, 2015), exert effects in the liver due to FGF15/19 signaling (Fu et al., 2016) and FXR-regulated gene expression (Trauner, Claudel, Fickert, Moustafa, & Wagner, 2010), and promote oxidative metabolism in BAT (Watanabe et al., 2006).

Circulating levels of the soluble form of the leptin receptor also are markedly elevated, and the leptin free index is suppressed in LIRKO mice (Cohen et al., 2007); this effect also is reversed when FoxO1 is knocked out in the liver of LIRKO mice (I. O-Sullivan and T. Unterman, unpublished observation), suggesting that peripheral effects of leptin on metabolism also may be impacted in a FoxO1-dependent fashion in the setting of hepatic insulin resistance. In addition, a preliminary report in liver-specific IRS-1 and -2 knockout mice indicates that increased hepatic production of follistatin, which binds activin and other members of the transforming growth factor-β family, may promote insulin resistance in WAT and effects on HGP in a FoxO1-dependent fashion (Tao, Stohr, & White, 2017). We find that follistatin expression also is increased in the liver of LIRKO mice in a FoxO1-dependent fashion (I. O-Sullivan and T. Unterman, unpublished observations), supporting the concept that follistatin may contribute to FoxO1-dependent effects of hepatic insulin resistance, including effects on extrahepatic tissues. Clearly, additional studies are needed to identify the role of these and other hepatokines in mediating effects of FoxO proteins on extrahepatic tissues in the setting of hepatic insulin resistance.

10. SUMMARY

FoxO Forkhead transcription factors are major targets of insulin action that are highly conserved in evolution, from invertebrates to mammals, and play an important role in the adaptation to nutrient restriction. Signaling via the insulin/PI3K/Akt pathway disrupts the function of FoxO proteins through multiple mechanisms, and deacetylation by sirtuins and class IIa HDACs, in combination with HDAC3 also plays an important part in regulating the function of FoxO proteins. Regulation of FoxO1 in the liver is required for the ability of insulin to suppress HGP and maintain glucose homeostasis. In addition to direct effects on the expression of gluconeogenic genes, the effects of FoxO proteins on glycolytic, lipogenic, and lipid catabolic pathways also play an important role in mediating effects of FoxO proteins on HGP and glucose homeostasis. FoxO proteins in the liver also exert effects on extrahepatic targets of insulin action that are important in maintaining glucose homeostasis, including WAT, where the ability of insulin to suppress lipolysis is critical for the regulation of HGP, and BAT, where glucose utilization contributes to the maintenance of glucose homeostasis (Fig. 3). Taken together, these observations indicate that hepatic FoxO proteins function as part of a highly integrated system involved in maintaining metabolic homeostasis at the whole-body level.

ACKNOWLEDGMENTS

The author gratefully acknowledges the many helpful discussions with Dr. Stanley Blumenthal, input from InSug O-Sullivan and Wenwei Zhang, and ongoing support from the Department of Veterans Affairs Merit Review Program. The author also gratefully acknowledges Dr. Bernard Santarsiero at the University of Illinois at Chicago for preparing Fig. 1.

REFERENCES

Alessi, D. R., Caudwell, F. B., Andjelkovic, M., Hemmings, B. A., & Cohen, P. (1996). Molecular basis for the substrate specificity of protein kinase B; comparison with MAPKAP kinase-1 and p70 S6 kinase. *FEBS Letters*, *399*(3), 333–338.

Armoni, M., Harel, C., Karni, S., Chen, H., Bar-Yoseph, F., Ver, M. R., et al. (2006). FOXO1 represses peroxisome proliferator-activated receptor-gamma1 and -gamma2 gene promoters in primary adipocytes. A novel paradigm to increase insulin sensitivity. *The Journal of Biological Chemistry*, *281*(29), 19881–19891. https://doi.org/10.1074/jbc.M600320200.

Barthel, A., Schmoll, D., & Unterman, T. G. (2005). FoxO proteins in insulin action and metabolism. *Trends in Endocrinology and Metabolism*, *16*(4), 183–189. https://doi.org/10.1016/j.tem.2005.03.010.

Bastie, C. C., Nahle, Z., McLoughlin, T., Esser, K., Zhang, W., Unterman, T., et al. (2005). FoxO1 stimulates fatty acid uptake and oxidation in muscle cells through CD36-dependent and -independent mechanisms. *The Journal of Biological Chemistry*, *280*(14), 14222–14229. https://doi.org/10.1074/jbc.M413625200.

Biddinger, S. B., Haas, J. T., Yu, B. B., Bezy, O., Jing, E., Zhang, W., et al. (2008). Hepatic insulin resistance directly promotes formation of cholesterol gallstones. *Nature Medicine*, *14*(7), 778–782. https://doi.org/10.1038/nm1785.

Biggs, W. H., 3rd, Meisenhelder, J., Hunter, T., Cavenee, W. K., & Arden, K. C. (1999). Protein kinase B/Akt-mediated phosphorylation promotes nuclear exclusion of the winged helix transcription factor FKHR1. *Proceedings of the National Academy of Sciences of the United States of America*, *96*(13), 7421–7426.

BonDurant, L. D., Ameka, M., Naber, M. C., Markan, K. R., Idiga, S. O., Acevedo, M. R., et al. (2017). FGF21 regulates metabolism through adipose-dependent and -independent mechanisms. *Cell Metabolism*, *25*(4). 935–944, e934. https://doi.org/10.1016/j.cmet.2017.03.005.

Brucker, D. P., Maurer, G. D., Harter, P. N., Rieger, J., & Steinbach, J. P. (2016). FOXO3a orchestrates glioma cell responses to starvation conditions and promotes hypoxia-induced cell death. *International Journal of Oncology*, *49*(6), 2399–2410. https://doi.org/10.3892/ijo.2016.3760.

Brunet, A., Bonni, A., Zigmond, M. J., Lin, M. Z., Juo, P., Hu, L. S., et al. (1999). Akt promotes cell survival by phosphorylating and inhibiting a Forkhead transcription factor. *Cell*, *96*(6), 857–868.

Brunet, A., Park, J., Tran, H., Hu, L. S., Hemmings, B. A., & Greenberg, M. E. (2001). Protein kinase SGK mediates survival signals by phosphorylating the forkhead transcription factor FKHRL1 (FOXO3a). *Molecular and Cellular Biology*, *21*(3), 952–965. https://doi.org/10.1128/mcb.21.3.952-965.2001.

Brunet, A., Sweeney, L. B., Sturgill, J. F., Chua, K. F., Greer, P. L., Lin, Y., et al. (2004). Stress-dependent regulation of FOXO transcription factors by the SIRT1 deacetylase. *Science*, *303*(5666), 2011–2015. https://doi.org/10.1126/science.1094637.

Buteau, J., Shlien, A., Foisy, S., & Accili, D. (2007). Metabolic diapause in pancreatic beta-cells expressing a gain-of-function mutant of the forkhead protein Foxo1. *The Journal of Biological Chemistry*, *282*(1), 287–293. https://doi.org/10.1074/jbc.M606118200.

Campbell, P. J., Mandarino, L. J., & Gerich, J. E. (1988). Quantification of the relative impairment in actions of insulin on hepatic glucose production and peripheral glucose uptake in non-insulin-dependent diabetes mellitus. *Metabolism*, *37*(1), 15–21.

Chakrabarti, P., & Kandror, K. V. (2009). FoxO1 controls insulin-dependent adipose triglyceride lipase (ATGL) expression and lipolysis in adipocytes. *The Journal of Biological Chemistry*, *284*(20), 13296–13300. https://doi.org/10.1074/jbc.C800241200.

Cheng, Z., Guo, S., Copps, K., Dong, X., Kollipara, R., Rodgers, J. T., et al. (2009). Foxo1 integrates insulin signaling with mitochondrial function in the liver. *Nature Medicine*, *15*(11), 1307–1311. https://doi.org/10.1038/nm.2049.

Cichy, S. B., Uddin, S., Danilkovich, A., Guo, S., Klippel, A., & Unterman, T. G. (1998). Protein kinase B/Akt mediates effects of insulin on hepatic insulin-like growth factor-binding protein-1 gene expression through a conserved insulin response sequence. *The Journal of Biological Chemistry*, *273*(11), 6482–6487.

Cohen, S. E., Kokkotou, E., Biddinger, S. B., Kondo, T., Gebhardt, R., Kratzsch, J., et al. (2007). High circulating leptin receptors with normal leptin sensitivity in liver-specific insulin receptor knock-out (LIRKO) mice. *The Journal of Biological Chemistry*, *282*(32), 23672–23678. https://doi.org/10.1074/jbc.M704053200.

Consoli, A., Nurjhan, N., Capani, F., & Gerich, J. (1989). Predominant role of gluconeogenesis in increased hepatic glucose production in NIDDM. *Diabetes*, *38*(5), 550–557.

Danial, N. N., Gramm, C. F., Scorrano, L., Zhang, C. Y., Krauss, S., Ranger, A. M., et al. (2003). BAD and glucokinase reside in a mitochondrial complex that integrates glycolysis and apoptosis. *Nature*, *424*(6951), 952–956. https://doi.org/10.1038/nature01825.

Defronzo, R. A. (2009). Banting lecture. From the triumvirate to the ominous octet: A new paradigm for the treatment of type 2 diabetes mellitus. *Diabetes*, *58*(4), 773–795. https://doi.org/10.2337/db09-9028.

Delisle, G., & Fritz, I. B. (1967). Interrelations between hepatic fatty acid oxidation and gluconeogenesis: A possible regulatory role of carnitine palmityltransferase. *Proceedings of the National Academy of Sciences of the United States of America*, *58*(2), 790–797.

Diaz-Ruiz, R., Rigoulet, M., & Devin, A. (2011). The Warburg and Crabtree effects: On the origin of cancer cell energy metabolism and of yeast glucose repression. *Biochimica et Biophysica Acta*, *1807*(6), 568–576. https://doi.org/10.1016/j.bbabio.2010.08.010.

Dong, X. C., Copps, K. D., Guo, S., Li, Y., Kollipara, R., DePinho, R. A., et al. (2008). Inactivation of hepatic Foxo1 by insulin signaling is required for adaptive nutrient homeostasis and endocrine growth regulation. *Cell Metabolism*, *8*(1), 65–76. https://doi.org/10.1016/j.cmet.2008.06.006.

Dong, X., Park, S., Lin, X., Copps, K., Yi, X., & White, M. F. (2006). Irs1 and Irs2 signaling is essential for hepatic glucose homeostasis and systemic growth. *The Journal of Clinical Investigation*, *116*(1), 101–114. https://doi.org/10.1172/jci25735.

Dowell, P., Otto, T. C., Adi, S., & Lane, M. D. (2003). Convergence of peroxisome proliferator-activated receptor gamma and Foxo1 signaling pathways. *The Journal of Biological Chemistry*, *278*(46), 45485–45491. https://doi.org/10.1074/jbc.M309069200.

Emanuelli, B., Vienberg, S. G., Smyth, G., Cheng, C., Stanford, K. I., Arumugam, M., et al. (2014). Interplay between FGF21 and insulin action in the liver regulates metabolism. *The Journal of Clinical Investigation*, *124*(2), 515–527. https://doi.org/10.1172/jci67353.

Fan, W., Imamura, T., Sonoda, N., Sears, D. D., Patsouris, D., Kim, J. J., et al. (2009). FOXO1 transrepresses peroxisome proliferator-activated receptor gamma transactivation, coordinating an insulin-induced feed-forward response in adipocytes. *The Journal of Biological Chemistry*, *284*(18), 12188–12197. https://doi.org/10.1074/jbc.M808915200.

Fisher, S. J., & Kahn, C. R. (2003). Insulin signaling is required for insulin's direct and indirect action on hepatic glucose production. *The Journal of Clinical Investigation*, *111*(4), 463–468. https://doi.org/10.1172/jci16426.

Freedman, A. D., & Kohn, L. (1964). Pyruvate metabolism and control: Factors affecting pyruvic carboxylase activity. *Science, 145*(3627), 58–60.

Frescas, D., Valenti, L., & Accili, D. (2005). Nuclear trapping of the forkhead transcription factor FoxO1 via Sirt-dependent deacetylation promotes expression of glucogenetic genes. *The Journal of Biological Chemistry, 280*(21), 20589–20595. https://doi.org/10.1074/jbc.M412357200.

Fu, T., Kim, Y. C., Byun, S., Kim, D. H., Seok, S., Suino-Powell, K., et al. (2016). FXR primes the liver for intestinal FGF15 signaling by transient induction of beta-Klotho. *Molecular Endocrinology, 30*(1), 92–103. https://doi.org/10.1210/me.2015-1226.

Fukuoka, M., Daitoku, H., Hatta, M., Matsuzaki, H., Umemura, S., & Fukamizu, A. (2003). Negative regulation of forkhead transcription factor AFX (Foxo4) by CBP-induced acetylation. *International Journal of Molecular Medicine, 12*(4), 503–508.

Furuyama, T., Nakazawa, T., Nakano, I., & Mori, N. (2000). Identification of the differential distribution patterns of mRNAs and consensus binding sequences for mouse DAF-16 homologues. *The Biochemical Journal, 349*(Pt. 2), 629–634.

Ganjam, G. K., Dimova, E. Y., Unterman, T. G., & Kietzmann, T. (2009). FoxO1 and HNF-4 are involved in regulation of hepatic glucokinase gene expression by resveratrol. *The Journal of Biological Chemistry, 284*(45), 30783–30797. https://doi.org/10.1074/jbc.M109.045260.

Garland, P. B., Randle, P. J., & Newsholme, E. A. (1963). Citrate as an intermediary in the inhibition of phosphofructokinase in rat heart muscle by fatty acids, ketone bodies, pyruvate, diabetes, and starvation. *Nature, 200*, 169–170.

Gerhart-Hines, Z., Dominy, J. E., Jr., Blattler, S. M., Jedrychowski, M. P., Banks, A. S., Lim, J. H., et al. (2011). The cAMP/PKA pathway rapidly activates SIRT1 to promote fatty acid oxidation independently of changes in NAD(+). *Molecular Cell, 44*(6), 851–863. https://doi.org/10.1016/j.molcel.2011.12.005.

Giménez-Cassina, A., Garcia-Haro, L., Choi, C. S., Osundiji, M. A., Lane, E. A., Huang, H., et al. (2014). Regulation of hepatic energy metabolism and gluconeogenesis by BAD. *Cell Metabolism, 19*(2), 272–284. https://doi.org/10.1016/j.cmet.2013.12.001.

Goswami, R., Lacson, R., Yang, E., Sam, R., & Unterman, T. (1994). Functional analysis of glucocorticoid and insulin response sequences in the rat insulin-like growth factor-binding protein-1 promoter. *Endocrinology, 134*(2), 736–743. https://doi.org/10.1210/endo.134.2.7507835.

Gross, D. N., Wan, M., & Birnbaum, M. J. (2009). The role of FOXO in the regulation of metabolism. *Current Diabetes Reports, 9*(3), 208–214.

Guo, S., Rena, G., Cichy, S., He, X., Cohen, P., & Unterman, T. (1999). Phosphorylation of serine 256 by protein kinase B disrupts transactivation by FKHR and mediates effects of insulin on insulin-like growth factor-binding protein-1 promoter activity through a conserved insulin response sequence. *The Journal of Biological Chemistry, 274*(24), 17184–17192.

Haeusler, R. A., Hartil, K., Vaitheesvaran, B., Arrieta-Cruz, I., Knight, C. M., Cook, J. R., et al. (2014). Integrated control of hepatic lipogenesis versus glucose production requires FoxO transcription factors. *Nature Communications, 5*, 5190. https://doi.org/10.1038/ncomms6190.

Haeusler, R. A., Kaestner, K. H., & Accili, D. (2010). FoxOs function synergistically to promote glucose production. *The Journal of Biological Chemistry, 285*(46), 35245–35248. https://doi.org/10.1074/jbc.C110.175851.

Haeusler, R. A., Pratt-Hyatt, M., Welch, C. L., Klaassen, C. D., & Accili, D. (2012). Impaired generation of 12-hydroxylated bile acids links hepatic insulin signaling with dyslipidemia. *Cell Metabolism, 15*(1), 65–74. https://doi.org/10.1016/j.cmet.2011.11.010.

Hall, R. K., Yamasaki, T., Kucera, T., Waltner-Law, M., O'Brien, R., & Granner, D. K. (2000). Regulation of phosphoenolpyruvate carboxykinase and insulin-like growth factor-binding protein-1 gene expression by insulin. The role of winged helix/forkhead proteins. *The Journal of Biological Chemistry*, *275*(39), 30169–30175. https://doi.org/10.1074/jbc.M004898200.

Hirota, K., Daitoku, H., Matsuzaki, H., Araya, N., Yamagata, K., Asada, S., et al. (2003). Hepatocyte nuclear factor-4 is a novel downstream target of insulin via FKHR as a signal-regulated transcriptional inhibitor. *The Journal of Biological Chemistry*, *278*(15), 13056–13060. https://doi.org/10.1074/jbc.C200553200.

Hue, L., & Taegtmeyer, H. (2009). The Randle cycle revisited: A new head for an old hat. *American Journal of Physiology. Endocrinology and Metabolism*, *297*(3), E578–591. https://doi.org/10.1152/ajpendo.00093.2009.

Imai, S., Armstrong, C. M., Kaeberlein, M., & Guarente, L. (2000). Transcriptional silencing and longevity protein Sir2 is an NAD-dependent histone deacetylase. *Nature*, *403*(6771), 795–800. https://doi.org/10.1038/35001622.

Jacobs, F. M., van der Heide, L. P., Wijchers, P. J., Burbach, J. P., Hoekman, M. F., & Smidt, M. P. (2003). FoxO6, a novel member of the FoxO class of transcription factors with distinct shuttling dynamics. *The Journal of Biological Chemistry*, *278*(38), 35959–35967. https://doi.org/10.1074/jbc.M302804200.

Jing, E., Gesta, S., & Kahn, C. R. (2007). SIRT2 regulates adipocyte differentiation through FoxO1 acetylation/deacetylation. *Cell Metabolism*, *6*(2), 105–114. https://doi.org/10.1016/j.cmet.2007.07.003.

Kaestner, K. H., Knochel, W., & Martinez, D. E. (2000). Unified nomenclature for the winged helix/forkhead transcription factors. *Genes & Development*, *14*(2), 142–146.

Kamei, Y., Miura, S., Suzuki, M., Kai, Y., Mizukami, J., Taniguchi, T., et al. (2004). Skeletal muscle FOXO1 (FKHR) transgenic mice have less skeletal muscle mass, down-regulated Type I (slow twitch/red muscle) fiber genes, and impaired glycemic control. *The Journal of Biological Chemistry*, *279*(39), 41114–41123. https://doi.org/10.1074/jbc.M400674200.

Katsuma, S., Hirasawa, A., & Tsujimoto, G. (2005). Bile acids promote glucagon-like peptide-1 secretion through TGR5 in a murine enteroendocrine cell line STC-1. *Biochemical and Biophysical Research Communications*, *329*(1), 386–390. https://doi.org/10.1016/j.bbrc.2005.01.139.

Khan, S. A., Sathyanarayan, A., Mashek, M. T., Ong, K. T., Wollaston-Hayden, E. E., & Mashek, D. G. (2015). ATGL-catalyzed lipolysis regulates SIRT1 to control PGC-1alpha/PPAR-alpha signaling. *Diabetes*, *64*(2), 418–426. https://doi.org/10.2337/db14-0325.

Kim, T. H., Kim, H., Park, J. M., Im, S. S., Bae, J. S., Kim, M. Y., et al. (2009). Interrelationship between liver X receptor alpha, sterol regulatory element-binding protein-1c, peroxisome proliferator-activated receptor gamma, and small heterodimer partner in the transcriptional regulation of glucokinase gene expression in liver. *The Journal of Biological Chemistry*, *284*(22), 15071–15083. https://doi.org/10.1074/jbc.M109.006742.

Kim-Muller, J. Y., Zhao, S., Srivastava, S., Mugabo, Y., Noh, H. L., Kim, Y. R., et al. (2014). Metabolic inflexibility impairs insulin secretion and results in MODY-like diabetes in triple FoxO-deficient mice. *Cell Metabolism*, *20*(4), 593–602. https://doi.org/10.1016/j.cmet.2014.08.012.

Kliewer, S. A., & Mangelsdorf, D. J. (2015). Bile acids as hormones: The FXR-FGF15/19 pathway. *Digestive Diseases*, *33*(3), 327–331. https://doi.org/10.1159/000371670.

Kops, G. J., de Ruiter, N. D., De Vries-Smits, A. M., Powell, D. R., Bos, J. L., & Burgering, B. M. (1999). Direct control of the Forkhead transcription factor AFX by protein kinase B. *Nature*, *398*(6728), 630–634. https://doi.org/10.1038/19328.

Langlet, F., Haeusler, R. A., Lindén, D., Ericson, E., Norris, T., Johansson, A., et al. (2017). Selective inhibition of FOXO1 activator/repressor balance modulates hepatic glucose handling. *Cell*, *171*(4), 824–835, e18. https://doi.org/10.1016/j.cell.2017.09.045. (Epub 2017 Oct 19).

Lin, K., Dorman, J. B., Rodan, A., & Kenyon, C. (1997). daf-16: An HNF-3/forkhead family member that can function to double the life-span of Caenorhabditis elegans. *Science*, *278*(5341), 1319–1322.

Lu, M., Wan, M., Leavens, K. F., Chu, Q., Monks, B. R., Fernandez, S., et al. (2012). Insulin regulates liver metabolism in vivo in the absence of hepatic Akt and Foxo1. *Nature Medicine*, *18*(3), 388–395. https://doi.org/10.1038/nm.2686.

Mashek, D. G., Khan, S. A., Sathyanarayan, A., Ploeger, J. M., & Franklin, M. P. (2015). Hepatic lipid droplet biology: Getting to the root of fatty liver. *Hepatology*, *62*, 964–967. https://doi.org/10.1002/hep.27839.

Matsumoto, M., Pocai, A., Rossetti, L., Depinho, R. A., & Accili, D. (2007). Impaired regulation of hepatic glucose production in mice lacking the forkhead transcription factor Foxo1 in liver. *Cell Metabolism*, *6*(3), 208–216. https://doi.org/10.1016/j.cmet.2007.08.006.

Matsuzaki, H., Daitoku, H., Hatta, M., Aoyama, H., Yoshimochi, K., & Fukamizu, A. (2005). Acetylation of Foxo1 alters its DNA-binding ability and sensitivity to phosphorylation. *Proceedings of the National Academy of Sciences of the United States of America*, *102*(32), 11278–11283. https://doi.org/10.1073/pnas.0502738102.

McGarry, J. D., Leatherman, G. F., & Foster, D. W. (1978). Carnitine palmitoyltransferase I. The site of inhibition of hepatic fatty acid oxidation by malonyl-CoA. *The Journal of Biological Chemistry*, *253*(12), 4128–4136.

Meex, R. C. R., & Watt, M. J. (2017). Hepatokines: Linking nonalcoholic fatty liver disease and insulin resistance. *Nature Reviews. Endocrinology*, *13*(9), 509–520. https://doi.org/10.1038/nrendo.2017.56.

Michael, M. D., Kulkarni, R. N., Postic, C., Previs, S. F., Shulman, G. I., Magnuson, M. A., et al. (2000). Loss of insulin signaling in hepatocytes leads to severe insulin resistance and progressive hepatic dysfunction. *Molecular Cell*, *6*(1), 87–97.

Mihaylova, M. M., Vasquez, D. S., Ravnskjaer, K., Denechaud, P. D., Yu, R. T., Alvarez, J. G., et al. (2011). Class IIa histone deacetylases are hormone-activated regulators of FOXO and mammalian glucose homeostasis. *Cell*, *145*(4), 607–621. https://doi.org/10.1016/j.cell.2011.03.043.

Nakae, J., Barr, V., & Accili, D. (2000). Differential regulation of gene expression by insulin and IGF-1 receptors correlates with phosphorylation of a single amino acid residue in the forkhead transcription factor FKHR. *The EMBO Journal*, *19*(5), 989–996. https://doi.org/10.1093/emboj/19.5.989.

Nakae, J., Cao, Y., Oki, M., Orba, Y., Sawa, H., Kiyonari, H., et al. (2008). Forkhead transcription factor FoxO1 in adipose tissue regulates energy storage and expenditure. *Diabetes*, *57*(3), 563–576. https://doi.org/10.2337/db07-0698.

Nakae, J., Kitamura, T., Ogawa, W., Kasuga, M., & Accili, D. (2001). Insulin regulation of gene expression through the forkhead transcription factor Foxo1 (Fkhr) requires kinases distinct from Akt. *Biochemistry*, *40*(39), 11768–11776.

Nakae, J., Kitamura, T., Silver, D. L., & Accili, D. (2001). The forkhead transcription factor Foxo1 (Fkhr) confers insulin sensitivity onto glucose-6-phosphatase expression. *The Journal of Clinical Investigation*, *108*(9), 1359–1367. https://doi.org/10.1172/jci12876.

Nakae, J., Park, B. C., & Accili, D. (1999). Insulin stimulates phosphorylation of the forkhead transcription factor FKHR on serine 253 through a Wortmannin-sensitive pathway. *The Journal of Biological Chemistry*, *274*(23), 15982–15985.

Noriega, L. G., Feige, J. N., Canto, C., Yamamoto, H., Yu, J., Herman, M. A., et al. (2011). CREB and ChREBP oppositely regulate SIRT1 expression in response to energy availability. *EMBO Reports*, *12*(10), 1069–1076. https://doi.org/10.1038/embor.2011.151.

O'Brien, R. M., Lucas, P. C., Forest, C. D., Magnuson, M. A., & Granner, D. K. (1990). Identification of a sequence in the PEPCK gene that mediates a negative effect of insulin on transcription. *Science*, *249*(4968), 533–537.

O'Brien, R. M., Noisin, E. L., Suwanichkul, A., Yamasaki, T., Lucas, P. C., Wang, J. C., et al. (1995). Hepatic nuclear factor 3- and hormone-regulated expression of the phosphoenolpyruvate carboxykinase and insulin-like growth factor-binding protein 1 genes. *Molecular and Cellular Biology*, *15*(3), 1747–1758.

Obsil, T., & Obsilova, V. (2008). Structure/function relationships underlying regulation of FOXO transcription factors. *Oncogene*, *27*(16), 2263–2275. https://doi.org/10.1038/onc.2008.20.

Obsilova, V., Vecer, J., Herman, P., Pabianova, A., Sulc, M., Teisinger, J., et al. (2005). 14-3-3 Protein interacts with nuclear localization sequence of forkhead transcription factor FoxO4. *Biochemistry*, *44*(34), 11608–11617. https://doi.org/10.1021/bi050618r.

Ogg, S., Paradis, S., Gottlieb, S., Patterson, G. I., Lee, L., Tissenbaum, H. A., et al. (1997). The Fork head transcription factor DAF-16 transduces insulin-like metabolic and longevity signals in C. elegans. *Nature*, *389*(6654), 994–999. https://doi.org/10.1038/40194.

Ooi, G. T., Orlowski, C. C., Brown, A. L., Becker, R. E., Unterman, T. G., & Rechler, M. M. (1990). Different tissue distribution and hormonal regulation of messenger RNAs encoding rat insulin-like growth factor-binding proteins-1 and -2. *Molecular Endocrinology*, *4*(2), 321–328. https://doi.org/10.1210/mend-4-2-321.

O-Sullivan, I., Zhang, W., Wasserman, D. H., Liew, C. W., Liu, J., Paik, J., et al. (2015). FoxO1 integrates direct and indirect effects of insulin on hepatic glucose production and glucose utilization. *Nature Communications*, *6*, 7079. https://doi.org/10.1038/ncomms8079.

Pajvani, U. B., & Accili, D. (2015). The new biology of diabetes. *Diabetologia*, *58*(11), 2459–2468. https://doi.org/10.1007/s00125-015-3722-5.

Perry, R. J., Camporez, J. P., Kursawe, R., Titchenell, P. M., Zhang, D., Perry, C. J., et al. (2015). Hepatic acetyl CoA links adipose tissue inflammation to hepatic insulin resistance and type 2 diabetes. *Cell*, *160*(4), 745–758. https://doi.org/10.1016/j.cell.2015.01.012.

Phillips, L. S., & Unterman, T. G. (1984). Somatomedin activity in disorders of nutrition and metabolism. *Clinics in Endocrinology and Metabolism*, *13*(1), 145–189.

Plum, L., Lin, H. V., Dutia, R., Tanaka, J., Aizawa, K. S., Matsumoto, M., et al. (2009). The obesity susceptibility gene Cpe links FoxO1 signaling in hypothalamic pro-opiomelanocortin neurons with regulation of food intake. *Nature Medicine*, *15*(10), 1195–1201. https://doi.org/10.1038/nm.2026.

Powell, D. R., Lee, P. D., & Suwanichkul, A. (1993). Multihormonal regulation of IGFBP-1 promoter activity. *Advances in Experimental Medicine and Biology*, *343*, 205–214.

Puigserver, P., Rhee, J., Donovan, J., Walkey, C. J., Yoon, J. C., Oriente, F., et al. (2003). Insulin-regulated hepatic gluconeogenesis through FOXO1-PGC-1alpha interaction. *Nature*, *423*(6939), 550–555. https://doi.org/10.1038/nature01667.

Qiang, L., Banks, A. S., & Accili, D. (2010). Uncoupling of acetylation from phosphorylation regulates FoxO1 function independent of its subcellular localization. *The Journal of Biological Chemistry*, *285*(35), 27396–27401. https://doi.org/10.1074/jbc.M110.140228.

Rausa, F. M., Tan, Y., Zhou, H., Yoo, K. W., Stolz, D. B., Watkins, S. C., et al. (2000). Elevated levels of hepatocyte nuclear factor 3beta in mouse hepatocytes influence expression of genes involved in bile acid and glucose homeostasis. *Molecular and Cellular Biology*, *20*(21), 8264–8282.

Rebrin, K., Steil, G. M., Getty, L., & Bergman, R. N. (1995). Free fatty acid as a link in the regulation of hepatic glucose output by peripheral insulin. *Diabetes*, *44*(9), 1038–1045.

Rena, G., Prescott, A. R., Guo, S., Cohen, P., & Unterman, T. G. (2001). Roles of the forkhead in rhabdomyosarcoma (FKHR) phosphorylation sites in regulating 14-3-3 binding, transactivation and nuclear targetting. *The Biochemical Journal*, *354*(Pt. 3), 605–612.

Rena, G., Woods, Y. L., Prescott, A. R., Peggie, M., Unterman, T. G., Williams, M. R., et al. (2002). Two novel phosphorylation sites on FKHR that are critical for its nuclear exclusion. *The EMBO Journal, 21*(9), 2263–2271. https://doi.org/10.1093/emboj/21.9.2263.

Rodgers, J. T., Lerin, C., Haas, W., Gygi, S. P., Spiegelman, B. M., & Puigserver, P. (2005). Nutrient control of glucose homeostasis through a complex of PGC-1alpha and SIRT1. *Nature, 434*(7029), 113–118. https://doi.org/10.1038/nature03354.

Roth, U., Curth, K., Unterman, T. G., & Kietzmann, T. (2004). The transcription factors HIF-1 and HNF-4 and the coactivator p300 are involved in insulin-regulated glucokinase gene expression via the phosphatidylinositol 3-kinase/protein kinase B pathway. *The Journal of Biological Chemistry, 279*(4), 2623–2631. https://doi.org/10.1074/jbc.M308391200.

Ruderman, N. B., Toews, C. J., & Shafrir, E. (1969). Role of free fatty acids in glucose homeostasis. *Archives of Internal Medicine, 123*(3), 299–313.

Sandri, M., Sandri, C., Gilbert, A., Skurk, C., Calabria, E., Picard, A., et al. (2004). Foxo transcription factors induce the atrophy-related ubiquitin ligase atrogin-1 and cause skeletal muscle atrophy. *Cell, 117*(3), 399–412.

Schmoll, D., Walker, K. S., Alessi, D. R., Grempler, R., Burchell, A., Guo, S., et al. (2000). Regulation of glucose-6-phosphatase gene expression by protein kinase Balpha and the forkhead transcription factor FKHR. Evidence for insulin response unit-dependent and -independent effects of insulin on promoter activity. *The Journal of Biological Chemistry, 275*(46), 36324–36333. https://doi.org/10.1074/jbc.M003616200.

Shanik, M. H., Xu, Y., Skrha, J., Dankner, R., Zick, Y., & Roth, J. (2008). Insulin resistance and hyperinsulinemia: Is hyperinsulinemia the cart or the horse? *Diabetes Care, 31*(Suppl. 2), S262–268. https://doi.org/10.2337/dc08-s264.

Silhan, J., Vacha, P., Strnadova, P., Vecer, J., Herman, P., Sulc, M., et al. (2009). 14-3-3 Protein masks the DNA binding interface of forkhead transcription factor FOXO4. *The Journal of Biological Chemistry, 284*(29), 19349–19360. https://doi.org/10.1074/jbc.M109.002725.

Stoehr, O. W. M. (2017). In *Hepatic insulin resistance causing metabolic dysfunctions in brown adipose tissue and skeletal muscle 77th Scientific sessions of the American Diabetes Association, San Diego, CA*.

Sun, L., Fan, G., Shan, P., Qiu, X., Dong, S., Liao, L., et al. (2016). Regulation of energy homeostasis by the ubiquitin-independent REGγ proteasome. *Nature Communications, 7*, 12497. https://doi.org/10.1038/ncomms12497.

Suwanichkul, A., Allander, S. V., Morris, S. L., & Powell, D. R. (1994). Glucocorticoids and insulin regulate expression of the human gene for insulin-like growth factor-binding protein-1 through proximal promoter elements. *The Journal of Biological Chemistry, 269*(49), 30835–30841.

Talchai, C., Xuan, S., Kitamura, T., DePinho, R. A., & Accili, D. (2012). Generation of functional insulin-producing cells in the gut by Foxo1 ablation. *Nature Genetics, 44*(4), 406–412. s401, https://doi.org/10.1038/ng.2215.

Talchai, C., Xuan, S., Lin, H. V., Sussel, L., & Accili, D. (2012). Pancreatic beta cell dedifferentiation as a mechanism of diabetic beta cell failure. *Cell, 150*(6), 1223–1234. https://doi.org/10.1016/j.cell.2012.07.029.

Tang, E. D., Nunez, G., Barr, F. G., & Guan, K. L. (1999). Negative regulation of the forkhead transcription factor FKHR by Akt. *The Journal of Biological Chemistry, 274*(24), 16741–16746.

Tao, R. C. K., Stohr, O., & White, M. F. (2017). In *Hepatic insulin signaling regulates follistatin to coordinate hepatic glucose production 77th Annual scientific sessions of the American Diabetes Association, San Diego, CA*.

Tao, R., Wei, D., Gao, H., Liu, Y., DePinho, R. A., & Dong, X. C. (2011). Hepatic FoxOs regulate lipid metabolism via modulation of expression of the nicotinamide phosphoribosyltransferase gene. *The Journal of Biological Chemistry*, *286*(16), 14681–14690. https://doi.org/10.1074/jbc.M110.201061.

Titchenell, P. M., Chu, Q., Monks, B. R., & Birnbaum, M. J. (2015). Hepatic insulin signalling is dispensable for suppression of glucose output by insulin in vivo. *Nature Communications*, *6*, 7078. https://doi.org/10.1038/ncomms8078.

Titchenell, P. M., Quinn, W. J., Lu, M., Chu, Q., Lu, W., Li, C., et al. (2016). Direct hepatocyte insulin signaling is required for lipogenesis but is dispensable for the suppression of glucose production. *Cell Metabolism*, *23*(6), 1154–1166. https://doi.org/10.1016/j.cmet.2016.04.022.

Trauner, M., Claudel, T., Fickert, P., Moustafa, T., & Wagner, M. (2010). Bile acids as regulators of hepatic lipid and glucose metabolism. *Digestive Diseases*, *28*(1), 220–224. https://doi.org/10.1159/000282091.

Tsai, K. L., Sun, Y. J., Huang, C. Y., Yang, J. Y., Hung, M. C., & Hsiao, C. D. (2007). Crystal structure of the human FOXO3a-DBD/DNA complex suggests the effects of post-translational modification. *Nucleic Acids Research*, *35*(20), 6984–6994. https://doi.org/10.1093/nar/gkm703.

Unterman, T. G., Fareeduddin, A., Harris, M. A., Goswami, R. G., Porcella, A., Costa, R. H., et al. (1994). Hepatocyte nuclear factor-3 (HNF-3) binds to the insulin response sequence in the IGF binding protein-1 (IGFBP-1) promoter and enhances promoter function. *Biochemical and Biophysical Research Communications*, *203*(3), 1835–1841. https://doi.org/10.1006/bbrc.1994.2401.

Unterman, T. G., Oehler, D. T., Murphy, L. J., & Lacson, R. G. (1991). Multihormonal regulation of insulin-like growth factor-binding protein-1 in rat H4IIE hepatoma cells: The dominant role of insulin. *Endocrinology*, *128*(6), 2693–2701. https://doi.org/10.1210/endo-128-6-2693.

Unterman, T. G., Patel, K., Mahathre, V. K., Rajamohan, G., Oehler, D. T., & Becker, R. E. (1990). Regulation of low molecular weight insulin-like growth factor binding proteins in experimental diabetes mellitus. *Endocrinology*, *126*(5), 2614–2624. https://doi.org/10.1210/endo-126-5-2614.

Vander Kooi, B. T., Streeper, R. S., Svitek, C. A., Oeser, J. K., Powell, D. R., & O'Brien, R. M. (2003). The three insulin response sequences in the glucose-6-phosphatase catalytic subunit gene promoter are functionally distinct. *The Journal of Biological Chemistry*, *278*(14), 11782–11793. https://doi.org/10.1074/jbc.M212570200.

Wang, Q., Liu, C., Uchida, A., Chuang, J. C., Walker, A., Liu, T., et al. (2014). Arcuate AgRP neurons mediate orexigenic and glucoregulatory actions of ghrelin. *Molecular Metabolism*, *3*(1), 64–72. https://doi.org/10.1016/j.molmet.2013.10.001.

Watanabe, M., Houten, S. M., Mataki, C., Christoffolete, M. A., Kim, B. W., Sato, H., et al. (2006). Bile acids induce energy expenditure by promoting intracellular thyroid hormone activation. *Nature*, *439*(7075), 484–489. https://doi.org/10.1038/nature04330.

Williamson, J. R., Kreisberg, R. A., & Felts, P. W. (1966). Mechanism for the stimulation of gluconeogenesis by fatty acids in perfused rat liver. *Proceedings of the National Academy of Sciences of the United States of America*, *56*(1), 247–254.

Woods, Y. L., Rena, G., Morrice, N., Barthel, A., Becker, W., Guo, S., et al. (2001). The kinase DYRK1A phosphorylates the transcription factor FKHR at Ser329 in vitro, a novel in vivo phosphorylation site. *The Biochemical Journal*, *355*(Pt. 3), 597–607.

Xia, S. J., & Barr, F. G. (2004). Analysis of the transforming and growth suppressive activities of the PAX3-FKHR oncoprotein. *Oncogene*, *23*(41), 6864–6871. https://doi.org/10.1038/sj.onc.1207850.

Xiao, X., Chen, C., Guo, P., Zhang, T., Fischbach, S., Fusco, J., et al. (2017). Forkhead box protein 1 (FoxO1) inhibits accelerated beta cell aging in pancreas-specific SMAD7 mutant mice. *The Journal of Biological Chemistry*, *292*(8), 3456–3465. https://doi.org/10.1074/jbc.M116.770032.

Xiong, X., Tao, R., DePinho, R. A., & Dong, X. C. (2013). Deletion of hepatic FoxO1/3/4 genes in mice significantly impacts on glucose metabolism through downregulation of gluconeogenesis and upregulation of glycolysis. *PLoS One*, *8*(8), e74340. https://doi.org/10.1371/journal.pone.0074340.

Yakar, S., Setser, J., Zhao, H., Stannard, B., Haluzik, M., Glatt, V., et al. (2004). Inhibition of growth hormone action improves insulin sensitivity in liver IGF-1-deficient mice. *The Journal of Clinical Investigation*, *113*(1), 96–105. https://doi.org/10.1172/jci17763.

Yang, X., Lu, X., Lombes, M., Rha, G. B., Chi, Y. I., Guerin, T. M., et al. (2010). The G(0)/G(1) switch gene 2 regulates adipose lipolysis through association with adipose triglyceride lipase. *Cell Metabolism*, *11*(3), 194–205. https://doi.org/10.1016/j.cmet.2010.02.003.

Yeagley, D., Guo, S., Unterman, T., & Quinn, P. G. (2001). Gene- and activation-specific mechanisms for insulin inhibition of basal and glucocorticoid-induced insulin-like growth factor binding protein-1 and phosphoenolpyruvate carboxykinase transcription. Roles of forkhead and insulin response sequences. *The Journal of Biological Chemistry*, *276*(36), 33705–33710. https://doi.org/10.1074/jbc.M101215200.

Yeoh, S. I., & Baxter, R. C. (1988). Metabolic regulation of the growth hormone independent insulin-like growth factor binding protein in human plasma. *Acta Endocrinologica*, *119*(4), 465–473.

Yoshida, K., Murao, K., Imachi, H., Cao, W. M., Yu, X., Li, J., et al. (2007). Pancreatic glucokinase is activated by insulin-like growth factor-I. *Endocrinology*, *148*(6), 2904–2913. https://doi.org/10.1210/en.2006-1149.

Yoshimochi, K., Daitoku, H., & Fukamizu, A. (2010). PCAF represses transactivation function of FOXO1 in an acetyltransferase-independent manner. *Journal of Receptor and Signal Transduction Research*, *30*(1), 43–49. https://doi.org/10.3109/10799890903517947.

Zhang, W., Bu, S. Y., Mashek, M. T., O-Sullivan, I., Sibai, Z., Khan, S. A., et al. (2016). Integrated regulation of hepatic lipid and glucose metabolism by adipose triacylglycerol lipase and FoxO proteins. *Cell Reports*, *15*(2), 349–359. https://doi.org/10.1016/j.celrep.2016.03.021.

Zhang, X., Gan, L., Pan, H., Guo, S., He, X., Olson, S. T., et al. (2002). Phosphorylation of serine 256 suppresses transactivation by FKHR (FOXO1) by multiple mechanisms. Direct and indirect effects on nuclear/cytoplasmic shuttling and DNA binding. *The Journal of Biological Chemistry*, *277*(47), 45276–45284. https://doi.org/10.1074/jbc.M208063200.

Zhang, K., Li, L., Qi, Y., Zhu, X., Gan, B., DePinho, R. A., et al. (2012). Hepatic suppression of Foxo1 and Foxo3 causes hypoglycemia and hyperlipidemia in mice. *Endocrinology*, *153*(2), 631–646. https://doi.org/10.1210/en.2011-1527.

Zhang, W., Patil, S., Chauhan, B., Guo, S., Powell, D. R., Le, J., et al. (2006). FoxO1 regulates multiple metabolic pathways in the liver: Effects on gluconeogenic, glycolytic, and lipogenic gene expression. *The Journal of Biological Chemistry*, *281*(15), 10105–10117. https://doi.org/10.1074/jbc.M600272200.

Zhao, X., Gan, L., Pan, H., Kan, D., Majeski, M., Adam, S. A., et al. (2004). Multiple elements regulate nuclear/cytoplasmic shuttling of FOXO1: Characterization of phosphorylation- and 14-3-3-dependent and -independent mechanisms. *The Biochemical Journal*, *378*(Pt. 3), 839–849. https://doi.org/10.1042/BJ20031450.

Zimmermann, R., Strauss, J. G., Haemmerle, G., Schoiswohl, G., Birner-Gruenberger, R., Riederer, M., et al. (2004). Fat mobilization in adipose tissue is promoted by adipose triglyceride lipase. *Science*, *306*(5700), 1383–1386. https://doi.org/10.1126/science.1100747.

CHAPTER SIX

The Role of FoxOs in Bone Health and Disease

Ha-Neui Kim*, Srividhya Iyer†, Rebecca Ring*, Maria Almeida*,†,1

*Center for Osteoporosis and Metabolic Bone Diseases, University of Arkansas for Medical Sciences, Little Rock, AR, United States

†University of Arkansas for Medical Sciences, Little Rock, AR, United States

[1]Corresponding author: e-mail address: schullermaria@uams.edu

Contents

Abstract

Recent studies with murine models of cell-specific loss- or gain-of-function of FoxOs have provided novel insights into the function and signaling of these transcription factors on the skeleton. They have revealed that FoxO actions in chondrocytes are critical for normal skeletal development, and FoxO actions in cells of the osteoclast or osteoblast lineage greatly influence bone resorption and formation and, consequently, bone mass. FoxOs also act in osteoblast progenitors to inhibit Wnt signaling and bone formation. Additionally, FoxOs decrease bone resorption via direct antioxidant effects on osteoclasts and upregulation of the antiosteoclastogenic cytokine OPG in cells of the osteoblast lineage. Deacetylation of FoxOs by the NAD-dependent histone deacetylase Sirt1 in both osteoblasts and osteoclasts stimulates bone formation and inhibits bone resorption, making Sirt1 activators promising therapeutic agents for diseases of low bone mass. In this chapter, we review these advances and discuss their implications for the pathogenesis and treatment of estrogen deficiency-, Type 1 diabetes-, and age-related osteoporosis.

Current Topics in Developmental Biology, Volume 127
ISSN 0070-2153
https://doi.org/10.1016/bs.ctdb.2017.10.004

1. BONE GROWTH, REMODELING, AND DISEASE

The skeleton is a dynamic tissue that serves many functions. It protects vital internal organs, acts as a framework for muscular attachment and locomotion, houses the bone marrow and influences hematopoiesis, and is the main repository for calcium and phosphate. Bone is constantly degraded and replaced by teams of osteoclasts and osteoblasts operating in synchrony to renew the skeleton throughout life. In the adult skeleton bone mass is maintained because the bone resorbed by the osteoclasts is fully replaced by new bone formed by the osteoblasts. Aging and several pathologies cause an imbalance between bone resorption and formation, leading to loss of bone mass and strength and increased risk of osteoporotic fractures (Manolagas, 2000).

Osteoclasts are short-lived giant multinucleated cells, derived from the fusion of myeloid lineage precursors (Teitelbaum & Ross, 2003). Macrophage colony-stimulating factor (M-CSF) and receptor activator of nuclear factor κ-B ligand (RANKL) and their respective receptors on osteoclast progenitors are essential for osteoclast differentiation. Osteoclasts break down the mineralized matrix of bone via polarized secretion of proteolytic enzymes and hydrogen ions into a sealed microenvironment formed by a "podosome belt" that tightly adheres to the bone area targeted for removal. Understanding what drives osteoclast formation and resorption activity is particularly important for diseases characterized by outpaced osteoclast-mediated bone resorption relative to bone deposition by osteoblasts, including postmenopausal osteoporosis.

Osteoblasts are terminally differentiated cells of the mesenchymal lineage responsible for the deposition of bone matrix into the spaces excavated by osteoclasts (Long, 2012). Because osteoblasts have a short life span, they need to be constantly replenished with new ones originating from osteoprogenitors that express the osteoblast determining transcriptions factors Runx2 and Osterix1. Wnt signaling is critical for osteoblast development, bone formation, and the overall balance of bone remodeling (Baron & Kneissel, 2013). Extracellular Wnt proteins bind to receptor frizzled and coreceptor lipoprotein-like receptor protein (LRP)5/6 present on the cell surface of mesenchymal progenitors. This binding triggers an intracellular cascade of events whereby the coactivator β-catenin is released from a proteasomal degradation complex, translocates into the nucleus, and binds of the T-cell factor/lymphoid enhancer factor (TCF/LEF) transcription

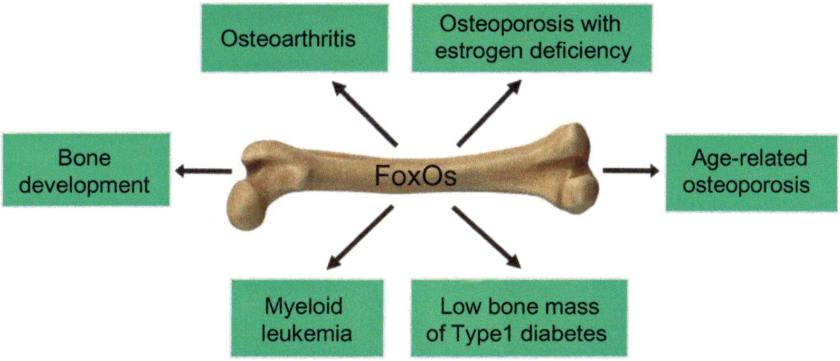

Fig. 1 Actions of FoxOs in bone cells alter bone mass and contribute to several skeletal diseases.

factors to activate or suppress the expression of Wnt target genes (Clevers & Nusse, 2012). The significance of Wnt signaling for bone formation is highlighted by evidence that humans with loss- or gain-of-function mutations in LRP5 have low or high bone mass, respectively (Lara-Castillo & Johnson, 2015). Genetic defects in the synthesis of sclerostin—a secreted Wnt signaling inhibitor—also cause high bone mass dysplasias (van Lierop, Appelman-Dijkstra, & Papapoulos, 2017). At the end of their secretory life most osteoblast die by apoptosis (Manolagas, 2000). A subset of the remaining osteoblasts is embedded in the bone matrix and becomes long-lived, dendritic osteocytes. Osteocytes control both bone resorption and formation by producing RANKL and sclerostin, respectively (Jilka & O'Brien, 2016). As in several other mammalian cell types, FoxO1, 3, and 4 expression overlaps in bone cells such as chondrocytes, osteoclasts, and osteoblasts. Actions of FoxOs in these cells are critical for bone development, and the control of bone mass under physiological and pathological conditions (Fig. 1).

2. FoxO EFFECTS IN CHONDROCYTES

Longitudinal bone growth occurs at the growth plate through endochondral ossification, a process in which cartilage is formed and then replaced by bone (Kronenberg, 2003). During endochondral ossification growth plate chondrocytes switch from a resting state to active proliferation and to hypertrophy. Hypertrophic chondrocytes secrete Type 10 collagen and eventually undergo apoptosis leaving behind a cartilaginous matrix scaffold which is resorbed by the osteoclasts and replaced with collagen 1-rich bone matrix by the osteoblasts. Endochondral ossification is tightly controlled by hormones and growth factors including sex steroids, growth hormone, thyroid hormones, and insulin growth factor I (van der Eerden, Karperien, & Wit, 2003). Long bones continue to grow in length throughout childhood until the cartilaginous epiphyseal growth plate is fully replaced by mineralized bone. This growth plate closure occurs after puberty due to increased serum levels of estrogen in both males and females (Almeida et al., 2017; Parfitt, 2002).

Mice with chondrocyte-specific loss of function of FoxO1, 3, and 4 have increased overall body and tail length and suffer from skeletal deformities (Eelen et al., 2016). These features are associated with a distinct elongation of the hypertrophic zone of the growth plate. The PI3K/Akt signaling pathway promotes chondrocyte maturation, proliferation, and cartilage matrix production (Ford-Hutchinson et al., 2007). Organ culture experiments have implicated FoxOs in the actions of Akt in chondrocytes (Rokutanda et al., 2009). Consistent with these observations, stimulation of PI3K/Akt signaling by deletion of the tumor suppressor gene phosphatase and tensin homolog deleted from chromosome 10 (PTEN) in chondrocytes causes bone defects which are similar to the ones seen with deletion of FoxOs, including increased hypertrophic zone length, increased total body length, and hyperkyphosis (Ford-Hutchinson et al., 2007). Together, this work indicates that signaling by the Akt/FoxO axis is crucial for normal endochondral ossification and bone development.

2.1 Osteoarthritis

Osteoarthritis (OA) is the most common musculoskeletal disease and a leading cause of disability. OA is characterized by degeneration of articular chondrocytes and synovial tissues and loss of joint cartilage with age. The phosphorylation of FoxOs is increased in articular chondrocytes and synovial tissues from aged humans and OA patients (Akasaki, Hasegawa, et al., 2014; Ludikhuize et al., 2007), suggesting that changes in FoxO activation are involved in the pathogenesis of OA. FoxOs stimulate the expression of antioxidant genes in both murine and human chondrocytes and promote resistance against oxidative stress-induced cell death (Akasaki, Alvarez-Garcia, et al., 2014; Eelen et al., 2016). Autophagy is a critical energy recycling mechanism in which cellular components, including protein aggregates and organelles, are targeted to the lysosomes for degradation (Cuervo, 2008). Suppression of autophagy contributes to cellular and organismal aging, while activation of autophagy may protect against certain pathologies of aging including OA (Lotz & Carames, 2012). FoxO3 promotes autophagic activity and cell viability in chondrocytes by decreasing ROS (Shen, Cai, Peng, & Chen, 2015). Moreover, inhibition of Akt and activation of FoxOs are associated with the stimulatory actions of glucosamine—a cartilage protective compound—on autophagy of articular chondrocytes in mice, further supporting a protective role of FoxOs in articular cartilage (Carames et al., 2013).

3. FoxO EFFECTS IN OSTEOCLASTS

Work performed during the last two decades has elucidated that RANKL increases ROS in osteoclast progenitor cells and ROS potentiates osteoclast differentiation (Garrett et al., 1990; Ha et al., 2004; Lee et al., 2005). However, until recently the mechanism responsible for the increase in ROS and the functional relevance of these mechanisms to bone homeostasis have remained unclear. Studies using mouse models of loss- and gain-of-function of FoxOs or attenuation of mitochondrial ROS in osteoclasts have elucidated that FoxOs are important regulators of osteoclast differentiation and bone resorption by decreasing ROS (Bartell et al., 2014). Indeed, combined loss of FoxO1, 3, and 4 in osteoclast progenitors increases proliferation, osteoclast formation, and bone resorption leading to reduced trabecular and cortical bone mass. Conversely, gain-of-function of FoxO3 inhibits osteoclast differentiation and bone resorption, and increases bone mass. These effects are due to increased expression of antioxidant enzymes like catalase to inhibit H_2O_2 production (Bartell et al., 2014). RANKL decreases the levels and activity of FoxO1, 3, and 4 via Akt-mediated phosphorylation and proteosomal degradation, thereby allowing a buildup of ROS (Bartell et al., 2014; Liu, Wang, Wei, Sun, & Feng, 2005; Tan et al., 2015). FoxOs also stimulate the expression of hemeoxygenase-1 (HO-1) in osteoclast progenitors (Bartell et al., 2014). HO-1 catabolizes heme and attenuates mitochondrial oxidative phosphorylation and ATP production in macrophages (Jais et al., 2014). These findings, along with the evidence that HO-1 decreases the number of osteoclasts and bone resorption in mice (Ke et al., 2015), support the contention that this enzyme contributes to the antiosteoclastogenic effects of FoxOs.

3.1 Bone Loss Caused by Estrogen Deficiency

Loss of estrogens at menopause increases osteoclast number and bone resorption leading to loss of bone mass and strength and increased risk of osteoporotic fractures (Almeida et al., 2017). Mice do not experience menopause, but following ovariectomy are a faithful model of the adverse effects of estrogen deficiency on both trabecular and cortical bone (Manolagas, O'Brien, & Almeida, 2013). Loss of bone mass caused by sex steroid deficiency in mice is associated with an increase in ROS levels in bone cells (Almeida, Han, Martin-Millan, Plotkin, et al., 2007; Lean et al., 2003). Mice with restrained H_2O_2 generation in the mitochondria of osteoclasts are protected from the effects of estrogen deficiency, indicating that ROS is required for the increase in bone resorption and loss of bone mass (Ucer et al., 2017). FoxOs in osteoclast precursors are targets of Sirt1 (Fig. 2). Sirt1-mediated deacetylation stimulates the expression of the FoxO-target genes catalase and HO-1 (Kim et al., 2015). Like FoxOs, Sirt1 in osteoclast attenuates bone resorption (Edwards et al., 2013; Hah et al., 2014). Sirt1 activators, such as the natural polyphenol resveratrol or synthetic Sirt1-activating compounds (STAC), inhibit osteoclastogenesis in a FoxO-dependent manner (Kim et al., 2015). Importantly, administration of resveratrol to rats prevents the loss of bone mass caused by estrogen deficiency (Su, Yang, Zhao, Kuo, & Yen, 2007). Likewise, the STAC SRT3025 ameliorates the skeletal effects of estrogen deficiency in mice (Artsi et al., 2014). Together, these results indicate that the actions of FoxOs on the skeleton can be harnessed with the use of Sirt1 activators. These compounds, therefore, may be a rationale pharmacotherapeutic approach to the treatment of diseases associated with increased bone resorption.

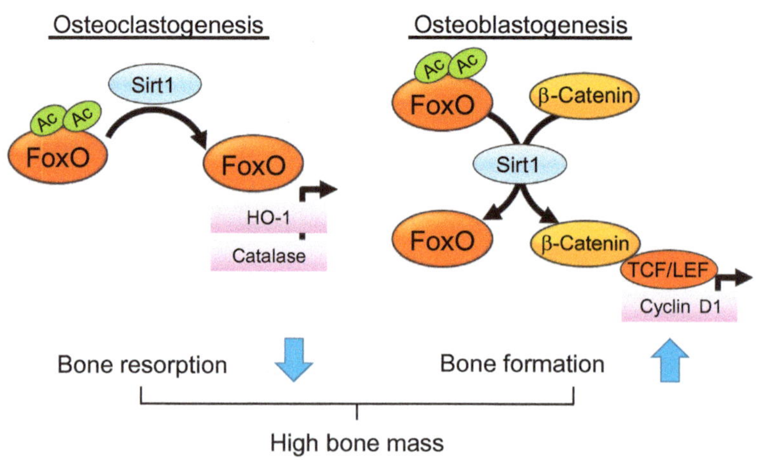

Fig. 2 Deacetylation of FoxOs by Sirt1 reduces osteoclast formation and increases osteoblast formation via direct actions in osteoclast and osteoblast progenitor cells.

4. FoxO EFFECTS IN OSTEOBLASTS

Mice with loss- or gain- of-function of individual or combined floxed alleles for FoxOs 1, 3, and 4 have revealed that FoxOs act directly in distinct cell types within the osteoblast lineage. In osteoblast progenitors, FoxOs reduce osteoblastogenesis by restraining Wnt signaling (Iyer et al., 2013). Specifically, FoxOs bind to and divert β-catenin away from TCF/LEF-mediated transcription and, thereby, attenuate the levels of cyclin D1 and proliferation (Fig. 2). Via this mechanism, FoxOs in osteoblast progenitors decrease bone mass. In difference to the suppressive effect in osteoprogenitors, FoxOs promote survival of osteoblasts by increasing the expression of antioxidant enzymes like catalase and superoxide dismutase and preventing oxidative stress (Ambrogini et al., 2010; Rached et al., 2010). In addition, FoxO1 promotes the accumulation of glutathione, a peptide with redox-active sulfhydryl moieties which reduces ROS. The increase in glutathione is due to stimulation of protein synthesis caused by FoxO1 interaction with ATF4, a transcription factor that promotes amino acid import (Rached et al., 2010). Besides the direct antiosteoclastogenic actions, described earlier, FoxOs in osteoblasts inhibit bone resorption via paracrine mechanisms (Ambrogini et al., 2010; Rached et al., 2010). Most likely, these antiosteoclastogenic actions of FoxOs are due to stimulation of osteoprotegerin (OPG), the decoy receptor for RANKL (Ferron et al., 2010; Iyer et al., 2013).

4.1 Low Bone Mass in Type 1 Diabetes

Type 1 diabetes is a complex metabolic disease caused by reduced insulin secretion in the pancreatic beta cells. Children and adolescents with Type 1 diabetes have low bone mass (Hui, Epstein, & Johnston, 1985). An increased risk of fragility fractures and delayed fracture healing are among the multiple comorbidities in patients with Type 1 diabetes (Compston, Smith, Matthews, & Schofield, 1994; Hamann, Kirschner, Gunther, & Hofbauer, 2012; Khan & Fraser, 2015; Nicodemus & Folsom, 2001; Vestergaard, 2007; Vestergaard, Rejnmark, & Mosekilde, 2005; Weber, Haynes, Leonard, Willi, & Denburg, 2015). Rodent models of genetic or pharmacologic Type 1 diabetes, similar to patients, exhibit low bone mass and reduced expression of osteoblast markers associated with impaired bone formation (Botolin et al., 2005; Bouillon et al., 1995; McCabe, 2007; Thrailkill et al., 2005). In several organs, for example, liver and muscle,

FoxOs are major mediators of the actions of insulin. Insulin stimulates Akt-mediated FoxO phosphorylation and nuclear exclusion. Lack of insulin, on the other hand, promotes FoxO-mediated transcription. Mouse models with deletion of insulin receptor in osteoblasts and their progenitors have revealed that insulin promotes osteoblast formation and increases bone mass (Fulzele et al., 2010; Thrailkill et al., 2014). Additionally, insulin promotes bone resorption by inhibiting FoxO-mediated transcription and thereby the expression of the antiosteoclastogenic cytokine OPG (Ferron et al., 2010). Mice lacking FoxO1, 3, and 4 in osteoblasts are protected, at least in part, from the skeletal effects of diabetes caused by streptozotocin—a drug that eliminates pancreatic beta cells (Iyer et al., 2017). Although FoxO deletion did not alter the effects of diabetes on bone formation, it did prevent the decrease in OPG caused by hyperglycemia and, thereby, the increase in bone resorption. This work suggests that inactivation of FoxOs in osteoblast progenitors might ameliorate the adverse effects of Type 1 diabetes on the skeleton.

4.2 Myeloid Leukemia

Hematopoietic stem cells (HSCs) reside in specialized niches within the bone marrow in close association with the cells of mesenchymal origin. Multiple genetic studies along with advances in imaging techniques have elucidated that distinct subpopulations of mesenchymal lineage cells can affect the maintenance, homing, and self-renewal capacity of the HSC via paracrine effects (Kfoury & Scadden, 2015). Furthermore, osteolineage cells can induce myeloid malignancies (Kode et al., 2014; Raaijmakers et al., 2010). Constitutive activation of β-catenin in osteoblasts increases expression of the Notch ligand, Jagged-1, and stimulation of Notch signaling in HSCs. This causes the expansion of myeloid lineage cells concomitantly with a reduction in erythroid- and B-cells, in a manner that is reminiscent of acute myeloid leukemia in humans (Kode et al., 2014). FoxO1 interacts with β-catenin in osteoblasts to induce expression of Jagged-1 and, thereby, mediates the leukemogenic properties of β-catenin (Kode et al., 2016). It should be noted, however, that FoxO1 insufficiency in osteoblasts per se does not alter the hematopoietic lineage in mice, suggesting that the effects of FoxO1 on HSCs are restricted to certain pathologic conditions.

4.3 Age-Related Bone Loss

Old age is the most important risk factor for osteoporotic fractures. The adverse effect of aging bone homeostasis results, to a large extent, from the loss of bone mass and deterioration of the bone microarchitecture (Roux et al., 2012; Zebaze et al., 2010). These changes occur in both sexes, but are greater in women due to the loss of estrogen at menopause (Nirody et al., 2015; Shanbhogue, Brixen, & Hansen, 2016). Like humans, both female and male mice lose bone mass with old age. Loss of bone mass with aging is due to a decrease in the number of osteoblasts. This decrease is caused, at least in part, by an increase in mitochondrial ROS in cells of the osteoblast lineage (Ucer et al., 2017). ROS-mediated activation of FoxOs leads to a reduction in the β-catenin required for Wnt signaling (Almeida, Han, Martin-Millan, O'Brien, & Manolagas, 2007; Hoogeboom et al., 2008). Lipid oxidation, inflammatory cytokines, and glucocorticoids increase with old age and cause a ROS-mediated increase in FoxO activity in osteoblastic cells (Almeida, Ambrogini, Han, Manolagas, & Jilka, 2009; Almeida, Han, Ambrogini, Weinstein, & Manolagas, 2011; Weinstein et al., 2010). Mice lacking FoxOs in osteoprogenitors exhibit high bone mass throughout life, indicating that stimulation of FoxOs contributes to the decreased bone formation that occurs with aging (Iyer et al., 2013). In contrast, Sirt1 actions in osteoblast progenitors stimulate osteoblast formation and increase bone mass (Iyer et al., 2014; Simic et al., 2013). Deacetylation of FoxOs by Sirt1 prevents FoxO association with β-catenin and potentiates Wnt signaling, leading to increased osteoblast proliferation (Iyer et al., 2014) (Fig. 2). These findings suggest that the actions of Sirt1 on bone formation are mediated by FoxOs. Sirt1 also deacetylates β-catenin, thereby increasing its nuclear localization and osteoblast differentiation (Simic et al., 2013). Importantly, long-term administration of SRT2104 to mice attenuates the loss of bone mass with age (Merken et al., 2014).

A decline in the number of osteoprogenitors due to cell senescence might contribute to the age-related loss of bone mass (Kim et al., 2017). Osteoprogenitor senescence is associated with activation of p53 and elevated expression of the cell cycle inhibitor p21. Senescence is a cellular process in which damaged cells are withdrawn from the cell cycle, avoid apoptosis, and develop an altered secretory activity, known as the senescence-associated secretory phenotype (Campisi & d'Adda di, 2007). Accumulation of senescent cells causes age-related diseases and curtails longevity (Baker et al., 2016; Chang et al., 2016). FoxO4 is critical for senescent cell viability. This effect of FoxO4 results from its binding to active p53, thereby precluding p53-mediated apoptosis and promoting p21 expression and cell cycle arrest (Baar et al., 2017). This mechanism contributes to the loss of hair, renal function, and activity in mouse models of accelerated aging and in naturally aged wild-type mice. Interestingly, signaling via a FoxO3/β-catenin complex in response to glucose-induced oxidative stress promotes p21 expression, strengthening the connection between FoxOs and senescence pathways (McClelland Descalzo et al., 2016). Future studies should elucidate whether FoxOs are involved in the senescence of osteoprogenitors.

5. FUTURE CONSIDERATIONS

Sirt1-mediated posttranslational modifications of FoxOs inhibit bone resorption and stimulate bone formation, making this signaling axis an ideal therapeutic target to counteract the loss of bone. Sirt1/FoxOs are well-established contributors to longevity in nematodes, flies, and mammals and represent a critical node for several degenerative diseases of aging. This evidence further highlights the significance of targeting this pathway (Longo & Kennedy, 2006; Oellerich & Potente, 2012). Recent studies by Baar et al. have shown that specific interference with protein–protein interaction between FoxO4 and p53 using cell-penetrating peptides can ameliorate several age-associated pathologies in a mouse model of accelerated aging as well as in naturally aged mice (Baar et al., 2017). Cell-penetrating peptides can mimic surface-exposed protein-binding motifs, bind to these motifs in target proteins, and thereby compete for one of the endogenous-interacting partners. These peptides are normally fused to a moiety capable of translocating cargoes of different molecular sizes across the cell membrane (Herce & Garcia, 2007; Jones & Sayers, 2012). This work raises the exciting possibility that the interaction of FoxOs with β-catenin is a potential therapeutic target. A major advantage of the use of cell-penetrating peptides is that only part of the entire range of FoxO functions is inhibited. In other words, they can prevent undesirable actions of a particular protein, while leaving the remainder actions unaffected. For example, inhibition of all effects of FoxOs in the osteoblast lineage is not desirable in view of their antistress effects in osteoblasts. Future work to elucidate the cellular pathways that are regulated by FoxOs will improve our understanding on the role of these proteins in bone formation and resorption. The role of FoxOs as therapeutic targets for skeletal diseases may also become clearer.

ACKNOWLEDGMENTS

The research of the authors is supported by the National Institutes of Health (R01 AR56679) and the University of Arkansas for Medical Sciences (UAMS) Translational Research Institute. We thank Stavros Manolagas for critically reviewing the manuscript.

REFERENCES

Akasaki, Y., Alvarez-Garcia, O., Saito, M., Carames, B., Iwamoto, Y., & Lotz, M. K. (2014). FoxO transcription factors support oxidative stress resistance in human chondrocytes. *Arthritis & Rhematology, 66*, 3349–3358.

Akasaki, Y., Hasegawa, A., Saito, M., Asahara, H., Iwamoto, Y., & Lotz, M. K. (2014). Dysregulated FOXO transcription factors in articular cartilage in aging and osteoarthritis. *Osteoarthritis and Cartilage, 22*, 162–170.

Almeida, M., Ambrogini, E., Han, L., Manolagas, S. C., & Jilka, R. L. (2009). Increased lipid oxidation causes oxidative stress, increased PPAR-gamma expression and diminished pro-osteogenic Wnt signaling in the skeleton. *The Journal of Biological Chemistry, 284*, 27438–27448.

Almeida, M., Han, L., Ambrogini, E., Weinstein, R. S., & Manolagas, S. C. (2011). Glucocorticoids and tumor necrosis factor (TNF) alpha increase oxidative stress and suppress WNT signaling in osteoblasts. *The Journal of Biological Chemistry, 286*, 44326–44335.

Almeida, M., Han, L., Martin-Millan, M., O'Brien, C. A., & Manolagas, S. C. (2007). Oxidative stress antagonizes Wnt signaling in osteoblast precursors by diverting beta-catenin from T cell factor- to forkhead box O-mediated transcription. *The Journal of Biological Chemistry, 282*, 27298–27305.

Almeida, M., Han, L., Martin-Millan, M., Plotkin, L. I., Stewart, S. A., Roberson, P. K., et al. (2007). Skeletal involution by age-associated oxidative stress and its acceleration by loss of sex steroids. *The Journal of Biological Chemistry, 282*, 27285–27297.

Almeida, M., Laurent, M. R., Dubois, V., Claessens, F., O'Brien, C. A., Bouillon, R., et al. (2017). Estrogens and androgens in skeletal physiology and pathophysiology. *Physiological Reviews, 97*, 135–187.

Ambrogini, E., Almeida, M., Martin-Millan, M., Paik, J., dePinho, R., Han, L., et al. (2010). FoxO-mediated defense against oxidative stress in osteoblasts is indispensable for skeletal homeostasis in mice. *Cell Metabolism, 11*, 136–146.

Artsi, H., Cohen-Kfir, E., Gurt, I., Shahar, R., Bajayo, A., Kalish, N., et al. (2014). The Sirtuin1 activator SRT3025 down-regulates sclerostin and rescues ovariectomy-induced bone loss and biomechanical deterioration in female mice. *Endocrinology, 155*, 3508–3515.

Baar, M. P., Brandt, R. M., Putavet, D. A., Klein, J. D., Derks, K. W., Bourgeois, B. R., et al. (2017). Targeted apoptosis of senescent cells restores tissue homeostasis in response to chemotoxicity and aging. *Cell, 169*, 132–147.e16.

Baker, D. J., Childs, B. G., Durik, M., Wijers, M. E., Sieben, C. J., Zhong, J., et al. (2016). Naturally occurring p16(Ink4a)-positive cells shorten healthy lifespan. *Nature, 530*, 184–189.

Baron, R., & Kneissel, M. (2013). WNT signaling in bone homeostasis and disease: From human mutations to treatments. *Nature Medicine, 19*, 179–192.

Bartell, S. M., Kim, H. N., Ambrogini, E., Han, L., Iyer, S., Serra, U. S., et al. (2014). FoxO proteins restrain osteoclastogenesis and bone resorption by attenuating H2O2 accumulation. *Nature Communications, 5*, 3773.

Botolin, S., Faugere, M. C., Malluche, H., Orth, M., Meyer, R., & McCabe, L. R. (2005). Increased bone adiposity and peroxisomal proliferator-activated receptor-gamma2 expression in type I diabetic mice. *Endocrinology, 146*, 3622–3631.

Bouillon, R., Bex, M., Van Herck, E., Laureys, J., Dooms, L., Lesaffre, E., et al. (1995). Influence of age, sex, and insulin on osteoblast function: Osteoblast dysfunction in diabetes mellitus. *The Journal of Clinical Endocrinology and Metabolism, 80*, 1194–1202.

Campisi, J., & d'Adda di, F. F. (2007). Cellular senescence: When bad things happen to good cells. *Nature Reviews. Molecular Cell Biology, 8*, 729–740.

Carames, B., Kiosses, W. B., Akasaki, Y., Brinson, D. C., Eap, W., Koziol, J., et al. (2013). Glucosamine activates autophagy in vitro and in vivo. *Arthritis and Rheumatism*, *65*, 1843–1852.

Chang, J., Wang, Y., Shao, L., Laberge, R. M., Demaria, M., Campisi, J., et al. (2016). Clearance of senescent cells by ABT263 rejuvenates aged hematopoietic stem cells in mice. *Nature Medicine*, *22*, 78–83.

Clevers, H., & Nusse, R. (2012). Wnt/beta-catenin signaling and disease. *Cell*, *149*, 1192–1205.

Compston, J. E., Smith, E. M., Matthews, C., & Schofield, P. (1994). Whole body composition and regional bone mass in women with insulin-dependent diabetes mellitus. *Clinical Endocrinology*, *41*, 289–293.

Cuervo, A. M. (2008). Autophagy and aging: Keeping that old broom working. *Trends in Genetics*, *24*, 604–612.

Edwards, J. R., Perrien, D. S., Fleming, N., Nyman, J. S., Ono, K., Connelly, L., et al. (2013). Silent information regulator (Sir)T1 inhibits NF-kappaB signaling to maintain normal skeletal remodeling. *Journal of Bone and Mineral Research*, *28*, 960–969.

Eelen, G., Verlinden, L., Maes, C., Beullens, I., Gysemans, C., Paik, J. H., et al. (2016). Forkhead box O transcription factors in chondrocytes regulate endochondral bone formation. *The Journal of Steroid Biochemistry and Molecular Biology*, *164*, 337–343.

Ferron, M., Wei, J., Yoshizawa, T., Del Fattore, A., DePinho, R. A., Teti, A., et al. (2010). Insulin signaling in osteoblasts integrates bone remodeling and energy metabolism. *Cell*, *142*, 296–308.

Ford-Hutchinson, A. F., Ali, Z., Lines, S. E., Hallgrimsson, B., Boyd, S. K., & Jirik, F. R. (2007). Inactivation of Pten in osteo-chondroprogenitor cells leads to epiphyseal growth plate abnormalities and skeletal overgrowth. *Journal of Bone and Mineral Research*, *22*, 1245–1259.

Fulzele, K., Riddle, R. C., DiGirolamo, D. J., Cao, X., Wan, C., Chen, D., et al. (2010). Insulin receptor signaling in osteoblasts regulates postnatal bone acquisition and body composition. *Cell*, *142*, 309–319.

Garrett, I. R., Boyce, B. F., Oreffo, R. O., Bonewald, L., Poser, J., & Mundy, G. R. (1990). Oxygen-derived free radicals stimulate osteoclastic bone resorption in rodent bone in vitro and in vivo. *The Journal of Clinical Investigation*, *85*, 632–639.

Ha, H., Kwak, H. B., Lee, S. W., Jin, H. M., Kim, H. M., Kim, H. H., et al. (2004). Reactive oxygen species mediate RANK signaling in osteoclasts. *Experimental Cell Research*, *301*, 119–127.

Hah, Y. S., Cheon, Y. H., Lim, H. S., Cho, H. Y., Park, B. H., Ka, S. O., et al. (2014). Myeloid deletion of SIRT1 aggravates serum transfer arthritis in mice via nuclear factor-kappaB activation. *PLoS One*, *9*, e87733.

Hamann, C., Kirschner, S., Gunther, K. P., & Hofbauer, L. C. (2012). Bone, sweet bone—Osteoporotic fractures in diabetes mellitus. *Nature Reviews. Endocrinology*, *8*, 297–305.

Herce, H. D., & Garcia, A. E. (2007). Cell penetrating peptides: How do they do it? *Journal of Biological Physics*, *33*, 345–356.

Hoogeboom, D., Essers, M. A., Polderman, P. E., Voets, E., Smits, L. M., & Burgering, B. M. (2008). Interaction of FOXO with beta-catenin inhibits beta-catenin/T cell factor activity. *The Journal of Biological Chemistry*, *283*, 9224–9230.

Hui, S. L., Epstein, S., & Johnston, C. C., Jr. (1985). A prospective study of bone mass in patients with type I diabetes. *The Journal of Clinical Endocrinology and Metabolism*, *60*, 74–80.

Iyer, S., Ambrogini, E., Bartell, S. M., Han, L., Roberson, P. K., Cabo, R., et al. (2013). FoxOs attenuate bone formation by suppressing Wnt signaling. *The Journal of Clinical Investigation*, *123*, 3404–3419.

Iyer, S., Han, L., Ambrogini, E., Yavropoulou, M., Fowlkes, J., & Manolagas, S. C. (2017). Deletion of FoxO1, 3 and 4 in osteoblast progenitors attenuates the loss of bone mass in a model of type 1 diabetes. *Journal of Bone and Mineral Research*, *32*, 60–69.

Iyer, S., Han, L., Bartel, S. M., Kim, H. N., Gubrij, I., de Cabo, R., et al. (2014). Sirtuin1 (Sirt1) promotes cortical bone formation by preventing beta (β)-catenin sequestration by FoxO transcription factors in osteoblast progenitors. *The Journal of Biological Chemistry*, *289*, 24069–24078.

Jais, A., Einwallner, E., Sharif, O., Gossens, K., Lu, T. T., Soyal, S. M., et al. (2014). Heme oxygenase-1 drives metaflammation and insulin resistance in mouse and man. *Cell*, *158*, 25–40.

Jilka, R. L., & O'Brien, C. A. (2016). The role of osteocytes in age-related bone loss. *Current Osteoporosis Reports*, *14*, 16–25.

Jones, A. T., & Sayers, E. J. (2012). Cell entry of cell penetrating peptides: Tales of tails wagging dogs. *Journal of Controlled Release*, *161*, 582–591.

Ke, K., Safder, M. A., Sul, O. J., Kim, W. K., Suh, J. H., Joe, Y., et al. (2015). Hemeoxygenase-1 maintains bone mass via attenuating a redox imbalance in osteoclast. *Molecular and Cellular Endocrinology*, *409*, 11–20.

Kfoury, Y., & Scadden, D. T. (2015). Mesenchymal cell contributions to the stem cell niche. *Cell Stem Cell*, *16*, 239–253.

Khan, T. S., & Fraser, L. A. (2015). Type 1 diabetes and osteoporosis: From molecular pathways to bone phenotype. *Journal of Osteoporosis*, *2015*, 174186.

Kim, H. N., Chang, J., Shao, L., Han, L., Iyer, S., Manolagas, S. C., et al. (2017). DNA damage and senescence in osteoprogenitors expressing Osx1 may cause their decrease with age. *Aging Cell*, *16*, 693–703.

Kim, H. N., Han, L., Iyer, S., de, C. R., Zhao, H., O'Brien, C. A., et al. (2015). Sirtuin1 suppresses osteoclastogenesis by deacetylating FoxOs. *Molecular Endocrinology*, *29*, 1498–1509.

Kode, A., Manavalan, J. S., Mosialou, I., Bhagat, G., Rathinam, C. V., Luo, N., et al. (2014). Leukaemogenesis induced by an activating beta-catenin mutation in osteoblasts. *Nature*, *506*, 240–244.

Kode, A., Mosialou, I., Manavalan, S. J., Rathinam, C. V., Friedman, R. A., Teruya-Feldstein, J., et al. (2016). FoxO1-dependent induction of acute myeloid leukemia by osteoblasts in mice. *Leukemia*, *30*, 1–13.

Kronenberg, H. M. (2003). Developmental regulation of the growth plate. *Nature*, *423*, 332–336.

Lara-Castillo, N., & Johnson, M. L. (2015). LRP receptor family member associated bone disease. *Reviews in Endocrine & Metabolic Disorders*, *16*, 141–148.

Lean, J. M., Davies, J. T., Fuller, K., Jagger, C. J., Kirstein, B., Partington, G. A., et al. (2003). A crucial role for thiol antioxidants in estrogen-deficiency bone loss. *The Journal of Clinical Investigation*, *112*, 915–923.

Lee, N. K., Choi, Y. G., Baik, J. Y., Han, S. Y., Jeong, D. W., Bae, Y. S., et al. (2005). A crucial role for reactive oxygen species in RANKL-induced osteoclast differentiation. *Blood*, *106*, 852–859.

Liu, W., Wang, S., Wei, S., Sun, L., & Feng, X. (2005). Receptor activator of NF-kappaB (RANK) cytoplasmic motif, 369PFQEP373, plays a predominant role in osteoclast survival in part by activating Akt/PKB and its downstream effector AFX/FOXO4. *The Journal of Biological Chemistry*, *280*, 43064–43072.

Long, F. (2012). Building strong bones: Molecular regulation of the osteoblast lineage. *Nature Reviews. Molecular Cell Biology*, *13*, 27–38.

Longo, V. D., & Kennedy, B. K. (2006). Sirtuins in aging and age-related disease. *Cell*, *126*, 257–268.

Lotz, M., & Carames, B. (2012). Autophagy: A new therapeutic target in cartilage injury and osteoarthritis. *The Journal of the American Academy of Orthopaedic Surgeons*, *20*, 261–262.

Ludikhuize, J., de Launay, D., Groot, D., Smeets, T. J., Vinkenoog, M., Sanders, M. E., et al. (2007). Inhibition of forkhead box class O family member transcription factors in rheumatoid synovial tissue. *Arthritis and Rheumatism*, *56*, 2180–2191.

Manolagas, S. C. (2000). Birth and death of bone cells: Basic regulatory mechanisms and implications for the pathogenesis and treatment of osteoporosis. *Endocrine Reviews*, *21*, 115–137.

Manolagas, S. C., O'Brien, C. A., & Almeida, M. (2013). The role of estrogen and androgen receptors in bone health and disease. *Nature Reviews. Endocrinology*, *9*, 699–712.

McCabe, L. R. (2007). Understanding the pathology and mechanisms of type I diabetic bone loss. *Journal of Cellular Biochemistry*, *102*, 1343–1357.

McClelland Descalzo, D. L., Satoorian, T. S., Walker, L. M., Sparks, N. R., Pulyanina, P. Y., & Zur Nieden, N. I. (2016). Glucose-induced oxidative stress reduces proliferation in embryonic stem cells via FOXO3A/beta-catenin-dependent transcription of p21 (cip1). *Stem Cell Reports*, 7, 55–68.

Merken, E. M., Mitchell, S. J., Martin-Montalvo, A., Minor, R. K., Almeida, M., Gomes, A. P., et al. (2014). SRT2104 extends survival of male mice on a standard diet and preserves bone and muscle mass. *Aging Cell*, *13*, 787–796.

Nicodemus, K. K., & Folsom, A. R. (2001). Type 1 and type 2 diabetes and incident hip fractures in postmenopausal women. *Diabetes Care*, *24*, 1192–1197.

Nirody, J. A., Cheng, K. P., Parrish, R. M., Burghardt, A. J., Majumdar, S., Link, T. M., et al. (2015). Spatial distribution of intracortical porosity varies across age and sex. *Bone*, *75*, 88–95.

Oellerich, M. F., & Potente, M. (2012). FOXOs and sirtuins in vascular growth, maintenance, and aging. *Circulation Research*, *110*, 1238–1251.

Parfitt, A. M. (2002). Misconceptions (1): Epiphyseal fusion causes cessation of growth. *Bone*, *30*, 337–339.

Raaijmakers, M. H., Mukherjee, S., Guo, S., Zhang, S., Kobayashi, T., Schoonmaker, J. A., et al. (2010). Bone progenitor dysfunction induces myelodysplasia and secondary leukaemia. *Nature*, *464*, 852–857.

Rached, M. T., Kode, A., Xu, L., Yoshikawa, Y., Paik, J. H., DePinho, R. A., et al. (2010). FoxO1 is a positive regulator of bone formation by favoring protein synthesis and resistance to oxidative stress in osteoblasts. *Cell Metabolism*, *11*, 147–160.

Rokutanda, S., Fujita, T., Kanatani, N., Yoshida, C. A., Komori, H., Liu, W., et al. (2009). Akt regulates skeletal development through GSK3, mTOR, and FoxOs. *Developmental Biology*, *328*, 78–93.

Roux, C., Wyman, A., Hooven, F. H., Gehlbach, S. H., Adachi, J. D., Chapurlat, R. D., et al. (2012). Burden of non-hip, non-vertebral fractures on quality of life in postmenopausal women: The global longitudinal study of osteoporosis in women (GLOW). *Osteoporosis International*, *23*, 2863–2871.

Shanbhogue, V. V., Brixen, K., & Hansen, S. (2016). Age- and sex-related changes in bone microarchitecture and estimated strength: A three-year prospective study using HRpQCT. *Journal of Bone and Mineral Research*, *31*, 1541–1549.

Shen, C., Cai, G. Q., Peng, J. P., & Chen, X. D. (2015). Autophagy protects chondrocytes from glucocorticoids-induced apoptosis via ROS/Akt/FOXO3 signaling. *Osteoarthritis and Cartilage*, *23*, 2279–2287.

Simic, P., Zainabadi, K., Bell, E., Sykes, D. B., Saez, B., Lotinun, S., et al. (2013). SIRT1 regulates differentiation of mesenchymal stem cells by deacetylating beta-catenin. *EMBO Molecular Medicine*, *5*, 430–440.

Su, J. L., Yang, C. Y., Zhao, M., Kuo, M. L., & Yen, M. L. (2007). Forkhead proteins are critical for bone morphogenetic protein-2 regulation and anti-tumor activity of resveratrol. *The Journal of Biological Chemistry*, *282*, 19385–19398.

Tan, P., Guan, H., Xie, L., Mi, B., Fang, Z., Li, J., et al. (2015). FOXO1 inhibits osteoclastogenesis partially by antagnozing MYC. *Scientific Reports*, *5*, 16835.

Teitelbaum, S. L., & Ross, F. P. (2003). Genetic regulation of osteoclast development and function. *Nature Reviews. Genetics*, *4*, 638–649.

Thrailkill, K., Bunn, R. C., Lumpkin, C., Jr., Wahl, E., Cockrell, G., Morris, L., et al. (2014). Loss of insulin receptor in osteoprogenitor cells impairs structural strength of bone. *Journal of Diabetes Research*, *2014*, 703589.

Thrailkill, K. M., Liu, L., Wahl, E. C., Bunn, R. C., Perrien, D. S., Cockrell, G. E., et al. (2005). Bone formation is impaired in a model of type 1 diabetes. *Diabetes*, *54*, 2875–2881.

Ucer, S., Iyer, S., Kim, H. N., Han, L., Rutlen, C., Allison, K., et al. (2017). The effects of aging and sex steroid deficiency on the murine skeleton are independent and mechanistically distinct. *Journal of Bone and Mineral Research*, *32*, 560–574.

van der Eerden, B. C., Karperien, M., & Wit, J. M. (2003). Systemic and local regulation of the growth plate. *Endocrine Reviews*, *24*, 782–801.

van Lierop, A. H., Appelman-Dijkstra, N. M., & Papapoulos, S. E. (2017). Sclerostin deficiency in humans. *Bone*, *96*, 51–62.

Vestergaard, P. (2007). Discrepancies in bone mineral density and fracture risk in patients with type 1 and type 2 diabetes—A meta-analysis. *Osteoporosis International*, *18*, 427–444.

Vestergaard, P., Rejnmark, L., & Mosekilde, L. (2005). Relative fracture risk in patients with diabetes mellitus, and the impact of insulin and oral antidiabetic medication on relative fracture risk. *Diabetologia*, *48*, 1292–1299.

Weber, D. R., Haynes, K., Leonard, M. B., Willi, S. M., & Denburg, M. R. (2015). Type 1 diabetes is associated with an increased risk of fracture across the life span: A population-based cohort study using the health improvement network (THIN). *Diabetes Care*, *38*, 1913–1920.

Weinstein, R. S., Wan, C., Liu, Q., Wang, Y., Almeida, M., O'Brien, C. A., et al. (2010). Endogenous glucocorticoids decrease skeletal angiogenesis, vascularity, hydration, and strength in aged mice. *Aging Cell*, *9*, 147–161.

Zebaze, R. M., Ghasem-Zadeh, A., Bohte, A., Iuliano-Burns, S., Mirams, M., Price, R. I., et al. (2010). Intracortical remodelling and porosity in the distal radius and post-mortem femurs of women: A cross-sectional study. *Lancet*, *375*, 1729–1736.

CHAPTER SEVEN

Regulation of FOXO Factors in Mammalian Cells

Abigail K. Brown, Ashley E. Webb[1]
Brown University, Providence, RI, United States
[1]Corresponding author: e-mail address: ashley_webb@brown.edu

Contents

Abstract

Forkhead box O (FOXO) transcription factors are central regulators of cellular homeostasis. FOXOs respond to a wide range of external stimuli, including growth factor signaling, oxidative stress, genotoxic stress, and nutrient deprivation. These signaling inputs regulate FOXOs through a number of posttranslational modifications, including phosphorylation, acetylation, ubiquitination, and methylation. Covalent modifications

Current Topics in Developmental Biology, Volume 127
ISSN 0070-2153
https://doi.org/10.1016/bs.ctdb.2017.10.006

can affect localization, DNA binding, and interactions with other cofactors in the cell. FOXOs integrate the various modifications to regulate cell type-specific gene expression programs that are essential for metabolic homeostasis, redox balance, and the stress response. Together, these functions are critical for coordinating a response to environmental fluctuations in order to maintain cellular homeostasis during development and to support healthy aging.

1. INTRODUCTION: THE FOXO FAMILY

FOXO transcription factors are conserved regulators of cellular homeostasis, the stress response, and longevity. To carry out these functions, FOXOs regulate a variety of cellular processes, including cell cycle regulation, redox balance, proteostasis, apoptosis, metabolism, and repair of DNA damage (Brunet et al., 1999; Dijkers, Medema, Lammers, Koenderman, & Coffer, 2000; Kops et al., 2002; Lee, Kennedy, Tolonen, & Ruvkun, 2003; Mammucari et al., 2007; Medema, Kops, Bos, & Burgering, 2000; Nemoto & Finkel, 2002; Tran et al., 2002). In humans, the FOXO family comprises *FOXO1*, *FOXO3*, *FOXO4*, and *FOXO6*. *Drosophila* and *C. elegans* each contain a single FOXO gene, known as dFOXO and *daf-16*, respectively. All FOXO family members contain a forkhead box DNA binding domain that is highly conserved across species and nearly identical among mammalian FOXOs. FOXOs usually function as transcriptional activators, although repression of transcription has been observed in some cases (Ramaswamy, Nakamura, Sansal, Bergeron, & Sellers, 2002; Webb et al., 2013). In all species, FOXO activity is tightly regulated at the posttranslational level. FOXOs receive inputs from various signaling pathways in the form of covalent modifications and protein–protein interactions. Thus, FOXOs are influenced by a combination of external stimuli and intrinsic factors, which together direct transcriptional responses. In this review, we discuss our current understanding of how FOXOs are regulated, with a focus on mammalian systems.

1.1 FOXO Domain Structure and Posttranslational Modifications

FOXO transcription factors are characterized by a "forkhead" DNA binding domain of approximately 110 amino acids that recognizes the consensus sequence TGTTTAC (Furuyama, Nakazawa, Nakano, & Mori, 2000; Xuan & Zhang, 2005). In addition to the DNA binding domain, FOXOs contain short conserved sequences near their N- and C- termini, as well as conserved nuclear localization and export sequences (NLS and NES, respectively; Fig. 1) (Obsil & Obsilova, 2008). In all FOXO family members the NLS overlaps with the C-terminal end of the DNA binding domain and can be modified posttranslationally. Intriguingly, outside of these conserved domains, FOXOs contain long stretches of putative disordered domains. The intrinsically disordered domains of FOXO make up approximately 75% of the total protein and include the predicted transactivation domains (Wang, Marshall, & Ikura, 2015). The function of these domains is not fully understood, but they are likely to be more flexible regions involved in protein–protein interactions and chromatin remodeling, or serve as linkers to facilitate intra- and intermolecular interactions.

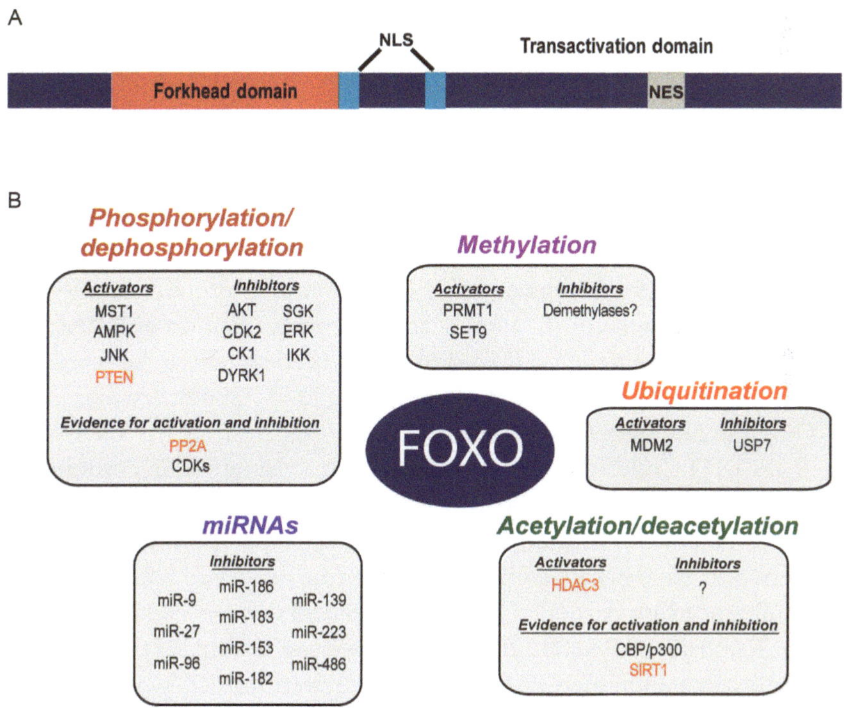

Fig. 1 FOXO regulation through posttranslational modifications. (A) General domain structure of the FOXO family of transcription factors. The conserved DNA binding domain is termed the Forkhead box. NES, nuclear export signal; NLS, nuclear localization signal. (B) FOXOs are posttranslationally modified by a number of different enzymes that affect their transcriptional activity. FOXOs are also regulated at the posttranscriptional level by several different miRNAs. Proteins and miRNAs that have been identified as direct regulators of FOXOs in mammalian cells are listed. Note that proteins in *red* perform the opposite function than the categorized covalent modification (i.e., PTEN and PP2A are phosphatases and HDAC3 and SIRT1 are deacetylases).

FOXO transcription factors are extensively posttranslationally modified. Covalent modifications to FOXOs include phosphorylation, acetylation, methylation, and ubiquitination (Fig. 1). These modifications activate or inhibit FOXO activity by affecting localization, stability, DNA binding, or protein–protein interactions. For example, phosphorylation of FOXO transcription factors has been observed at multiple sites throughout the protein, and a number of kinases have been implicated in their regulation. The first evidence of direct regulation by a protein kinase was phosphorylation at three sites by the serine/threonine kinase AKT (also known as PKB) (Brunet et al., 1999; Kops et al., 1999). Additional kinases including SGK, AMPK, JNK, ERK, and MST1 have been observed to modify FOXOs (Asada et al., 2007; Brunet et al., 2001; Essers et al., 2004; Greer et al., 2007; Lehtinen et al., 2006). Activity of these kinases can inhibit (e.g., AKT, SGK) or augment (e.g., JNK, MST1) FOXO transcriptional activity. FOXOs also contain a number of lysine residues that are targeted by protein acetyltransferases, deacetylases, ubiquitin ligases, and methyltransferases (Brunet et al., 2004; Calnan et al., 2012; Fukuoka et al., 2003; Motta et al., 2004; Yamagata et al., 2008). Intriguingly, a number of these lysine modifications occur in and around the DNA binding domain and nuclear localization sequence. Covalent modification at these sites alters the activity of FOXOs through the regulation of DNA binding, protein stability, and protein–protein interactions. Mono- and polyubiquitination of FOXO factors have been observed and linked to transcriptional activity and stability, respectively (van der Horst et al., 2006; Yang et al., 2008).

The large number of modifications to the FOXO family has led to a proposed "FOXO Code," whereby modifications act singly or in combination to selectively recruit protein partners to direct specific programs of gene expression (Calnan & Brunet, 2008). Indeed, cell type-specific gene expression networks have been identified at the chromatin level, which are likely to be influenced by particular extrinsic cues and lineage-specific cofactors (Webb, Kundaje, & Brunet, 2016). Here, we discuss the upstream regulatory inputs that influence FOXO activity in various cell types to ensure appropriate downstream responses.

2. REGULATION BY INSULIN/IGF SIGNALING AND OTHER GROWTH FACTORS

The insulin/insulin-like growth factor (IGF) signaling pathway functions in an evolutionarily conserved manner to regulate FOXO activity

(Brunet et al., 1999; Lin, Dorman, Rodan, & Kenyon, 1997; Ogg et al., 1997). Insulin and IGF proteins regulate a variety of processes across tissues, including metabolism, growth, cognition, and proteostasis. Insulin, IGF, and several other growth factors (e.g., EGF, NGF, PDGF) activate a common downstream signaling pathway, which includes activation of AKT and MAPK. This signaling pathway is conserved from worms to humans and is a major regulatory input for FOXOs.

2.1 Inhibition by AKT and Other Kinases

Insulin and IGF receptors function as membrane-bound receptors with intracellular tyrosine kinase domains (Siddle, 2011). Once stimulated, the receptors autophosphorylate and recruit the intracellular insulin receptor substrate proteins (IRS; IRS1–3 in mammals). IRS proteins function as adapters that organize downstream signaling events, in particular through recruitment of phosphoinositol-3-kinase (PI3K) (Shaw, 2011). PI3K is a lipid kinase that generates phosphatidylinositol triphosphate (PIP_3) and subsequent activation of phosphoinositide protein kinase (PDK) and AKT (Fig. 2). The protein kinase AKT is a key direct regulator of FOXO activity. Mammals have three AKT genes, named AKT1, 2, and 3, which have tissue-specific functions (Gonzalez & McGraw, 2009). Full activation of AKT involves phosphorylation by PDK-1 and mTORC2 at Thr308 and Ser473, respectively (Alessi et al., 1996; Sarbassov, Guertin, Ali, & Sabatini, 2005). Once activated, AKT directly phosphorylates FOXOs 1, 3, and 4 at three sites, resulting in sequestration in the cytoplasm (Brunet et al., 1999; Kops et al., 1999). The same sites that are targeted by AKT can also be phosphorylated by serum and glucocorticoid-induced kinase (SGK) (Brunet et al., 2001). The first two phosphosites are binding sites for the 14-3-3 proteins, which are localized to both the cytoplasm and the nucleus (Brunet et al., 1999; Obsil, Ghirlando, Anderson, Hickman, & Dyda, 2003). 14-3-3 binding in the cytoplasm prevents the cytoplasmic pool of FOXO from entering the nucleus, whereas 14-3-3-FOXO interaction in the nucleus is thought to expose the nuclear export sequence and increase transport back to the cytoplasm (Brunet et al., 2002; Obsilova et al., 2005). Once they have been shuttled to the cytoplasm, FOXOs can be degraded by the proteasome. The multisubunit E3 ligase SKP1-CUL-F-box (SCF) has been shown to recognize phosphorylated Ser256 on FOXO1 and polyubiquitinate the N-terminal half of the protein, tagging it for proteasomal degradation (Huang et al., 2005).

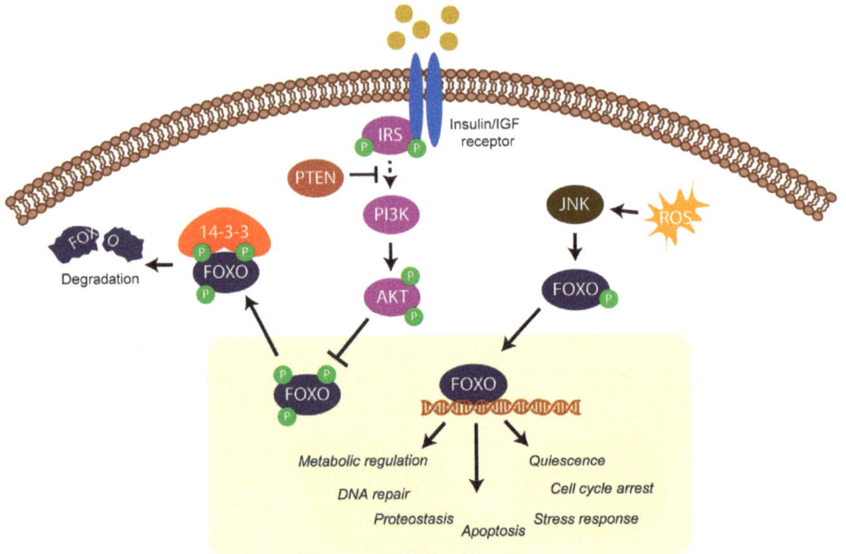

Fig. 2 FOXO activity, stability, and localization are regulated by the insulin/IGF signaling pathway and reactive oxygen species. Insulin/insulin-like growth factors activate transmembrane receptors and in turn activate the insulin receptor substrate (IRS) proteins. IRS proteins then stimulate downstream signaling events, resulting in the phosphorylation and activation of AKT. Activated AKT phosphorylates FOXO proteins on three sites and triggers their inactivation and nuclear exclusion. FOXO phosphorylation by AKT enhances binding to 14-3-3 proteins and either cytoplasmic sequestration and/or degradation by the proteasome. Phosphatase and tensin homolog (PTEN) promotes FOXO activation and nuclear localization by inhibiting the activation of PI3K. In contrast to inhibition by insulin/IGF signaling, reactive oxygen species (ROS) promote nuclear localization and activation of FOXO by activating the c-Jun N-terminal kinase (JNK). JNK directly phosphorylates FOXOs, thereby stimulating activation of transcriptional regulation. Once activated, FOXOs regulate several cellular processes that promote cellular homeostasis.

14-3-3-independent regulation of FOXO localization has been observed in the case of FOXO1 (Rena, Prescott, Guo, Cohen, & Unterman, 2001). Although the precise mechanism has not been defined in all cases, additional protein kinases can potentiate the nuclear export of FOXOs. For example, cyclin-dependent kinase 2 (CDK2) phosphorylates FOXO1 at Ser249, which resides in the NLS (Huang, Regan, Lou, Chen, & Tindall, 2006). Casein kinase 1 (CK1) recognizes FOXO1 phosphorylated by AKT at Ser319, and subsequently phosphorylates Ser322 and Ser325 (Rena, Bain, Elliott, & Cohen, 2004; Rena et al., 2002). Modification of these sites enhances binding to the Ran-dependent export machinery, thereby reducing nuclear FOXO levels by accelerating nuclear export.

Thus, tight regulation of the nuclear levels of FOXO proteins involves multiple mechanisms and phosphorylation events in addition to interaction with 14-3-3 proteins (Zhao et al., 2004).

AKT also phosphorylates and inhibits the transcriptional activity of FOXO6. However, FOXO6 stands apart from the other mammalian FOXOs because it is only phosphorylated by AKT at two sites, Thr26 and Ser184, and it does not appear to shuttle between the nucleus and the cytoplasm and instead is constitutively nuclear (van der Heide, Jacobs, Burbach, Hoekman, & Smidt, 2005).

2.2 Regulation by Phosphatases

Phosphatases function both upstream of FOXOs and directly on the FOXOs themselves to attenuate and fine-tune regulation by insulin/IGF signaling. The phosphatase and tensin homolog (PTEN) is a lipid/protein phosphatase that inhibits PI3 kinase signaling by dephosphorylating PIP_3 to $PI(4,5)P_2$ (Maehama & Dixon, 1998). PIP_3 is critical for AKT activation; therefore, high PTEN activity reduces the active pool of AKT and increases FOXO activity. Genetic evidence supports an important role for PTEN upstream of FOXOs since genetic ablation of the PTEN and FOXO factors results in similar phenotypes in the mouse. For example, PTEN deficiency causes increased neural progenitor proliferation in vitro and larger brain size in vivo, similar to FOXO-deficient mice (Amiri et al., 2012; Groszer et al., 2006; Paik et al., 2009; Renault et al., 2009). Similarly, ablation of either PTEN or FOXOs in hematopoietic stem cells causes enhanced cell cycle entry and eventual exhaustion of the stem cell pool (Miyamoto et al., 2007; Tothova et al., 2007; Yilmaz et al., 2006; Zhang et al., 2006). PTEN is a tumor suppressor commonly mutated in cancers, including prostate, brain, and breast cancer (Hollander, Blumenthal, & Dennis, 2011). PTEN-negative prostate and renal carcinoma cells have elevated cytoplasmic FOXOs (Modur, Nagarajan, Evers, & Milbrandt, 2002; Nakamura et al., 2000), and expression of constitutively active FOXO in these cells can decrease tumorigenesis (Ramaswamy et al., 2002). However, the degree of FOXO dysregulation in most PTEN-associated cancers is not known, and the extent to which reduced FOXO activity in the absence of PTEN contributes to tumorigenicity will be an important question in the future. Nevertheless, PTEN is a critical upstream activator of FOXO transcription factors in healthy mammalian cells, and reduced FOXO activity in PTEN-negative tumors enhances tumorigenesis.

In addition to regulation by PTEN, FOXOs are regulated both directly and indirectly by protein phosphatase 2A (PP2A). Direct interactions between both FOXO1 and FOXO3 and PP2A have been observed, and pharmacological inhibition or ablation of PP2A enhances phosphorylation at the AKT phosphosites on FOXO1 (Yan et al., 2008). Dephosphorylation of the AKT phosphosites T32 and S253 by PP2A releases FOXOs from 14-3-3 binding and increases nuclear entry (Singh et al., 2010). The model that has emerged from these observations is one in which inhibition by AKT or SGK is tightly regulated in mammalian cells, whereas dephosphorylation by PP2A is constitutive and nonspecific. Under this model, the kinases are the important inhibitory nodes, and in their absence FOXOs are dephosphorylated and activated by default. However, upstream of FOXOs, AKT is targeted by specific PP2A regulatory subunits. In *C. elegans*, a key regulatory subunit of the PP2A holoenzyme, B56β/*pptr-1*, is actively regulated to balance FOXO stability with degradation (Padmanabhan et al., 2009). The B56β subunit specifically targets PP2A to dephosphorylate AKT-1, but not AKT-2 or AKT-3. Modulation of B56β/*pptr-1* activity in the worm influences FOXO/DAF-16 localization and affects FOXO/DAF-16-regulated phenotypes. Similarly, the *Drosophila* ortholog of B56β/*pptr-1*, *widerborst*, was identified as a regulator of the AKT–FOXO pathway in flies (Vereshchagina, Ramel, Bitoun, & Wilson, 2008). To what extent particular phosphatases target specific AKTs or FOXOs in mammalian cells remains largely unknown. Knockdown of B56β, but not the related factor B56α, in cultured adipocyte cell lines resulted in increased AKT-1 phosphorylation, suggesting that this interaction is conserved from worms to humans (Padmanabhan et al., 2009). However, whether this alters FOXO activity is not known. The B55α PP2A regulatory subunit has also been found to specifically target AKT (Kuo et al., 2008), but factors specific for FOXOs remain unknown. Thus, future studies will be required to dissect the context-specific regulation of FOXO activity by additional phosphatases.

3. NUTRIENT SIGNALING AND METABOLIC STRESS

FOXOs are activated by upstream signals under conditions of low nutrient availability or starvation. For example, in the liver, low nutrient status and thus low levels of insulin signaling activate FOXOs to restore glucose levels via glycogenolysis and gluconeogenesis. In this context, FOXOs regulate a number of target genes involved in energy metabolism, including the gluconeogenic enzymes glucose-6-phosphatase (*G6Pase*), phosphoenolpyruvate carboxykinase (*Pepck*), and (PGC1a) (Altomonte et al., 2003; Matsumoto, Pocai, Rossetti, Depinho, & Accili, 2007; Puigserver et al., 2003). FOXOs also maintain metabolic balance through regulation of autophagy genes, including *Atg5*, *Atg12*, and *Becn1* (Webb & Brunet, 2014). Restoration of metabolic balance in response to starvation or limited nutrients is also a key function of FOXOs in beta cells of the pancreas and skeletal muscle. In addition to regulation by insulin/IGF/AKT signaling, FOXOs are also regulated by AMP-activated protein kinase (AMPK), a central intracellular regulator of energy homeostasis.

3.1 Regulation by AMPK

AMPK is an energy-sensing kinase that is activated in response to high AMP/ATP ratios as a result of ATP depletion (Hardie & Carling, 1997). AMPK functions as a trimeric complex that is allosterically activated by AMP binding. When cellular ATP levels are low, AMP-bound AMPK is directly phosphorylated at Thr172 by either LKB1 or CAMKK2 (Hawley et al., 2005; Shaw et al., 2004; Woods et al., 2005). Phospho-AMPK phosphorylates a number of enzymes and transcriptional regulators involved in metabolic regulation, including lipid metabolism, protein synthesis, and glucose uptake, in order to replenish the ATP supply. AMPK directly phosphorylates FOXO3 at six serine/threonine residues that are distinct from the AKT phosphosites (Thr179, Ser399, Ser413, Ser555, Ser588, and Ser626) (Greer et al., 2007). Unlike regulation by other kinases (e.g., AKT, SGK, MST1), modification by AMPK does not appear to alter the localization of FOXO3. Thus, the role of AMPK in the regulation of FOXOs may primarily be to promote interaction between cofactors and FOXO3 to affect specific target genes. Consistent with this notion, mutation of the six AMPK phosphorylation sites disrupted expression of novel targets involved in energy homoeostasis, including acetyl-CoA thioesterase (*Acot12*), ferroportin (*Slc40a1*), and the mitochondrial proton carrier uncoupling protein 2 (*Ucp2*). Alternatively, activation by AMPK may increase the overall efficiency of FOXO-mediated transcriptional activation. Indeed, AMPK phosphorylation of FOXO3 has been observed to enhance binding to CBP/p300 and transactivation in luciferase assays (Wang et al., 2012).

Some, but not all, of the AMPK sites are found in other mammalian and nonmammalian FOXO family members. Regulation of FOXOs by AMPK has been detected in *C. elegans*, and this pathway regulates life span extension in response to dietary restriction in worms (Greer et al., 2007). Thus, the AMPK–FOXO regulatory axis is conserved from worms to mammals.

Fasting conditions can activate Class IIa histone deacetylases (HDACs) to deacetylate and activate FOXOs in the liver. Specifically, in response to food deprivation, HDAC3 binds to target genes involved in gluconeogenesis and deacetylates FOXOs to stimulate FOXO-mediated gene activation (Mihaylova et al., 2011). This cellular context involves dual regulation by both phosphorylation and acetylation: low insulin levels drive FOXOs into the nucleus, which are then further activated by deacetylation by HDACs. Activation of *G6pc* and *Pck1* (PEPCK) restores glucose homeostasis under these conditions, and the activation of FOXOs is reversed upon refeeding. In summary, FOXOs are critical for maintaining metabolic balance in conditions of nutrient deprivation. FOXOs integrate signals from insulin/IGF signaling, AMPK, HDACs, and possibly other inputs to respond to changing nutrient availability.

4. OXIDATIVE AND GENOTOXIC STRESS STIMULI

Regulation of the stress response is a key function of FOXO transcription factors. This important function was first observed in *C. elegans* where long-lived *daf-2* mutants display resistance to oxidative stress (paraquat), a phenotype which is dependent on FOXO/DAF-16 (Honda & Honda, 1999). FOXOs are able to respond to stress signals independent of growth factor signaling since stress can induce nuclear entry in high growth factor conditions (Brunet et al., 2004; Frescas, Valenti, & Accili, 2005). Once they have been activated, FOXOs activate a number of downstream target genes encoding antioxidant enzymes that regulate redox homeostasis, including catalase, sestrins, and superoxide dismutase (SOD) (Klotz et al., 2015).

4.1 Regulation by the MAPK Family

Activation of FOXOs in response to stress is mediated in part by Jun-N-terminal kinase (JNK). JNK is activated by different cellular stresses, including UV irradiation and oxidative stress. In the presence of low-level oxidative stress JNK directly phosphorylates and activates FOXO4 at Thr447 and Thr451 (Fig. 2) (Essers et al., 2004). Modification of these sites disrupts the interaction with 14-3-3 proteins, allowing FOXO nuclear entry. JNK also directly phosphorylates 14-3-3, perturbing the interaction with FOXO and releasing it into the nucleus. Intriguingly, this regulatory axis is conserved across species and impacts life span in model organisms.

JNK activation enhances dFOXO nuclear entry and extends life span in a FOXO-dependent manner in *Drosophila* (Wang, Bohmann, & Jasper, 2005). However, it remains unclear if FOXO is directly phosphorylated by JNK in this context, particularly because JNK has been found to phosphorylate other proteins in the insulin/IGF pathway, including IRS-1 (Hirosumi et al., 2002). Evidence from *C. elegans* suggests that JNK directly phosphorylates FOXO/DAF-16 and, similar to *Drosophila*, overexpression of JNK in worms extends life span in a FOXO/DAF-16-dependent manner (Oh et al., 2005). Together, these lines of evidence indicate that JNK can activate FOXOs via direct and indirect mechanisms. In *Drosophila*, activation of the JNK–FOXO axis culminates in expression of a program of genes that function in protection against oxidative stress (Wang, Bohmann, & Jasper, 2003). Similar targets are likely to reside downstream of JNK-stimulated FOXO in mammalian cells, but they have yet to be identified. In addition to JNK, p38 MAPK activates FOXO3 in response to genotoxic and environmental stress. p38 phosphorylates FOXO3 at Ser294 and Ser425, which are JNK target sites, as well as at a site distinct from JNK (Ser7) (Ho et al., 2012). Whether activation by p38 induces a response that completely overlaps with the JNK-induced response or is distinct from JNK-mediated stimulation remains unknown. However, mutation of Ser7 to Ala was sufficient to impair nuclear localization in response to genotoxic stress (doxorubicin treatment), suggesting that phosphorylation by p38 is a key input for full activation of FOXO in this context.

In contrast to the stimulation by JNK and p38, the MAPK family member ERK is a potent inhibitor of FOXO activity. In tumor cells, stimulation with growth factors activates ERK, which directly phosphorylates FOXO3 at three sites: Ser294, Ser344, and Ser425 (Asada et al., 2007; Yang et al., 2008). Phosphorylation at these sites promotes cytoplasmic retention, ubiquitination by the E3-ligase MDM2, and subsequent degradation by the proteasome. Mutation of the three ERK target sites causes resistance to MDM2-mediated degradation and can block tumor growth (Yang et al., 2008). This finding suggests parallel regulation of FOXOs downstream of growth factor signaling since the same stimuli simultaneously inhibit PI3K/AKT signaling and ERK activity.

4.2 Activation by Ste20-Like Kinase MST1

In addition to the MAPK family, the Ste20-like kinase (MST1) has been shown to phosphorylate and activate FOXOs in response to oxidative stress

(Lehtinen et al., 2006; Yuan et al., 2009). MST1-mediated activation of FOXOs was first discovered in primary neurons, where treatment with hydrogen peroxide induced phosphorylation of FOXOs, followed by FOXO-dependent apoptosis. Similar to other activating modifications, the MST1 phosphosite was found to be within the forkhead domain (Ser212 and Ser207 in FOXO1 and FOXO3, respectively), and phosphorylation at this site disrupted 14-3-3 binding. This regulatory axis has been observed in other mammalian cell types and is evolutionarily conserved in *C. elegans* (Lehtinen et al., 2006). For example, *Mst1*$^{-/-}$ T cells are highly sensitive to oxidative stress, which correlates with reduced FOXO1 and FOXO3 stability and activity (Choi et al., 2009; Du et al., 2014). However, in contrast to what has been observed in neurons, MST1 does not regulate FOXO1 or FOXO3 localization in Treg cells. The MST1–FOXO interaction may have implications for FOXO's tumor suppressor activity since treatment of lymphoma cell lines with the anticancer drug α-tocopheryl succinate, which causes accumulation of ROS, also induced apoptosis via the MST1–FOXO1 axis (Valis et al., 2011). The downstream effects of activation of the MST1–FOXO pathway are not fully understood, but appear to be cell type and species specific. In neurons, FOXO3 induces cell death in response to activation of MST1, and the proapoptotic gene BIM was identified as the key factor mediating this response (Lehtinen et al., 2006). In contrast, FOXOs have been proposed to activate ROS-detoxifying genes (e.g., MnSOD and catalase) and cell survival (e.g., Il7r) in naïve T cells (Choi et al., 2009). Activation of FOXOs by MST1 in Treg cells occurs during normal development, independent of exogenously induced stress (Du et al., 2014). In this situation, FOXOs are critical for regulating expression of *Foxp3*, which functions as a master regulator of Treg identity. Moreover, overexpression of the worm MST1 (*cst-1*) can extend life span in nematodes in a FOXO/DAF-16-dependent manner. Thus, the cellular and organismal context of FOXO activation is likely to direct the appropriate downstream response in each case.

The extent to which activation of FOXOs in this context differs from activation by JNK has not been determined. It is possible that JNK and MST1 function as redundant activators of the same gene programs that mediate the stress response. Alternatively, modification by each kinase may provide a binding site for particular binding partners that guide activation of specific gene networks. Additional work will be required to determine the precise downstream readouts of FOXO in each case and the extent to which they function as separate and overlapping networks.

4.3 Ubiquitination/Deubiquitination in Response to Stress

Monoubiquitination has also been observed to regulate FOXO activity in response to oxidative stress (peroxide treatment) (van der Horst et al., 2006). In this context, MDM2 has been observed to monoubiquitinate FOXO4 and increase its nuclear localization and transcriptional activity (Brenkman, de Keizer, van den Broek, Jochemsen, & Burgering, 2008). The deubiquitinating enzyme herpes virus-associated ubiquitin-specific protease (USP7) binds monoubiquitinated FOXO4 to attenuate the response, suggesting that, in this context, MDM2 alters the localization but not the stability of FOXOs. Interestingly, the authors observe that low levels of MDM2 potentiate FOXO4 activity, whereas high levels of MDM2 overexpression have a repressive effect. This result contrasts the observation that MDM2 induces FOXO degradation upon growth factor stimulation (Yang et al., 2008), and suggests that the effects of ubiquitination are context specific, depending either on the levels and/or nature of the stimulus.

4.4 Acetylation and Deacetylation Modulates FOXO Activity

Early studies from *C. elegans* identified an interaction between the nicotinamide adenine dinucleotide (NAD)-dependent histone deacetylase Sir2 and DAF-16/FOXO. Studies in mammalian cells confirmed that FOXOs are acetylated by p300 and CBP (CREB binding protein) in response to oxidative stress. The same conditions induce a physical association between various FOXO family members and the mammalian ortholog of Sir2, SIRT1. Knockout of SIRT1 enhances FOXO acetylation in fibroblasts and embryonic stem cells. SIRT1 is only present in the nucleus, and only associates with and deacetylates nuclear FOXOs in the presence of oxidative stress. Acetylation of FOXOs has been observed to both increase and decrease its transcriptional activity, depending on the target.

Motta et al. reported that acetylation of FOXO3 by p300 (E1A binding protein 300) enhanced transcriptional activity, and the effect could be reversed by the SIRT1 deacetylase (Motta et al., 2004). The authors extended their findings to other FOXOs and found that FOXOs 1 and 4 could also be deacetylated and inhibited by SIRT1. Moreover, knockout of SIRT1 correlated with increased transcription of two previously identified FOXO targets, *Pepck* (phosphoenolpyruvate carboxylase) and *Igfpb1* (insulin-like growth factor binding protein 1). In contrast, others have observed that acetylation of FOXOs by CBP (CREB binding protein) inhibits transcriptional activity in the context of oxidative stress, and SIRT1 can relieve this inhibition (Daitoku et al., 2004; Fukuoka et al., 2003; van der Horst et al., 2004). In this case, overexpression of SIRT1 promoted FOXO-mediated cell cycle arrest and induction of p27/*Cdkn1b* and MnSOD. Brunet et al. observed that the effect of SIRT1–FOXO3 association was target dependent: SIRT1 enhanced induction of p27/CDKN1B and GADD45, but inhibited BIM activation (Brunet et al., 2004). This observation suggests that the acetyltransferases and deacetylases targeting FOXOs are modulators of FOXO activity, instead of simple on/off switches such as kinases (e.g., AKT and MAPK). Subsequent studies investigating the nature of the FOXO–acetyltransferase interaction supports this notion. Interestingly, the FOXO4–CBP interaction was found to be stabilized by the formation of a cysteine disulfide bridge that forms under conditions of high oxidative stress. In this case, the cysteine residues were observed to mediate the inhibition of FOXO4-induced cell cycle exit by p300, while promoting FOXO4-induced apoptosis (Dansen et al., 2009). Altogether, these findings suggest that acetylation of FOXOs has dual effects on modulating the response to oxidative stress: acetylation enhances apoptosis while reducing stress resistance. The precise mechanism by which acetylation tips the balance toward apoptosis is unknown, but the location of the acetylation sites, which primarily reside in the DNA binding domain, suggests that acetylation may affect binding specificity at different targets. Studies have demonstrated that acetylation of FOXOs can disrupt their ability to bind DNA (Brent, Anand, & Marmorstein, 2008; Matsuzaki et al., 2005; Tsai et al., 2007), but how this modification might alter target selection remains unclear. It must also be considered that histone acetylation and deacetylation, which activate and repress transcription, respectively, are performed by the same set of enzymes. Thus, it remains a challenge to unravel transcriptional effects due to alterations in histone acetylation vs FOXO activity in some experiments. Nevertheless, acetylation is a key covalent modification in the regulation of FOXO activity in mammalian systems.

4.5 Regulation by Cyclin-Dependent Kinases Under Genotoxic Stress

FOXOs function in the response to genotoxic stress, by integrating with the repair machinery and activating DNA damage response genes (Tran et al., 2002; Tsai, Chung, Takahashi, Xu, & Hu, 2008). CDK2 can directly phosphorylate FOXO1 in response to DNA damage. CDK2 plays an important role in the G1/S DNA damage checkpoint. CDK2 normally complexes with cyclin E to promote the G1/S transition. Under genotoxic stress, p21/CDKN1A inhibits CDK2, thereby inducing cell cycle arrest and preventing replication of damaged DNA (Satyanarayana & Kaldis, 2009). Huang et al. observed that CDK2 phosphorylates FOXO1 on Ser249, causing nuclear export and inhibition of FOXO1. DNA damage induced by a topoisomerase I inhibitor (camptothecin) inhibited CDK2 phosphorylation of FOXO1 and resulted in FOXO-induced apoptosis (Huang et al., 2006). CDK1, which regulates entry into mitosis, has been observed to phosphorylate the same site on FOXO1 (Ser249) in neurons. In this case, phosphorylation was observed to induce nuclear localization and activate FOXO1 by disrupting FOXO1-14-3-3 binding. CDK1 can induce apoptosis in neurons, and does so through the activation of FOXO1 in this context (Yuan et al., 2008). How phosphorylation at the same site can inhibit FOXO activity in one context (DNA damage/CDK2 activation) and enhance activity in another (postmitotic neurons/CDK1 activation) remains unclear. It is unlikely to be a difference between proliferating and terminally differentiated cells since Yuan et al. also observed activation of FOXO1 by CDK1 in dividing cells. An interesting possibility is that the same phosphosite is recognized by different cellular machinery in different phases of the cell cycle, resulting in differential regulation of FOXO1. Future work will be necessary to define the precise mechanism underlying this difference.

5. REGULATION OF FOXOS IN CANCER AND THE IMMUNE RESPONSE

FOXOs have emerged as critical regulators of the immune response. FOXOs generally function to restrain the immune response and inhibit tumor growth through the activation of genes that inhibit the cell cycle (e.g., *Cdkn1b*/p27), induce apoptosis (e.g., Bim), and limit proinflammatory cytokine production (e.g., TNFα). Mitogenic signals alter FOXO activity through both AKT-dependent and -independent signaling pathways, which we review below.

5.1 AKT-Independent Regulation of FOXO Localization by IkB Kinase in Tumors

In the tumor setting, regulation of FOXO activity can be entirely independent of AKT activity. In this context FOXOs can instead be inhibited by the IkB kinase/NK-kB signaling pathway. IkB kinase (IKK) is a trimeric complex that is activated by proinflammatory cytokines and viral infections. Once activated, IKK phosphorylates IkB proteins and targets them for degradation. IkB proteins bind and inhibit the NF-kB family of transcriptional regulators. Thus, high levels of cytokines and IKK signaling induce an NF-kB-mediated transcriptional program, promoting cell survival and the immune response (Luo, Kamata, & Karin, 2005). Hu et al. observed that some breast cancer tumors that are negative for activated (phosphorylated) AKT can have high levels of cytoplasmic FOXO protein, indicating an AKT-independent inhibition of FOXO activity in tumors (Hu et al., 2004). In such tumors, the authors found elevated IKK activity, indicating reciprocal activation states for FOXOs and IKK signaling. Interestingly, tumors with high nuclear FOXO levels and low IKK corresponded to improved survival rates. The authors went on to explore the mechanism underlying this correlation and discovered that the kinase subunits of IKK, IKKα, and IKKβ could inhibit the transcriptional activity of FOXO3 and induce export to the cytoplasm for degradation by the proteasome. The attenuation of FOXO3 activity was also observed using a version of FOXO3 with mutated AKT phosphosites, suggesting that inhibition by IKK signaling is independent of AKT. IKK was found to directly bind FOXO3 and phosphorylate it at Ser644, which is not present on FOXO1 or FOXO4, and direct regulation of these FOXOs by IKK has not been reported.

Regulation of FOXO3 by IKK has been observed in the context of leukemic cells as well. Similar to the observations in breast cancer cells, inhibition of IKK activity in acute myeloid leukemia blast cells induced FOXO3-mediated cell cycle arrest and apoptosis (Chapuis et al., 2010). Notably, AKT can also phosphorylate and stimulate IKK directly (Dan et al., 2008). Thus, regulation of FOXO3 by IKK can be independent of AKT in the case of activation by cytokines (TNF-α), but it may also work through AKT-induced IKK activity.

5.2 Regulation by T Cell Receptors, B Cell Receptors, and Cytokine Signaling

FOXO transcription factors play an important role in maintenance of the immune system, autoimmunity, and in the response to pathogens. Immune cell-specific ablation of FOXOs causes immune dysfunction phenotypes, depending on the particular FOXO and cell type that is targeted. For example, knockout of FOXO3 in mouse T cells causes hyperactivation of T cells and inflammatory disease.

T cells are stimulated in response to contact with antigen-presenting cells (APCs) and exposure to IL-2. IL-2 binds a trimeric cell surface receptor (IL-2R) and promotes T cell survival and expansion by activating the PI3K–AKT pathway (Minami, Kono, Miyazaki, & Taniguchi, 1993).

Similar to the effect downstream of insulin/IGF signaling, signaling through AKT in this context inhibits FOXO activity (Stahl et al., 2002). Likewise, T cell receptor (TCR/CD28) stimulation through APC contact activates the PI3K–AKT pathway, which in turn drives T cell proliferation required for the immune response. Antigen-induced activation of PI3K in T cells inhibits FOXO1 activity through AKT-mediated phosphorylation and nuclear exclusion (Fabre et al., 2005). Moreover, introduction of a mutant form of FOXO1 that cannot be phosphorylated by AKT was constitutively nuclear and inhibited T cell proliferation even in the presence of APC contact. In the context of T cell stimulation, the Vav family of guanine exchange factors (GEFs) are activated by TCR/CD28 stimulation and regulate PI3K/AKT activity upstream of FOXO1 (Charvet et al., 2006).

Cytokine withdrawal can also induce FOXO activation in primary B cells. Similar to T cells, cytokines stimulate PI3K/AKT to promote B cell survival, and cytokine deprivation induces FOXO-mediated apoptosis via activation of *Bim* and loss of mitochondrial integrity (Dijkers et al., 2002). In addition to inducing B cell death, FOXO1 has also emerged as a critical regulator of B cell development through the regulation of a target network that guides differentiation and homeostasis in the periphery (Szydlowski, Jablonska, & Juszczynski, 2014). In this context, FOXO1 and FOXO3 regulate B cell specialization through the regulation of *Rag1* and *Rag2* transcription, which induce genomic editing of the B cell receptor. Thus, in immature B cells PI3K/AKT signaling inhibits FOXO activity. Signaling through the B cell receptor involving the SLP-65 adapter protein inhibits PI3K/AKT, thereby stabilizing FOXOs and inducing genomic rearrangements that drive B cell maturation (Amin & Schlissel, 2008; Herzog et al., 2008).

In summary, FOXOs are important regulators of immune function and are under tight regulation by PI3K/AKT in T cells, B cells, and other immune cell types. This signaling network functions downstream of a variety of stimuli, including cytokines and antigen presentation through T cell and B cell receptors. FOXOs in turn regulate apoptosis, differentiation, cell cycle exit, and maturation to promote healthy immune system function and development.

6. OTHER INPUTS

6.1 Regulation by Methyltransferases

Methylation of FOXO1 and FOXO3 has been reported to alter their stability. Two arginine residues on FOXO1 (Arg248 and 250) can be methylated by protein arginine methyltransferase 1 (PRMT1). Methylation at these sites reduced phosphorylation by AKT at Ser253, thereby promoting FOXO1 nuclear entry and reducing degradation by the proteasome (Yamagata et al., 2008). Similarly, the Set9 methyltransferase can methylate Lys271 on FOXO3 to alter stability and transcriptional activity (Calnan et al., 2012). Similarly, Lys270 on FOXO3 was found to be methylated by Set9 in a different cell type (neurons) in response to oxidative stress (Xie et al., 2012). Interestingly, the Lys271 is a conserved residue among the FOXO family, and was previously identified as an acetylation site that is targeted by the SIRT1 deacetylase (Brunet et al., 2004). Although the precise interplay between these modifications has yet to be determined, these findings suggest that competition between different modifications may help balance the response to stress. Alternatively, the same residues may be modified differently in different cell types used to modulate FOXO activity.

6.2 Regulation of Feeding Behavior by DYRK1A

Dual-specificity tyrosine-phosphorylated and regulated kinase 1A (DYRK1A) is a nuclear-localized protein kinase that functions in the nervous system (Smith et al., 1997). Woods et al. demonstrated that DYRK1A binds and phosphorylates FOXO1 on Ser329, a site that is conserved in FOXO3, FOXO4, and DAF-16 (Woods et al., 2001). Phosphorylation by DYRK1 reduces the nuclear pool of FOXO1 and inhibits its transcriptional activity. Regulation of FOXOs by DYRK1 appears to be independent of insulin/IGF signaling, at least in 293T cells, because stimulation of the PI3K/AKT pathway did not affect phosphorylation at Ser329. More recent work in flies and mice established an in vivo function for FOXO regulation by DYRK1. In *Drosophila*, Dyrk1a can activate Sirt1, which in turn deacetylates and activates dFOXO to regulate food intake. This circuitry appears to be intact in the mouse hypothalamus since Dyrk1 transgenic mice displayed decreased FOXO acetylation and increased NPY transcription (Hong et al., 2012). Importantly, the initial study identifying a DYRK1–FOXO1 interaction reported direct phosphorylation and inhibition of FOXO1, whereas the more recent study reported that DYRK1 increases FOXO1 transcriptional indirectly through the activation of SIRT1. The reason for this difference is not clear, but it suggests that neurons regulating food intake may have specialized signaling networks to regulate FOXO activity to control neuropeptide expression and feeding behavior.

6.3 Modification of FOXOs by Additional Protein Kinases

A number of additional protein kinases have been reported to phosphorylate FOXO transcription factors and modulate their activity, including nemo-like kinase (NLK), glycogen synthase kinase 3 (GSK3), PERK, and protein kinase A (PKA) (Huo et al., 2014; Kim, Kim, Lee, & Chung, 2010; Zhang et al., 2013). Interestingly, PKA can phosphorylate FOXO1 at the three AKT target sites (Thr24, Ser256, and Ser319) and inhibit its transcriptional activity in endothelial cells (Lee, Chen, Pullikotil, & Quon, 2011). The PKA consensus site is nearly identical to the AKT consensus sequence ((R/K)XX(S/T) and (R/K)XRXX(S/T), respectively) (Huang, Tsai, Chen, Wu, & Chen, 2007; Kane et al., 2002). Thus, it remains possible that PKA targets FOXOs in a variety of cellular contexts. However, the extent to which PKA targets these sites in vivo remains unclear.

6.4 Regulation by miRNAs

Although most studies have focused on posttranslational regulation of FOXO factors, several groups have reported posttranscriptional regulation by miRNAs. For example, the FOXO1 3′ untranslated region is targeted by miR-27a, miR-96, and miR-182 in breast cancer cells (MCF7 cells). Expression of these microRNAs maintains FOXO1 protein at a low level in breast cancer cells compared to normal breast tissue, thereby promoting an oncogenic state (Guttilla & White, 2009). Similarly, FOXO1 is maintained at a low level in endometrial cancer cells (HEC-1B and Ishikawa cells) through direct regulation by miR-9, miR-27, miR-96, miR-153, miR-182, miR-183, and miR-186 (Myatt et al., 2010). Other examples of miRNAs that target FOXOs include miR-139, miR-223, and miR-486 (Hasseine et al., 2009; Small et al., 2010; Wu et al., 2012). Thus, inhibition of FOXOs by miRNAs adds an additional layer of regulation, which is frequently used in the cancer cells to promote tumorigenesis.

7. CONCLUDING REMARKS

The FOXO family of transcription factors are key integrators of environmental cues and stimuli that maintain cellular homeostasis. Downstream functions of FOXOs are conserved across species, and include regulation of the stress response, nutrient signaling, DNA repair, and cell cycle status. Since the initial discovery that AKT is a potent inhibitor of FOXO nuclear entry and transcriptional activity, a wide range of stimuli and enzymes have been reported to impinge on FOXO activity. However, a number of important questions remain to be answered. For example, precisely how different enzymes and modifications are integrated to affect FOXO localization and fine-tune its transcriptional activity remains unclear. In addition, little is known about how different modifications or stimuli affect target selection at the chromatin level. It is likely that FOXOs bind and activate different target genes depending on the particular environmental cues involved, but this has yet to be explored. FOXOs are likely to interact with specific cofactors depending on their modification state, but to date the number of cofactors interacting with FOXOs remains limited. This will be important to clarify since binding partners are likely to be integral to binding site selection, and to help direct cell type-specific transcriptional responses. These studies would provide important insight into how FOXOs guide development, maintain cellular homeostasis, and promote healthy aging.

ACKNOWLEDGMENTS

Work in our laboratory is supported by NIH R01 AG053268, the American Federation for Aging Research, the Glenn Foundation for Research on the Biology of Aging, and a subproject award from the NIH/NIA U19 AG023122.

REFERENCES

Alessi, D. R., Andjelkovic, M., Caudwell, B., Cron, P., Morrice, N., Cohen, P., et al. (1996). Mechanism of activation of protein kinase B by insulin and IGF-1. *The EMBO Journal*, *15*(23), 6541–6551.

Altomonte, J., Richter, A., Harbaran, S., Suriawinata, J., Nakae, J., Thung, S. N., et al. (2003). Inhibition of Foxo1 function is associated with improved fasting glycemia in diabetic mice. *American Journal of Physiology. Endocrinology and Metabolism*, *285*(4), E718–728. https://doi.org/10.1152/ajpendo.00156.2003.

Amin, R. H., & Schlissel, M. S. (2008). Foxo1 directly regulates the transcription of recombination-activating genes during B cell development. *Nature Immunology*, *9*(6), 613–622. https://doi.org/10.1038/ni.1612.

Amiri, A., Cho, W., Zhou, J., Birnbaum, S. G., Sinton, C. M., McKay, R. M., et al. (2012). Pten deletion in adult hippocampal neural stem/progenitor cells causes cellular abnormalities and alters neurogenesis. *The Journal of Neuroscience*, *32*(17), 5880–5890. https://doi.org/10.1523/JNEUROSCI.5462-11.2012.

Asada, S., Daitoku, H., Matsuzaki, H., Saito, T., Sudo, T., Mukai, H., et al. (2007). Mitogen-activated protein kinases, Erk and p38, phosphorylate and regulate Foxo1. *Cellular Signalling*, *19*(3), 519–527. https://doi.org/10.1016/j.cellsig.2006.08.015.

Brenkman, A. B., de Keizer, P. L., van den Broek, N. J., Jochemsen, A. G., & Burgering, B. M. (2008). Mdm2 induces mono-ubiquitination of FOXO4. *PLoS One*, *3*(7), e2819. https://doi.org/10.1371/journal.pone.0002819.

Brent, M. M., Anand, R., & Marmorstein, R. (2008). Structural basis for DNA recognition by FoxO1 and its regulation by posttranslational modification. *Structure*, *16*(9), 1407–1416. https://doi.org/10.1016/j.str.2008.06.013.

Brunet, A., Bonni, A., Zigmond, M. J., Lin, M. Z., Juo, P., Hu, L. S., et al. (1999). Akt promotes cell survival by phosphorylating and inhibiting a Forkhead transcription factor. *Cell, 96*(6), 857–868.

Brunet, A., Kanai, F., Stehn, J., Xu, J., Sarbassova, D., Frangioni, J. V., et al. (2002). 14-3-3 transits to the nucleus and participates in dynamic nucleocytoplasmic transport. *The Journal of Cell Biology, 156*(5), 817–828. https://doi.org/10.1083/jcb.200112059.

Brunet, A., Park, J., Tran, H., Hu, L. S., Hemmings, B. A., & Greenberg, M. E. (2001). Protein kinase SGK mediates survival signals by phosphorylating the forkhead transcription factor FKHRL1 (FOXO3a). *Molecular and Cellular Biology, 21*(3), 952–965. https://doi.org/10.1128/MCB.21.3.952-965.2001.

Brunet, A., Sweeney, L. B., Sturgill, J. F., Chua, K. F., Greer, P. L., Lin, Y., et al. (2004). Stress-dependent regulation of FOXO transcription factors by the SIRT1 deacetylase. *Science, 303*(5666), 2011–2015. https://doi.org/10.1126/science.1094637.

Calnan, D. R., & Brunet, A. (2008). The FoxO code. *Oncogene, 27*(16), 2276–2288. https://doi.org/10.1038/onc.2008.21.

Calnan, D. R., Webb, A. E., White, J. L., Stowe, T. R., Goswami, T., Shi, X., et al. (2012). Methylation by Set9 modulates FoxO3 stability and transcriptional activity. *Aging (Albany NY), 4*(7), 462–479.

Chapuis, N., Park, S., Leotoing, L., Tamburini, J., Verdier, F., Bardet, V., et al. (2010). IkappaB kinase overcomes PI3K/Akt and ERK/MAPK to control FOXO3a activity in acute myeloid leukemia. *Blood, 116*(20), 4240–4250. https://doi.org/10.1182/blood-2009-12-260711.

Charvet, C., Canonigo, A. J., Becart, S., Maurer, U., Miletic, A. V., Swat, W., et al. (2006). Vav1 promotes T cell cycle progression by linking TCR/CD28 costimulation to FOXO1 and p27kip1 expression. *Journal of Immunology, 177*(8), 5024–5031.

Choi, J., Oh, S., Lee, D., Oh, H. J., Park, J. Y., Lee, S. B., et al. (2009). Mst1-FoxO signaling protects naive T lymphocytes from cellular oxidative stress in mice. *PLoS One, 4*(11), e8011. https://doi.org/10.1371/journal.pone.0008011.

Daitoku, H., Hatta, M., Matsuzaki, H., Aratani, S., Ohshima, T., Miyagishi, M., et al. (2004). Silent information regulator 2 potentiates Foxo1-mediated transcription through its deacetylase activity. *Proceedings of the National Academy of Sciences of the United States of America, 101*(27), 10042–10047. https://doi.org/10.1073/pnas.0400593101.

Dan, H. C., Cooper, M. J., Cogswell, P. C., Duncan, J. A., Ting, J. P., & Baldwin, A. S. (2008). Akt-dependent regulation of NF-{kappa}B is controlled by mTOR and Raptor in association with IKK. *Genes & Development, 22*(11), 1490–1500. https://doi.org/10.1101/gad.1662308.

Dansen, T. B., Smits, L. M., van Triest, M. H., de Keizer, P. L., van Leenen, D., Koerkamp, M. G., et al. (2009). Redox-sensitive cysteines bridge p300/CBP-mediated acetylation and FoxO4 activity. *Nature Chemical Biology, 5*(9), 664–672. https://doi.org/10.1038/nchembio.194.

Dijkers, P. F., Birkenkamp, K. U., Lam, E. W., Thomas, N. S., Lammers, J. W., Koenderman, L., et al. (2002). FKHR-L1 can act as a critical effector of cell death induced by cytokine withdrawal: Protein kinase B-enhanced cell survival through maintenance of mitochondrial integrity. *The Journal of Cell Biology, 156*(3), 531–542. https://doi.org/10.1083/jcb.200108084.

Dijkers, P. F., Medema, R. H., Lammers, J. W., Koenderman, L., & Coffer, P. J. (2000). Expression of the pro-apoptotic Bcl-2 family member Bim is regulated by the forkhead transcription factor FKHR-L1. *Current Biology, 10*(19), 1201–1204.

Du, X., Shi, H., Li, J., Dong, Y., Liang, J., Ye, J., et al. (2014). Mst1/Mst2 regulate development and function of regulatory T cells through modulation of Foxo1/Foxo3 stability in autoimmune disease. *Journal of Immunology, 192*(4), 1525–1535. https://doi.org/10.4049/jimmunol.1301060.

Essers, M. A., Weijzen, S., de Vries-Smits, A. M., Saarloos, I., de Ruiter, N. D., Bos, J. L., et al. (2004). FOXO transcription factor activation by oxidative stress mediated by the small GTPase Ral and JNK. *The EMBO Journal*, *23*(24), 4802–4812. https://doi.org/10.1038/sj.emboj.7600476.

Fabre, S., Lang, V., Harriague, J., Jobart, A., Unterman, T. G., Trautmann, A., et al. (2005). Stable activation of phosphatidylinositol 3-kinase in the T cell immunological synapse stimulates Akt signaling to FoxO1 nuclear exclusion and cell growth control. *Journal of Immunology*, *174*(7), 4161–4171.

Frescas, D., Valenti, L., & Accili, D. (2005). Nuclear trapping of the forkhead transcription factor FoxO1 via Sirt-dependent deacetylation promotes expression of glucogenetic genes. *The Journal of Biological Chemistry*, *280*(21), 20589–20595. https://doi.org/10.1074/jbc.M412357200.

Fukuoka, M., Daitoku, H., Hatta, M., Matsuzaki, H., Umemura, S., & Fukamizu, A. (2003). Negative regulation of forkhead transcription factor AFX (Foxo4) by CBP-induced acetylation. *International Journal of Molecular Medicine*, *12*(4), 503–508.

Furuyama, T., Nakazawa, T., Nakano, I., & Mori, N. (2000). Identification of the differential distribution patterns of mRNAs and consensus binding sequences for mouse DAF-16 homologues. *The Biochemical Journal*, *349*(Pt. 2), 629–634.

Gonzalez, E., & McGraw, T. E. (2009). The Akt kinases: Isoform specificity in metabolism and cancer. *Cell Cycle*, *8*(16), 2502–2508. https://doi.org/10.4161/cc.8.16.9335.

Greer, E. L., Dowlatshahi, D., Banko, M. R., Villen, J., Hoang, K., Blanchard, D., et al. (2007). An AMPK-FOXO pathway mediates longevity induced by a novel method of dietary restriction in C. elegans. *Current Biology*, *17*(19), 1646–1656. https://doi.org/10.1016/j.cub.2007.08.047.

Greer, E. L., Oskoui, P. R., Banko, M. R., Maniar, J. M., Gygi, M. P., Gygi, S. P., et al. (2007). The energy sensor AMP-activated protein kinase directly regulates the mammalian FOXO3 transcription factor. *The Journal of Biological Chemistry*, *282*(41), 30107–30119. https://doi.org/10.1074/jbc.M705325200.

Groszer, M., Erickson, R., Scripture-Adams, D. D., Dougherty, J. D., Le Belle, J., Zack, J. A., et al. (2006). PTEN negatively regulates neural stem cell self-renewal by modulating G0-G1 cell cycle entry. *Proceedings of the National Academy of Sciences of the United States of America*, *103*(1), 111–116. https://doi.org/10.1073/pnas.0509939103.

Guttilla, I. K., & White, B. A. (2009). Coordinate regulation of FOXO1 by miR-27a, miR-96, and miR-182 in breast cancer cells. *The Journal of Biological Chemistry*, *284*(35), 23204–23216. https://doi.org/10.1074/jbc.M109.031427.

Hardie, D. G., & Carling, D. (1997). The AMP-activated protein kinase—Fuel gauge of the mammalian cell? *European Journal of Biochemistry*, *246*(2), 259–273.

Hasseine, L. K., Hinault, C., Lebrun, P., Gautier, N., Paul-Bellon, R., & Van Obberghen, E. (2009). miR-139 impacts FoxO1 action by decreasing FoxO1 protein in mouse hepatocytes. *Biochemical and Biophysical Research Communications*, *390*(4), 1278–1282. https://doi.org/10.1016/j.bbrc.2009.10.135.

Hawley, S. A., Pan, D. A., Mustard, K. J., Ross, L., Bain, J., Edelman, A. M., et al. (2005). Calmodulin-dependent protein kinase kinase-beta is an alternative upstream kinase for AMP-activated protein kinase. *Cell Metabolism*, *2*(1), 9–19. https://doi.org/10.1016/j.cmet.2005.05.009.

Herzog, S., Hug, E., Meixlsperger, S., Paik, J. H., DePinho, R. A., Reth, M., et al. (2008). SLP-65 regulates immunoglobulin light chain gene recombination through the PI(3)K-PKB-Foxo pathway. *Nature Immunology*, *9*(6), 623–631. https://doi.org/10.1038/ni.1616.

Hirosumi, J., Tuncman, G., Chang, L., Gorgun, C. Z., Uysal, K. T., Maeda, K., et al. (2002). A central role for JNK in obesity and insulin resistance. *Nature*, *420*(6913), 333–336. https://doi.org/10.1038/nature01137.

Ho, K. K., McGuire, V. A., Koo, C. Y., Muir, K. W., de Olano, N., Maifoshie, E., et al. (2012). Phosphorylation of FOXO3a on Ser-7 by p38 promotes its nuclear localization in response to doxorubicin. *The Journal of Biological Chemistry*, *287*(2), 1545–1555. https://doi.org/10.1074/jbc.M111.284224.

Hollander, M. C., Blumenthal, G. M., & Dennis, P. A. (2011). PTEN loss in the continuum of common cancers, rare syndromes and mouse models. *Nature Reviews Cancer*, *11*(4), 289–301. https://doi.org/10.1038/nrc3037.

Honda, Y., & Honda, S. (1999). The daf-2 gene network for longevity regulates oxidative stress resistance and Mn-superoxide dismutase gene expression in Caenorhabditis elegans. *The FASEB Journal*, *13*(11), 1385–1393.

Hong, S. H., Lee, K. S., Kwak, S. J., Kim, A. K., Bai, H., Jung, M. S., et al. (2012). Minibrain/Dyrk1a regulates food intake through the Sir2-FOXO-sNPF/NPY pathway in Drosophila and mammals. *PLoS Genetics*, *8*(8), e1002857. https://doi.org/10.1371/journal.pgen.1002857.

Hu, M. C., Lee, D. F., Xia, W., Golfman, L. S., Ou-Yang, F., Yang, J. Y., et al. (2004). IkappaB kinase promotes tumorigenesis through inhibition of forkhead FOXO3a. *Cell*, *117*(2), 225–237.

Huang, H., Regan, K. M., Lou, Z., Chen, J., & Tindall, D. J. (2006). CDK2-dependent phosphorylation of FOXO1 as an apoptotic response to DNA damage. *Science*, *314*(5797), 294–297. https://doi.org/10.1126/science.1130512.

Huang, H., Regan, K. M., Wang, F., Wang, D., Smith, D. I., van Deursen, J. M., et al. (2005). Skp2 inhibits FOXO1 in tumor suppression through ubiquitin-mediated degradation. *Proceedings of the National Academy of Sciences of the United States of America*, *102*(5), 1649–1654. https://doi.org/10.1073/pnas.0406789102.

Huang, S. Y., Tsai, M. L., Chen, G. Y., Wu, C. J., & Chen, S. H. (2007). A systematic MS-based approach for identifying in vitro substrates of PKA and PKG in rat uteri. *Journal of Proteome Research*, *6*(7), 2674–2684. https://doi.org/10.1021/pr070134c.

Huo, X., Liu, S., Shao, T., Hua, H., Kong, Q., Wang, J., et al. (2014). GSK3 protein positively regulates type I insulin-like growth factor receptor through forkhead transcription factors FOXO1/3/4. *The Journal of Biological Chemistry*, *289*(36), 24759–24770. https://doi.org/10.1074/jbc.M114.580738.

Kane, S., Sano, H., Liu, S. C., Asara, J. M., Lane, W. S., Garner, C. C., et al. (2002). A method to identify serine kinase substrates. Akt phosphorylates a novel adipocyte protein with a Rab GTPase-activating protein (GAP) domain. *The Journal of Biological Chemistry*, *277*(25), 22115–22118. https://doi.org/10.1074/jbc.C200198200.

Kim, S., Kim, Y., Lee, J., & Chung, J. (2010). Regulation of FOXO1 by TAK1-Nemo-like kinase pathway. *The Journal of Biological Chemistry*, *285*(11), 8122–8129. https://doi.org/10.1074/jbc.M110.101824.

Klotz, L. O., Sanchez-Ramos, C., Prieto-Arroyo, I., Urbanek, P., Steinbrenner, H., & Monsalve, M. (2015). Redox regulation of FoxO transcription factors. *Redox Biology*, *6*, 51–72. https://doi.org/10.1016/j.redox.2015.06.019.

Kops, G. J., Dansen, T. B., Polderman, P. E., Saarloos, I., Wirtz, K. W., Coffer, P. J., et al. (2002). Forkhead transcription factor FOXO3a protects quiescent cells from oxidative stress. *Nature*, *419*(6904), 316–321. https://doi.org/10.1038/nature01036.

Kops, G. J., de Ruiter, N. D., De Vries-Smits, A. M., Powell, D. R., Bos, J. L., & Burgering, B. M. (1999). Direct control of the Forkhead transcription factor AFX by protein kinase B. *Nature*, *398*(6728), 630–634. https://doi.org/10.1038/19328.

Kuo, Y. C., Huang, K. Y., Yang, C. H., Yang, Y. S., Lee, W. Y., & Chiang, C. W. (2008). Regulation of phosphorylation of Thr-308 of Akt, cell proliferation, and survival by the B55alpha regulatory subunit targeting of the protein phosphatase 2A holoenzyme to Akt. *The Journal of Biological Chemistry*, *283*(4), 1882–1892. https://doi.org/10.1074/jbc.M709585200.

Lee, J. W., Chen, H., Pullikotil, P., & Quon, M. J. (2011). Protein kinase A-alpha directly phosphorylates FoxO1 in vascular endothelial cells to regulate expression of vascular cellular adhesion molecule-1 mRNA. *The Journal of Biological Chemistry*, *286*(8), 6423–6432. https://doi.org/10.1074/jbc.M110.180661.

Lee, S. S., Kennedy, S., Tolonen, A. C., & Ruvkun, G. (2003). DAF-16 target genes that control C. elegans life-span and metabolism. *Science*, *300*(5619), 644–647. https://doi.org/10.1126/science.1083614.

Lehtinen, M. K., Yuan, Z., Boag, P. R., Yang, Y., Villen, J., Becker, E. B., et al. (2006). A conserved MST-FOXO signaling pathway mediates oxidative-stress responses and extends life span. *Cell*, *125*(5), 987–1001. https://doi.org/10.1016/j.cell.2006.03.046.

Lin, K., Dorman, J. B., Rodan, A., & Kenyon, C. (1997). daf-16: An HNF-3/forkhead family member that can function to double the life-span of Caenorhabditis elegans. *Science*, *278*(5341), 1319–1322.

Luo, J. L., Kamata, H., & Karin, M. (2005). IKK/NF-kappaB signaling: Balancing life and death—A new approach to cancer therapy. *The Journal of Clinical Investigation*, *115*(10), 2625–2632. https://doi.org/10.1172/JCI26322.

Maehama, T., & Dixon, J. E. (1998). The tumor suppressor, PTEN/MMAC1, dephosphorylates the lipid second messenger, phosphatidylinositol 3,4,5-trisphosphate. *The Journal of Biological Chemistry*, *273*(22), 13375–13378.

Mammucari, C., Milan, G., Romanello, V., Masiero, E., Rudolf, R., Del Piccolo, P., et al. (2007). FoxO3 controls autophagy in skeletal muscle in vivo. *Cell Metabolism*, *6*(6), 458–471. https://doi.org/10.1016/j.cmet.2007.11.001.

Matsumoto, M., Pocai, A., Rossetti, L., Depinho, R. A., & Accili, D. (2007). Impaired regulation of hepatic glucose production in mice lacking the forkhead transcription factor Foxo1 in liver. *Cell Metabolism*, *6*(3), 208–216. https://doi.org/10.1016/j.cmet.2007.08.006.

Matsuzaki, H., Daitoku, H., Hatta, M., Aoyama, H., Yoshimochi, K., & Fukamizu, A. (2005). Acetylation of Foxo1 alters its DNA-binding ability and sensitivity to phosphorylation. *Proceedings of the National Academy of Sciences of the United States of America*, *102*(32), 11278–11283. https://doi.org/10.1073/pnas.0502738102.

Medema, R. H., Kops, G. J., Bos, J. L., & Burgering, B. M. (2000). AFX-like Forkhead transcription factors mediate cell-cycle regulation by Ras and PKB through p27kip1. *Nature*, *404*(6779), 782–787. https://doi.org/10.1038/35008115.

Mihaylova, M. M., Vasquez, D. S., Ravnskjaer, K., Denechaud, P. D., Yu, R. T., Alvarez, J. G., et al. (2011). Class IIa histone deacetylases are hormone-activated regulators of FOXO and mammalian glucose homeostasis. *Cell*, *145*(4), 607–621. https://doi.org/10.1016/j.cell.2011.03.043.

Minami, Y., Kono, T., Miyazaki, T., & Taniguchi, T. (1993). The IL-2 receptor complex: Its structure, function, and target genes. *Annual Review of Immunology*, *11*, 245–268. https://doi.org/10.1146/annurev.iy.11.040193.001333.

Miyamoto, K., Araki, K. Y., Naka, K., Arai, F., Takubo, K., Yamazaki, S., et al. (2007). Foxo3a is essential for maintenance of the hematopoietic stem cell pool. *Cell Stem Cell*, *1*(1), 101–112. https://doi.org/10.1016/j.stem.2007.02.001.

Modur, V., Nagarajan, R., Evers, B. M., & Milbrandt, J. (2002). FOXO proteins regulate tumor necrosis factor-related apoptosis inducing ligand expression. Implications for PTEN mutation in prostate cancer. *The Journal of Biological Chemistry*, *277*(49), 47928–47937. https://doi.org/10.1074/jbc.M207509200.

Motta, M. C., Divecha, N., Lemieux, M., Kamel, C., Chen, D., Gu, W., et al. (2004). Mammalian SIRT1 represses forkhead transcription factors. *Cell*, *116*(4), 551–563.

Myatt, S. S., Wang, J., Monteiro, L. J., Christian, M., Ho, K. K., Fusi, L., et al. (2010). Definition of microRNAs that repress expression of the tumor suppressor gene FOXO1 in endometrial cancer. *Cancer Research*, *70*(1), 367–377. https://doi.org/10.1158/0008-5472.CAN-09-1891.

Nakamura, N., Ramaswamy, S., Vazquez, F., Signoretti, S., Loda, M., & Sellers, W. R. (2000). Forkhead transcription factors are critical effectors of cell death and cell cycle arrest downstream of PTEN. *Molecular and Cellular Biology, 20*(23), 8969–8982.

Nemoto, S., & Finkel, T. (2002). Redox regulation of forkhead proteins through a p66shc-dependent signaling pathway. *Science, 295*(5564), 2450–2452. https://doi.org/10.1126/science.1069004.

Obsil, T., Ghirlando, R., Anderson, D. E., Hickman, A. B., & Dyda, F. (2003). Two 14-3-3 binding motifs are required for stable association of Forkhead transcription factor FOXO4 with 14-3-3 proteins and inhibition of DNA binding. *Biochemistry, 42*(51), 15264–15272. https://doi.org/10.1021/bi0352724.

Obsil, T., & Obsilova, V. (2008). Structure/function relationships underlying regulation of FOXO transcription factors. *Oncogene, 27*(16), 2263–2275. https://doi.org/10.1038/onc.2008.20.

Obsilova, V., Vecer, J., Herman, P., Pabianova, A., Sulc, M., Teisinger, J., et al. (2005). 14-3-3 Protein interacts with nuclear localization sequence of forkhead transcription factor FoxO4. *Biochemistry, 44*(34), 11608–11617. https://doi.org/10.1021/bi050618r.

Ogg, S., Paradis, S., Gottlieb, S., Patterson, G. I., Lee, L., Tissenbaum, H. A., et al. (1997). The Fork head transcription factor DAF-16 transduces insulin-like metabolic and longevity signals in C. elegans. *Nature, 389*(6654), 994–999. https://doi.org/10.1038/40194.

Oh, S. W., Mukhopadhyay, A., Svrzikapa, N., Jiang, F., Davis, R. J., & Tissenbaum, H. A. (2005). JNK regulates lifespan in Caenorhabditis elegans by modulating nuclear translocation of forkhead transcription factor/DAF-16. *Proceedings of the National Academy of Sciences of the United States of America, 102*(12), 4494–4499. https://doi.org/10.1073/pnas.0500749102.

Padmanabhan, S., Mukhopadhyay, A., Narasimhan, S. D., Tesz, G., Czech, M. P., & Tissenbaum, H. A. (2009). A PP2A regulatory subunit regulates C. elegans insulin/IGF-1 signaling by modulating AKT-1 phosphorylation. *Cell, 136*(5), 939–951. https://doi.org/10.1016/j.cell.2009.01.025.

Paik, J. H., Ding, Z., Narurkar, R., Ramkissoon, S., Muller, F., Kamoun, W. S., et al. (2009). FoxOs cooperatively regulate diverse pathways governing neural stem cell homeostasis. *Cell Stem Cell, 5*(5), 540–553. https://doi.org/10.1016/j.stem.2009.09.013.

Puigserver, P., Rhee, J., Donovan, J., Walkey, C. J., Yoon, J. C., Oriente, F., et al. (2003). Insulin-regulated hepatic gluconeogenesis through FOXO1-PGC-1alpha interaction. *Nature, 423*(6939), 550–555. https://doi.org/10.1038/nature01667.

Ramaswamy, S., Nakamura, N., Sansal, I., Bergeron, L., & Sellers, W. R. (2002). A novel mechanism of gene regulation and tumor suppression by the transcription factor FKHR. *Cancer Cell, 2*(1), 81–91.

Rena, G., Bain, J., Elliott, M., & Cohen, P. (2004). D4476, a cell-permeant inhibitor of CK1, suppresses the site-specific phosphorylation and nuclear exclusion of FOXO1a. *EMBO Reports, 5*(1), 60–65. https://doi.org/10.1038/sj.embor.7400048.

Rena, G., Prescott, A. R., Guo, S., Cohen, P., & Unterman, T. G. (2001). Roles of the forkhead in rhabdomyosarcoma (FKHR) phosphorylation sites in regulating 14-3-3 binding, transactivation and nuclear targetting. *The Biochemical Journal, 354*(Pt. 3), 605–612.

Rena, G., Woods, Y. L., Prescott, A. R., Peggie, M., Unterman, T. G., Williams, M. R., et al. (2002). Two novel phosphorylation sites on FKHR that are critical for its nuclear exclusion. *The EMBO Journal, 21*(9), 2263–2271. https://doi.org/10.1093/emboj/21.9.2263.

Renault, V. M., Rafalski, V. A., Morgan, A. A., Salih, D. A., Brett, J. O., Webb, A. E., et al. (2009). FoxO3 regulates neural stem cell homeostasis. *Cell Stem Cell, 5*(5), 527–539. https://doi.org/10.1016/j.stem.2009.09.014.

Sarbassov, D. D., Guertin, D. A., Ali, S. M., & Sabatini, D. M. (2005). Phosphorylation and regulation of Akt/PKB by the rictor-mTOR complex. *Science*, *307*(5712), 1098–1101. https://doi.org/10.1126/science.1106148.

Satyanarayana, A., & Kaldis, P. (2009). A dual role of Cdk2 in DNA damage response. *Cell Division*, *4*, 9. https://doi.org/10.1186/1747-1028-4-9.

Shaw, L. M. (2011). The insulin receptor substrate (IRS) proteins: At the intersection of metabolism and cancer. *Cell Cycle*, *10*(11), 1750–1756. https://doi.org/10.4161/cc.10.11.15824.

Shaw, R. J., Kosmatka, M., Bardeesy, N., Hurley, R. L., Witters, L. A., DePinho, R. A., et al. (2004). The tumor suppressor LKB1 kinase directly activates AMP-activated kinase and regulates apoptosis in response to energy stress. *Proceedings of the National Academy of Sciences of the United States of America*, *101*(10), 3329–3335. https://doi.org/10.1073/pnas.0308061100.

Siddle, K. (2011). Signalling by insulin and IGF receptors: Supporting acts and new players. *Journal of Molecular Endocrinology*, *47*(1), R1–10. https://doi.org/10.1530/JME-11-0022.

Singh, A., Ye, M., Bucur, O., Zhu, S., Tanya Santos, M., Rabinovitz, I., et al. (2010). Protein phosphatase 2A reactivates FOXO3a through a dynamic interplay with 14-3-3 and AKT. *Molecular Biology of the Cell*, *21*(6), 1140–1152. https://doi.org/10.1091/mbc.E09-09-0795.

Small, E. M., O'Rourke, J. R., Moresi, V., Sutherland, L. B., McAnally, J., Gerard, R. D., et al. (2010). Regulation of PI3-kinase/Akt signaling by muscle-enriched microRNA-486. *Proceedings of the National Academy of Sciences of the United States of America*, *107*(9), 4218–4223. https://doi.org/10.1073/pnas.1000300107.

Smith, D. J., Stevens, M. E., Sudanagunta, S. P., Bronson, R. T., Makhinson, M., Watabe, A. M., et al. (1997). Functional screening of 2 Mb of human chromosome 21q22.2 in transgenic mice implicates minibrain in learning defects associated with Down syndrome. *Nature Genetics*, *16*(1), 28–36. https://doi.org/10.1038/ng0597-28.

Stahl, M., Dijkers, P. F., Kops, G. J., Lens, S. M., Coffer, P. J., Burgering, B. M., et al. (2002). The forkhead transcription factor FoxO regulates transcription of p27Kip1 and Bim in response to IL-2. *Journal of Immunology*, *168*(10), 5024–5031.

Szydlowski, M., Jablonska, E., & Juszczynski, P. (2014). FOXO1 transcription factor: A critical effector of the PI3K-AKT axis in B-cell development. *International Reviews of Immunology*, *33*(2), 146–157. https://doi.org/10.3109/08830185.2014.885022.

Tothova, Z., Kollipara, R., Huntly, B. J., Lee, B. H., Castrillon, D. H., Cullen, D. E., et al. (2007). FoxOs are critical mediators of hematopoietic stem cell resistance to physiologic oxidative stress. *Cell*, *128*(2), 325–339. https://doi.org/10.1016/j.cell.2007.01.003.

Tran, H., Brunet, A., Grenier, J. M., Datta, S. R., Fornace, A. J., Jr., DiStefano, P. S., et al. (2002). DNA repair pathway stimulated by the forkhead transcription factor FOXO3a through the Gadd45 protein. *Science*, *296*(5567), 530–534. https://doi.org/10.1126/science.1068712.

Tsai, W. B., Chung, Y. M., Takahashi, Y., Xu, Z., & Hu, M. C. (2008). Functional interaction between FOXO3a and ATM regulates DNA damage response. *Nature Cell Biology*, *10*(4), 460–467. https://doi.org/10.1038/ncb1709.

Tsai, K. L., Sun, Y. J., Huang, C. Y., Yang, J. Y., Hung, M. C., & Hsiao, C. D. (2007). Crystal structure of the human FOXO3a-DBD/DNA complex suggests the effects of post-translational modification. *Nucleic Acids Research*, *35*(20), 6984–6994. https://doi.org/10.1093/nar/gkm703.

Valis, K., Prochazka, L., Boura, E., Chladova, J., Obsil, T., Rohlena, J., et al. (2011). Hippo/Mst1 stimulates transcription of the proapoptotic mediator NOXA in a FoxO1-dependent manner. *Cancer Research*, *71*(3), 946–954. https://doi.org/10.1158/0008-5472.CAN-10-2203.

van der Heide, L. P., Jacobs, F. M., Burbach, J. P., Hoekman, M. F., & Smidt, M. P. (2005). FoxO6 transcriptional activity is regulated by Thr26 and Ser184, independent of nucleo-cytoplasmic shuttling. *The Biochemical Journal*, *391*(Pt. 3), 623–629. https://doi.org/10.1042/BJ20050525.

van der Horst, A., de Vries-Smits, A. M., Brenkman, A. B., van Triest, M. H., van den Broek, N., Colland, F., et al. (2006). FOXO4 transcriptional activity is regulated by monoubiquitination and USP7/HAUSP. *Nature Cell Biology*, *8*(10), 1064–1073. https://doi.org/10.1038/ncb1469.

van der Horst, A., Tertoolen, L. G., de Vries-Smits, L. M., Frye, R. A., Medema, R. H., & Burgering, B. M. (2004). FOXO4 is acetylated upon peroxide stress and deacetylated by the longevity protein hSir2(SIRT1). *The Journal of Biological Chemistry*, *279*(28), 28873–28879. https://doi.org/10.1074/jbc.M401138200.

Vereshchagina, N., Ramel, M. C., Bitoun, E., & Wilson, C. (2008). The protein phosphatase PP2A-B' subunit Widerborst is a negative regulator of cytoplasmic activated Akt and lipid metabolism in Drosophila. *Journal of Cell Science*, *121*(Pt. 20), 3383–3392. https://doi.org/10.1242/jcs.035220.

Wang, M. C., Bohmann, D., & Jasper, H. (2003). JNK signaling confers tolerance to oxidative stress and extends lifespan in Drosophila. *Developmental Cell*, *5*(5), 811–816.

Wang, M. C., Bohmann, D., & Jasper, H. (2005). JNK extends life span and limits growth by antagonizing cellular and organism-wide responses to insulin signaling. *Cell*, *121*(1), 115–125. https://doi.org/10.1016/j.cell.2005.02.030.

Wang, F., Marshall, C. B., & Ikura, M. (2015). Forkhead followed by disordered tail: The intrinsically disordered regions of FOXO3a. *Intrinsically Disordered Proteins*, *3*(1), e1056906. https://doi.org/10.1080/21690707.2015.1056906.

Wang, F., Marshall, C. B., Yamamoto, K., Li, G. Y., Gasmi-Seabrook, G. M., Okada, H., et al. (2012). Structures of KIX domain of CBP in complex with two FOXO3a transactivation domains reveal promiscuity and plasticity in coactivator recruitment. *Proceedings of the National Academy of Sciences of the United States of America*, *109*(16), 6078–6083. https://doi.org/10.1073/pnas.1119073109.

Webb, A. E., & Brunet, A. (2014). FOXO transcription factors: Key regulators of cellular quality control. *Trends in Biochemical Sciences*, *39*(4), 159–169. https://doi.org/10.1016/j.tibs.2014.02.003.

Webb, A. E., Kundaje, A., & Brunet, A. (2016). Characterization of the direct targets of FOXO transcription factors throughout evolution. *Aging Cell*, *15*(4), 673–685. https://doi.org/10.1111/acel.12479.

Webb, A. E., Pollina, E. A., Vierbuchen, T., Urban, N., Ucar, D., Leeman, D. S., et al. (2013). FOXO3 shares common targets with ASCL1 genome-wide and inhibits ASCL1-dependent neurogenesis. *Cell Reports*, *4*(3), 477–491. https://doi.org/10.1016/j.celrep.2013.06.035.

Woods, A., Dickerson, K., Heath, R., Hong, S. P., Momcilovic, M., Johnstone, S. R., et al. (2005). Ca2+/calmodulin-dependent protein kinase kinase-beta acts upstream of AMP-activated protein kinase in mammalian cells. *Cell Metabolism*, *2*(1), 21–33. https://doi.org/10.1016/j.cmet.2005.06.005.

Woods, Y. L., Rena, G., Morrice, N., Barthel, A., Becker, W., Guo, S., et al. (2001). The kinase DYRK1A phosphorylates the transcription factor FKHR at Ser329 in vitro, a novel in vivo phosphorylation site. *The Biochemical Journal*, *355*(Pt. 3), 597–607.

Wu, L., Li, H., Jia, C. Y., Cheng, W., Yu, M., Peng, M., et al. (2012). MicroRNA-223 regulates FOXO1 expression and cell proliferation. *FEBS Letters*, *586*(7), 1038–1043. https://doi.org/10.1016/j.febslet.2012.02.050.

Xie, Q., Hao, Y., Tao, L., Peng, S., Rao, C., Chen, H., et al. (2012). Lysine methylation of FOXO3 regulates oxidative stress-induced neuronal cell death. *EMBO Reports*, *13*(4), 371–377. https://doi.org/10.1038/embor.2012.25.

Xuan, Z., & Zhang, M. Q. (2005). From worm to human: Bioinformatics approaches to identify FOXO target genes. *Mechanisms of Ageing and Development*, *126*(1), 209–215. https://doi.org/10.1016/j.mad.2004.09.021.

Yamagata, K., Daitoku, H., Takahashi, Y., Namiki, K., Hisatake, K., Kako, K., et al. (2008). Arginine methylation of FOXO transcription factors inhibits their phosphorylation by Akt. *Molecular Cell*, *32*(2), 221–231. https://doi.org/10.1016/j.molcel.2008.09.013.

Yan, L., Lavin, V. A., Moser, L. R., Cui, Q., Kanies, C., & Yang, E. (2008). PP2A regulates the pro-apoptotic activity of FOXO1. *The Journal of Biological Chemistry*, *283*(12), 7411–7420. https://doi.org/10.1074/jbc.M708083200.

Yang, J. Y., Zong, C. S., Xia, W., Yamaguchi, H., Ding, Q., Xie, X., et al. (2008). ERK promotes tumorigenesis by inhibiting FOXO3a via MDM2-mediated degradation. *Nature Cell Biology*, *10*(2), 138–148. https://doi.org/10.1038/ncb1676.

Yilmaz, O. H., Valdez, R., Theisen, B. K., Guo, W., Ferguson, D. O., Wu, H., et al. (2006). Pten dependence distinguishes haematopoietic stem cells from leukaemia-initiating cells. *Nature*, *441*(7092), 475–482. https://doi.org/10.1038/nature04703.

Yuan, Z., Becker, E. B., Merlo, P., Yamada, T., DiBacco, S., Konishi, Y., et al. (2008). Activation of FOXO1 by Cdk1 in cycling cells and postmitotic neurons. *Science*, *319*(5870), 1665–1668. https://doi.org/10.1126/science.1152337.

Yuan, Z., Lehtinen, M. K., Merlo, P., Villen, J., Gygi, S., & Bonni, A. (2009). Regulation of neuronal cell death by MST1-FOXO1 signaling. *The Journal of Biological Chemistry*, *284*(17), 11285–11292. https://doi.org/10.1074/jbc.M900461200.

Zhang, J., Grindley, J. C., Yin, T., Jayasinghe, S., He, X. C., Ross, J. T., et al. (2006). PTEN maintains haematopoietic stem cells and acts in lineage choice and leukaemia prevention. *Nature*, *441*(7092), 518–522. https://doi.org/10.1038/nature04747.

Zhang, W., Hietakangas, V., Wee, S., Lim, S. C., Gunaratne, J., & Cohen, S. M. (2013). ER stress potentiates insulin resistance through PERK-mediated FOXO phosphorylation. *Genes & Development*, *27*(4), 441–449. https://doi.org/10.1101/gad.201731.112.

Zhao, X., Gan, L., Pan, H., Kan, D., Majeski, M., Adam, S. A., et al. (2004). Multiple elements regulate nuclear/cytoplasmic shuttling of FOXO1: Characterization of phosphorylation- and 14-3-3-dependent and -independent mechanisms. *The Biochemical Journal*, *378*(Pt. 3), 839–849. https://doi.org/10.1042/BJ20031450.

CHAPTER EIGHT

FOXO3 and Exceptional Longevity: Insights From Hydra to Humans

Philip M.C. Davy*, Richard C. Allsopp*, Timothy A. Donlon[†,‡], Brian J. Morris[†,§,¶], Donald Craig Willcox[†,§,||], Bradley J. Willcox[†,§,1]
*Institute for Biogenesis Research, University of Hawaii, Honolulu, HI, United States
[†]Honolulu Heart Program/Honolulu-Asia Aging Study, Kuakini Medical Center, Honolulu, HI, United States
[‡]Ohana Genetics, Honolulu, HI, United States
[§]John A. Burns School of Medicine, University of Hawaii, Honolulu, HI, United States
[¶]School of Medical Sciences and Bosch Institute, University of Sydney, Sydney, NSW, Australia
[||]Okinawa International University, Ginowan, Okinawa, Japan
[1]Corresponding authors: e-mail address: willcox@hawaii.edu

Contents

Abstract

Aging is a complex, multifactorial process with significant plasticity. While several biological pathways appear to influence aging, few genes have been identified that are both evolutionarily conserved and have a strong impact on aging and age-related phenotypes. The FoxO3 gene (*FOXO3*), and its homologs in model organisms, appears especially important, forming a key gene in the insulin/insulin-like growth factor-signaling pathway, and influencing life span across diverse species. We highlight some of the key findings that are associated with FoxO3 protein, its gene and homologs in relation to lifespan in different species, and the insights these findings might provide about the molecular, cellular, and physiological processes that modulate aging and longevity in humans.

1. FoxO GENES AS DETERMINANTS OF LONGEVITY

Biological aging is a process that can be difficult to define outside of a specific context. However, for individuals, the process of aging over time consistently incurs the loss of homeostasis that eventually results in death. The lifespan of an individual in relation to other members of their species

Current Topics in Developmental Biology, Volume 127
ISSN 0070-2153
https://doi.org/10.1016/bs.ctdb.2017.10.001

can be expressed as longevity or "exceptional longevity" if it exceeds some specified threshold (Willcox, Willcox, & Ferrucci, 2008). Many extrinsic factors interact with intrinsic genetic factors to influence longevity, resulting in significant plasticity of the life span of multicellular organisms from invertebrates to humans (Kenyon, 2005). Both entropic loss of order and genetic programs are deterministic factors in the rate of aging and overall longevity of living organisms (Hayflick, 2007). Although we have limited tools for influencing the second law of thermodynamics in cells, we have many tools for studying the impact of genetic variations on longevity (Deelen, Beekman, Capri, Franceschi, & Slagboom, 2013). The exact contribution of the lottery of genetics to longevity varies between individuals, but has been consistently shown to exceed 20% in average-lived populations (Budovsky et al., 2013; Shadyab & LaCroix, 2015) and approaches 50% in select long-lived populations, such as centenarians (Willcox, Willcox, Hsueh, & Suzuki, 2006; Yashin, Iachine, & Harris, 1999).

The first candidate genes to be associated with exceptional longevity (100+ years of age) in humans were *HLA-DRW9* ($P=0.0001$) and *HLA-DR1* ($P=0.037$), which were found in human leukocyte antigen (HLA) complex region genes in long-lived Okinawans in lower and higher frequency, respectively, by the Okinawa Centenarian Study (Takata, Suzuki, Ishii, Sekiguchi, & Iri, 1987). The first widely replicated longevity-associated gene was *APOE*, a lipid-binding protein involved in important transport functions in the lymphatic and vascular systems (Schächter et al., 1994). The frequency of the *APOE* ε4 allele, considered a risk allele for cardiovascular disease (CVD) and Alzheimer's disease (AD), was originally observed to be lower in centenarians than in adult controls (age 20–70) by 6% ($P<0.001$), whereas the frequency of the protective ε2 allele was 6% higher ($P<0.01$), leading to an apparent association with longevity (Schächter et al., 1994). This initial finding in Caucasians (French) has since been confirmed in multiple Caucasian and Asian populations (Garatachea et al., 2015). Carriers of the ε2 allele have an approximately 20% lower prevalence of coronary heart disease (Bennet et al., 2007) and significantly lower cognitive decline in extreme old age (97+, odds ratio 0.58) (Lindahl-Jacobsen et al., 2013). Attrition of ε4 allele carriers from mortality at earlier ages may explain the association of the ε2 allele with longevity (Sebastiani et al., 2017).

Unlike ApoE (Davignon, Gregg, & Sing, 1988), interest in the role of the forkhead box O (FoxO) family of transcription factors with lifespan began not in humans—but in nematodes. Kenyon and colleagues reported that mutations in the *daf-2* gene of *Caenorhabditis elegans* resulted in a 2.3-fold increase in their lifespan, with the majority of mutants also retaining their health after their wild-type counterparts had already died (Kenyon, Chang, Gensch, Rudner, & Tabtiang, 1993). The morbidity compression (i.e., healthspan) and longevity-conferring (i.e., lifespan) action of the *daf-2* mutant were dependent on a related gene, *daf-16* (Kenyon et al., 1993), which proved to be the *C. elegans* ortholog of human FoxO genes (Ogg et al., 1997).

We carried out the first investigation into the possible association of variants in the 125 kb FoxO gene (*FOXO*) with human longevity in a nested case–control candidate gene study targeting insulin signaling and glucose homeostasis genes, which are associated with the longevity benefits of caloric restriction (Willcox, Donlon, et al., 2008). Utilizing a large, longitudinal cohort of Japanese-American men well characterized for aging phenotypes—the Honolulu Heart Program—we found the first association of several single nucleotide polymorphisms (SNPs) in human *FOXO3* (and to a lesser extent *FOXO1*) with longevity. Heterozygotes and homozygotes of the protective variant of the *rs2802292* SNP in particular had a 1.9-fold ($P < 0.0003$), or 2.8-fold ($P < 0.0007$) increased chance, respectively, of living to ≥95 years compared with homozygote carriers of the nonprotective variant (Willcox, Donlon, et al., 2008). This significant association between *FOXO3* SNPs and longevity in humans has been independently replicated in multiple populations (Morris et al., 2014), including a genome-wide association study (GWAS) (Broer et al., 2015).

A GWAS follows the same broad design concept as a candidate gene study in that each use a defined outcome within the sample population to investigate the association of genetic variation with a specific phenotype. GWAS are considered to produce more rigorous associations because they assess polymorphisms across the entire genome, and because they employ a Bonferroni correction for multiple testing that is at least three orders of magnitude greater than those used to assess the statistical significance of candidate gene hypotheses. The drawback of requiring a strength of association such that the likelihood of the null hypothesis being disproved is less than 1 in 20 million ($P < 5 \times 10^{-8}$) is that a large cohort must be recruited for a GWAS. Studies from the CHARGE consortium exemplify this. A metaanalysis of potential longevity associations performed in 2010 involved 1836 longevity cases (90+ years of age) and 1955 controls (aged 55–80 years), but *FOXO3* SNP(s) did not appear in the 24 most statistically significant candidates despite consistent replication with longevity in independent studies of Japanese-American (Willcox, Donlon, et al., 2008),

Italian (Anselmi et al., 2009), German (Flachsbart et al., 2009), Han Chinese (Li et al., 2009), and Danish (Soerensen et al., 2010), among other populations. In fact, none of those 24 candidates from the original CHARGE metaanalysis achieved genome-wide significance (Newman et al., 2010). Just 5 years later, however, the CHARGE consortium had expanded its metaanalysis to 6036 longevity cases and 3757 controls and confirmed the candidacy of SNPs in *APOE* (odds ratio = 1.20; corrected P value = 4.8×10^{-4}) and *FOXO3* (odds ratio = 1.17; uncorrected P value = 1.85×10^{-10}) as potential longevity genes (Broer et al., 2015). Although the Japanese-American and Italian cohorts were exclusively male, female participants were included in the studies of German, Danish, Han-Chinese, and various Caucasian populations included in the CHARGE consortium of cohorts. This is important since women have a greater average lifespan than men, and because gender-specific effects are observed in studies of association of polymorphisms and can affect the power of genetic association studies (Magi, Lindgren, & Morris, 2010). Both males and females display an association between different *FOXO3* SNPs and longevity, with some variability between each when studied separately (Li et al., 2009; Soerensen et al., 2010). Given the high linkage disequilibrium between *FOXO3* SNPs (Soerensen et al., 2010), we can reasonably assume that effects observed for one SNP apply to another, and therefore that the "protective" alleles of *FOXO3* may confer longevity to both men and women.

The exact physiological mechanism(s) by which *FOXO3* SNPs confer longevity were not immediately clear from the genetic association studies. The effect does not appear to be mediated by a coding variant in an exon but rather is associated with a ~100 kb region in or near intron 2, which contains noncoding SNPs in high LD (Donlon et al., 2012). To quantify the effect size of the protective *FOXO3* allele(s) on risk of mortality, we studied the association of *rs2802292* SNP alleles with major age-associated clinical causes of death—i.e., coronary heart disease (CHD), cancer, hypertension, diabetes, and stroke, among others. We investigated the effect size on all-cause mortality and cause-specific mortality for the *FOXO3* protective allele across 17 years of prospective data in two cohorts that represented three distinct ethnic groups. The risk for all-cause mortality was 10% lower in carriers of the protective allele (hazard ratio = 0.90; 95% CI 0.84–0.95; $P = 0.001$), with CHD being the only cause of death significantly associated with lack of the protective allele. A metaanalysis of results from three ethnically different populations (Asian, Black, and Caucasian) revealed a 26%

reduction in mortality from CHD mortality (combined hazard ratio = 0.74; 95% CI 0.64–0.86; $P = 0.00004$) for carriers of the protective allele (Willcox et al., 2016). Thus, the protective alleles of both *APOE* and *FOXO3* each reduce the risk of death from CHD (Bennet et al., 2007; Willcox et al., 2016), albeit likely via different principal mechanisms.

2. THE INSULIN SIGNALING PATHWAY AND LIFE SPAN

The link between the insulin/IGF-1 signaling (IIS) pathway and longevity was first apparent from research involving *C. elegans* in the 1990s when *daf-2* was identified as the homolog of the insulin receptor, and its downstream effector, *daf-16*, was found to be the homolog of *FOXO* (Dorman, Albinder, Shroyer, & Kenyon, 1995; Kenyon et al., 1993; Ogg et al., 1997). The IIS pathway begins with the transmembrane DAF-2/insulin receptor (invertebrates/mammals). Binding of insulin or growth hormones initiates a phosphorylation cascade that involves AGE-1/PI3K and results in the AKT (protein kinase B)-mediated phosphorylation of DAF-16/FoxO3, causing the repression of FoxO3 translocation to the nucleus (Fig. 1). In this manner, the energy-sensing pathway regulates the FoxO3 protein by blocking its activity as a transcription factor in the presence of the appropriate stimulus. In *C. elegans*, the extension of lifespan associated with repression of the IIS pathway, for example, as a result of genetic mutation(*s*) (Kenyon et al., 1993), is entirely dependent on a functional copy of *DAF-16*.

The functions of FoxO3 and its homologs in the IIS pathway are highly conserved (O'Neill, Kiely, Coakley, Manning, & Long-Smith, 2012). The link between the IIS pathway and longevity in mice (Blüher, Kahn, & Kahn, 2003; Shimokawa et al., 2015), *Drosophila melanogaster* (Clancy et al., 2001; Giannakou et al., 2004; Tatar et al., 2001), and *Saccharomyces cerevisiae* (Fabrizio, Pozza, Pletcher, Gendron, & Longo, 2001; Postnikoff, Malo, Wong, & Harkness, 2012), formed the basis for our studies of insulin signaling genes in candidate gene studies of human longevity (Willcox, Donlon, et al., 2008). Interestingly, although we observed a strong association of *FOXO3* variants with longevity, SNPs in other genes in the IIS pathway (specifically, *ATF4*, *CBL*, *CDKN2*, *EXO1*, and *JUN*) did not exhibit an association (Morris et al., 2014). Nevertheless, a previous study found an association between an *AKT* SNP and longevity (Pawlikowska et al., 2009). The range of species across which the IIS pathway affects

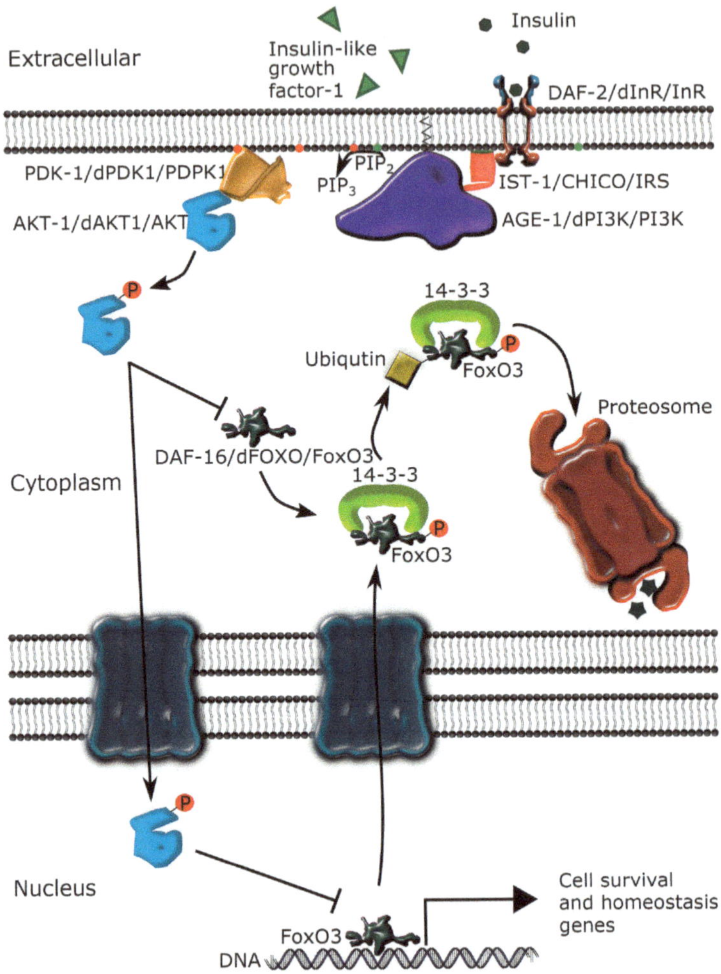

Fig. 1 The role of the insulin/IGF-1 signaling pathway in regulating FoxO3 activity is conserved between *C. elegans/D. melanogaster*/humans. Following extracellular binding of the insulin receptor (InR) by a substrate (e.g., insulin or insulin-like growth factor-1), the insulin receptor substrate (IRS) associates with InR on the cytoplasmic side of the cellular membrane. IRS in turn recruits PI3K which phosphorylates the membrane lipid phosphatidylinositol-2-phosphate (PIP_2), creating phosphatidylinositol-3-phosphate (PIP_3). PIP_3 is bound by phosphoinositide-dependent protein kinase-1 (PDPK1) which activates AKT via phosphorylation. Active AKT inhibits the nuclear translocation of FoxO3 and promotes its exclusion from the nucleus via a phosphorylation event. Phosphorylated FoxO3 is bound by the 14-3-3 regulator protein, leading to ubiquitination and proteasomal degradation.

longevity supports the hypothesis that a highly conserved longevity mechanism is centered around *FOXO3*. Given the key role of FoxO3 in translating IIS pathway signals into changes in the pattern of gene expression, and a significant difference in longevity between carriers of *FOXO3* alleles, it seems reasonable to conclude that FoxO3 forms part of an evolutionarily conserved control node with considerable influence on longevity.

Caloric restriction (CR) is the single strongest intervention for extension of lifespan, with a 30% CR resulting in approximately 30% increased life span in rodents and more across the spectrum of other model organisms (Bonkowski, Rocha, Masternak, Regaiey, & Bartke, 2006; Harrison, Archer, & Astle, 1984; Lakowski & Hekimi, 1998; McCay, Crowell, & Maynard, 1935). The effects of CR are strongly connected to the IIS pathway. Mutations in the insulin receptor of various species lead to longer life spans (Bartke et al., 2001; Blüher et al., 2003; Clancy et al., 2001; Kimura, Tissenbaum, Liu, & Ruvkun, 1997; Lakowski & Hekimi, 1998). Adipose tissue seems to be key to the effects of FoxO and IIS on lifespan. In *D. melanogaster*, induced expression of the native FoxO gene (*dFOXO*) in the fat body of adults produced a sex-specific increase in lifespan of 20%–50% favoring female flies (Giannakou et al., 2004). In mice, fat-specific knockout of the insulin receptor results in an 18% increase in lifespan, accompanied by reduced adiposity (50%–70% less fat mass) (Blüher et al., 2003). Additionally, without functional FoxO3, mice cannot benefit from the lifespan extension effects of CR (Shimokawa et al., 2015). Studies of nonhuman primates (Rhesus monkeys) have shown that CR improves healthspan and may increase lifespan (Mattison et al., 2017). The possibility of an effect in humans is supported by improvements in various biomarkers of healthy aging, cardiovascular and metabolic health in particular (Das, Balasubramanian, & Weerasekara, 2017; Fontana, Meyer, Klein, & Holloszy, 2004; Most, Tosti, Redman, & Fontana, 2017; Ravussin et al., 2015; Walford, Mock, Verdery, & MacCallum, 2002). The traditional Okinawan diet, which results in a 10%–15% CR vs the average diet of other Japanese, may contribute to these people having the world's longest lifespan and a remarkable healthspan (Bernstein et al., 2004; Willcox & Willcox, 2014; Willcox et al., 2007). There is no question that most people in Western society consume an excess of calories and that this poses a significant hazard to healthspan and lifespan. On the other hand, diets that push the limits of caloric restriction are associated with negative effects on lifespan and healthspan (Walford et al., 2002; Willcox et al., 2004).

3. TRANSCRIPTIONAL IMPACTS OF FoxO GENES ACROSS SPECIES

FoxOs are transcriptional regulators that sit at the interface between multiple signaling pathways and gene expression. FoxO3 influences expression of a wide variety of genes and plays a key role in numerous cellular processes (Morris, Willcox, Donlon, & Willcox, 2015). These include redox regulation, autophagy, energy homeostasis, DNA repair, cell cycle arrest, and stem cell homeostasis. The role of FoxO3 can be accurately summarized as being a central component of cellular stress resistance, be it oxidative, metabolic, or replicative (Morris et al., 2015). FoxO3 binds a 5′-TTGTTTAC-3′ sequence found at relatively high frequency throughout the genome. This sequence is recognized by a highly conserved DNA-binding domain (DBD) that is shared across the forkhead box family (Furuyama, Nakazawa, Nakano, & Mori, 2000). There is thus a need for additional levels of variation in the mode of action between FoxO transcription factors besides DNA–protein interactions; e.g., posttranslational modifications (PTM) and interaction with coregulators, including microRNAs (Eijkelenboom & Burgering, 2013). FoxO proteins range in size from ~400 to 700 amino acids in length. Besides sharing a similar DBD, they also contain two nuclear localization sequences and a single nuclear export sequence. FoxO proteins across species share highly conserved PTM sites of phosphorylation, acetylation, and ubiquitination (Smith & Shanley, 2010).

Following activation, the IIS pathway represses FoxO3 activity by increasing the levels of the lipid phosphatidylinositol-3-phosphate, causing phosphoinositide-3-kinase (PI3K) to activate the serine/threonine kinase AKT, which then phosphorylates FoxO3 (Brunet et al., 1999). These phosphorylation events create binding sites for the regulator protein 14-3-3, which disrupts FoxO3/DNA interactions and promotes exclusion of FoxO3 from the nucleus, thus effectively blocking its transcriptional activity and permitting degradation of FoxO3 in the cytoplasm (Dobson et al., 2011). However, in the presence of reactive oxygen species (ROS), phosphorylation of FoxO3 by MST1 or JNK at alternative PTM sites dissociates 14-3-3, leading to increased expression of oxidative stress response genes such as catalase, thioredoxin-dependent peroxide reductase, sentrin-specific peptidase 1, and superoxide dismutase (SOD) 2 (Eijkelenboom & Burgering, 2013; Morris et al., 2015). In response to nutrient deprivation, FoxO3 is upregulated by the energy sensor AMP-activated protein kinase

(AMPK), which phosphorylates FoxO3 at several serine residues, leading to activation of FoxO3, which in turn activates specific target genes involved in stress resistance and energy metabolism. In addition to affecting nuclear translocation, activation of FoxO3 by AMPK increases its binding affinity to CBP–p300 (Wang, Chan, et al., 2012; Wang, Marshall, et al., 2012), a transcription coactivation complex that functions as a histone acetyltransferase to relax chromatin. In an apparent contradiction, the CBP–p300 complex mediates the acetylation of FoxO proteins, which appears to inhibit their DNA binding (van der Heide & Smidt, 2005). Sirtuins 1 and 2—which have been linked to longevity in mice and nematodes—deacetylate FoxO3, inhibiting its transcriptional activity and promoting ubiquitin-mediated degradation (Wang, Chan, et al., 2012; Wang, Marshall, et al., 2012). The extension of lifespan in *C. elegans* in response to oxidative stress requires the FoxO3 gene homolog *daf-16* (Heidler, Hartwig, Daniel, & Wenzel, 2010). The daf-16 protein upregulates ROS-neutralizing SOD (Honda & Honda, 2002) via an oxidative stress response conserved in *Drosophila* (Curtis et al., 2007). Thus, multiple interdependent mechanisms continuously adjust the subcellular location and transcriptional regulatory capacity of FoxO3 to influence its role in energy homeostasis and redox regulation.

Autophagy is a critical cell process that maintains cellular homeostasis by regulating the turnover and degradation of damaged cellular components by packaging and delivering them to the lysosome for degradation. In skeletal myocytes, FoxO3 transcriptionally activates autophagy via the LC3 and BNIP3 genes (Mammucari et al., 2007). In human and mouse embryonic cell lines the prelude to this is the upregulation of PI3K, causing AKT1 to exclude FoxO1 from the nucleus (Zhou et al., 2012). This process impacts skeletal muscle atrophy (Mammucari et al., 2007) and cardiomyocyte size in rodents (Sengupta, Molkentin, & Yutzey, 2009), hinting at a possible influence of FoxO3-mediated autophagy on longevity. FoxO3-mediated activation of autophagy during osteoblast differentiation in mesenchymal stem cells to help maintain redox homeostasis (Gómez-Puerto et al., 2016) represents another cellular and physiological mechanism by which FoxO3 may contribute to longevity, namely, stem cell homeostasis.

The stem cell theory of aging postulates that gradual loss of the ability of adult stem cell populations to contribute to tissue homeostasis is a prime driver of organismal aging (Schultz & Sinclair, 2016). This theory is supported by evidence of reductions in the number and/or function of a variety of stem cell populations across the range of organismal complexity from *C. elegans* to humans (Schultz & Sinclair, 2016). Therefore, mechanisms by which stem cell homeostasis is maintained are of great relevance to the longevity of different organisms, and the essential role of FoxO3 in this process has been demonstrated repeatedly. The Cnidarian, *Hydra vulgaris*, is considered to be immortal because it shows no evidence of senescence (Martı´nez, 1998). The constant renewal of its tissues is achieved by highly efficient maintenance of its stem cells (Bosch, Anton-Erxleben, Hemmrich, & Khalturin, 2010) in a FoxO-dependent manner (Boehm et al., 2012). *C. elegans* has only one stem cell population, the germline stem cells. The regulatory mechanism controlling proliferation of these stem cells has also been shown to be dependent on the activity of *daf-16* (Michaelson, Korta, Capua, & Hubbard, 2010). Similarly, dFoxO regulates germline stem cells in *Drosophila* by upregulating the Notch-suppressing gene, *fng*, in the niche cells supporting this stem cell population (Yang et al., 2013).

Mice present the best evidence that FoxO3-mediated regulation of stem cell homeostasis could have an impact on human longevity. In stem cells (satellite cells) of mouse skeletal muscle, FoxO3-mediated Notch signaling maintains their quiescence and thereby the long-term self-renewal capacity of the tissue (Gopinath, Webb, Brunet, & Rando, 2014). Hematopoietic stem cells (HSC) in *Foxo3* knockout mice have an impaired oxidative stress response and there is a significantly greater decline in their frequency in bone marrow with age (Miyamoto et al., 2007). Independent of ROS management, *Foxo3* knockout mice have defects in HSC mitochondrial metabolism that affect their long-term homeostasis, suggesting a broader role for FoxO3 in stem cell maintenance (Rimmelé et al., 2015). Indeed, the role of FoxO3 in autophagy is important to HSC maintenance, particularly for old HSC (Warr et al., 2013). In the absence of functional FoxO3, murine HSC demonstrated significant downregulation of proautophagic genes in response to external stress. This FoxO3-dependent mechanism is essential for the continued survival of old HSC (Warr et al., 2013). FoxO3 knockout mice also exhibit fewer neural stem cells (NSC), with impaired self-renewal and differentiation capacity (Renault et al., 2009). This increases significantly with age compared to wild-type controls and may be attributable to the loss FoxO3-mediated redox balance (Yeo et al., 2013). FoxO3 also plays a role in another important regulator of NSC self-renewal and differentiation control, Wnt. A mouse knockout model in which *Foxo1*, *Foxo3*, and *Foxo4* were deleted demonstrated that negative regulation of Wnt played a role in the early loss of NSC (Paik et al., 2009). In humans, both circulating

HSC counts and differentiation capacity (Moresi et al., 2005), as well as NSC frequency (Maslov, Barone, Plunkett, & Pruitt, 2004) decline with age. HSC aging in the human thymus results in loss of T cell diversity (Naylor et al., 2005) which may partly explain the decline in adaptive immune response in the elderly (Ginaldi et al., 1999).

DNA repair and cell cycle control are essential to the maintenance of genome stability in replicating cells, and the increase in genome instability with age (Burhans & Weinberger, 2007) suggests a relation between these processes and longevity. ROS directly and indirectly cause DNA damage, which induces FoxO3-dependent cell cycle arrest via the mitogen-activated protein kinase signaling pathway, leading to downregulation of the proliferation regulatory gene, *Myc* (Kress et al., 2011). The main target of the DNA damage-sensing factor ataxia telangiectasia mutated is H2AX. H2AX is a specialized histone whose mouse knockout model exhibits decreased *FoxO3* expression and changes in expression of FoxO3-regulated genes in response to irradiative damage (Tarrade et al., 2015). The changes to FoxO3-regulated gene expression in *H2ax* knockout mice result in decreased DNA repair response (involving genes *Ddb2* and *Hspa1b*), weaker assertion of cell cycle control (via Cdkn1a), and increased apoptosis signaling (via Bcl2l11) (Tarrade et al., 2015). Together, these results indicate a significant role for FoxO3 in the DNA damage response and repair process independent of redox homeostasis.

4. EXTRINSIC INFLUENCES AND THE FUTURE OF FoxO3-MEDIATED LONGEVITY

All of the earlier evidence supporting the many mechanisms by which FoxO3 influences longevity is good news for those lucky enough to receive a protective genotype of *FOXO3* from the genetic lottery that is reproduction. Identification of SNPs that affect phenotype, particularly one as complex as longevity, is vital to the investigation of genetic contributions to aging. The knowledge we have gained regarding the role of FoxO3 in human longevity can be leveraged to benefit healthspan and perhaps lifespan for everyone, irrespective of whether they carry the protective haplotype of *FOXO3* or the common haplotype. An important part of this process will be the identification of extrinsic modifiers of *FOXO3* expression—diet, lifestyle, and small molecules—and assessment of their effect, both short-term and long-term, on age-related diseases and longevity. The highly conserved nature of the *FOXO*/longevity relationship should yield a rich translation of beneficial effects from model organisms to humans.

Energy restricted diets improve biomarkers and other phenotypes of healthy aging in humans (Civitarese et al., 2007; Fontana et al., 2004; Ravussin et al., 2015; Walford et al., 2002; Willcox & Willcox, 2014; Willcox et al., 2007; Witte, Fobker, Gellner, Knecht, & Flöel, 2009). In the general population, however, adherence to diet-management plans is generally poor. In contrast, nutraceutical and dietary supplement usage, which is on the rise, is more likely to be effective. Identification of FoxO3 activators is in its infancy. We know from *C. elegans* that under conditions of stress, the AGE-1 hypertonic stress resistance mechanism includes the DAF-16-mediated upregulation of trehalose synthesis enzymes TSP-1 and TPS-2 (Lamitina & Strange, 2005). Trehalose is a glucose disaccharide linked to cytoprotection across multiple species. When fed to young worms, trehalose extends their life span by 30%; even when the feeding started in late adulthood, trehalose compressed morbidity and extended remaining lifespan in *C. elegans* (Honda, Tanaka, & Honda, 2010). Trehalose also stimulates autophagy—a FoxO3-mediated process—in multiple human cells (Belzile et al., 2016; Chen et al., 2016). When fed to older mice, trehalose improved arterial endothelium-dependent dilation (LaRocca et al., 2012). Endothelium-dependent dilation of arteries occurs when the endothelial cells produce nitric oxide (NO), which signals relaxation in the underlying vascular smooth muscle cells. Trehalose treatment of aged mice reduces vascular oxidative stress and normalizes cytokine expression (LaRocca et al., 2012). In a similar manner, oral supplementation with trehalose in humans results in improvement in endothelium-dependent microvascular function, which is considered a major indicator of CVD risk (Kaplon et al., 2016). While it remains to be demonstrated directly that trehalose activates FoxO protein(s) in humans, given these associations, trehalose may offer an approach to nutraceutical activation of FoxO3. It also suggests that a FoxO3-dependent mechanism in endothelial cells may play a role in the impact that protective variant(s) in linkage disequilibrium with *FOXO3* SNP *rs2802292* has on CHD mortality.

Supplementation with the potent polyphenol curcumin improves endothelial function of resistance arteries by 37% ($P=0.03$) via the same kind of increase in NO bioavailability and sensitivity as trehalose supplementation (Kaplon et al., 2016; Santos-Parker et al., 2017). However, curcumin may also have a more direct link to FoxO3 regulatory pathways. Curcumin can reduce phosphorylation of FoxO3, leading to its nuclear translocation and resultant upregulation of expression of stress resistance and lipid transport genes (Zingg et al., 2012). This may explain the dFOXO-dependent 21% increase in *Drosophila* lifespan following treatment with the curcumin metabolite tetrahydrocurcumin (Xiang et al., 2011). Green tea also contains potent polyphenolic compounds, one of which—epigallocatechin-3-gallate—has been shown to activate FoxO3 (Belguise, Guo, & Sonenshein, 2007). In breast cancer cells, this natural compound, via activation of FoxO3, had the added benefit of inhibiting their invasive phenotype by upregulation of estrogen receptor alpha (Belguise et al., 2007). The microalgae *Haematococcus pluvialis* produces astaxanthin (a xanthophyll carotenoid), under conditions of stress. This compound is a potent activator of FoxO3 and increases *C. elegans* life span by up to 30% (Yazaki, Yoshikoshi, Oshiro, & Yanase, 2011). The effect is abrogated in a *daf-16* null mutant.

Although these natural compounds are promising options for nutraceutical-mediated activation of FoxO3, pharmaceutical compounds approved for other uses have been shown to activate FoxO3. Notably, bepridil (now withdrawn in the United States) and trifluoperazine (an antipsychotic) promote FoxO3 translocation to the nucleus by inhibiting AKT phosphorylation (Park et al., 2016). The drug metformin induces AMPK-mediated phosphorylation and nuclear translocation of FoxO3 (Sato et al., 2012). It should be noted that metformin is being used in the first clinical trial approved to study the effect of a drug on longevity (Barzilai, Crandall, Kritchevsky, & Espeland, 2016). The screening of novel and existing libraries of small molecular compounds for FoxO nuclear translocation capabilities utilizing cells modified with green fluorescent protein-tagged FoxO has produced a wealth of potential activators (Cautain et al., 2016; Zanella, Rosado, García, Carnero, & Link, 2008). When measuring nuclear translocation of FoxO as the endpoint, compounds may act by influencing the activity of other components of the IIS pathway (Zanella et al., 2008), or the exclusion of active FoxO proteins from the nucleus (Cautain et al., 2016). These successful small-scale investigations of pharmaceutical activation of FoxO(s) suggest that high-throughput screening may reveal a veritable library of old drugs that can be put to new uses for individuals with age-related morbidities modifiable by FoxO3. Those seeking a better chance of achieving exceptional longevity may soon have the option of utilizing methods that take advantage of the positive effect of FoxO3 on lifespan and healthspan, regardless of their luck in the genetic lottery.

REFERENCES

Anselmi, C. V., Malovini, A., Roncarati, R., Novelli, V., Villa, F., Condorelli, G., et al. (2009). Association of the FOXO3A locus with extreme longevity in a southern Italian centenarian study. *Rejuvenation Research*, *12*, 95–104.

Bartke, A., Wright, J. C., Mattison, J. A., Ingram, D. K., Miller, R. A., & Roth, G. S. (2001). Longevity: Extending the lifespan of long-lived mice. *Nature*, *414*, 412.

Barzilai, N., Crandall, J. P., Kritchevsky, S. B., & Espeland, M. A. (2016). Metformin as a tool to target aging. *Cell Metabolism*, *23*, 1060–1065.

Belguise, K., Guo, S., & Sonenshein, G. E. (2007). Activation of FOXO3a by the green tea polyphenol epigallocatechin-3-gallate induces estrogen receptor expression reversing invasive phenotype of breast cancer cells. *Cancer Research*, *67*, 5763–5770.

Belzile, J.-P., Sabalza, M., Craig, M., Clark, E., Morello, C. S., & Spector, D. H. (2016). Trehalose, an mTOR-independent inducer of autophagy, inhibits human cytomegalovirus infection in multiple cell types. *Journal of Virology*, *90*, 1259–1277.

Bennet, A. M., Angelantonio, E. D., Ye, Z., Wensley, F., Dahlin, A., Ahlbom, A., et al. (2007). Association of apolipoprotein E genotypes with lipid levels and coronary risk. *JAMA*, *298*, 1300–1311.

Bernstein, A. M., Willcox, B. J., Tamaki, H., Kunishima, N., Suzuki, M., Willcox, D. C., et al. (2004). First autopsy study of an Okinawan centenarian: Absence of many age-related diseases. *The Journals of Gerontology. Series A, Biological Sciences and Medical Sciences*, *59*, 1195–1199.

Blüher, M., Kahn, B. B., & Kahn, C. R. (2003). Extended longevity in mice lacking the insulin receptor in adipose tissue. *Science*, *299*, 572–574.

Boehm, A.-M., Khalturin, K., Anton-Erxleben, F., Hemmrich, G., Klostermeier, U. C., Lopez-Quintero, J. A., et al. (2012). FoxO is a critical regulator of stem cell maintenance in immortal hydra. *Proceedings of the National Academy of Sciences*, *109*, 19697–19702.

Bonkowski, M. S., Rocha, J. S., Masternak, M. M., Regaiey, K. A. A., & Bartke, A. (2006). Targeted disruption of growth hormone receptor interferes with the beneficial actions of calorie restriction. *Proceedings of the National Academy of Sciences*, *103*, 7901–7905.

Bosch, T. C. G., Anton-Erxleben, F., Hemmrich, G., & Khalturin, K. (2010). The hydra polyp: Nothing but an active stem cell community. *Development, Growth & Differentiation*, *52*, 15–25.

Broer, L., Buchman, A. S., Deelen, J., Evans, D. S., Faul, J. D., Lunetta, K. L., et al. (2015). GWAS of longevity in CHARGE consortium confirms APOE and FOXO3 candidacy. *The Journals of Gerontology. Series A, Biological Sciences and Medical Sciences*, *70*, 110–118.

Brunet, A., Bonni, A., Zigmond, M. J., Lin, M. Z., Juo, P., Hu, L. S., et al. (1999). Akt promotes cell survival by phosphorylating and inhibiting a forkhead transcription factor. *Cell*, *96*, 857–868.

Budovsky, A., Craig, T., Wang, J., Tacutu, R., Csordas, A., Lourenço, J., et al. (2013). LongevityMap: A database of human genetic variants associated with longevity. *Trends in Genetics*, *29*, 559–560.

Burhans, W. C., & Weinberger, M. (2007). DNA replication stress, genome instability and aging. *Nucleic Acids Research*, *35*, 7545–7556.

Cautain, B., Castillo, F., Musso, L., Ferreira, B. I., de Pedro, N., Rodriguez Quesada, L., et al. (2016). Discovery of a novel, isothiazolonaphthoquinone-based small molecule activator of FOXO nuclear-cytoplasmic shuttling. *PLoS One*, *11*, e0167491.

Chen, X., Li, M., Li, L., Xu, S., Huang, D., Ju, M., et al. (2016). Trehalose, sucrose and raffinose are novel activators of autophagy in human keratinocytes through an mTOR-independent pathway. *Scientific Reports*, *6*, 28423.

Civitarese, A. E., Carling, S., Heilbronn, L. K., Hulver, M. H., Ukropcova, B., Deutsch, W. A., et al. (2007). Calorie restriction increases muscle mitochondrial biogenesis in healthy humans. *PLoS Medicine*, *4*, e76.

Clancy, D. J., Gems, D., Harshman, L. G., Oldham, S., Stocker, H., Hafen, E., et al. (2001). Extension of life-span by loss of CHICO, a drosophila insulin receptor substrate protein. *Science, 292*, 104–106.
Curtis, C., Landis, G. N., Folk, D., Wehr, N. B., Hoe, N., Waskar, M., et al. (2007). Transcriptional profiling of MnSOD-mediated lifespan extension in Drosophila reveals a species-general network of aging and metabolic genes. *Genome Biology, 8*, R262.
Das, S. K., Balasubramanian, P., & Weerasekara, Y. K. (2017). Nutrition modulation of human aging: The calorie restriction paradigm. *Molecular and Cellular Endocrinology, 455*, 148–157, S0303-7207, 30224–1.
Davignon, J., Gregg, R. E., & Sing, C. F. (1988). Apolipoprotein E polymorphism and atherosclerosis. *Arteriosclerosis, Thrombosis, and Vascular Biology, 8*, 1–21.
Deelen, J., Beekman, M., Capri, M., Franceschi, C., & Slagboom, P. E. (2013). Identifying the genomic determinants of aging and longevity in human population studies: Progress and challenges. *BioEssays, 35*, 386–396.
Dobson, M., Ramakrishnan, G., Ma, S., Kaplun, L., Balan, V., Fridman, R., et al. (2011). Bimodal regulation of FoxO3 by AKT and 14-3-3. *Biochimica et Biophysica Acta, 1813*, 1453–1464.
Donlon, T. A., Curb, J. D., He, Q., Grove, J. S., Masaki, K. H., Rodriguez, B., et al. (2012). FOXO3 gene variants and human aging: Coding variants may not be key players. *The Journals of Gerontology. Series A, Biological Sciences and Medical Sciences, 67*, 1132–1139.
Dorman, J. B., Albinder, B., Shroyer, T., & Kenyon, C. (1995). The age-1 and daf-2 genes function in a common pathway to control the lifespan of *Caenorhabditis elegans*. *Genetics, 141*, 1399–1406.
Eijkelenboom, A., & Burgering, B. M. T. (2013). FOXOs: Signalling integrators for homeostasis maintenance. *Nature Reviews Molecular Cell Biology, 14*, 83–97.
Fabrizio, P., Pozza, F., Pletcher, S. D., Gendron, C. M., & Longo, V. D. (2001). Regulation of longevity and stress resistance by Sch9 in yeast. *Science, 292*, 288–290.
Flachsbart, F., Caliebe, A., Kleindorp, R., Blanché, H., von Eller-Eberstein, H., Nikolaus, S., et al. (2009). Association of FOXO3A variation with human longevity confirmed in German centenarians. *Proceedings of the National Academy of Sciences of the United States of America, 106*, 2700–2705.
Fontana, L., Meyer, T. E., Klein, S., & Holloszy, J. O. (2004). Long-term calorie restriction is highly effective in reducing the risk for atherosclerosis in humans. *Proceedings of the National Academy of Sciences of the United States of America, 101*, 6659–6663.
Furuyama, T., Nakazawa, T., Nakano, I., & Mori, N. (2000). Identification of the differential distribution patterns of mRNAs and consensus binding sequences for mouse DAF-16 homologues. *The Biochemical Journal, 349*, 629–634.
Garatachea, N., Marín, P. J., Santos-Lozano, A., Sanchis-Gomar, F., Emanuele, E., & Lucia, A. (2015). The ApoE gene is related with exceptional longevity: A systematic review and meta-analysis. *Rejuvenation Research, 18*, 3–13.
Giannakou, M. E., Goss, M., Jünger, M. A., Hafen, E., Leevers, S. J., & Partridge, L. (2004). Long-lived drosophila with overexpressed dFOXO in adult fat body. *Science, 305*, 361.
Ginaldi, L., De Martinis, M., D'Ostilio, A., Marini, L., Loreto, M. F., Martorelli, V., et al. (1999). The immune system in the elderly: II. Specific cellular immunity. *Immunologic Research, 20*, 109–115.
Gómez-Puerto, M. C., Verhagen, L. P., Braat, A. K., Lam, E. W.-F., Coffer, P. J., & Lorenowicz, M. J. (2016). Activation of autophagy by FOXO3 regulates redox homeostasis during osteogenic differentiation. *Autophagy, 12*, 1804–1816.
Gopinath, S. D., Webb, A. E., Brunet, A., & Rando, T. A. (2014). FOXO3 promotes quiescence in adult muscle stem cells during the process of self-renewal. *Stem Cell Reports, 2*, 414–426.

Harrison, D. E., Archer, J. R., & Astle, C. M. (1984). Effects of food restriction on aging: Separation of food intake and adiposity. *Proceedings of the National Academy of Sciences, 81*, 1835–1838.

Hayflick, L. (2007). Entropy explains aging, genetic determinism explains longevity, and undefined terminology explains misunderstanding both. *PLoS Genetics, 3*(12), 2351–2354.

Heidler, T., Hartwig, K., Daniel, H., & Wenzel, U. (2010). Caenorhabditis elegans lifespan extension caused by treatment with an orally active ROS-generator is dependent on DAF-16 and SIR-2.1. *Biogerontology, 11*, 183–195.

Honda, Y., & Honda, S. (2002). Oxidative stress and life span determination in the nematode *Caenorhabditis elegans*. *Annals of the New York Academy of Sciences, 959*, 466–474.

Honda, Y., Tanaka, M., & Honda, S. (2010). Trehalose extends longevity in the nematode *Caenorhabditis elegans*. *Aging Cell, 9*, 558–569.

Kaplon, R. E., Hill, S. D., Bispham, N. Z., Santos-Parker, J. R., Nowlan, M. J., Snyder, L. L., et al. (2016). Oral trehalose supplementation improves resistance artery endothelial function in healthy middle-aged and older adults. *Aging, 8*, 1167–1183.

Kenyon, C. (2005). The plasticity of aging: Insights from long-lived mutants. *Cell, 120*, 449–460.

Kenyon, C., Chang, J., Gensch, E., Rudner, A., & Tabtiang, R. (1993). A *C. elegans* mutant that lives twice as long as wild type. *Nature, 366*, 461–464.

Kimura, K. D., Tissenbaum, H. A., Liu, Y., & Ruvkun, G. (1997). daf-2, an insulin receptor-like gene that regulates longevity and diapause in *Caenorhabditis elegans*. *Science, 277*, 942–946.

Kress, T. R., Cannell, I. G., Brenkman, A. B., Samans, B., Gaestel, M., Roepman, P., et al. (2011). The MK5/PRAK kinase and Myc form a negative feedback loop that is disrupted during colorectal tumorigenesis. *Molecular Cell, 41*, 445–457.

Lakowski, B., & Hekimi, S. (1998). The genetics of caloric restriction in *Caenorhabditis elegans*. *Proceedings of the National Academy of Sciences of the United States of America, 95*, 13091–13096.

Lamitina, S. T., & Strange, K. (2005). Transcriptional targets of DAF-16 insulin signaling pathway protect *C. elegans* from extreme hypertonic stress. *American Journal of Physiology Cell Physiology, 288*, C467–C474.

LaRocca, T. J., Henson, G. D., Thorburn, A., Sindler, A. L., Pierce, G. L., & Seals, D. R. (2012). Translational evidence that impaired autophagy contributes to arterial ageing. *The Journal of Physiology, 590*, 3305–3316.

Li, Y., Wang, W.-J., Cao, H., Lu, J., Wu, C., Hu, F.-Y., et al. (2009). Genetic association of FOXO1A and FOXO3A with longevity trait in Han Chinese populations. *Human Molecular Genetics, 18*, 4897–4904.

Lindahl-Jacobsen, R., Tan, Q., Mengel-From, J., Christensen, K., Nebel, A., & Christiansen, L. (2013). Effects of the APOE ε2 allele on mortality and cognitive function in the oldest old. *The Journals of Gerontology Series A: Biological Sciences and Medical Sciences, 68*, 389–394.

Magi, R., Lindgren, C. M., & Morris, A. P. (2010). Meta-analysis of sex-specific genome-wide association studies. *Genetic Epidemiology, 34*, 846–853.

Mammucari, C., Milan, G., Romanello, V., Masiero, E., Rudolf, R., Piccolo, P. D., et al. (2007). FoxO3 controls autophagy in skeletal muscle in vivo. *Cell Metabolism, 6*, 458–471.

Martínez, D. E. (1998). Mortality patterns suggest lack of senescence in hydra. *Experimental Gerontology, 33*, 217–225.

Maslov, A. Y., Barone, T. A., Plunkett, R. J., & Pruitt, S. C. (2004). Neural stem cell detection, characterization, and age-related changes in the subventricular zone of mice. *The Journal of Neuroscience, 24*, 1726–1733.

Mattison, J. A., Colman, R. J., Beasley, T. M., Allison, D. B., Kemnitz, J. W., Roth, G. S., et al. (2017). Caloric restriction improves health and survival of rhesus monkeys. *Nature Communications*, *8*, 14063.

McCay, C. M., Crowell, M. F., & Maynard, L. A. (1935). The effect of retarded growth upon the length of life span and upon the ultimate body size one figure. *The Journal of Nutrition*, *10*, 63–79.

Michaelson, D., Korta, D. Z., Capua, Y., & Hubbard, E. J. A. (2010). Insulin signaling promotes germline proliferation in *C. elegans*. *Development (Cambridge, England)*, *137*, 671–680.

Miyamoto, K., Araki, K. Y., Naka, K., Arai, F., Takubo, K., Yamazaki, S., et al. (2007). Foxo3a is essential for maintenance of the hematopoietic stem cell pool. *Cell Stem Cell*, *1*, 101–112.

Moresi, R., Tesei, S., Costarelli, L., Viticchi, C., Stecconi, R., Bernardini, G., et al. (2005). Age- and gender-related alterations of the number and clonogenic capacity of circulating CD34+ progenitor cells. *Biogerontology*, *6*, 185–192.

Morris, B. J., Donlon, T. A., He, Q., Grove, J. S., Masaki, K. H., Elliott, A., et al. (2014). Association analyses of insulin signaling pathway gene polymorphisms with healthy aging and longevity in Americans of Japanese ancestry. *The Journals of Gerontology Series A, Biological Sciences and Medical Sciences*, *69A*, 270–273.

Morris, B. J., Willcox, D. C., Donlon, T. A., & Willcox, B. J. (2015). FOXO3: A major gene for human longevity—A mini-review. *Gerontology*, *61*, 515–525.

Most, J., Tosti, V., Redman, L. M., & Fontana, L. (2017). Calorie restriction in humans: An update. *Ageing Research Reviews*, *39*, 36–45S1568-1637, 30183-0.

Naylor, K., Li, G., Vallejo, A. N., Lee, W.-W., Koetz, K., Bryl, E., et al. (2005). The influence of age on T cell generation and TCR diversity. *Journal of Immunology*, *174*, 7446–7452.

Newman, A. B., Walter, S., Lunetta, K. L., Garcia, M. E., Slagboom, P. E., Christensen, K., et al. (2010). A meta-analysis of four genome-wide association studies of survival to age 90 years or older: The cohorts for heart and aging research in genomic epidemiology consortium. *The Journals of Gerontology Series A, Biological Sciences and Medical Sciences*, *65A*, 478–487.

O'Neill, C., Kiely, A. P., Coakley, M. F., Manning, S., & Long-Smith, C. M. (2012). Insulin and IGF-1 signalling: Longevity, protein homoeostasis and Alzheimer's disease. *Biochemical Society Transactions*, *40*, 721–727.

Ogg, S., Paradis, S., Gottlieb, S., Patterson, G. I., Lee, L., Tissenbaum, H. A., et al. (1997). The fork head transcription factor DAF-16 transduces insulin-like metabolic and longevity signals in *C. elegans*. *Nature*, *389*, 994–999.

Paik, J., Ding, Z., Narurkar, R., Ramkissoon, S., Muller, F., Kamoun, W. S., et al. (2009). FoxOs cooperatively regulate diverse pathways governing neural stem cell homeostasis. *Cell Stem Cell*, *5*, 540–553.

Park, S.-H., Chung, Y. M., Ma, J., Yang, Q., Berek, J. S., & Hu, M. C.-T. (2016). Pharmacological activation of FOXO3 suppresses triple-negative breast cancer in vitro and in vivo. *Oncotarget*, 7, 42110–42125.

Pawlikowska, L., Hu, D., Huntsman, S., Sung, A., Chu, C., Chen, J., et al. (2009). Association of common genetic variation in the insulin/IGF1 signaling pathway with human longevity. *Aging Cell*, *8*, 460–472.

Postnikoff, S. D. L., Malo, M. E., Wong, B., & Harkness, T. A. A. (2012). The yeast Forkhead transcription factors Fkh1 and Fkh2 regulate lifespan and stress response together with the anaphase-promoting complex. *PLoS Genetics*, *8*, e1002583.

Ravussin, E., Redman, L. M., Rochon, J., Das, S. K., Fontana, L., Kraus, W. E., et al. (2015). A 2-year randomized controlled trial of human caloric restriction: Feasibility and effects on predictors of health span and longevity. *The Journals of Gerontology. Series A, Biological Sciences and Medical Sciences*, *70*, 1097–1104.

Renault, V. M., Rafalski, V. A., Morgan, A. A., Salih, D. A. M., Brett, J. O., Webb, A. E., et al. (2009). FoxO3 regulates neural stem cell homeostasis. *Cell Stem Cell*, *5*, 527–539.
Rimmelé, P., Liang, R., Bigarella, C. L., Kocabas, F., Xie, J., Serasinghe, M. N., et al. (2015). Mitochondrial metabolism in hematopoietic stem cells requires functional FOXO3. *EMBO Reports*, *16*, 1164–1176.
Santos-Parker, J. R., Strahler, T. R., Bassett, C. J., Bispham, N. Z., Chonchol, M. B., & Seals, D. R. (2017). Curcumin supplementation improves vascular endothelial function in healthy middle-aged and older adults by increasing nitric oxide bioavailability and reducing oxidative stress. *Aging*, *9*, 187–205.
Sato, A., Sunayama, J., Okada, M., Watanabe, E., Seino, S., Shibuya, K., et al. (2012). Glioma-initiating cell elimination by metformin activation of FOXO3 via AMPK. *Stem Cells Translational Medicine*, *1*, 811–824.
Schächter, F., Faure-Delanef, L., Guénot, F., Rouger, H., Froguel, P., Lesueur-Ginot, L., et al. (1994). Genetic associations with human longevity at the APOE and ACE loci. *Nature Genetics*, *6*, 29–32.
Schultz, M. B., & Sinclair, D. A. (2016). When stem cells grow old: Phenotypes and mechanisms of stem cell aging. *Development*, *143*, 3–14.
Sebastiani, P., Bae, H., Gurinovich, A., Soerensen, M., Puca, A., & Perls, T. T. (2017). Limitations and risks of meta-analyses of longevity studies. *Mechanisms of Ageing and Development*, *165*, 139–146.
Sengupta, A., Molkentin, J. D., & Yutzey, K. E. (2009). FoxO transcription factors promote autophagy in cardiomyocytes. *The Journal of Biological Chemistry*, *284*, 28319–28331.
Shadyab, A. H., & LaCroix, A. Z. (2015). Genetic factors associated with longevity: A review of recent findings. *Ageing Research Reviews*, *19*, 1–7.
Shimokawa, I., Komatsu, T., Hayashi, N., Kim, S.-E., Kawata, T., Park, S., et al. (2015). The life-extending effect of dietary restriction requires Foxo3 in mice. *Aging Cell*, *14*, 707–709.
Smith, G. R., & Shanley, D. P. (2010). Modelling the response of FOXO transcription factors to multiple post-translational modifications made by ageing-related signalling pathways. *PLoS One*, *5*, e11092.
Soerensen, M., Dato, S., Christensen, K., McGue, M., Stevnsner, T., Bohr, V. A., et al. (2010). Replication of an association of variation in the FOXO3A gene with human longevity using both case-control and longitudinal data. *Aging Cell*, *9*, 1010–1017.
Takata, H., Suzuki, M., Ishii, T., Sekiguchi, S., & Iri, H. (1987). Influence of major histocompatibility complex region genes on human longevity among Okinawan-Japanese centenarians and Monagenarians. *Lancet*, *2*, 824–826.
Tarrade, S., Bhardwaj, T., Flegal, M., Bertrand, L., Velegzhaninov, I., Moskalev, A., et al. (2015). Histone H2AX is involved in FoxO3a-mediated transcriptional responses to ionizing radiation to maintain genome stability. *International Journal of Molecular Sciences*, *16*, 29996–30014.
Tatar, M., Kopelman, A., Epstein, D., Tu, M.-P., Yin, C.-M., & Garofalo, R. S. (2001). A mutant drosophila insulin receptor homolog that extends life-span and impairs neuroendocrine function. *Science*, *292*, 107–110.
van der Heide, L. P., & Smidt, M. P. (2005). Regulation of FoxO activity by CBP/p300-mediated acetylation. *Trends in Biochemical Sciences*, *30*, 81–86.
Walford, R. L., Mock, D., Verdery, R., & MacCallum, T. (2002). Calorie restriction in biosphere 2Alterations in physiologic, hematologic, hormonal, and biochemical parameters in humans restricted for a 2-year period. *The Journals of Gerontology Series A: Biological Sciences and Medical Sciences*, *57*, B211–B224.
Wang, F., Chan, C.-H., Chen, K., Guan, X., Lin, H.-K., & Tong, Q. (2012). Deacetylation of FOXO3 by SIRT1 or SIRT2 leads to Skp2-mediated FOXO3 ubiquitination and degradation. *Oncogene*, *31*, 1546–1557.

Wang, F., Marshall, C. B., Yamamoto, K., Li, G.-Y., Gasmi-Seabrook, G. M. C., Okada, H., et al. (2012). Structures of KIX domain of CBP in complex with two FOXO3a transactivation domains reveal promiscuity and plasticity in coactivator recruitment. *Proceedings of the National Academy of Sciences of the United States of America, 109*, 6078–6083.

Warr, M. R., Binnewies, M., Flach, J., Reynaud, D., Garg, T., Malhotra, R., et al. (2013). FoxO3a directs a protective autophagy program in hematopoietic stem cells. *Nature, 494*, 323–327.

Willcox, B. J., Donlon, T. A., He, Q., Chen, R., Grove, J. S., Yano, K., et al. (2008). FOXO3A genotype is strongly associated with human longevity. *Proceedings of the National Academy of Sciences of the United States of America, 105*, 13987–13992.

Willcox, B. J., Tranah, G. J., Chen, R., Morris, B. J., Masaki, K. H., He, Q., et al. (2016). The FoxO3 gene and cause-specific mortality. *Aging Cell, 15*, 617–624.

Willcox, B. J., & Willcox, D. C. (2014). Caloric restriction, caloric restriction mimetics, and healthy aging in Okinawa: Controversies and clinical implications. *Current Opinion in Clinical Nutrition and Metabolic Care, 17*, 51–58.

Willcox, B. J., Willcox, D. C., & Ferrucci, L. (2008). Secrets of healthy aging and longevity from exceptional survivors around the globe: Lessons from octogenarians to supercentenarians. *The Journals of Gerontology Series A, Biological Sciences and Medical Sciences, 63*, 1181–1185.

Willcox, D. C., Willcox, B. J., Hsueh, W. C., & Suzuki, M. (2006). Genetic determinants of exceptional human longevity: Insights from the Okinawa centenarian study. *Age (Dordrecht, Netherlands), 28*, 313–332.

Willcox, B. J., Willcox, D. C., Todoriki, H., Fujiyoshi, A., Yano, K., He, Q., et al. (2007). Caloric restriction, the traditional Okinawan diet, and healthy aging. *Annals of the New York Academy of Sciences, 1114*, 434–455.

Willcox, B. J., Yano, K., Chen, R., Willcox, D. C., Rodriguez, B. L., Masaki, K. H., et al. (2004). How much should we eat? The association between energy intake and mortality in a 36-year follow-up study of Japanese-American men. *The Journals of Gerontology. Series A, Biological Sciences and Medical Sciences, 59*, 789–795.

Witte, A. V., Fobker, M., Gellner, R., Knecht, S., & Flöel, A. (2009). Caloric restriction improves memory in elderly humans. *Proceedings of the National Academy of Sciences of the United States of America, 106*, 1255–1260.

Xiang, L., Nakamura, Y., Lim, Y.-M., Yamasaki, Y., Kurokawa-Nose, Y., Maruyama, W., et al. (2011). Tetrahydrocurcumin extends life span and inhibits the oxidative stress response by regulating the FOXO forkhead transcription factor. *Aging, 3*, 1098–1109.

Yang, S.-A., Wang, W.-D., Chen, C.-T., Tseng, C.-Y., Chen, Y.-N., & Hsu, H.-J. (2013). FOXO/fringe is necessary for maintenance of the germline stem cell niche in response to insulin insufficiency. *Developmental Biology, 382*, 124–135.

Yashin, A. I., Iachine, I. A., & Harris, J. R. (1999). Half of the variation in susceptibility to mortality is genetic: Findings from Swedish twin survival data. *Behavior Genetics, 29*, 11–19.

Yazaki, K., Yoshikoshi, C., Oshiro, S., & Yanase, S. (2011). Supplemental cellular protection by a carotenoid extends lifespan via Ins/IGF-1 signaling in Caenorhabditis elegans. *Oxidative Medicine and Cellular Longevity, 2011*, 596240.

Yeo, H., Lyssiotis, C. A., Zhang, Y., Ying, H., Asara, J. M., Cantley, L. C., et al. (2013). FoxO3 coordinates metabolic pathways to maintain redox balance in neural stem cells. *The EMBO Journal, 32*, 2589–2602.

Zanella, F., Rosado, A., García, B., Carnero, A., & Link, W. (2008). Chemical genetic analysis of FOXO nuclear-cytoplasmic shuttling by using image-based cell screening. *Chembiochem, 9*, 2229–2237.

Zhou, J., Liao, W., Yang, J., Ma, K., Li, X., Wang, Y., et al. (2012). FOXO3 induces FOXO1-dependent autophagy by activating the AKT1 signaling pathway. *Autophagy*, *8*, 1712–1723.

Zingg, J.-M., Hasan, S. T., Cowan, D., Ricciarelli, R., Azzi, A., & Meydani, M. (2012). Regulatory effects of curcumin on lipid accumulation in monocytes/macrophages. *Journal of Cellular Biochemistry*, *113*, 833–840.

MIX
Papier | Fördert gute Waldnutzung
FSC® C083411

Kolibri 360 GmbH
Ferdinand-Jühlke-Straße 7
99095 Erfurt, Deutschland
produktsicherheit@kolibri360.de

VOLUME ONE HUNDRED AND TWENTY NINE

CURRENT TOPICS IN DEVELOPMENTAL BIOLOGY

Human Embryonic Stem Cells in Development

Edited by

ALI H. BRIVANLOU, Ph.D

Robert & Harriet Heilbrunn Professor,
Laboratory of Stem cell Biology and Molecular Embryology,
The Rockefeller University; Graduate School of Architecture,
Planning and Preservation, Columbia University,
New York, NY, United States

ELSEVIER

ACADEMIC PRESS

An imprint of Elsevier

Academic Press is an imprint of Elsevier
50 Hampshire Street, 5th Floor, Cambridge, MA 02139, United States
525 B Street, Suite 1650, San Diego, CA 92101, United States
The Boulevard, Langford Lane, Kidlington, Oxford OX5 1GB, United Kingdom
125 London Wall, London, EC2Y 5AS, United Kingdom

First edition 2018

ISBN: 978-0-12-804251-9
ISSN: 0070-2153

For information on all Academic Press publications
visit our website at https://www.elsevier.com/books-and-journals

Publisher: Zoe Kruze
Acquisition Editor: Zoe Kruze
Editorial Project Manager: Shellie Bryant
Production Project Manager: Denny Mansingh
Cover Designer: Greg Harris

Typeset by SPi Global, India

CONTENTS

CONTRIBUTORS

Ziad Al Tanoury
Brigham and Women's Hospital; Harvard Medical School; Harvard Stem Cell Institute, Boston, MA, United States

Paola Arlotta
Harvard University, Cambridge, MA, United States

Marc Astick
Université Libre de Bruxelles (U.L.B.), Institut de Recherches en Biologie Humaine et Moléculaire (IRIBHM), and ULB Neuroscience Institute (UNI), Brussels; VIB-KU Leuven Center for Brain & Disease Research; Department of Neurosciences, Leuven Brain Institute, KUL, Leuven, Belgium

Juliana Brown
Harvard University, Cambridge, MA, United States

Jérome Chal
Brigham and Women's Hospital; Harvard Medical School; Harvard Stem Cell Institute, Boston, MA, United States

Carla A.C. Gonçalves
Novo Nordisk Foundation Center for Stem Cell Biology (DanStem), University of Copenhagen, Copenhagen, Denmark

Anne Grapin-Botton
Novo Nordisk Foundation Center for Stem Cell Biology (DanStem), University of Copenhagen, Copenhagen, Denmark

Naoko Irie
Wellcome Trust/Cancer Research UK Gurdon Institute; University of Cambridge, Cambridge, United Kingdom

Yung Hae Kim
Novo Nordisk Foundation Center for Stem Cell Biology (DanStem), University of Copenhagen, Copenhagen, Denmark

Maja B.K. Petersen
Novo Nordisk Foundation Center for Stem Cell Biology (DanStem), University of Copenhagen, Copenhagen; Department of Islet and Stem Cell Biology, Novo Nordisk, Måløv, Denmark

Olivier Pourquié
Brigham and Women's Hospital; Harvard Medical School; Harvard Stem Cell Institute, Boston, MA, United States

Giorgia Quadrato
Harvard University, Cambridge, MA, United States

Eric D. Siggia
Center for Studies in Physics and Biology, The Rockefeller University, New York, NY, United States

M. Azim Surani
Wellcome Trust/Cancer Research UK Gurdon Institute; University of Cambridge, Cambridge, United Kingdom

Anastasiya Sybirna
Wellcome Trust/Cancer Research UK Gurdon Institute; Wellcome Trust Medical Research Council Stem Cell Institute, University of Cambridge, Cambridge, United Kingdom

Pierre Vanderhaeghen
Université Libre de Bruxelles (U.L.B.), Institut de Recherches en Biologie Humaine et Moléculaire (IRIBHM), and ULB Neuroscience Institute (UNI); Welbio, Université Libre de Bruxelles (U.L.B.), Brussels; VIB-KU Leuven Center for Brain & Disease Research; Department of Neurosciences, Leuven Brain Institute, KUL, Leuven, Belgium

Aryeh Warmflash
Departments of Biosciences and Bioengineering, Rice University, Houston, TX, United States

PREFACE

For centuries, and regardless of their cultural origins, humans have been mesmerized by the process of early development. The sequence of events that transforms an amphibian egg into a tadpole and then a tadpole to a frog, or the inherent majesty of a bird hatching from the shell of its egg have always ignited fascination and curiosity. The first written description of animal embryonic development dates back more than 2300 years ago by Aristotle. The field of embryology, under the global umbrella of Developmental Biology, constitutes one of the oldest branches of biology. The study of "model systems" such as the fruit fly, fish, frogs, and mice has shed light on the molecular and cellular processes that underlie emergence of shape and contours that distinguish animals of different species. Elegant approaches of classical and modern experimental embryology, as well as genetic tools, have been used to provide answers to the most basic questions such as cell fate decisions, morphogenetic movements, and axis determination. The animals used in these studies are, however, supposed to "model" human development where early embryos are not easily accessible, mostly due to ethical concerns and the paucity of source of biological material. The emergence of human pluripotent stem cells in general and embryonic stem cells (hESCs) in particular has provided a new window and an excellent platform for the study of the early human embryo, and the validation of knowledge acquired from model systems and their relevance for our own development. Thus, human developmental biology has gradually morphed into hESC studies. In the beginning, utilitarian rationale and the benefit for clinical applications were used as a rationale to undertake these kinds of controversial work, natural human curiosity about its own origins has gradually taken over, and hESCs are frequently used to understand the origins of cell fate decisions. The recent appreciation of self-organization of hESCs into discrete and functional "organoids" and "gastruloids" which can be followed dynamically and measured with subcellular quantitative tools has revolutionized not only our understanding of the relevance of knowledge obtained from model systems but also highlighted the existence of human-specific traits, which, with hindsight, should have been expected. Now self-organizing artificial human embryos are modeling human development.

In this book I have gathered a few examples from a fast-ever-growing literature centered around lessons learned from hESCs in the context of human developmental biology. This knowledge provides a road map toward the understanding of the natural process of cell and organ formation which inevitably will have a very strong impact in clinical applications and the rational design of hESC differentiation for regenerative medicine, a process already on the horizon. I have decided to organize the six different chapters of this book in chronological order. Eric Siggia and Aryeh Warmflash describe the physical and molecular processes that allow hESC colonies grown on confined geometry to self-organize into 2D and 3D artificial human embryos (or "gastruloids"). Naoko Iries, Anastasiya Sioberm, and Azim Surani ask: What can stem cell models tell us about human germ cell biology? I have organized the subsequent chapters in the order of tissues derived from each of the three embryonic germ layers: ectoderm, mesoderm, and endoderm. Two chapters will be dedicated to models of brain development. Marc Astick and Pierre Vanderhaeghen explore the path of human pluripotent stem cells on their journey toward corticogenesis and the emergence of cortical circuitry. Juliana Brown, Giorgio Quatrado, and Paola Arlotta discuss how the basic understanding of 3D models of human brain development can be used to unravel the aspects of brain disease, taking a slight translational point of view. Olivier Pourquié, Ziad Al Tanoury, and Jerome Chal take us to the emergence of human embryonic muscles, mesodermal derivatives. Finally, Maja Borup Kjaer Peteren, Carla Alexandra Carvalho Goncalves, Yung Hae Kim, and Anne Grapin-Botton discuss how human pluripotent stem cells can recapitulate the development of an endodermal derivative: the pancreas. Together these six chapters provide an overview of how hESCs, rather than traditional animal models, can be used to decipher the normal process of cell and tissue specification, starting from embryonic germ layers, and how ultimately this basic knowledge can guide rationale design of clinical strategies.

This book represents a collective effort to provide a snapshot of how hESCs have influenced our current understanding of human development. I started my studies of developmental biology originally focusing on the early development of the frog *Xenopus laevis* in the mid-1980s. Both my graduate advisor, Richard Harland, and my postdoctoral mentor, Doug Melton, had a transformative role in my education. If someone had told me back then that one day I would even contemplate studying human embryology directly, rather than modeling it, I would have accused them of being lunatics. Today, natural human embryos can be studied post in vitro attachment for 14 days, and self-organizing artificial embryos, gastruloids, and organoids model human development with human cells. The quantitative resolution of these studies has now surpassed those of traditional model systems and acts as a mirror reflecting our past, which will continuously enlighten us about our own origins, a process that is long overdue.

These kinds of projects would have not self-assembled if it was not due to a collective effort. I am indebted to all the authors who, despite the tremendous investment of time and energy, and many other priorities, still decided to accept my offer and contributed to the different chapters. I am also grateful to the members of my laboratory who, through many discussions, helped sharpen my own knowledge and point of views, and to Jean-Marx Santel, who has been instrumental and extremely patient in correcting my English and helping to organize the many aspects of the puzzle and collection of chapters. Finally, I am grateful to my parents, Yahya and Mahin, who raised me in Iran, invested, and firmly planted the seed of education that has since informed my life in my journey from Tehran to France to the United States; and to my two beautiful American sons, Amir and Nima, to whom this book is dedicated. I hope that it will serve as a source of inspiration for students, postdocs, and colleagues inside and outside the field of human developmental biology.

Ali H. Brivanlou Ph.D
Robert & Harriet Heilbrunn Professor,
Laboratory of Stem cell Biology and Molecular Embryology,
The Rockefeller University; Graduate School of Architecture,
Planning and Preservation, Columbia University,
New York, NY, United States

CHAPTER ONE

Modeling Mammalian Gastrulation With Embryonic Stem Cells

Eric D. Siggia*[,1], Aryeh Warmflash†[,1]
*Center for Studies in Physics and Biology, The Rockefeller University, New York, NY, United States
†Departments of Biosciences and Bioengineering, Rice University, Houston, TX, United States
[1]Corresponding author: e-mail address: siggiae@rockefeller.edu; aryeh.warmflash@rice.edu

Contents

Abstract

Understanding cell fate patterning and morphogenesis in the mammalian embryo remains a formidable challenge. Recently, in vivo models based on embryonic stem cells (ESCs) have emerged as complementary methods to quantitatively dissect the physical and molecular processes that shape the embryo. Here we review recent developments in using ESCs to create both two- and three-dimensional culture models that shed light on mammalian gastrulation.

1. INTRODUCTION

During embryogenesis, the processes of cell differentiation, growth, division, and movement all occur simultaneously in a three-dimensional environment. For the mammalian embryo, this occurs in utero. The complexity of studying this process makes it crucial to develop simplified

Current Topics in Developmental Biology, Volume 129
ISSN 0070-2153
https://doi.org/10.1016/bs.ctdb.2018.03.001

systems where these processes can be separated, readily observed, and studied in a controlled manner. Further, when considering human embryogenesis, the nearly completely lack of access to actual embryos means that developing synthetic systems may be the best route to understanding uniquely human features of early development. In this review, we focus on modeling mammalian development at gastrulation stages with systems derived from embryonic stem cells (ESCs). We will consider both two- and three-dimensional culture systems and will focus on the most recent developments. The reader is referred to other reviews for further discussion of earlier work (Heemskerk & Warmflash, 2016; Sasai, 2013; Simunovic & Brivanlou, 2017; Turner, Baillie-Johnson, & Martinez Arias, 2016).

Gastrulation occurs in the posterior region of the embryo under the control of signals emanating from two extraembryonic tissues, the visceral endoderm and the trophectoderm (Arnold & Robertson, 2009). BMP signals from the trophectoderm initiate gastrulation at the proximal end of the embryo, while inhibitors to BMP, Nodal, and Wnt are secreted from the anterior visceral endoderm and ensure that the site of gastrulation, known as the primitive streak, is confined to the posterior side of the embryo (Arnold & Robertson, 2009; Perea-Gomez et al., 2002). The BMP signals are triggered by Nodal signals initiating from the epiblast, and in turn activate Wnt signals in the epiblast which further activate Nodal (Ben-Haim et al., 2006). These high levels of Wnt and Nodal in the primitive streak are essential for gastrulation, and Nodal is thought to pattern the resulting mesendoderm in a dose-dependent manner with the highest levels of Nodal being required for endoderm and axial mesoderm and lower levels giving rise to paraxial and lateral mesoderm (Dunn, Vincent, Oxburgh, Robertson, & Bikoff, 2004). These facts have largely been inferred from the patterns of expression and knockout phenotypes of pathway components, and understanding the relationship between BMP, Wnt, and Nodal signals and the resulting cell fates remains a challenge. Further, how the potential gradients of Wnt and Nodal activity are established and interpreted remains largely obscure.

ESCs offer an exciting window into mammalian development and have been used to model a wide variety of cell fate decisions and differentiation programs that take place in early embryogenesis (e.g., Chambers et al., 2009; Lippmann et al., 2015; Loh et al., 2016; Teo et al., 2011). Until recently, there were no methods to generate reproducible patterns from ESCs, and most protocols have either been highly optimized to produce a single cell type (e.g., Chambers et al., 2009; Loh et al., 2016; Pagliuca et al., 2014) or else yield a unpatterned mixture of different derivatives (Bernardo et al., 2011; Tang et al., 2012; Xu et al., 2002; Yu, Pan, Yu, & Thomson, 2011).

Typically, stem cells are grown in colonies of variable size and shapes, and the position of a cell within the colony as well as the local cell density has a profound effect on the outcomes of differentiation. Recently, these challenges have been overcome by adopting techniques which control the size and shape of stem cell colonies. In these methods, complimentary patterns of extracellular matrix (ECM) proteins such as laminin and passivating materials that prevent cell and protein adhesion such as poly-L-lysine-grafted-polyethylene glycol (PLL-PEG) (Azioune, Storch, Bornens, Théry, & Piel, 2009) are deposited on the culture surface. Features on the scale of hundreds of microns can be imposed by microcontact printing where shapes are cast in polydimethylsiloxane (PDMS) elastomer, which are then used as stamps to transfer a pattern onto a slide (Qin, Xia, & Whitesides, 2010; Théry & Piel, 2009). Alternatives use photolithography, which allows for creating features on the micron scale. The slide is first coated with either the cell attractive or repellant coating, and ultraviolet light shined through a mask is used to burn away the coating in selected regions (Aziou et al., 2009). When the cells are seeded onto such coverslips, they adhere only where the surface has been coated with ECM proteins and then remain confined to those areas.

The first papers to examine hESC differentiation in micropatterned colonies relied in general on spontaneous rather than morphogen-induced differentiation and did not observe spatial patterning (Bauwens et al., 2008; Peerani et al., 2007). Warmflash et al. were the first to consider micropatterns as surrogates for embryonic patterning (Warmflash, Sorre, Etoc, Siggia, & Brivanlou, 2014), and it is useful to recapitulate their reasoning and principle results. The mammalian embryo derives from the epiblast, which prior to gastrulation is an apical–basal polarized pseudostratified epithelium, sharing a basement membrane with the visceral endoderm. Stem cells grown in microcolonies easily reach densities of $2–6 \times 10^3$ cells/mm^2 similar to the epiblast (Etoc et al., 2016), remain uniform, pluripotent, and display a similar epithelial morphology. The coated surface on which they are grown supplies the basement membrane. Colonies of 0.5–1 mm have cell numbers comparable to the mammalian epiblast just prior to gastrulation. Thus, microcolonies are a reasonable platform on which to assay the signals that lead to gastrulation and axis formation in the embryo. When cells are treated with BMP4 ligand, which mimics the gastrulation initiating signal from trophoblast, they initiate a patterning process that allocates cells to all three germ layers along the radial axis of the colony (Fig. 1A) (Warmflash et al., 2014). The outermost cells become trophoblast-like (see discussion of their fate below), the innermost cells differentiate to ectoderm, and rings of mesoderm and endoderm form in between.

While these micropatterned systems represent good models for the epiblastic disk, they do not recapitulate the morphogenesis that occurs in the embryo beginning at gastrulation. They also do not break the radial symmetry of the colony geometry, and the region corresponding to the primitive streak occupies a ring around the colony. Three-dimensional culture systems allow for more complex morphogenesis and symmetry breaking, including the formation of apical–basal polarized cysts (Bedzhov & Zernicka-Goetz, 2014; Harrison, Sozen, Christodoulou, Kyprianou, & Zernicka-Goetz, 2017; Shao et al., 2016), the elongation of the aggregate (van den Brink et al., 2014), and the emergence of primitive streak like regions with more natural geometries (Harrison et al., 2017; Shao et al., 2017). This more complex morphogenesis comes at a price, as these systems are not quantitatively reproducible in the sense that micropatterned two-dimensional colonies are. All synthetic systems have the potential to break the rigid connections between gastrulation, primitive streaks, and germ layers that we know from embryos. Here we review recent progress on both two- and three-dimensional systems that model gastrulation events and their fidelity to the embryo, with a focus on the emergence of both physical structure and cell fate patterns.

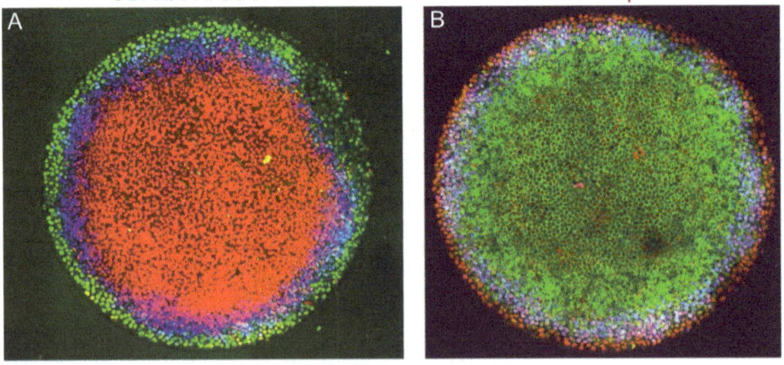

Fig. 1 Self-organized patterning in 2D micropatterning colonies. (A) A micropatterned two-dimensional colony of hESCs differentiates to trophoblast-like (CDX2 expressing; *green*), mesoderm (BRA expressing; *blue*), and ectoderm (SOX2 expressing; *red*). (B) Self-organized signaling in a micropatterned colony. Shown are pSmad1 (*red*), Smad2 (*green*), and BRA (*blue*) (Heemskerk et al., 2017).

2. MATHEMATICAL PRELIMINARIES

Subsequent discussion will be enhanced if we impose sharp definitions on terms whose meaning sometimes drifts in biological reviews. Any set of equations, describing systems where species can spread by diffusion and undergo a chemical reaction, will fall into the general category of *reaction–diffusion*. It is understood that diffusion in this context is merely a phenomenological approximation to some form of local transfer between cells, and the molecular mechanisms are debated and variable. Diffusion does preserve the material being transported, and when this is not the case, say due to molecular traps, one adds an effective decay rate to the system.

We reserve the term *Turing* system to a particular reaction–diffusion system for which the spatially uniform state is unstable and the system evolves toward a periodic pattern, whose wavelength scales as the square root of the diffusion constant divided by a rate (Turing, 1952). This definition excludes the case where there is a localized source of some activator or preferential signaling at the boundary of a tissue, which then propagates away from the source. The Bicoid gradient in Drosophila is not a Turing system, but does qualify as reaction–diffusion. The pair rule stripes once suggested a Turing mechanism but the reality is the antithesis. A Turing system is capable of spontaneous symmetry breaking, though in development there is almost always some bias that locks the pattern into a particular orientation so that the symmetry is always broken the same way with respect to the body axes. Thus, it is difficult to prove a Turing mechanism for pattern formation purely on the basis of experiments, and arguments in favor typically show the mathematical prerequisites are met and the phenomena resemble what is expected from a Turing model. The so-called activator–inhibitor systems are a particular type of reaction–diffusion model in which a diffusible species activates both its own production and that of a diffusible inhibitor. Under certain conditions, most notably that the inhibitor diffuses faster than the activator, activator–inhibitor systems display Turing properties (Meinhardt, 2008).

A *morphogen* is a signal whose levels can define more than two fates, i.e., we exclude bistable systems from the category of morphogens. Classic examples are Bicoid in Drosophila and Activin/Nodal and BMP in the context of isolated Xenopus animal cap cells (Green, New, & Smith, 1992; Wilson, Lagna, Suzuki, & Hemmati-Brivanlou, 1997). Note that the demonstration that a molecule can function as a morphogen in isolated cells does not mean that it necessarily does so in vivo. Putative morphogens including Activin/Nodal, BMP, and Wnt are not static in the vertebrate embryo in contrast to Bicoid (Schohl & Fagotto, 2002), and so the interpretation of these signals can be complex. Nonetheless, they can still convey positional information, i.e., distance from a defined source. For example, if the signal transduction pathway is adaptive, that is returns to its prestimulus baseline after a step increase in morphogen concentration, then its quantitative response is proportional to the time rate of change of the morphogen. If a morphogen turns on at a defined time and spreads, points near the source will experience a more abrupt change than points further away. Thus, a dynamic signal can convey positional information to an adaptive receiver. This is all easy to demonstrate mathematically, and the sensitivity of signaling outputs to the rate of change of TGFb ligands has been shown in a cell culture system (Sorre, Warmflash, Brivanlou, & Siggia, 2014).

Some signals, notably Wnt (Farin et al., 2016), operate at short distances or only by cell contacts and alone cannot coordinately pattern an embryo with a diameter of hundreds of microns. Nonetheless, the inhibitors are often longer range (as required for an activator–inhibitor Turing system) and can impose a pattern on a background of constant activator production (Meinhardt, 2008). Thus, it is of interest to study the movement of the inhibitors and micropatterned colonies could serve as an attractive platform for evaluating their range and mechanisms of action. One potential objection is that in cell culture secreted signals may escape into the bulk media and therefore not be relevant to patterning. Experimentally, this appears not to be the case, as knockdown or knockout of secreted inhibitors has clear patterning phenotypes in micropatterned colonies (Etoc et al., 2016; Warmflash et al., 2014). In cell culture, it is generally true that some secreted signals escape into the media and are homogeneous; nevertheless, autocrine signaling can occur even when the conditioned media transferred to naïve cells do not elicit paracrine signaling arguing that local signaling is still possible in cell culture.

To understand the distribution of inhibitors on micropatterns, a related effect should be noted. Assume an inhibitor is made uniformly, secreted, and adsorbed back onto the cell layer (perhaps to be endocytosed, but for whatever reason remains attached). If the inhibitor is released a distance z_0 above the disk away from the edges, then in a time of order of z_0^2/D all the inhibitor will be readsorbed on the surface, where D is the diffusion constant in the media. At the edge of the colony the inhibitor can mix into the volume.

The net result is that the profile of inhibitor can be described by two-dimensional diffusion within the layer (either directly cell to cell, or via secretion and local uptake from the media) and fixed at a low value at the edge (Etoc et al., 2016). This will restrict the activity of the activator to the colony edge with a range depending on the concentration of supplied activator.

3. COLONY ARCHITECTURE

In both the embryo and synthetic systems, the morphology of cells and tissues has a large impact on how signals are transmitted and ultimately how fates are acquired. We thus consider common physical aspects of stem cell systems before turning to fate determination.

3.1 Growth in Two-Dimensional Micropatterns

The apical–basal structure of micropatterned colonies in the pluripotent state was investigated in Etoc et al. (2016). They show that the apical tight junction marker ZO-1 and the centrioles were positioned on the apical side of the nucleus. In common with other polarized epithelia (Nallet-Staub et al., 2015), both the Activin/Nodal and BMP receptors are localized to the basolateral sides of the cells. This is more pronounced at high cell densities with the result that colonies become insensitive to apically applied morphogens. At the colony boundaries, however, the apical–basal axis becomes more radial perhaps associated with the stress fibers one finds there (Rosowski, Mertz, Norcross, Dufresne, & Horsley, 2015), with the result that the receptors remain apically exposed. The most compelling data for the receptor polarization arise from confluent cell colonies grown on filters. These are ~10 μm thick transparent membranes with 10–200 ~0.4 μm pores per cell. They are sealed into wells so that different media can be placed on the two sides. The strong asymmetry in response between apically and basally applied BMP or Activin ligands argues for basolateral receptor localization. There are anecdotal observations that colony edges tended to differentiate before the bulk, but that was not connected with receptor occlusion. Growth on filters with TGFβ ligands supplied from below is a simple technique to insure uniform application of cytokines when uniform signaling is desired. It is yet to be widely adapted in the stem cell field.

Any of the technologies used to make micropatterns can also make arbitrary shapes. This fact was exploited by Blin et al. to make lozenge-shaped domains and examine the effect of the corners on cell fate (Blin, Picart, Théry, & Puceat, 2017). Growing mouse ESCs under pluripotent conditions, they observed some spontaneous differentiation to Bra+ cells and controlled the fraction of such cells by adjusting colony density prior to replating. They observed the Bra + cells preferentially moved to the corners. This is consistent with old ideas that tissues behave as if endowed with a surface tension, so in this case we would infer that the Bra+ cells optimize their contact with media in preference to the undifferentiated cells by occupying the corners. Since there are no supplied morphogens or patterns of signaling, and the colonies appear to be somewhat layered from the start, the embryological relevance is unclear.

3.2 Three-Dimensional Culture Systems

There is a long history of papers tracing the influence of extracellular matrix on cancer as regards its chemical composition, mechanical properties, and dimensionality. The reconstitution of breast acinar networks from normal and cancerous endothelial cells is particularly revealing about the importance of the 3D physical environment of the cells (Lee, Kenny, Lee, & Bissell, 2007). These methods have slowly found their way to the stem cell field.

An interesting illustration of their potential is described in Shao et al. (2016). Cells are first seeded on a soft matrigel layer, allowed to form colonies for a day, and then embedded in a dilute matrigel solution, that favors the formation of closed epithelial cysts. Presumably the matrigel solution encourages cells to place their basal sides out, but nothing is known about the transition intermediate between the layer and the cyst, perhaps it resembles a neural rosette with the apical surfaces grouped into a circle and the basal sides radially extended. A combination of matrigel in the media and soft substrate for growth is required for the colony to spontaneously differentiate to squamous epithelial morphology and display a gene signature indicative of human amnion. Three-dimensional cysts form with either a soft culture substrate and standard growth media or on a hard surface with matrigel added to the culture media, but they remain columnar and pluripotent. Thus, cell contacts and the physical environment of the colony can have a profound effect on cell fates in the absence of supplied morphogens. We still know very little about outcomes when morphogens and the physical environment compete, or the extent to which cells in a cyst will reconstruct their own basal membrane de novo once some global cue establishes their collective polarity.

For the purposes of massively expanding human stem cell numbers Lei et al. used a PEG-based hydrogel that solidified when the temperature was raised to 37°C, that together with a chemically defined growth media, allowed single cells to expand to ~350 μm diameter balls that remained fully pluripotent (Lei & Schaffer, 2013). This should be contrasted with large 3D aggregates, called embryoid bodies, made from cells first grown on surfaces and then placed in suspension in differentiation media which differentiate in a mostly disorganized fashion and sometimes show apoptosis in the center (Coucouvanis & Martin, 1995).

For 3D stem cell culture with outcomes more relevant to the embryo, we have mostly data from mouse. In conjunction with their study of how the inner cell mass reorganizes to form the epiblast and amniotic cavity, the Zernicka-Goetz lab put mouse ESCs directly into matrigel (Bedzhov & Zernicka-Goetz, 2014). They formed a polarized epithelial cyst once more than a few cells were present, whose formation required the ECM components of matrigel. It is not yet clear how long the cysts remain pluripotent under these conditions, and whether by measures of gene expression the cells successively transit from their inner cell mass state to the pluripotent epiblast state.

Much larger cysts, again starting from mouse ESC, were induced by a neural differentiation protocol to form a dense polarized epithelium resembling the neural plate (Meinhardt et al., 2014). They respond to signals that regulated their fates along the anterior–posterior axis. In common with a neural epithelium, cells move to the apical surface prior to division. A similar culture has not yet been reported for human cells, but if the patterning mechanism follows that in vivo, it could prove to be a useful assay for the interaction of SHH and BMP signaling.

The extracellular matrix is generally consigned to a supporting role in morphogenesis, necessary but otherwise ignored. In a follow up to Meinhardt et al. (2014), Ranga et al. used synthetic hydrogels where they could control the ECM components (as well as mechanics) and systematically screened for the properties of the neural cysts and their propensity for spontaneous DV axis formation (Ranga et al., 2016). Their system revealed the generation of ECM by the cyst itself and how the basement membrane remodeled as the cyst grows. This may prove to be a feasible route to resolving how the specific components of the ECM contribute to morphogenesis.

Mouse is an appealing system in which to explore coculture of different cell types since stable cell lines exist for the lineages that derive from the blastocyst: primitive endoderm, trophoblast, and the standard ICM-derived ESC. In Harrison et al., trophoblast cell colonies were mixed with the cultures that generated the epiblast cysts and led to the formation of structures resembling the egg cylinder with the trophoblast ball capping the epiblast epithelial shell. The two cell populations established a common luminal compartment as in the embryo and then showed asymmetric expression of Bra and Wnt activity (Harrison et al., 2017). Cell fate patterning in this system is discussed later.

There is a considerable literature on the influence of substrate stiffness on the fate of stem cell colonies undergoing spontaneous differentiation (Engler, Sen, Sweeney, & Discher, 2006). A recent paper using hESC shows that a soft substrate can enhance a mesoderm induction (Przybyla, Lakins, & Weaver, 2016). In this case, cells on soft substrates preferred E-Cadherin-dependent cell–cell contacts to integrin-dependent contacts with the culture surface. This led to upregulation of β-Catenin and greater sensitivity to a mesoderm induction protocol. However, this protocol did not include Wnt, and they showed that cells cultured on stiff surfaces and supplemented with Wnt gave results similar to soft surfaces. Since multiple papers produce various mesoderm-derived fates on glass with good efficiency (Loh et al., 2016; Mendjan et al., 2014), we conclude that substantial doses of morphogens can override the effects of mechanics. An earlier paper however demonstrated enhanced yields of neural progenitors when subject to dual smad inhibition on soft substrates as compared to stiff ones (Sun et al., 2014). More generally, even if supplied morphogens can override the effects of mechanics in culture, mechanics may still play an important role in influencing differentiation outcomes at physiological concentrations in vivo.

4. SPATIAL PATTERNING OF CELL FATES

As embryogenesis proceeds, the cells of the embryo differentiate to appropriate fates depending on their spatial position. A complex network of ligands and their inhibitors is used to instruct these fate decisions. In vivo, these decisions are entwined with the processes of growth, cell division, and morphogenesis making quantitative study difficult. Moreover, while it is relatively straightforward to determine the patterns of gene expression for the mRNAs encoding the ligands and inhibitors, determining the spatial distributions of the proteins themselves as well as the signaling responses has proved much more difficult. Studying these processes in stem cells offers a potential alternative as imaging is considerably more straightforward in stem cell cultures than in mammalian embryos, and ligands can be applied in a controlled fashion making it possible to determine quantitative dynamic input–output relationships for each signaling pathway. In this section, we focus on recent progress studying cell fate patterning associated with gastrulation in 2D and 3D cultures of mouse and human ESCs.

4.1 Micropatterned Two-Dimensional Culture Systems

The signals governing patterning within micropatterned colonies treated with BMP4 are the same as those governing gastrulation in the mouse embryo (Arnold & Robertson, 2009). The externally supplied BMP4 activates transcription of Wnt ligands which in turn activate Nodal. Both Wnt and Nodal signals are required for the differentiation of the mesendoderm. The ligands themselves are not sufficient to generate the spatial pattern, and the Nodal inhibitors Lefty and Cerberus restrict the mesoderm to the rings. Without these inhibitors, the mesoderm will spread to fill the colony. In the absence of either Nodal or Wnt signals, mesendodermal fates are lost and the colony is divided between trophectodermal fates at the colony border and ectodermal fates at the center (Warmflash et al., 2014).

While the patterning of the germ layers is similar to the embryo, the differentiation of the outer cells from epiblast-like hESCs to trophoectoderm is quite different from the situation in vivo where the epiblast derives from the ICM only after it has split from the trophectodermal lineages. As a consequence, the identity of these cells has remained controversial, with some suggesting that they represent extraembryonic mesoderm rather than trophoblast (Bernardo et al., 2011). More recently, a substantial amount of data has been obtained showing similar transcriptional profiles, hormone secretion, and physiological responses between BMP4 differentiated hESCs and trophoblast (Amita et al., 2013; Horii et al., 2016; Li et al., 2013; Yang et al., 2015). Nonetheless, data showing that these cells can actually function in vivo are lacking. Thus, it remains unclear whether these cells represent true trophoblast that are differentiated by a different path than their in vivo counterparts, a different but molecularly similar cell type, or possibly a culture artifact which bears a resemblance to trophoblast, but does not correspond to any cell occurring in the embryo.

Whatever the precise identity of these cells, it is clear that their differentiation is dependent on BMP signaling. In micropatterns, an initially broad response to the added BMP ligand is refined over time so that only the trophectoderm-like cells at the border show sustained BMP signaling

(Warmflash et al., 2014) (Fig. 1A). In experiments with sparsely seeded cells in standard culture, it was shown that this sustained BMP response is required for cells to adopt this fate (Nemashkalo, Ruzo, Heemskerk, & Warmflash, 2017b), and terminating signaling early also prevented differentiation. During patterning in larger colonies, the restriction of these signals to the colony border is dependent on two factors, the prepattern in apical–basal receptor localization discussed earlier (Etoc et al., 2016), and inhibition by the secreted inhibitor Noggin. As discussed earlier, uniform production of Noggin within the colony together with diffusion is sufficient to create a gradient with the highest levels of Noggin at the colony center and the lowest levels at the edge. As Noggin is a direct BMP target in hESCs, as well as a Nodal target, it is likely that the patterns of BMP and Nodal signaling induce patterns of Noggin expression and that these play a role in shaping the resulting cell fate patterns.

As noted earlier, a cascade of signaling events is responsible for initiating the gastrulation-like processes in micropatterned colonies. The exogenously supplied BMP activates Wnt signaling which in turn activates Nodal (Fig. 1A). Both extracellular Nodal and Wnt inhibitors are required for limiting the spread of these signaling activities and the resulting mesendoderm differentiation. The architecture of these signaling circuits is reminiscent of the theoretically well-studied activator–inhibitor systems originally proposed by Meinhardt (2008) which are examples of Turing systems (see discussion earlier). Nodal activates both itself and its extracellular inhibitors Lefty1/2 and Cerberus. Similarly, Wnt signaling activates both the Wnt3 ligand and its extracellular inhibitor Dkk1. Nodal and its inhibitors lefty have also been proposed to act as these type of Turing systems in other contexts (Müller et al., 2012; Nakamura et al., 2006).

If Wnt and Nodal do indeed function as Turing systems in this context, they would be capable of generating similar patterns even in the absence of induction by BMP at the edge but these would be variable within the colony. That is, a stripe or patch of high-signaling cells would form stochastically at a particular position and inhibit further signaling and mesendoderm differentiation in the region around it. The function of the upstream BMP signaling is to bias this process so that the pattern is always the same from the edge of the colony inward. Similar mechanisms have been suggested to take place in other patterning systems. For example, the ventral neural tube is patterned under the control of the morphogen Sonic hedgehog (Shh). Shh expressed from the neural tube itself could create patterns in a self-organized fashion, but which side of the neural tube adopted a ventral fate would be random. Shh from the notochord, which lies ventral to the neural tube can bias the patterning process so that the ventral side of the neural patterning always aligns with the ventral side of the embryo (Turner et al., 2016).

Taken together, these results suggest a two-step patterning process. First, a combination of high Noggin concentrations and inaccessible receptors at the center of the colony restricts the response to exogenous BMP4 to the colony edge. BMP4 then activates two potential Turing systems, Wnt signaling and Nodal signaling which position stripes of these activities to the primitive streak like region where mesendoderm differentiation occurs. The finding that BMP signaling activates the inhibitor Noggin (Etoc et al., 2016) raises the possibility that BMP–Noggin also acts as a Turing system in patterning. In the future, it will be interesting to rigorously examine this possibility as well as the Wnt and Nodal patterning systems to better understand the relationships between these three and the patterns they generate.

Recently, another study has confirmed these experimental findings but proposed an alternative explanation for the observed pattern of cell fates in micropatterned colonies (Tewary, Ostblom, Shakiba, & Zandstra, 2017). Tewary et al. grew micropatterned colonies in a defined medium containing recombinant Nodal, differentiated these by adding BMP4, and found identical patterns to those in the micropatterning studies reviewed earlier. They proposed that a Turing system involving BMP4 and Noggin creates a gradient of BMP signaling as reflected in the activated signal transducer Smad1. The levels of pSmad1 are then proposed to determine cell fates in a concentration-dependent manner. As evidence for this model, they show that in smaller colonies, which typically differentiate entirely to the trophoectodermal fates found at the edges of large colonies, expression of mesoderm and ectodermal is induced by lower doses of BMP.

A number of experimental observations argue against this model. Experiments on BMP signaling in hESCs both in the context of micropatterned culture and in standard culture suggest that BMP cannot function as a classic morphogen in this context. First, while pSmad1 is highest at the colony edge, there is not a clear gradient of activity. In fact, the distribution of pSmad1 can be effectively modeled as a binary distribution with cells either on or off (Etoc et al., 2016). Three different studies show that pSmad1 is restricted to within about 100 μm of the edge of the colony

(Etoc et al., 2016; Tewary et al., 2017; Warmflash et al., 2014), which is too narrow a range to pattern all the cell fates within the colony in a concentration-dependent manner. The gradient is broader earlier in patterning, possibly suggesting a duration-dependent interpretation of BMP signaling with longer exposure needed for trophoectodermal than mesodermal fates; however, this is contradicted by the absolute requirement for both Nodal and Wnt signaling in forming the mesendoderm in these colonies (Tewary et al., 2017; Warmflash et al., 2014). Finally, experiments examining the dose-dependent response to BMP4 in very small colonies, which lack secondary signals, show that only a single fate is generated. That is, cells switch from pluripotent to trophectodermal fates above a threshold concentration without any alternative fates generated, supporting a binary model of cell fate decisions induced by BMP. In larger colonies, mesodermal fates are generated but these require secondary signals, and consistently, are only observed at particular cell densities (Nemashkalo, Ruzo, Heemskerk, & Warmflash, 2017a).

Further, the evidence in Tewary et al. is not consistent with a BMP4 and Noggin forming a Turing system in the sense defined earlier. As noted earlier, Turing systems generate self-organized patterns with fixed length scales determined by diffusion and decay constants. If BMP–Noggin formed such a system, the role of the exogenous BMP4 would be to trigger the formation of these self-organized patterns with a bias toward the edge, and the resulting patterns would be independent of BMP4 dose once the self-organizing system had been activated. The fact that the patterns can be rescued with lower BMP4 doses suggests that such a Turing system is not operating. If instead, there were a gradient of Noggin that is highest in the center, then the range over which BMP4 could overcome the Noggin repression would be dependent on the BMP4 dose. Further, the existence of doses of BMP4 that do not show the BMP4-dependent CDX2 fates at the colony border but do show mesendoderm differentiation is also consistent with the two-step model proposed earlier. Experiments show that CDX2 fates require sustained high levels BMP signaling, significantly beyond the times shown to be required to activate Nodal and Wnt signals (Nemashkalo et al., 2017a). Thus, at some doses, the level or duration of BMP signaling will not be high enough to give rise to CDX2 fate but will be sufficient to activate the Wnt and Nodal patterning systems giving rise to the mesendodermal fates at the colony edge.

During cell fate patterning, coherent territories of a single fate are generated, and signals between the cells of the territory may be required for differentiation. Gurdon originally demonstrated that groups of Xenopus animal cap cells, but not individual cells, are induced to form muscle by interaction with vegetal cells (Gurdon, 1988). Positive feedback in which signaling pathway activity enhances transcription of the genes encoding the pathway ligands has been proposed to generate coherent signaling and fate responses within a group of cells (Bolouri & Davidson, 2010), while negative feedback might be required to limit the extent of this territory and create cells fate patterns as discussed earlier (Saka, Lhoussaine, Kuttler, Ullner, & Thiel, 2011).

Recently, micropatterning approaches have been used to investigate the mechanisms underlying these phenomena at the single cell level (Nemashkalo et al., 2017a). When hESCs grown in small colonies of 1–8 cells are treated with BMP4, cells within each colony coordinate their response so that at intermediate doses where both pluripotent and trophoectodermal fates are present, each colony is typically composed of entirely CDX2+ or SOX2+ positive fates. These trends are strengthened as the colony size increases. Further, this trend toward uniformity within the colony reinforces the fates instructed by exogenously supplied signals, so that compared to smaller colonies, those with four or more cells retain pluripotency better in pluripotency supporting media and differentiate more sensitively and homogenously in response to BMP4. At the level of signaling, live cell imaging showed that larger colonies are better able to sustain the response to the BMP signal and therefore differentiate more homogenously. Smaller colonies show more variable signaling and differentiation. Correlating fates with signaling at the level of single cells shows that it is the cells with sustained signaling that differentiate to the trophectodermal fates. Positive feedback between BMP signaling and transcription of BMP ligands is a plausible molecular mechanism for these observations, but this remains to be tested.

4.2 Three-Dimensional Culture Systems

Early embryonic events have also been investigated in three-dimensional cultures of mESCs. In initial experiments, it was shown that embryoid bodies made from mESCs show spontaneous polarization of a Wnt signaling reporter and mesodermal gene expression suggestive of an anterior–posterior axis (ten Berge et al., 2008). It was also shown that the hierarchy of signaling from BMP to Wnt to Nodal is preserved, so that treatment with any of BMP, Wnt, or Activin can lead to activation of polarized Wnt signaling in these aggregates, but that BMP inhibition only blocks the polarization induced by BMP. Wnt and Activin are downstream of BMP and so activate polarized Wnt activity in a BMP-independent fashion. More recently, it was shown that when these aggregates are made from relatively small numbers of cells, the polarization is accompanied by elongation along this axis, with the posterior markers on one end. This effect can be enhanced by Wnt activation during a particular period in the culture (van den Brink et al., 2014). Moreover, it was observed in some aggregates that neural markers such as Sox1 and Sox2 are not expressed opposite the region of Bra expression on the long axis of the aggregate but instead on the shorter axis (Fig. 2A), and it was suggested that this represents a second axis in the aggregate, akin to the DV axis of the embryo (Turner et al., 2017).

While intriguing, further experiments will be required to support these claims. First, outside the context of the embryo, localized expression of germ layer markers such as Bra or Sox2 may result from the process of germ layer differentiation either under the spatial control of ligands or through more stochastic processes followed by cell sorting (as in Bli et al., 2017), rather than the formation of an axis equivalent to the AP axis of the embryo. Additional markers specific to particular AP positions such as Otx2 for anterior fates or particular Hox genes for more posterior ones could support these conclusions. It is also possible that more elaborate protocols will be required to define the AP position. For example, it was recently shown that neural/mesodermal progenitors can be maintained in a combination of Wnt and FGF signaling and during this time acquire a progressively more posterior

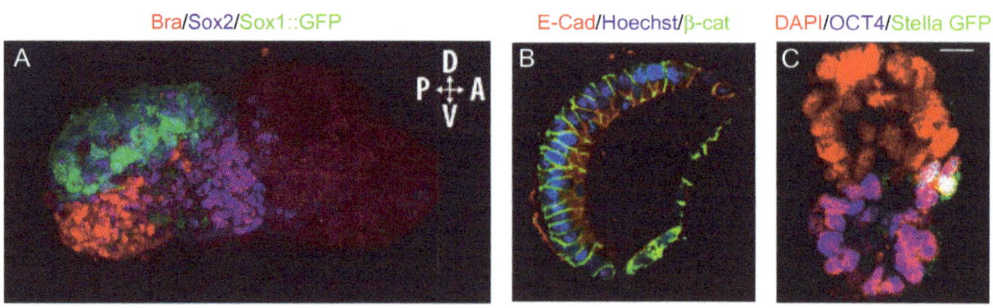

Fig. 2 Examples of self-patterning in 3D. (A) An embryoid body from mESC stained for Bra (*red*) and Sox2 (*blue*), and with Sox1::GFP in *green* (Turner et al., 2017). (B) A structurally asymmetric amniotic cyst from hESC, with a thick pseudostratified epithelium on one side and a thin amnion layer on the other, stained for nuclei (*blue*), βCAT (*green*), and E-Cad (*red*) (Shao et al., 2017). (C) The juxtaposition of mouse trophoblast cells (*top*) with an epiblast epithelium (*bottom*) results in an incipient primitive streak (*right*) breaking the azimuthal symmetry in the ring where the two types of cells are in contact, nuclei (*red*), Oct4 (*blue*), Stella-GFP (*green*). *Image courtesy of Magdalena Zernicka-Goetz.*

identity as defined by Hox gene expression. Treatment with retinoic acid at any time during this protocol induces differentiation to neural fates and freezes the AP identity of the cells (Lippmann et al., 2015). The claim of two independent axes requires multiple markers to be assayed and simultaneously visualized. In the absence of this, it is equally possible that axial elongation and AP axis formation can be decoupled in aggregates so that the Bra–Sox1/2 axis apparent in Fig. 2A corresponds to an AP axis or, as noted earlier, to germ layer differentiation without a clear correspondence to one of the major body axes. The DV axis would most clearly be demonstrated by visualizing ventral and dorsal fates within the same germ layer, for example, neural and epidermal fates within the ectoderm. These issues are also complicated by the variability seen within aggregates. All aggregates form a long axis with Bra and Wnt signaling on one end, allowing for quantification of these markers relative to this axis, but other aspects such as the Sox2 expression appear variable making it difficult to have an external reference by which all markers can be compared.

As discussed earlier, Harrison et al. (2017) developed a three-dimensional culture system which combines mESCs and trophoblast stem cells (TSCs) into a structure called an ETS embryo (for ESC- and TSC-derived embryo). In addition to recapitulating egg cylinder stage morphogenesis, ETS embryos also show asymmetric expression of primitive streak markers such as Brachyury and germ cell markers such as Stella, Fig. 2C. Thus, these embryos have two orthogonal axes, the proximal–distal axis, defined by the relative position of the ESCs and TSCs, and an AP-like axis in which the positioning of Brachyury and germ cell markers defines the posterior side. The development of the AP axis is particularly interesting as it occurs in the absence of the visceral endoderm, while in vivo, secreted signals from the anterior visceral endoderm are required to position the primitive streak in the posterior of the embryo (Perea-Gomez et al., 2002). An attractive model is that the generation of the primitive streak is under the control of a Turing system so that it stochastically forms on one side of the ETS embryo, and then the longer-range inhibitors prevent further Wnt/Nodal signaling and primitive streak formation on the opposite side. If this model is correct, an open question is what prevents similar mechanisms from operating in real embryos lacking the secreted inhibitors in the AVE (Perea-Gomez et al., 2002), in embryos in which the AVE does not form (Migeotte, Omelchenko, Hall, & Anderson, 2010; Nowotschin et al., 2013), or in the micropatterned human ESC colonies discussed earlier (Warmflash et al., 2014). In the former case, multiple primitive streaks form, while in the latter two cases, the radial symmetry of the embryo or colony is never broken resulting a ring of mesodermal differentiation rather than a streak on one side. The patterning by sorting rather than morphogens is not excluded in this system either.

Both these 3D systems lose essential aspects of in vivo gastrulation. A cell aggregate does not undergo the epithelial to mesenchymal transition (EMT), which is a necessary step in primitive streak formation. The ETS embryos do not have a well-characterized EMT or a mesenchymal layer covering the remaining epiblast epithelium. Many protocols exist to make mesendo derivatives from stem cells without obvious intermediate spatial organization. Would mixtures of cells fated to different germ layers sort and look so different from synthetic systems?

An example of a self-patterning three-dimensional system in human is the amniotic cysts discussed earlier. In most cases, these create relatively homogeneous aggregates of amniotic ectoderm; however, it was recently shown that in a minority of cases, polarized cysts form consisting of an amniotic half and an epiblastic half in which the epiblastic half retains its columnar epithelial morphology while the amniotic half differentiates to a squamous epithelium expressing markers of amnion such GATA3 and CDX2 (Shao et al., 2017) (Fig. 2B). As with the fully differentiated cysts, the differentiation of the amniotic half in polarized cysts requires BMP signaling which autonomously becomes asymmetric in the cyst. It is hypothesized that in the polarized cysts, BMP induction of BMP inhibitors limits the spread of amniotic differentiation and allows for the stable retention of epiblast fates in half the aggregate. This hypothesis remains to be proven, and, in any event, it remains unclear what distinguishes the fully differentiated cysts where the BMP-mediated differentiation spreads to the entire aggregate and the polarized cysts in which it is limited. Interestingly, in polarized cysts, a primitive streak-like region often develops from the epiblastic part; however, it remains unknown whether the amniotic half of the cyst plays a role in inducing this event, as the extraembryonic tissue does in vivo, or whether it arises spontaneously from the epiblast cells. If the amnion plays a role, the primitive-streak like region should initiate at the border between the epiblast and amnion cells and extend from there toward the center of the epiblast region, and it will be interesting to determine whether this is the case.

Finally, many examples of systems that undergo patterning and morphogenesis and model the development of particular organs have recently been developed (reviewed in Gjorevski, Ranga, & Lutolf, 2014; Simunovic & Brivanlou, 2017; Turner et al., 2016). One of the most relevant to early development are the neural cysts discussed earlier

(Meinhardt et al., 2014). In addition to the morphogenesis discussed earlier, they also represent an interesting in vitro system for studying cell fates within the developing neural tube. Cyst grown in neural induction conditions was uniformly anterior and dorsal, and could be ventralized through activation of the Shh hedgehog pathway. Interestingly, treatment with RA induced more posterior fates and also led to spontaneous dorsal–ventral patterning as assayed by sonic hedgehog (SHH) in the putative floor plate and several early motor neuron fate markers in their correct relative positions.

5. CONCLUSIONS

The stem cell systems reviewed here represent promising avenues for making progress on difficult problems in mammalian embryogenesis. To date, most work has shown that these systems recapitulate already known features of mammalian embryogenesis such as the cascade of signaling from BMP to Wnt to Nodal; however, new insights which may be applicable to the embryo are also beginning to emerge. One example is the role of both secreted inhibitors and receptor localization in restricting the response of the epiblast to BMP. In vivo, it is possible that localizing the receptors to the basal side of the embryo restricts signaling to the epiblast—extraembryonic (trophoblast or amnion) boundary where this localization breaks down. It further prevents signaling from cavity, which is apical to the cells, from globally initiating gastrulation. The use of filter systems and micropatterned colonies has begun to unravel these interactions in culture, and it will be important to test their relevance in vivo in the future.

The culture systems can also be combined with mathematical modeling to investigate fundamental issues of symmetry breaking in development. In this regard, while several experiments suggest that Wnt–Dkk or Nodal–Lefty function as activator–inhibitor systems to generate Turing patterns, it is notable that they do not break the azimuthal symmetry of the colonies but instead generate rings of primitive streak formation. This is in contrast to the situation in vivo in which the primitive streak only occupies the posterior side of the embryo. It is possible that by treating these colonies with high levels of BMP, which they strongly respond to on the entire perimeter, they are constrained to adopt azimuthally symmetric organizations of signaling and fate. More natural ways of inducing the gastrulation might reveal whether the cells are intrinsically capable of breaking this symmetry or whether interactions with extraembryonic tissues which are lacking in these culture systems are required.

ACKNOWLEDGMENT

We thank Idse Heemskerk, Alfonso Martinez-Arias, and Mijo Simunovic for comments on an earlier draft of this review. E.D.S. was supported in part by NSF Grant PHY 1502151, and A.W. by NSF Grant MCB-1553228, Simons Foundation Grant 511079, and NIH Grant R01GM126122.

REFERENCES

Amita, M., Adachi, K., Alexenko, A. P., Sinha, S., Schust, D. J., Schulz, L. C., et al. (2013). Complete and unidirectional conversion of human embryonic stem cells to trophoblast by BMP4. *Proceedings of the National Academy of Sciences of the United States of America*, *110*, E1212–E1221.

Arnold, S. J., & Robertson, E. J. (2009). Making a commitment: Cell lineage allocation and axis patterning in the early mouse embryo. *Nature Reviews. Molecular Cell Biology*, *10*, 91–103.

Azioune, A., Storch, M., Bornens, M., Théry, M., & Piel, M. (2009). Simple and rapid process for single cell micro-patterning. *Lap Chip*, *9*, 1640–1642.

Bauwens, C. L., Peerani, R., Niebruegge, S., Woodhouse, K. A., Kumacheva, E., Husain, M., et al. (2008). Control of human embryonic stem cell colony and aggregate size heterogeneity influences differentiation trajectories. *Stem Cells*, *26*, 2300–2310.

Bedzhov, I., & Zernicka-Goetz, M. (2014). Self-organizing properties of mouse pluripotent cells initiate morphogenesis upon implantation. *Cell*, *156*, 1032–1044.

Ben-Haim, N., Lu, C., Guzman-Ayala, M., Pescatore, L., Mesnard, D., Bischofberger, M., et al. (2006). The nodal precursor acting via activin receptors induces mesoderm by maintaining a source of its convertases and BMP4. *Developmental Cell*, *11*, 313–323.

Bernardo, A. S., Faial, T., Gardner, L., Niakan, K. K., Ortmann, D., Senner, C. E., et al. (2011). BRACHYURY and CDX2 mediate BMP-induced differentiation of human and mouse pluripotent stem cells into embryonic and extraembryonic lineages. *Cell Stem Cell*, *9*, 144–155.

Blin, G., Picart, C., Théry, M., & Puceat, M. (2017). Geometrical confinement guides brachyury self-patterning in embryonic stem cells. *bioRxiv*, 1–46. https://doi.org/10.1101/138354.

Bolouri, H., & Davidson, E. H. (2010). The gene regulatory network basis of the "community effect," and analysis of a sea urchin embryo example. *Developmental Biology*, *340*, 170–178.

Chambers, S. M., Fasano, C. A., Papapetrou, E. P., Tomishima, M., Sadelain, M., & Studer, L. (2009). Highly efficient neural conversion of human ES and iPS cells by dual inhibition of SMAD signaling. *Nature Biotechnology*, *27*, 275–280.

Coucouvanis, E., & Martin, G. R. (1995). Signals for death and survival: A two-step mechanism for cavitation in the vertebrate embryo. *Cell*, *83*, 279–287.

Dunn, N. R., Vincent, S. D., Oxburgh, L., Robertson, E. J., & Bikoff, E. K. (2004). Combinatorial activities of Smad2 and Smad3 regulate mesoderm formation and patterning in the mouse embryo. *Development*, *131*, 1717–1728.

Engler, A. J., Sen, S., Sweeney, H. L., & Discher, D. E. (2006). Matrix elasticity directs stem cell lineage specification. *Cell*, *126*, 677–689.

Etoc, F., Metzger, J., Ruzo, A., Kirst, C., Yoney, A., Ozair, M. Z., et al. (2016). A balance between secreted inhibitors and edge sensing controls gastruloid self-organization. *Developmental Cell*, *39*, 302–315.

Farin, H. F., Jordens, I., Mosa, M. H., Basak, O., Korving, J., Tauriello, D. V. F., et al. (2016). Visualization of a short-range Wnt gradient in the intestinal stem-cell niche. *Nature*, *530*, 340–343.

Gjorevski, N., Ranga, A., & Lutolf, M. P. (2014). Bioengineering approaches to guide stem cell-based organogenesis. *Development*, *141*, 1794–1804.

Green, J. B., New, H. V., & Smith, J. C. (1992). Responses of embryonic Xenopus cells to activin and FGF are separated by multiple dose thresholds and correspond to distinct axes of the mesoderm. *Cell, 71*, 731–739.

Gurdon, J. B. (1988). A community effect in animal development. *Nature, 336*, 772–774.

Harrison, S. E., Sozen, B., Christodoulou, N., Kyprianou, C., & Zernicka-Goetz, M. (2017). Assembly of embryonic and extra-embryonic stem cells to mimic embryogenesis in vitro. *Science*, eaal1810.

Heemskerk, I., Burt, K., Miller, M., Chabra, S., Guerra, M. C., & Warmflash, A. (2017). Morphogen dynamics control patterning in a stem cell model of the human embryo. *bioRxiv*, 1–18. https://doi.org/10.1101/202366.

Heemskerk, I., & Warmflash, A. (2016). Pluripotent stem cells as a model for embryonic patterning: From signaling dynamics to spatial organization in a dish. *Developmental Dynamics, 245*, 976–990.

Horii, M., Li, Y., Wakeland, A. K., Pizzo, D. P., Nelson, K. K., Sabatini, K., et al. (2016). Human pluripotent stem cells as a model of trophoblast differentiation in both normal development and disease. *Proceedings of the National Academy of Sciences of the United States of America, 113*, 3882–3891.

Lee, G. Y., Kenny, P. A., Lee, E. H., & Bissell, M. J. (2007). Three-dimensional culture models of normal and malignant breast epithelial cells. *Nature Methods, 4*, 359–365.

Lei, Y., & Schaffer, D. V. A. (2013). Fully defined and scalable 3D culture system for human pluripotent stem cell expansion and differentiation. *Proceedings of the National Academy of Sciences of the United States of America, 110*, E5039–E5048.

Li, Y., Moretto-Zita, M., Soncin, F., Wakeland, A., Wolfe, L., Leon-Garcia, S., et al. (2013). BMP4-directed trophoblast differentiation of human embryonic stem cells is mediated through a ΔNp63+ cytotrophoblast stem cell state. *Development, 140*, 3965–3976.

Lippmann, E. S., Williams, C. E., Ruhl, D. A., Estevez-Silva, M. C., Chapman, E. R., Coon, J. J., et al. (2015). Deterministic HOX patterning in human pluripotent stem cell-derived neuroectoderm. *Stem Cell Reports, 4*, 632–644.

Loh, K. M., Chen, A., Koh, P. W., Deng, T. Z., Sinha, R., Tsai, J. M., et al. (2016). Mapping the pairwise choices leading from pluripotency to human bone, heart, and other mesoderm cell types. *Cell, 166*, 451–467.

Meinhardt, H. (2008). Models of biological pattern formation: From elementary steps to the organization of embryonic axes. *Current Topics in Developmental Biology, 81*, 1–63.

Meinhardt, A., Eberle, D., Tazaki, A., Ranga, A., Niesche, M., Wilsch-Bräuninger, M., et al. (2014). 3D reconstitution of the patterned neural tube from embryonic stem cells. *Stem Cell Reports, 3*, 987–999.

Mendjan, S., Mascetti, V. L., Ortmann, D., Ortiz, M., Karjosukarso, D. W., Ng, Y., et al. (2014). NANOG and CDX2 pattern distinct subtypes of human mesoderm during exit from pluripotency. *Cell Stem Cell, 15*, 310–325.

Migeotte, I., Omelchenko, T., Hall, A., & Anderson, K. V. (2010). Rac1-dependent collective cell migration is required for specification of the anterior-posterior body axis of the mouse. *PLoS Biology, 8*, e1000442.

Müller, P., Rogers, K. W., Jordan, B. M., Lee, J. S., Robson, D., Ramanathan, S., et al. (2012). Differential diffusivity of nodal and lefty underlies a reaction-diffusion patterning system. *Science, 336*, 721–724.

Nakamura, T., Mine, N., Nakaguchi, E., Mochizuki, A., Yamamoto, M., Yashiro, K., et al. (2006). Generation of robust left-right asymmetry in the mouse embryo requires a self-enhancement and lateral-inhibition system. *Developmental Cell, 11*, 495–504.

Nallet-Staub, F., Yin, X., Gilbert, C., Marsaud, V., Ben Mimoun, S., Javelaud, D., et al. (2015). Cell density sensing alters TGF-B signaling in a cell-type-specific manner, independent from hippo pathway activation. *Developmental Cell, 32*, 640–651.

Nemashkalo, A., Ruzo, A., Heemskerk, I., & Warmflash, A. (2017a). Morphogen and community effects determine cell fates in response to BMP4 signaling in human embryonic stem cells. *bioRxiv*, 1–30.

Nemashkalo, A., Ruzo, A., Heemskerk, I., & Warmflash, A. (2017b). Morphogen and community effects determine cell fates in response to BMP4 signaling in human embryonic stem cells. *Development*, *144*, 3042–3053. https://doi.org/10.1242/dev.153239.

Nowotschin, S., Costello, I., Piliszek, A., Kwon, G. S., Mao, C.-A., Klein, W. H., et al. (2013). The T-box transcription factor eomesodermin is essential for AVE induction in the mouse embryo. *Genes & Development*, *27*, 997–1002.

Pagliuca, F. W., Millman, J. R., Gürtler, M., Segel, M., Van Dervort, A., Ryu, J. H., et al. (2014). Generation of functional human pancreatic B cells in vitro. *Cell*, *159*, 428–439.

Peerani, R., Rao, B. M., Bauwens, C., Yin, T., Wood, G. A., Nagy, A., et al. (2007). Niche-mediated control of human embryonic stem cell self-renewal and differentiation. *The EMBO Journal*, *26*, 4744–4755.

Perea-Gomez, A., Vella, F. D. J., Shawlot, W., Oulad-Abdelghani, M., Chazaud, C., Meno, C., et al. (2002). Nodal antagonists in the anterior visceral endoderm prevent the formation of multiple primitive streaks. *Developmental Cell*, *3*, 745–756.

Przybyla, L., Lakins, J. N., & Weaver, V. M. (2016). Tissue mechanics orchestrate Wnt-dependent human embryonic stem cell differentiation. *Cell Stem Cell*, *19*, 462–475.

Qin, D., Xia, Y., & Whitesides, G. M. (2010). Soft lithography for micro- and nanoscale patterning. *Nature Protocols*, *5*, 491–502.

Ranga, A., Girgin, M., Meinhardt, A., Eberle, D., Caiazzo, M., Tanaka, E. M., et al. (2016). Neural tube morphogenesis in synthetic 3D microenvironments. *Proceedings of the National Academy of Sciences of the United States of America*, *113*, E6831–E6839.

Rosowski, K. A., Mertz, A. F., Norcross, S., Dufresne, E. R., & Horsley, V. (2015). Edges of human embryonic stem cell colonies display distinct mechanical properties and differentiation potential. *Scientific Reports*, *5*, 14218.

Saka, Y., Lhoussaine, C., Kuttler, C., Ullner, E., & Thiel, M. (2011). Theoretical basis of the community effect in development. *BMC Systems Biology*, *5*, 54.

Sasai, Y. (2013). Cytosystems dynamics in self-organization of tissue architecture. *Nature*, *493*, 318–326.

Schohl, A., & Fagotto, F. (2002). Beta-catenin, MAPK and Smad signaling during early Xenopus development. *Development*, *129*, 37–52.

Shao, Y., Taniguchi, K., Gurdziel, K., Townshend, R. F., Xue, X., Yong, K. M. A., et al. (2016). Self-organized amniogenesis by human pluripotent stem cells in a biomimetic implantation-like niche. *Nature Materials*, *16*, 419–425.

Shao, Y., Taniguchi, K., Townshend, R. F., Miki, T., Gumucio, D. L., & Fu, J. A. (2017). Pluripotent stem cell-based model for post-implantation human amniotic sac development. *Nature Communications*, *8*, 1–14.

Simunovic, M., & Brivanlou, A. H. (2017). Embryoids, organoids and gastruloids: New approaches to understanding embryogenesis. *Development*, *144*, 976–985.

Sorre, B., Warmflash, A., Brivanlou, A. H., & Siggia, E. D. (2014). Encoding of temporal signals by the TGF-B pathway and implications for embryonic patterning. *Developmental Cell*, *30*, 334–342.

Sun, Y., Yong, K. M. A., Villa-Diaz, L. G., Zhang, X., Chen, W., Philson, R., et al. (2014). Hippo/YAP-mediated rigidity-dependent motor neuron differentiation of human pluripotent stem cells. *Nature Materials*, *13*, 599–604.

Tang, C., Ardehali, R., Rinkevich, Y., Seita, J., Lee, A. S., Mosley, A. R., et al. (2012). Isolation of primitive endoderm, mesoderm, vascular endothelial, and trophoblast progenitors from human pluripotent stem cells. *Nature Biotechnology*, *30*, 531–542.

ten Berge, D., Koole, W., Fuerer, C., Fish, M., Eroglu, E., & Nusse, R. (2008). Wnt signaling mediates self-organization and axis formation in embryoid bodies. *Cell Stem Cell*, *3*, 508–518.

Teo, A. K. K., Arnold, S. J., Trotter, M. W. B., Brown, S., Ang, L. T., Chng, Z., et al. (2011). Pluripotency factors regulate definitive endoderm specification through eomesodermin. *Genes & Development*, *25*, 238–250.

Tewary, M., Ostblom, J. E., Shakiba, N., & Zandstra, P. W. A. (2017). Defined platform of human peri-gastrulation-like biological fate patterning reveals coordination between reaction-diffusion and positional-information. *bioRxiv*, 1–41.

Théry, M., & Piel, M. (2009). Adhesive micropatterns for cells: A microcontact printing protocol. *Cold Spring Harbor Protocols*, *2009*. pdb.prot5255.

Turing, A. (1952). The chemical basis of morphogenesis. *Philosophical Transactions of the Royal Society of London. Series B, Biological Sciences*, *237*, 37–72.

Turner, D., Alonso-Crisostomo, L., Girgin, M., Baillie-Johnson, P., Glodowski, C. R., Hayward, P. C., et al. (2017). Gastruloids develop the three body axes in the absence of extraembryonic tissues and spatially localised signalling. *bioRxiv*, 1–26.

Turner, D. A., Baillie-Johnson, P., & Martinez Arias, A. (2016). Organoids and the genetically encoded self-assembly of embryonic stem cells. *BioEssays*, *38*, 181–191.

van den Brink, S. C., Baillie-Johnson, P., Balayo, T., Hadjantonakis, A.-K., Nowotschin, S., Turner, D. A., et al. (2014). Symmetry breaking, germ layer specification and axial organisation in aggregates of mouse embryonic stem cells. *Development*, *141*, 4231–4242.

Warmflash, A., Sorre, B., Etoc, F., Siggia, E. D., & Brivanlou, A. H. (2014). A method to recapitulate early embryonic spatial patterning in human embryonic stem cells. *Nature Methods*, *11*, 847–854.

Wilson, P. A., Lagna, G., Suzuki, A., & Hemmati-Brivanlou, A. (1997). Concentration-dependent patterning of the Xenopus ectoderm by BMP4 and its signal transducer Smad1. *Development*, *124*, 3177–3184.

Xu, R.-H., Chen, X., Li, D. S., Li, R., Addicks, G. C., Glennon, C., et al. (2002). BMP4 initiates human embryonic stem cell differentiation to trophoblast. *Nature Biotechnology*, *20*, 1261–1264.

Yang, Y., Adachi, K., Sheridan, M. A., Alexenko, A. P., Schust, D. J., Schulz, L. C., et al. (2015). Heightened potency of human pluripotent stem cell lines created by transient BMP4 exposure. *Proceedings of the National Academy of Sciences of the United States of America*, *112*, E2337–E2346.

Yu, P., Pan, G., Yu, J., & Thomson, J. A. (2011). FGF2 sustains NANOG and switches the outcome of BMP4-induced human embryonic stem cell differentiation. *Cell Stem Cell*, *8*, 326–334.

CHAPTER TWO

What Can Stem Cell Models Tell Us About Human Germ Cell Biology?

Naoko Irie[*,†,1], Anastasiya Sybirna[*,†,‡], M. Azim Surani[*,†,1]
[*]Wellcome Trust/Cancer Research UK Gurdon Institute, University of Cambridge, Cambridge, United Kingdom
[†]University of Cambridge, Cambridge, United Kingdom
[‡]Wellcome Trust Medical Research Council Stem Cell Institute, University of Cambridge, Cambridge, United Kingdom
[1]Corresponding authors: e-mail addresses: n.irie@gurdon.cam.ac.uk; a.surani@gurdon.cam.ac.uk

Contents

Abstract

Fusion of sperm and egg generates a totipotent zygote that develops into a whole organism. Accordingly, the "immortal" germline transmits genetic and epigenetic information to subsequent generations with consequences for human health and disease. In mammals, primordial germ cells (PGCs) originate from peri-gastrulation embryos. While early human embryos are inaccessible for research, *in vitro* model systems using pluripotent stem cells have provided critical insights into human PGC specification, which differs from that in mice. This might stem from significant differences in early embryogenesis at the morphological and molecular levels, including pluripotency

ISSN 0070-2153
https://doi.org/10.1016/bs.ctdb.2018.02.010

networks. Here, we discuss recent advances and experimental systems used to study mammalian germ cell development. We also highlight key aspects of germ cell disorders, as well as mitochondrial and potentially epigenetic inheritance in humans.

1. INTRODUCTION

Primordial germ cells (PGCs) are embryonic precursors of sperm and egg. There are two primary mechanisms of PGC fate specification: instructive through signaling as in mammals (Chatfield et al., 2014; Magnúsdóttir & Surani, 2014), or the inheritance of germplasm containing deterministic factors as in *Drosophila*, teleost fish, and frogs (Extavour & Akam, 2003; Johnson & Alberio, 2015). Human PGCs (hPGCs) were first reported in week 4 (Wk 4) embryos in the early 1900s (De Felici, 2013; Felix, 1912; Fuss, 1911), with subsequent studies describing their migration from the yolk sac to the primitive gonadal fold (Politzer, 1930, 1933; Witschi, 1948). Mouse PGCs (mPGCs) were first reported as alkaline phosphatase (AP)-positive cells (Chiquoine, 1954) and later revealed to originate from the posterior proximal pregastrulation epiblast (Lawson & Hage, 1994).

The first molecular study to explore the genetic basis for mPGC specification was carried out using single-cell analysis (Saitou, Barton, & Surani, 2002) and identified *Prdm1* as the key determinant of mPGC fate (Oosterhuis & Looijenga, 2005), which established the foundation for subsequent advances in understanding the mechanism of mammalian PGC specification. On embryonic day (E) 6.25, the combined signaling of WNT, BMP4, BMP8b, and BMP2 induces founding PGCs in a subpopulation of equipotent epiblast cells (Lawson et al., 1999; Ohinata et al., 2009; Ying, Liu, Marble, Lawson, & Zhao, 2000; Ying & Zhao, 2001). PGC specification commences with the expression of transcription factor (TF) *Prdm1* (encodes BLIMP1) (Oosterhuis & Looijenga, 2005), followed by *Prdm14* and *Tfap2c* (encodes AP2γ) that are necessary and sufficient for mouse germ cell fate (Magnúsdóttir et al., 2013; Nakaki et al., 2013). Concomitantly, mPGCs re-express pluripotency-related markers, such as *Pou5f1* (encodes OCT4), *Nanog*, and *Sox2* (Magnúsdóttir & Surani, 2014; Saitou & Yamaji, 2012), as they migrate along the hindgut to colonize the developing genital ridges, and undergo epigenetic resetting, including global DNA demethylation and redistribution of certain histone marks (Hackett et al., 2013; Hajkova et al., 2008; Leitch, Tang, & Surani, 2013; Tang et al., 2015; Weick & Miska, 2014).

Recent studies revealed key differences in the transcription factor networks engaged in mPGC and hPGC specification that might be attributed to divergent early embryogenesis and pluripotency regulation (Boroviak & Nichols, 2017; Irie, Tang, & Azim Surani, 2014): rodents feature a cup-shaped egg cylinder, and non-rodent mammals form a flat embryonic disc epiblast. This might impact on signaling gradients and downstream molecular pathways for PGC specification (Irie et al., 2014; Tang et al., 2015), which complicates extrapolation of studies from mouse to human.

It is possible to investigate *in vivo* hPGCs from Wk 4 (migrating hPGCs), up to Wks 25/26 (fetal germ cells in the testis and ovary, respectively) (Gkountela et al., 2015; Guo et al., 2015; Li et al., 2017; Tang et al., 2015). However, peri-implantation human embryos at ~Wks 2–3, where hPGCs originate, are not available for research. This necessitates the development of *in vitro* model systems using human pluripotent stem cells (hPSCs), including embryonic stem cells (ESCs), derived from the inner cell mass (ICM) of the blastocyst, and induced pluripotent stem cells (iPSCs) obtained by reprogramming of differentiated cells. While initial attempts had limited success (see Hayashi, Saitou, & Yamanaka, 2012), later studies reported robust *in vitro* differentiation of human PGC-like cells (hPGCLCs) from hPSCs (Irie et al., 2015; Sasaki et al., 2015; Sugawa et al., 2015). Importantly, these experiments revealed a critical role of SOX17 TF in hPGC fate, which was hitherto primarily known as a regulator of endoderm (Irie et al., 2015; Seguin, Draper, Nagy, & Rossant, 2008). While BLIMP1 is a conserved regulator of mouse and human PGCs, it acts downstream of SOX17 in hPGCs (Irie et al., 2015). Furthermore, SOX2, important for mPGC fate, is absent from human PGCs (de Jong, Stoop, Gillis, van Gurp, et al., 2008; Irie et al., 2015). Notably, these key features of human PGC molecular network are also observed in monkey and pig PGCs which all develop as flat bilaminar embryonic discs (Kobayashi et al., 2017; Sasaki et al., 2016).

Here, we discuss the molecular control of early germline development in mammals and how PGC versus soma cell fate decision can be modeled using *in vitro* PSCs and non-rodent mammalian models. Furthermore, we highlight how misregulation of germline development may contribute to human diseases. In particular, we review the origins of germ cell tumors, as well as inheritance of mitochondrial DNA (mtDNA) mutations and potential epigenetic modifications.

2. SIGNALING FOR GERM CELL SPECIFICATION IN MAMMALIAN EMBRYOS

During human prenatal development, which lasts around 38 weeks, hPGCs are set aside from somatic cells at Wks 2–3 and initiate the molecular program to form unipotent gametes (Tang, Kobayashi, Irie, Dietmann, & Surani, 2016). Upon implantation, the mouse embryo turns into the so-called egg cylinder consisting of the cup-shaped epiblast, with extraembryonic ectoderm (ExE) sitting on top. These two layers are surrounded by the visceral endoderm (VE) featuring a recognizable region called the anterior VE (AVE) (Fig. 1). ExE, AVE, and the epiblast emit various signals required for primitive streak formation and PGC specification in the posterior region. In contrast to the mouse, the postimplantation human embryo forms a flat bilaminar structure consisting of the epiblast layer and the VE (also known as the hypoblast) lying underneath (Fig. 1). Of note, human embryos lack tissues equivalent to the ExE. Gastrulation initiates posteriorly and transforms the embryo into a trilaminar disc composed of three germ layers: ectoderm, mesoderm, and definitive endoderm (Tang et al., 2016). This type of peri-gastrulation embryo has been observed in other non-rodent mammals, such as rabbit, cow, pig, and primates (Boroviak & Nichols, 2017; Fig. 1). Another important difference between rodent and primate embryos is the origin and timing of the amnion formation. While in marmoset, rhesus macaque, and human, the amniotic epithelium arises from the pre-implantation epiblast before gastrulation, in mouse, it is derived mainly from the extraembryonic mesoderm and extraembryonic ectoderm and appears later, at the onset of gastrulation (Pereira et al., 2011). One report suggested that cynomolgus monkey PGCs (cmPGCs) might be specified in the amnion (Sasaki et al., 2016), which underscores the importance of further studying the amniotic development.

Pronounced morphological differences, in particular, the absence of signal-producing ExE, raise the question of how similar the patterning signals are in planar embryos compared to the mouse. Mouse studies showed that the components of WNT and BMP pathways, as well as their antagonists, are critical for the temporal and spatial control of PGC specification. In mouse embryos at E5.5–6.5 when PGCs are specified, WNT3 is detected in the posterior VE and in the epiblast (Kemp, Willems, Abdo, Lambiv, & Leyns, 2005; Liu et al., 1999), while WNT direct target, *Brachyury* (*T*), marks the formation of the primitive streak in the posterior epiblast (Yamaguchi, Takada, Yoshikawa, Wu, & McMahon, 1999). BMP4 and BMP8b are secreted from the ExE in E6.0 mouse embryos (Lawson et al., 1999; Ying, Qi, & Zhao, 2001). BMP4 is the main PGC-inducing signal, sufficient to specify PGCs from epiblast cells *ex vivo* and *in vitro* (Ohinata et al., 2009). BMP8b, on the other hand, restricts the expansion of the AVE, which is the source of inhibitory factors Cer1 and Dkk1, antagonizing BMP and WNT, respectively (Ohinata et al., 2009).

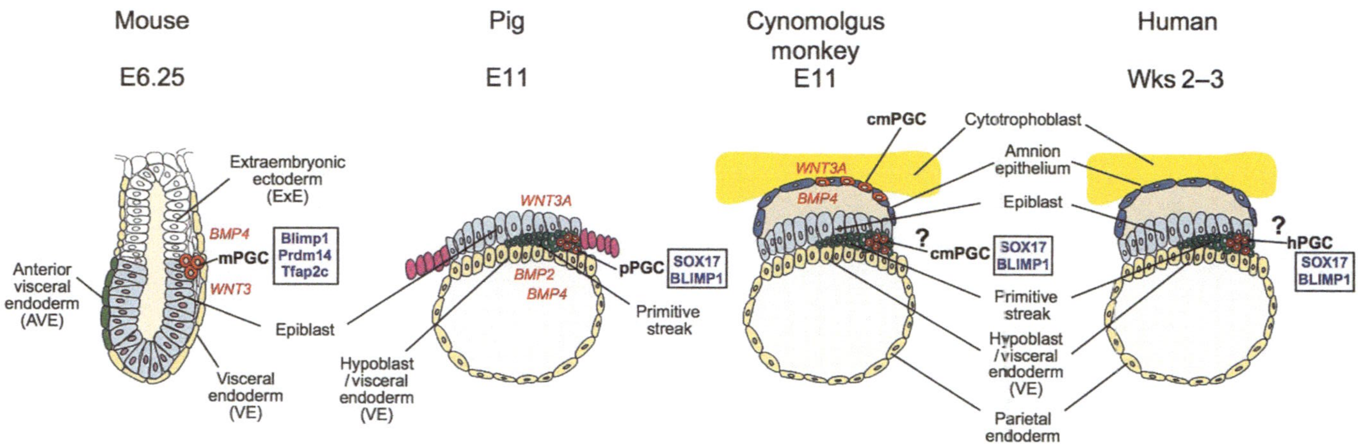

Fig. 1 Schematic model for PGC specification in early embryos of mouse, pig, cynomolgus monkey, and human. Mouse egg cylinder epiblast at E6.25 consists of an elongated cup-shaped epiblast, with trophectoderm-derived extraembryonic ectoderm (ExE) sitting on top; both are surrounded by the VE including the specialized signaling region called the anterior VE (AVE). Bone morphogenetic protein 4 (BMP4) from the ExE, together with WNT3 from the posterior VE, as well as inhibitory signals from the AVE, induces mPGCs expressing the key TFs, *Prdm1/Prdm14/Tfap2c*, in the posterior epiblast before primitive streak formation. On the other hand, pig (E11), cynomolgus monkey (E11), and human (Wks 2–3) embryos form a flat bilaminar structure consisting of the epiblast layer and the VE (also known as the hypoblast) lying underneath, with the primitive streak/nascent mesendoderm sitting between the two layers at the posterior end. PGCs are induced by WNT and BMP signaling from the surrounding tissues and upregulate PGC master TFs, SOX17 and BLIMP1. In pig, WNT and BMP2/4 signaling can be detected at the posterior epiblast. Cynomolgus monkey and human, but not mouse and pig, embryos at these stages exhibit amnion epithelium; however, the timing of the amnion formation relative to PGC specification differs between the two species. In the cynomolgus monkey embryo, both WNT3A/BMP4 expression and SOX17^{+}/TFAP2C^{+} presumable nascent cmPGCs can be observed in the amnion; cmPGCs later appear and expand in the posterior region. In the human embryo, hPGCs are likely specified around Wks 2–3, but their tissue of origin and inductive signals cannot be experimentally tested *in vivo*.

BMP2 is expressed in the VE from around E5.5 and is believed to enhance the BMP4 signal in the posterior region (Ying & Zhao, 2001). Indeed, BMP2 is also able to induce mouse and human PGC-like cells *in vitro* (Irie et al., 2015; Ohinata et al., 2009).

One way to visualize the patterning of non-rodent disc embryos is to "project" the E6.5 mouse egg cylinder (corresponding to the trilaminar human embryo around E17) onto a flat surface (Beddington & Robertson, 1999; Behringer, Wakamiya, Tsang, & Tam, 2000). Here, the epiblast would represent the upper layer, while the VE, the bottom layer of the disc; the ingressing mesoderm cells would form the inner layer. In such a model, the ExE would positionally correspond to the peripheral epiblast or the amnion epithelium, while the VE would be equivalent to the yolk sac epithelium. This would place the source of WNT and BMP2 signals in the posterior hypoblast and BMP4 would appear in a ring-like region around the epiblast. Such a hypothetical model is supported by experimental evidence in flat-disc epiblast mammals. Indeed, in rabbit and porcine embryos, WNT3 and its target *T* can be detected predominantly in the posterior rim of the embryonic disc when it elongates at the posterior pole, generating a posterior gastrula extension (Hassoun, Püschel, & Viebahn, 2010; Idkowiak, Weisheit, Plitzner, & Viebahn, 2004; Viebahn, Stortz, Mitchell, & Blum, 2002; Yoshida et al., 2016). Furthermore, prior to PGC specification, BMP2 is expressed in the posterior hypoblast, followed by BMP4 expression in the epiblast and hypoblast. Nascent PGCs, identified by BLIMP1 expression, appear in the posterior epiblast (Hopf, Viebahn, & Puschel, 2011). The region of PGC induction in rabbits and pigs is likely restricted by BMP and WNT inhibitors (CER1 and DKK1, respectively) emanating from the anterior hypoblast, as is the case in mice (Idkowiak et al., 2004).

A recent study proposed that cynomolgus monkey PGCs (cmPGCs) might originate from the nascent amnion and then migrate to posterior amnion and expand in the posterior yolk sac. However, the authors could not exclude the possibility of additional cmPGC specification in the posterior epiblast (Sasaki et al., 2016). At E11 *WNT3A*, but not *WNT3*, is expressed in the amnion and adjacent parts of the placenta including the cytotrophoblast layer, and its downstream target, *AXIN2*, can be detected in the amnion. *BMP4* is first observed in the E11 amnion, but then shifts to the proximal posterior region at E12. *BMP2*, on the other hand, is strongly expressed in the hypoblast at E11 and then spreads to the parietal endoderm and extraembryonic mesenchyme at E12. Similar to other models, WNT antagonist *DKK1* and BMP antagonist *CER1* are expressed in the AVE at E11, potentially ensuring the spatial restriction of PGC specification. At E12, a few cells in the amnion, posterior epiblast, as well as between the epiblast and hypoblast start expressing T protein; T^{+}-primitive streak cells appear at E13. T-positive cells that upregulate either SOX17 or SOX17 and TFAP2C are the presumptive nascent PGCs (Sasaki et al., 2016; Fig. 1).

Altogether, the comparative analysis across mammals points to conserved roles of WNT and BMP signaling in PGC specification. The localization of activating and inhibitory ligands for WNT and BMP pathways most likely determines the region of PGC specification. The cells that exhibit competency to become PGCs (PGC-competent cells) and nascent PGCs are marked by T expression, which appears to be lower than in the surrounding mesodermal cells (Kobayashi et al., 2017). Importantly, *in vitro* hPGC models show that PGCLCs also arise from T-positive cells (Irie et al., 2015; Kobayashi et al., 2017; Sasaki et al., 2015). However, while T is expressed in the majority of PGC-competent cells *in vitro*, roughly half the cells show upregulation of germ cell genes. One possibility is that the gene dosage of *T* might be critical for PGC specification. It would be of interest to identify additional markers for PGC-competent cells and address the molecular mechanisms of competence acquisition. In the following sections, we review the current knowledge on PGC competence *in vivo* and *in vitro*.

3. PERMISSIVE CELL STATE FOR GERM CELL SPECIFICATION IN THE EMBRYO

Cell fate decisions are tightly regulated, both temporally and spatially: a subpopulation of cells gain the ability to respond to the right extracellular cues in time and space in the developing embryo. Common signaling pathways often control different developmental processes in a highly parsimonious fashion (Barolo & Posakony, 2002). While BMP signaling is essential for germ cell fate in the epiblast cells, it is also required for differentiation of mesoderm and cardiac progenitors, as well as gastric patterning and bone organogenesis later in the development (Wang et al., 2014). The same is true for transcription factors, which often operate in context-dependent manners. This can be exemplified by SOX17, which, in humans, acts as both an endoderm specifier and a master regulator of the hPGC fate (Irie et al., 2015; Seguin et al., 2008). The question is how cell-type-specific transcriptional programs ensue from seemingly identical inputs. This likely reflects different states of the target cells, possessing or lacking competence to form certain lineages that are spatially and temporally regulated.

Developmental competence can be defined as the propensity of cells to differentiate to a particular cell type(s) in response to appropriate stimuli (Gunesdogan & Surani, 2016). This notion is not synonymous to stem cell potency. For example, mouse naïve PSCs are pluripotent: when injected into mouse blastocysts, they contribute to all three germ layers, as well as the germline (Ying et al., 2008). Mouse PSCs (mPSCs), however, are not competent to form PGCs directly; rather, they have to be pre-induced into a competent state (EpiLCs, see later) to be able to respond to BMP and form germ cell precursors (Hayashi, Ohta, Kurimoto, Aramaki, & Saitou, 2011). Competence can be associated with a number of factors, such as the presence of appropriate cell surface receptors, expression of certain transcription factors, and availability of intracellular signal transducers (Gunesdogan & Surani, 2016). In addition to chemical cues, the phase of the cell cycle, cell-to-cell contacts, and the stiffness of the extracellular matrix emerge as important players in cells' differentiation propensity (Bellas & Chen, 2014; Kaylan, Ermilova, Yada, & Underhill, 2016; Pauklin & Vallier, 2013). There is also increasing evidence suggesting that competent cells possess a unique epigenetic landscape that "poises" lineage-specific genes to be rapidly induced in response to the right stimulus; the regulatory elements of these genes are typically more accessible for TFs and transcriptional machinery (Long, Prescott, & Wysocka, 2016; Rada-Iglesias et al., 2011; Wang et al., 2015). This could be achieved by differential histone modifications and DNA methylation at these regulatory elements.

Both *in vitro* and *in vivo* studies in mouse, porcine, monkey, and human cells suggest that such temporal and spatial regulation exists for the gain of germline competency (Gunesdogan, Magnusdottir, & Surani, 2014; Gunesdogan & Surani, 2016; Kobayashi et al., 2017; Sasaki et al., 2016; Tang et al., 2016). In the mouse model, the founder PGC population arises exclusively in the posterior proximal epiblast and is very small (around 40 AP-positive cells at E7.25) (Ginsburg, Snow, & McLaren, 1990; Magnúsdóttir & Surani, 2014), raising the question of whether only a limited subpopulation of epiblast cells is PGC competent. Epiblasts before E5.5 are refractory to PGC induction, but acquire competence in response to WNT signaling followed by BMP to induce PGC fate (Aramaki et al., 2013; Ohinata et al., 2009). An *ex vivo* epiblast culture system shows that mouse epiblast cells maintain the competency between E5.5 and E6.25 and lose the responsiveness to BMP afterward (Ohinata et al., 2009). Interestingly, the isolated E5.5–6.25 epiblast cells without VE and ExE can be uniformly specified as PGCs in response to BMPs, suggesting that epiblast cells may be equipotent at this stage. Spatial regulation can be explained by the AVE inhibitory factors, Cer1 and Dkk1 antagonizing BMP and WNT, respectively, regulated by BMP8b from ExE (Ohinata et al., 2009). As mentioned earlier, most of the non-rodent mammalian epiblasts do not have the equivalent structure to the ExE.

Deletion of either *Wnt3* or WNT effector β-catenin prevents germ cell induction by BMP, confirming the role of WNT in conferring PGC competence (Aramaki et al., 2013; Ohinata et al., 2009). Interestingly, Wnt3-KO epiblast cells express BMP receptors as well as Smad1 and Smad5, which transduce the BMP signal (Ohinata et al., 2009). Therefore, the exact mechanism of WNT-mediated competence remains unclear. One possibility is that WNT/β-catenin pathway is required for the expression of *T*, which was shown to directly induce key germ cell TFs *Prdm1* and *Prdm14* in an *in vitro* system of mouse PGCLC specification (Aramaki et al., 2013). However, while T-deficient cells were unable to upregulate *Prdm14*, they could still express *Prdm1* in response to BMP signaling (Aramaki et al., 2013). This is in contrast to Wnt3-null epiblast cells that failed to induce *Prdm1* expression (Aramaki et al., 2013). This suggests that while T is critical for proper mPGC specification, it might not be the only Wnt target conferring PGC competence. Importantly, WNT is also required for gastrulation and mesoderm formation (Liu et al., 1999). Thus, the specificity of WNT in the context of germ cell lineage remains to be elucidated. It is possible that high dosage of BMP after a pulse of Wnt signaling promotes PGC fate over mesoderm.

Conserved roles of WNT signaling in PGC competence have recently been confirmed in non-rodent animals. In porcine embryos, WNT signaling is detected in pre-primitive streak stage (~E11) followed by the expression of BMP2 and BMP4 in the early primitive streak (~E11.5 to E12), when nascent PGCs appear (Kobayashi et al., 2017; Valdez Magana, Rodriguez, Zhang, Webb, & Alberio, 2014; Yoshida et al., 2016). In line with this, in *ex vivo* cultured porcine epiblasts PGCs are specified in response to BMP4 at the pre-primitive streak (~E11) and early primitive streak (~E11.5 to E12) stages, but not at the earlier bilaminar disc stage (~E9.5 to 10) (Kobayashi et al., 2017). Interestingly, WNT inhibition abrogates PGC competency at the preprimitive streak cells (~E11), but not in the early primitive streak (~E11.5 to E12). We might speculate that once competence-related WNT targets are upregulated, sustained activity of WNT signaling is dispensable for porcine PGC (pPGC) specification. Indeed, WNT target, T, is induced in the pseudostratified posterior epiblast cells in the preprimitive streak embryo. This is followed by the pPGC specification identified by the expression of SOX17, BLIMP1,

TFAP2C, and NANOG in response to BMP signaling (Kobayashi et al., 2017). Of note, not all epiblast cells at the competent stages gave rise to pPGCs in the *ex vivo* culture (Kobayashi et al., 2017), which suggests that unlike E6 mouse epiblasts, pig epiblasts at E11–E12 may not be equipotent. Alternatively, the studied *ex vivo* pig epiblasts might have contained some sources of inhibitory signals.

Sasaki and colleagues reported that in cynomolgus monkey nascent PGCs might arise in the dorsal amnion from T-expressing competent cells at around E11, before gastrulation (see above) (Sasaki et al., 2016). Despite the potential difference in the location of PGC specification, their data confirmed the key roles of WNT and BMP signaling for PGC specification. Also, the authors discussed the possibility for the dual origin of cmPGCs from both the amnion and posterior epiblast. Indeed, *in vitro* cmPGCLCs were efficiently derived from cmPSCs in a similar manner to human PGCLC induction (Kobayashi et al., 2017). These *in vitro* studies suggest that the germline-competent state represents epiblast-derived incipient mesoderm/pre-mesendoderm (pre-ME) state in both human and cynomolgus monkey (Kobayashi et al., 2017; Sasaki et al., 2015). Additional studies, such as lineage tracing experiments, are important to confirm the amnion origin of cmPGCs, which can help clarify whether it is specific to cynomolgus monkeys or common in other species.

In primates, the amnion is derived directly from epiblast cells and has been reported to possess some pluripotent stem cell features, such as OCT4 and NANOG expression (Dobreva, Pereira, Deprest, & Zwijsen, 2010). Therefore, it is possible that *in vitro* PGC-competent state recapitulated some aspects of the nascent amnion development. A recent study reported differentiation of amnion-like cells from hPSCs *in vitro*, in the form of a three-dimensional (3D) human postimplantation amniotic sac embryoid (PASE) (Shao et al., 2017). PASE recapitulates a number of postimplantation events, including amniotic sac development, activation of BMP signaling, and primitive streak formation. Importantly, however, there was no SOX17/NANOG-positive PGC population observed in either the amnion or the epiblast part of the PASE (Shao et al., 2017).

Harrison and colleagues reported specification of mPGCs in 3D organoids formed by mESC cocultured with extraembryonic trophoblast stem cells (Harrison, Sozen, Christodoulou, Kyprianou, & Zernicka-Goetz, 2017). Similar approaches might not be applicable using human cells, because unlike in mice, where the epiblast is in direct contact with the ExE, human epiblast disc sits on the hypoblast and is surrounded by the amniotic cavity (Boroviak &

Nichols, 2017). The timing and mechanism of amnion formation vary among different amniotes (Boroviak & Nichols, 2017). Studies in marmoset, human, and rhesus monkey have shown that the process of amnion formation is equivalent in these primate species (Enders & Lopata, 1999; Luckett, 1975). The amniotic cavity however forms as early as E7–8 in the human embryo, compared to E10 in rhesus monkey (Luckett, 1975) and E11 in the cynomolgus monkey (Sasaki et al., 2016). These timing differences are important, considering that hPGC specification presumably occurs around Wks 2–3 (E14–21) (Tang et al., 2016), while nascent cmPGCs were observed in the amnion concomitantly with its formation at E11 (Sasaki et al., 2016). Other mammalian species have a distinct mode of amniogenesis compared to mice where the amniotic cavity is established by cavitation of the epiblast (Oestrup et al., 2009). The pig amnion is formed by dorsal fusion of the amniotic folds (derived from the trophectoderm and extraembryonic mesoderm) with the embryonic disc at E14–15, where pPGC specification occurs earlier at E11.5–12 (Kobayashi et al., 2017). Extended *in vitro* culture of early human embryos mimicking implantation might provide further insights on the origin of human PGCs (Deglincerti et al., 2016; Shahbazi et al., 2016).

4. INDUCING GERMLINE COMPETENCY IN PLURIPOTENT STEM CELLS

mPGC specification can be recapitulated *in vitro* using mPSCs (either ESCs or iPSCs) (Hayashi & Saitou, 2014). Notably, mPSCs cultured in so-called naïve conditions, which bear resemblance to pre-implantation epiblast (Boroviak, Loos, Bertone, Smith, & Nichols, 2014), have to be coaxed into germline-competent state termed epiblast-like cells (EpiLCs). EpiLCs can respond to BMP4 forming mPGCLCs with high efficiencies (Hayashi et al., 2011). *In vitro* induced mPGC-like cells (mPGCLCs) can develop into mature sperm if transplanted to mouse testes (Hayashi et al., 2011; Ishikura et al., 2016) or into oocytes if combined *in vitro* with gonadal somatic cells from the embryonic ovary (Hikabe et al., 2016).

EpiLCs are generated by a 48-h exposure of naïve mPSCs to activin A and FGF2. Notably, longer (>2 days) induction results in competence loss (Hayashi et al., 2011). The transient EpiLC identity is reminiscent of the short PGC competence time observed in the epiblasts (Hayashi et al., 2011; Ohinata et al., 2009). Furthermore, the transcriptome of EpiLCs is highly similar to that of mouse embryos at pre-gastrulation stage, when PGC specification occurs (Hayashi et al., 2011). Comparison of epigenetic profiles of naïve (mESCs) and PGC-competent (EpiLC) states revealed that lineage specification genes, including those involved in PGC fate, were primed for activation in EpiLCs, but not in naïve mESCs (Buecker et al., 2014; Kurimoto et al., 2015). Thus, PGC-specific enhancers were marked with both H3K27me3 and H3K4me1 (poised enhancers), while many developmental promoters were bivalent (marked by H3K27me3 and H3K4me3). This is consistent with the idea of competent cells possessing characteristic poised chromatin signatures (Rada-Iglesias et al., 2011; Wang et al., 2015).

Initial attempts to derive germ cells from hPSCs yielded PGCLCs at very low frequencies, largely representing spontaneous differentiation (Clark et al., 2004; Gkountela et al., 2013; Kee, Angeles, Flores, Nguyen, & Reijo Pera, 2009). When the protocol developed for mouse PGCLC induction is applied to conventional hPSCs (maintained with fibroblast growth factor 2, FGF2), the efficiency of hPGCLC induction is extremely low (<5%). Intriguingly, when cultured in an alternative culture condition, containing leukemia inhibitory factor (LIF), FGF2, and transforming growth factor β1 (TGFβ), as well as four small-molecule inhibitors of GSK3 and three MAP kinases (p38, JNK, and MEK) (named 4i), hPSCs acquire competence for germ cell fate (Gafni et al., 2013; Irie et al., 2015). 4i hPSCs self-renew in the competent state and can respond to BMP2 or BMP4 generating PGCLCs with high efficiencies (around 50%) (Fig. 2). The competency of 4i hPSCs is further confirmed by the finding that overexpression of *SOX17*, the master regulator of human germ cell fate, is sufficient to induce PGC genes in the absence of BMPs (Irie et al., 2015). In contrast, overexpression of *SOX17* in conventional hPSC context results in endoderm differentiation (Seguin et al., 2008). Importantly, PGCLC competence is reversible by alternating 4i and conventional conditions. This makes 4i hPSCs a tractable and robust system to study the dynamics of human PGC competence acquisition and maintenance.

Further, an alternative approach to hPGCLC specification was established (Kobayashi et al., 2017; Sasaki et al., 2015). It relies on initial differentiation of conventional hPSCs to incipient mesoderm-like cells (iMeLCs)/pre-ME via combined action of activin A and WNT signaling agonist Chiron (CH) (Fig. 2). iMeLCs/pre-ME temporarily gain competency for PGC lineage and can then acquire PGC fate in response to BMP2/4. Of note, the timing of maximal germline competency *in vitro* seems to depend on the initial conventional hPSC culture condition (42–48 h of competency induction for iMeLCs and 12 h for pre-ME) (Kobayashi et al., 2017; Sasaki et al., 2015). Based on transcriptional profiles, PGCLCs derived from both 4i hPSCs

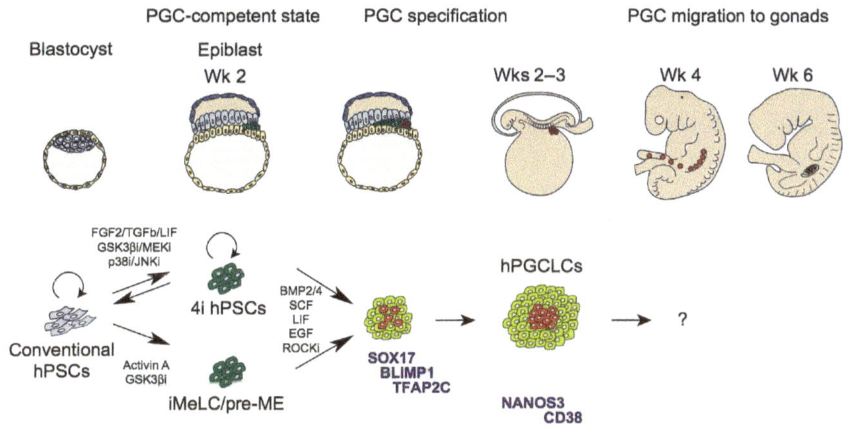

Fig. 2 Human PGC specification *in vivo* and from pluripotent stem cells. Conventional hPSCs cultured in FGF2-containing medium exhibit low germline competence, but can reversibly gain it when grown in so-called 4i medium, which contains TGFβ1, FGF2, and LIF, together with four inhibitors of MEK1/2 (MEKi), GSK3β (GSK3βi), JUN amino-terminal kinase (JNKi), and mitogen-activated protein kinase p38 (p38i). Cells in the 4i condition can-self-renew and maintain the ability to differentiate into PGCLCs. Alternative germline-competent state, called incipient mesoderm-like cells (iMeLCs)/premesoendoderm (pre-ME), can be transiently induced from conventional PSCs in the presence of activin A and GSK3β inhibitor. These germline-competent cells express some mesendodermal genes, which could be similar to nascent primitive streak/mesendoderm in human embryos around Wk 2. The 4i/iMeLC/pre-ME cells can differentiate into hPGCLCs efficiently by forming three-dimensional aggregates in the presence of BMP2 or BMP4, SCF, LIF, EGF, and Rho-associated protein kinase inhibitor (ROCKi) with around 50% efficiency. During hPGCLC differentiation, SOX17 expression is induced as early as day 1 which is followed by upregulation of BLIMP1 and TFAP2C. On days 4–5, hPGCLCs expressing NANOS3 and CD38 show the characteristic features of *in vivo* pre-migratory hPGCs in Wks 2–3 embryos. Culture conditions able to promote further development of hPGCLCs have not been fully established to date.

and iMeLCs/pre-ME are likely counterparts of *in vivo* pre-migratory PGCs (Irie et al., 2015; Kobayashi et al., 2017; Sasaki et al., 2015). Notably, the pre-ME induction described by Kobayashi and colleagues recapitulated the progressive gain and loss of germline competence *in vitro*. Thus, 12-h exposure to activin A and CH yields PGC-competent pre-ME, while longer (24- and 36-h) culture produces mesendoderm capable of endoderm and mesoderm but not PGC differentiation (Kobayashi et al., 2017).

The mechanisms underlying the gain of the competency in human 4i PSCs and iMeLC/pre-ME remain to be fully elucidated. The commonality between the two protocols is the activation of WNT signaling via supplementation with WNT agonist (CH, GSK3β inhibitor) resulting in T expression

(Irie et al., 2015; Sasaki et al., 2015). It is therefore probable that the involvement of WNT in PGC competence, described in mice, pigs, and cynomolgus monkeys (Aramaki et al., 2013; Kobayashi et al., 2017; Ohinata et al., 2009; Sasaki et al., 2016), could also be conserved in humans. To prove this, it would be informative to test if *WNT3* knockout or WNT signaling inhibition interferes with human PGCLC competence, similar to what was demonstrated by pig *ex vivo* epiblast culture experiments (Kobayashi et al., 2017). Furthermore, comparing the epigenetic landscape of conventional hPSCs versus competent 4i hPSCs and iMeLC/pre-ME could identify genes poised for activation upon PGC induction, similar to studies performed in the mouse system.

Based on transcriptome analysis, iMeLCs and pre-ME seem to represent a slightly later developmental stage than mouse-competent cells (EpiLCs) (Kobayashi et al., 2017; Sasaki et al., 2015). Interestingly, both 4i hPSCs and iMeLCs/pre-ME, but not conventional hPSCs, express a number of primitive-streak-related genes, such as *T*, *MIXL1*, and *EOMES*, suggesting that acquisition of PGC competence in humans could occur later than in mice and might coincide with the onset of gastrulation (Irie et al., 2015; Sasaki et al., 2015). Considering the crucial role of T for mPGC fate (Aramaki et al., 2013) and the timely expression of T for pPGC specification, it is noteworthy to address its role in human PGC development. Immunofluorescence studies showed that *T* expression is heterogeneous in 4i hPSCs (Irie et al., 2015). It is also possible that T marks competent cells, but is not required for competence acquisition per se. Gene knockout studies can provide conclusive answers to this question.

Importantly, human iMeLCs/pre-ME, similar to mPGC fate-competent EpiLCs, represent a transient entity, while 4i hPSCs can self-renew indefinitely in PGC-competent state. 4i hPSC adaptation to self-renewal *in vitro* could, at least partially, account for the differences in transcriptomes between the two cell types (Sasaki et al., 2015).

Recently, several groups succeeded in establishing "ground-state" hPSC, sharing many properties with naïve mPSCs and displaying similar transcriptome profiles and DNA methylation levels to human ICM (Takashima et al., 2014; Theunissen et al., 2014). It would be of interest to attempt recapitulating the human development using the naïve hPSC to PGC specification via a competent state (von Meyenn et al., 2016). While mouse EpiLCs can be induced in the presence of activin A and FGF2, human germline-competent state seems to be dependent on activin A and WNT signaling. It is still an open question whether these different signaling requirements represent fundamental differences between species or reflect distinct developmental stages of the source pluripotent stem cells.

Considering the inaccessibility of human peri-implantation embryos, the ability to mimic gastrulation *in vitro* is worthy of further investigation. Recently, several groups managed to recapitulate some aspects of mouse and human peri-implantation development *in vitro* (Bedzhov, Leung, Bialecka, & Zernicka-Goetz, 2014; Bedzhov & Zernicka-Goetz, 2014; Deglincerti et al., 2016; Shahbazi et al., 2016). While opening a number of exciting possibilities, this could pose a number of ethical challenges. When exactly should these experiments be terminated: at the onset of human gastrulation or when the nervous system starts to develop? Should the rules be identical when starting with human embryos and hPSCs? Even if this work does not achieve direct visualization of PGC specification in human embryos, as in the mouse (Harrison et al., 2017), it would allow closer investigation of peri-gastrulation epiblast, which could contain PGC-competent cells.

5. TRANSCRIPTION FACTORS FOR HUMAN GERM CELL SPECIFICATION AND EPIGENETIC RESETTING

The ability to model human PGC specification *in vitro* combined with genome editing techniques has been instrumental to study the genes involved in human PGC development. Among the key TFs conferring mPGC fate (*Prdm1*, *Tfap2c*, and *Prdm14*) (Magnúsdóttir et al., 2013; Nakaki et al., 2013), *PRDM1*/BLIMP1 is also required for human PGC specification (Irie et al., 2015; Sasaki et al., 2015). However, the roles of TFAP2C and PRDM14 in hPGC fate remain uncertain. They are both expressed in hPGCLCs/hPGCs *in vitro* and *in vivo* (Irie et al., 2015). Strikingly, however, PRDM14 levels are lower in PGCLCs than in hPSCs and it appears to be localized in the cytoplasm in gonadal hPGCs and E14 pig PGCs (Irie et al., 2015; Kobayashi et al., 2017). It remains to be elucidated whether, in addition to being a transcriptional regulator, PRDM14 has any roles in the cytoplasm. *Prdm14* is critical for mPGC fate: it is one of the first genes to be upregulated in this lineage and *Prdm14*-knockout mice are sterile (Yamaji et al., 2008). Also, overexpression of *Prdm14* in competent cells can induce mPGC fate *in vitro* even in the absence of cytokines (Nakaki et al., 2013). Considering these crucial roles of *Prdm14* in mPGCs, it is of great interest to address its function, if any, in human germ cells. Importantly, mouse PRDM14 has been implicated in epigenetic reprogramming in naïve pluripotency (ground-state mPSCs and the ICM), and the generation of mouse and human iPSCs and mPGCs (Chia et al., 2010; Grabole et al., 2013; Nakaki & Saitou, 2014; Payer et al., 2013; Yamaji et al., 2008, 2013). Since PRDM14 is expressed at low levels in hPGCLCs concomitantly with the initiation of DNA demethylation (Irie et al., 2015), it could serve a similar function in hPGCs.

Immunofluorescence analyses of human PGCLCs and gonadal PGCs revealed specific expression of some TFs absent from the mouse germline. Most prominently, a classic endoderm TF, SOX17, was shown to be critical for hPGCLC induction as it acts upstream and together with BLIMP1 to specify human germ cell fate (Irie et al., 2015). Interestingly, unlike mPGCs that express Sox2, human PGCs, and germ cell tumors, seminomas express SOX17 (de Jong, Stoop, Gillis, van Gurp, et al., 2008; Table 1). Further, while the lack of *SOX17* impairs human PGC specification, *Sox17*-null mice can specify and develop PGCs when endoderm formation is rescued by chimera experiments (Hara et al., 2009; Irie et al., 2015; Kanai-Azuma et al., 2002). Although Sox17 was temporarily detected in E6.5–E7.5 mPGCs, its expression is not maintained in later-stage mPGCs (Yabuta, Kurimoto, Ohinata, Seki, & Saitou, 2006). Crucially, SOX17 is also expressed in nascent and gonadal PGCs of other non-rodent mammals, such as pigs and cynomolgus monkeys (Kobayashi et al., 2017; Sasaki et al., 2016). Another report found *Sox17* expression in the posterior epiblast of perigastrulation rabbit embryos; this region is distinct from the anterior segment of the primitive streak, known to form definitive endoderm in mouse and chick (Hassoun et al., 2010). The involvement of SOX17 in PGC fate in non-rodent mammals might be related to the timing of PGC specification. It would be of interest to identify the ancestral mechanism of PGC induction and explore when, how, and why rodents acquired the unique molecular network for early PGC development. For this, it would be informative to find other mammals that do not rely on SOX17 for PGC specification.

In *in vitro* human PGC specification, the exposure of T-positive germline-competent cells (iMeLCs, pre-ME, or 4i hPSCs) to BMP4 or BMP2 along with supporting cytokines (stem cell factor (SCF), LIF, epidermal growth factor (EGF)) induces SOX17 expression in a subset of cells within 12h (Irie et al., 2015; Kobayashi et al., 2017; Sasaki et al., 2015). Once SOX17 is expressed in the nascent PGC precursors, BLIMP1 and then TFAP2C are upregulated sequentially. In addition, SOX17 knockout cells fail to induce BLIMP1 and TFAP2C, suggesting that SOX17 acts upstream of BLIMP1 and TFAP2C to establish the hPGC molecular network. Specified PGCLCs start expressing NANOS3, followed by cell surface protein CD38 (Table 1; Irie et al., 2015). CD38 is known as a hematopoietic lineage marker (Jackson & Bell, 1990; Malavasi et al., 2008), but is also specific to gonadal hPGCs. Of note, CD38 together with SOX17 is also expressed in

Table 1 Expression of Key Pluripotency and PGC Genes in Human Cells *in vitro* and *in vivo*

State	Gene	Naïve hPSC, ICM	Conventional hPSC	PGC-Competent 4i hPSC, iMeLC, Pre-ME	hPGCLC	hPGC	Seminoma	Embryonal Carcinoma	References
Naïve pluripotency	*TFCP2L1*	+	−	−	+	+	+	ND	Warrier et al. (2017)
	KLF4	+	−	−	+	+	+	ND	
	KLF17	+	−	−	−	−	−	ND	
Pluripotency	*OCT4*	+	+	+	+	+	+	+	Warrier et al. (2017), Irie et al. (2015), Sasaki et al. (2015), and Kobayashi et al. (2017)
	NANOG	+	+	+	+	+	+	+	
	SOX2	+	+	+	−	−	−	+	
	PRDM14	+	+	+	+	+	+	+	
	SOX3	ND	+	+[a]	−	−	−	ND	
	ZIC3	+	+	+	−	−	ND	ND	
	ESRRB	+	−	−	−	−	ND	ND	
	CD30	ND	+	−	−	−	−	+	
Early PGC	*SOX17*	−	−	−	+	+	+	−	Irie et al. (2015), Sasaki et al. (2015), and Kobayashi et al. (2017)
	PRDM1	−	−	−	+	+	+	−	
	TFAP2C	−	−	−	+	+	+	−	
	NANOS3	−	−	−	+	+	−	−	
	CD38	−	−	−	+	+	+	−	
	DND1	−	−	−	+	+	+	ND	

Continued

Table 1 Expression of Key Pluripotency and PGC Genes in Human Cells *in vitro* and *in vivo*—cont'd

State	Gene	Naïve hPSC, ICM	Conventional hPSC	PGC-Competent 4i hPSC, iMeLC, Pre-ME	hPGCLC	hPGC	Seminoma	Embryonal Carcinoma	References
Late PGC	*DAZL*	ND	–	–	–	+	ND	ND	Irie et al. (2015), Sasaki et al. (2015), and Kobayashi et al. (2017)
	DDX4	ND	–	–	–	+	ND	ND	
Mesoderm	*T*	–	–	+	+	+	–	–	Tang et al. (2015), Guo et al. (2015), Gkountela et al. (2015), and Li et al. (2017)
Endoderm	*GATA4*	–	–	+[a]	+	+	+	–	de Jong, Stoop, Gillis, van Gurp, et al. (2008) and Irie et al. (2015)
Trophectoderm	*TEAD4*	–	+	+[a]	+	+	+	ND	de Jong, Stoop, Gillis, van Gurp, et al. (2008) and Pallesen and Hamilton-Dutoit (1988)

[a]Data from iMeLC and 4i hPSCs.
ND: not determined.

seminomas, while embryonic carcinomas and hESCs are marked by CD30 and SOX2 (de Jong, Stoop, Gillis, van Gurp, et al., 2008; Pallesen & Hamilton-Dutoit, 1988). Thus, these markers can be relevant to the diagnosis of human germ cell tumors (see later) and the use of human stem cell models can help dissect the molecular mechanisms of pluripotent versus germ cell identity in normal development and disease. In addition to CD38, AP, KIT EpCAM, and Integrin α6 are useful cell surface markers allowing isolation of hPGCLCs and gonadal PGCs (Irie et al., 2015; Tang et al., 2015; Sasaki et al., 2015). Although AP is also expressed in hPSCs, it becomes restricted to PGCLCs in the aggregates supplemented with cytokines; it is also specific to PGCs in the gonads. Interestingly, KIT expression in *in vitro* induced hPGCLCs is not detected on the cell surface, but found at the RNA level. This is presumably due to the presence of the KIT ligand, SCF, in the PGC differentiation medium, which might induce KIT internalization (Broudy et al., 1999; Jahn et al., 2002). Indeed, KIT could be detected at the cell surface, when SCF was omitted from the cytokine cocktail for PGC induction (unpublished observations).

Human PGCLCs cannot be induced without SOX17, suggesting that the expression of SOX17 is critical for this cell fate decision (Irie et al., 2015). On the other hand, in the absence of BLIMP1, some nascent AP^+ PGCLCs could still be detected, albeit showing an aberrant transcriptional profile with lower expression of PGC genes and ectopic expression of somatic lineage genes, including endodermal markers likely induced by SOX17. In addition, BLIMP1-null cells cannot form NANOS3-positive PGCLCs. Thus, BLIMP1 seems to be critical to establish the nascent PGC program, inducing germ cell genes and repressing somatic genes to allow the initiation of epigenetic programming phase and further PGC development (Irie et al., 2015; Sasaki et al., 2015; Tang et al., 2015). Interestingly, *SOX17* overexpression can restore some PGC gene expression in SOX17-null cells but it also induces endodermal genes, presumably due to the wrong dosage of SOX17 (Kobayashi et al., 2017). Instead, co-expression of BLIMP1 together with SOX17 efficiently and rapidly induces the PGC molecular network even in the absence of BMPs. On the other hand, overexpression of TFAP2C alone or in combination with either SOX17 or BLIMP1 did not have the same effect (Kobayashi et al., 2017). This is in contrast to mouse PGCLCs, which can be induced by combined overexpression of the key TFs (*Prdm1*, *Prdm14*, and *Tfap2c*), any two of them or *Prdm14* alone (Nakaki et al., 2013).

Notably, unlike mPGCs, hPGCs also express some lineage genes, such as *GATA4* and *TEAD4*, even though BLIMP1 acts to suppress somatic markers, including *HOX*, mesodermal, endodermal, and neuronal genes (Table 1; Tang et al., 2015). This might be related to differences in the timing of the PGC specification between mouse and human, which presumably occurs later in humans, around the gastrulation stage.

A remarkable feature of PGCs is that they express a number of pluripotency-associated genes. Thus, *OCT4*, *NANOG*, and *PRDM14*, are detected in mouse, human, porcine, and monkey PGCs (Irie et al., 2015; Kobayashi et al., 2017; Sasaki et al., 2016, 2015). NANOG, which is expressed in competent cells, seems to be downregulated in SOX17-positive PGC precursors, but is upregulated in PGCLCs once BLIMP1 is expressed (Irie et al., 2015; Kobayashi et al., 2017). Interestingly, while *Nanog* overexpression can induce mPGCLCs from mPSCs even without cytokines (Murakami et al., 2016), *NANOG* alone cannot induce human PGCLCs (Kobayashi et al., 2017). Also, unlike mPGCs, SOX2 is rapidly downregulated during hPGCLC induction and is not re-expressed in gonadal PGCs (Irie et al., 2015; Sasaki et al., 2015), reminiscent of the mutually exclusive expression pattern of SOX17 and SOX2 described in other contexts (Aksoy et al., 2013; de Jong, Stoop, Gillis, van Gurp, et al., 2008). Moreover, hPGCs and mPGCs differ in the set of expressed "naïve" pluripotency factors, characteristic of the ICM, and ground-state PSCs. For example, mPGCs express *Klf2* and *Esrrb*, while human PGCs upregulate *KLF4* and *TFCP2L1* instead (Table 1; Tang et al., 2015). Not all naïve TFs are re-expressed in hPGCs; for instance, *KLF17* is specific to hypomethylated hPSCs and ICM, but is absent from hPGCs (Tang et al., 2015). The presence of *KLF4* instead of *KLF2* in human PGCs also warrants further investigation. Ectopic expression of *Klf2* together with *Prdm14* in the presence of LIF in mouse epiblast stem cells (EpiSCs) induces hallmarks of naïve pluripotency, such as DNA demethylation and X-chromosome reactivation (Gillich et al., 2012). This suggests that KLF2/PRDM14 play a role in epigenetic resetting during mPGC specification. Since KLF2 is absent from the human PGC lineage, it is interesting to explore whether KLF4 could play a similar role in this context.

The exact functions of the pluripotency genes in PGC biology are elusive, but they have been implicated in PGC proliferation and survival (Campolo et al., 2013; Chambers et al., 2007; Kehler et al., 2004; Yamaguchi et al., 2009). It is also possible to speculate that, similar to the process of iPSC generation, pluripotency regulators could facilitate epigenetic reprogramming (Leitch & Smith, 2013; Papp & Plath, 2013). As discussed earlier, *Prdm14* and *Nanog* are involved in PGC specification, at least in mice (Murakami et al., 2016; Yamaji et al., 2008). Pluripotency genes are not continuously expressed throughout gametogenesis: they are downregulated in male gonocytes undergoing mitotic arrest, but then some factors reappear in spermatogonial stem cells (SSCs), the tissue-specific stem cells fueling spermatogenesis (Phillips, Gassei, & Orwig, 2010). Accordingly, pluripotent stem cells can be derived from both mPGCs and SSCs, referred to as embryonic germ cells (EGCs, see later) and germline stem cells (GSCs), respectively (Kanatsu-Shinohara et al., 2004; Ko et al., 2009; Leitch, Nichols, et al., 2013; Shamblott et al., 1998; Tanaka, Kanatsu-Shinohara, Hirose, Ogura, & Shinohara, 2015). In females, the expression of pluripotency factors ceases at the time of arrest in prophase of meiosis I (Lopes et al., 2005).

Specified PGCs undergo DNA demethylation via both active and (mainly) passive routes. Passive DNA demethylation relies on downregulation of de novo DNA methyltransferases (DNMT3A and DNMT3B) as well as DNA methylation maintenance machinery (such as DNMT1 and UHRF1), which results in progressive dilution of 5-methylcytosine (5mC) with each round of DNA replication (Kagiwada, Kurimoto, Hirota, Yamaji, & Saitou, 2013; Seisenberger et al., 2012). Active DNA demethylation involves the conversion of 5mC to 5-hydroxymethylcytosine (5hmC) by TET enzymes (Ten-Eleven Translocation methylcytosine dioxygenases); 5hmC is poorly recognized by DNA methyltransferases and therefore cannot serve as a template for the mark maintenance (Hackett et al., 2013; Tang et al., 2016). The initiation of epigenetic resetting in *in vitro* nascent hPGCLCs can be observed by PGC-specific enrichment of TET1 expression concomitant with the upregulation of 5hmC (Irie et al., 2015). This is followed by suppression of DNA methylation maintenance through repression of DNMT3A, DNMT3B, and UHRF1 in hPGCs, allowing DNA demethylation. Interestingly, DNMT3B expression is misregulated in BLIMP1-mutant cells. These data demonstrate that the core molecular machinery for global DNA demethylation in PGC development is conserved between mouse and human (Tang et al., 2015).

Another conserved epigenetic player is arginine methyltransferase PRMT5, which shows nuclear localization in NANOS3-positive hPGCLCs (Irie et al., 2015) and TCam-2 (Eckert et al., 2008). In mouse E8 PGCs and naïve mPSCs, PRMT5 is localized in the nucleus and is responsible for arginine-3 dimethylation of histone H2A and H4, a repressive chromatin modification on LINE1 and IAP transposons (Kim et al., 2014). PRMT5 subsequently translocates to the cytoplasm, where it is involved in transposon silencing (Kim et al., 2014). Such roles of PRMT5 in preserving the genome integrity might be similar in human PGC development, which warrants further investigation.

In summary, despite some similarities, emerging evidence points to critical differences in the core PGC transcription factor networks in mouse and human. This might stem from the unique rodent embryogenesis, featuring the egg cylinder instead of the flat embryonic disc. In addition, the timing of PGC specification differs between the two species, with human PGCs presumably specified at a later developmental stage compared to mice. It would be of great interest to further explore these molecular differences in light of putative evolutionary adaptations.

6. IN VITRO GERM CELL INDUCTION BEYOND PGC SPECIFICATION

Some reports have claimed successful induction of haploid germ cells from mouse and human PSCs (Easley et al., 2012; Jung et al., 2017; Kee et al., 2009; Zhou et al., 2016). However, given the time frame of the procedure is surprisingly short (~10–14 days) relative to gametogenesis *in vivo*, it is unclear whether the induction faithfully recapitulates all the developmental steps.

Robust PGC specification from PSCs yielded PGCLCs with characteristics of migratory (~E9.5) and premigratory (~Wk 3) PGCs for mouse and human, respectively (Hayashi et al., 2011; Hayashi & Saitou, 2013; Irie et al., 2015; Sasaki et al., 2015). Prolonged culture of PGCLCs in the same condition does not stimulate their further development. However, this can be achieved by aggregation with relevant cell types. Thus, female mouse PGCLCs can develop into functional mature oocytes *in vitro* when aggregated with gonadal somatic cells from E12.5 female embryos in an appropriate culture medium (Hikabe et al., 2016). Similarly, male mouse PGCLCs in aggregates with gonadal somatic cells from E12.5 embryos can form spermatogonia-like cells that can be expanded like GSCs (Ishikura et al., 2016). GSC-like cells can develop into mature sperm when injected into adult mouse testes, whereas PGCs can undergo spermatogenesis only in neonatal testes (Ishikura et al., 2016). These reports indicate that somatic cells from mouse embryonic gonads at this stage can provide sufficient factor(s) to stimulate further PGCLC development; presumably these cells also differentiate in the aggregates, to match the developmental stage of germ cells they are mixed with. However, the factors produced by the gonadal somatic cells are not fully known. Retinoic acid (RA), SCF, LIF, TNFα, and cAMP-dependent protein kinase (PKA) have been identified as important players in the survival and proliferation of mPGCs after specification (Koshimizu, Watanabe, & Nakatsuji, 1995; Matsui et al., 1991; Matsui, Zsebo, & Hogan, 1992). However, PGCs in cultures supplemented with SCF and LIF only proliferate for a limited amount of time (Matsui et al., 1991). The addition of FGF extends their growth; however, instead of promoting further PGC differentiation, pluripotent stem cells, called EGCs, are established (Matsui et al., 1992). Ohta and colleagues reported that mouse PGCLCs can proliferate up to 7 more days on mouse embryonic fibroblasts expressing SCF by increasing cyclic AMP levels in the presence of RA agonists (Ohta et al., 2017). These expanded PGCLCs show erasure of DNA 5mC, including on imprints, which would make them equivalent to E13.5 PGCs. However, the expanded PGCs do not show sex-specific histone modification changes, suggesting that some additional factors, presumably from gonadal tissue, are required for their further development (Ohta et al., 2017). It is difficult to recapitulate the mouse protocol using human *in vitro* PGCLCs, since human gonadal tissue is a limited material. An alternative approach could be to use gonadal somatic cells from other mammals featuring flat-disc epiblasts and similar to human PGC TF network, such as pigs or rabbits.

7. MISREGULATION OF GERM CELL DEVELOPMENT

7.1 Germ Cell Tumors

Despite expressing a plethora of pluripotency factors, PGCs are unipotent since they can only form either sperm or egg and do not contribute to chimeras (Leitch et al., 2014). However, when placed into appropriate milieu *in vitro*, PGCs can give rise to self-renewing EGCs (Matsui et al., 1992; Shamblott et al., 1998). mEGCs very closely resemble ESCs (with the exception of the loss of imprints in EGCs derived from late PGCs) and extensively colonize chimeras (Labosky, Barlow, & Hogan, 1994; Leitch et al., 2010). It has been proposed that events similar to EGC formation could occur *in vivo* giving rise to teratomas—tumors that contain derivatives of all three embryonic germ layers (Cunningham, Ulbright, Pera, & Looijenga, 2012; Stevens, 1967). Malignant teratomas, teratocarcinomas, contain pluripotent cancer stem cells, called embryonic carcinoma cells (ECCs), which share many characteristics with ESCs (Cunningham et al., 2012; Damjanov & Andrews, 2007).

Teratomas represent a broader class of neoplasms called germ cell tumors (GCTs), presumably derived from germ cells stalled in their proper development (Oosterhuis & Looijenga, 2005). GCTs are recognized by characteristic histology and can be subdivided into adult GCTs and pediatric GCTs, occurring before puberty. Adult GCTs constitute the most common tumor type in young men in industrialized countries (Rajpert-De Meyts, McGlynn, Okamoto, Jewett, & Bokemeyer, 2016).

Initial experimental evidence for the germ cell origin of GCTs came from transplantation experiments in 129Sv strain of mice that show high incidence of spontaneous GCTs. When genital ridges from wild-type 129Sv were used, the recipients formed teratomas; in contrast, no tumors appeared after transplantation from isogenic steel-mutant mice that lack PGCs (Stevens, 1967). This identified PGCs as the tumor cells of origin. Molecular data from human patients further support the germ cell origin hypothesis. First, GCTs express many PGC- or pluripotency-related markers (e.g., OCT4 and NANOG), which are not normally detected in adult somatic tissues (Rajpert-De Meyts et al., 2016). Second, GCT cells are usually hypomethylated (unlike most somatic cancer types) and are devoid of imprints (Kawakami, Zhang, Okada, & Okamoto, 2006; Ushida et al., 2012). Also, noninvasive GCTs show a similar pattern of histone modifications as in gonadal PGCs (Almstrup et al., 2010). It has been speculated that testicular GCTs might arise from SSCs that reacquire pluripotent characteristics, potentially in response to abnormal SSC niche signaling (Silvan et al., 2013; Singh, Burnicka-Turek, Chauhan, & Hou, 2011). Further, molecular evidence is needed to support this hypothesis.

In addition to gonads, GCTs can also occur in extragonadal sites, such as the brain (intracranial), chest cavity (mediastinum), the base of the spine (sacrococcygeal region), and the abdomen (retroperitoneum). Such localization along the midline of the body recapitulates the PGC migration route and suggests that extragonadal germ cell tumors (EGGCTs) could originate from PGCs that failed to migrate to the gonad (Oosterhuis, Stoop, Honecker, & Looijenga, 2007). Normally these cells are eliminated by apoptosis, but it is possible that mutations in apoptosis-associated genes prevent clearance of ectopic PGCs (Cantú & Laird, 2017). In line with this, mutations in the proapoptotic gene *Bax* lead to increased number of ectopic germ cells in mice (Runyan, Gu, Shoemaker, Looijenga, & Wylie, 2008; Stallock, Molyneaux, Schaible, Knudson, & Wylie, 2003). Some studies have suggested that EGGCTs could represent metastases of the "burned-out" primary gonadal tumors (Scholz et al., 2002), but autopsy data identified most EGGCT as primary neoplasms (Luna & Valenzuela-Tamariz, 1976). Another hypothesis postulates that EGGCTs could arise from somatic cells that reactivate pluripotency genes in a permissive environment (Economou et al., 2015). However, this would not explain why EGGCTs show similar histology, marker gene expression, and imprint patterns to their gonadal counterparts (Fukushima et al., 2017; Oosterhuis & Looijenga, 2005; Oosterhuis et al., 2007; Rajpert-De Meyts et al., 2016; Schneider et al., 2001).

Many pediatric tumors show incomplete loss of imprints, which might point to their origin from migratory PGCs that have not fully erased their imprints (Rajpert-De Meyts et al., 2016). Adult GCTs, on the other hand, are thought to descend from a noninvasive precursor lesion termed carcinoma in situ (CIS), also referred to as intratubular germ cell neoplasia unclassified (Rajpert-De Meyts et al., 2016). The transcriptome and epigenome of CIS cells very closely resemble that of human gonadal PGCs (Almstrup et al., 2010). Acquisition of mutations, in particular 12p isochromosome (i(12p)), *NANOG* and *STELLA* locus, is the likely cause for transformation of CIS to invasive adult GCTs (Litchfield, Levy, Huddart, Shipley, & Turnbull, 2016). In almost 100% of cases dormant CIS progresses to invasive tumors, potentially triggered by puberty-associated hormonal changes (Rajpert-De Meyts & Skakkebaek, 1993).

Based on histology and expression of molecular markers, adult GCTs are classified into seminomas and nonseminomas, irrespective of their location (Oosterhuis & Looijenga, 2005; Rajpert-De Meyts et al., 2016). Seminomatous tumors in the ovaries are referred to as dysgerminomas and those in extragonadal sites are called germinomas. Nonseminomas include teratocarcinoma (teratoma with ECCs), yolk sac tumors, choriocarcinoma, and immature teratoma (Oosterhuis & Looijenga, 2005). Seminomas express many PGC-related markers, such as SOX17 and CD38 (Looijenga, Stoop, & Biermann, 2014). In contrast, embryonal carcinoma cells are reminiscent of hESCs and express SOX2 and CD30 (de Jong, Stoop, Gillis, van Gurp, et al., 2008; Irie et al., 2015; Pallesen & Hamilton-Dutoit, 1988). Of note, virtually all seminomas show mutations in KITL and c-Kit, while this is not always the case for nonseminomas. One study suggested possible interconversion between seminoma and EC. Thus, when human seminoma cell line TCam-2 was grafted into the central nervous system of immunodeficient mice, it seemingly differentiated to EC-like tumors (Nettersheim et al., 2012). Further experiments are required to validate the plasticity of these tumor types.

GCTs show extremely high heritability (around 50%) (Litchfield, Thomsen, et al., 2015). To date, genome-wide association studies have identified over 40 susceptibility loci, with mutations predominantly in noncoding regions (Litchfield et al., 2017). Crucially, many of the identified SNPs are located near genes implicated in PGC biology, such as *PRDM14* and *DAZL*, or physically interact with putative causal genes (Litchfield, Holroyd, et al., 2015; Litchfield et al., 2017; Ruark et al., 2013). It is possible to envisage that these sequences span regulatory elements such as enhancers or insulators. Indeed, some mutations disrupted TF-binding motifs for some PGC-related factors, including PRDM1, NANOG, DMRT1, OCT4, and KLF4 (Litchfield et al., 2017).

Molecular events leading to PGC transformation remain poorly understood. Data from mice suggest that testicular teratomas could arise from PGCs that do not undergo mitotic arrest in G0 in preparation for meiosis (Heaney et al., 2012). Failed mitotic–meiotic switch would explain why these cells continue to proliferate and express pluripotency-related genes. Importantly, however, the available mouse models of GCTs can only recapitulate some aspects of pediatric teratomas and adult nonseminomas (Nettersheim et al., 2012). Only a handful of seminoma cell lines (e.g., TCam-2; de Jong, Stoop, Gillis, Hersmus, et al., 2008) have been established; therefore, there is a great need for development of new experimental models of GCTs. One way to study the transition of unipotent PGCs toward CIS and EC cells could be derivation of EG cells from human PGCs. While establishment of human EGCs has been reported, these cells were unstable and not fully characterized (Shamblott et al., 1998). Further optimization of culture conditions for EGC derivation and maintenance might be required. Notably, 2i/LIF conditions promoting mouse ground-state pluripotency drive efficient EGC derivation from both mouse and rat PGCs (Leitch et al., 2010). It would be of interest to try establishing human EGC lines using recent formulations supporting human naïve pluripotency (Theunissen et al., 2016).

Most GCTs, both gonadal and extragonadal, respond well to treatment, such as cisplatin, and the cure rate reaches 90% (Rajpert-De Meyts et al., 2016). However, many testicular cancer patients must undergo partial or complete orchiectomy and GCT survivors suffer from numerous chemotherapy side effects, including infertility (Oosterhuis & Looijenga, 2005; Rajpert-De Meyts et al., 2016). While adult patients can consider sperm or oocyte collection, this is not an option in the case of infantile GCTs. It is therefore of great importance to further understand the molecular etiology of various types of GCTs to be able to prevent or treat them at earlier stages and with less invasive methods. Using patient-specific iPSCs or hESCs with introduced disease-associated SNPs in conjunction with *in vitro* PGC differentiation could help to address the mechanisms of disease onset.

7.2 *Inheritance of Epigenetic Mutations Through the Germline*

In addition to passing on the genetic blueprint to the offspring, germ cells can also transmit epigenetic information in the form of DNA methylation, histone modifications, and noncoding RNAs (Heard & Martienssen, 2014; Miska & Ferguson-Smith, 2016). Most of these nongenetic components get reconfigured during early embryogenesis and thus constitute intragenerational inheritance. If the changes persist for more than one generation, however, they are referred to as transgenerational epigenetic inheritance (TEI). TEI is well established in plants, where environmentally induced epimutations can be stable over multiple generations and confer selectable traits (Heard & Martienssen, 2014). By contrast, the existence of TEI in mammals remains uncertain due to the comprehensive erasure of epigenetic information in two waves of reprogramming: first in developing PGCs and then in pre-implantation embryos (Hemberger, Dean, & Reik, 2009). Epigenetic reprogramming involves global DNA demethylation, as well as redistribution of histone marks and chromatin remodeling. However, there are some noteworthy exceptions. For example, PGCs reacquire methylation at imprinted loci, which are resistant to reprogramming in the ICM to ensure parent-of-origin expression (McGrath & Solter, 1984; Surani, Barton, & Norris, 1984). However, since imprints are erased and re-established in every generation, they do not constitute TEI.

Surprisingly, a significant proportion of loci retains DNA methylation during epigenetic reprogramming in human PGCs, with some of the regions also evading the second wave of reprogramming (Tang et al., 2015). The majority of these so-called escapees are repeat elements, suggesting that certain transposons have to be constitutively silenced by DNA methylation. Furthermore, some repeat-poor differentially methylated regions (DMRs) were found in the gene bodies and regulatory regions of genes predominantly expressed in the brain (Tang et al., 2015). Some of these genes have been implicated in neural development and their misregulation has been associated with neurological diseases. Of note, while escapees were also described in mPGCs (Guibert, Forne, & Weber, 2012; Hackett et al., 2013; Seisenberger et al., 2012), most of them are not conserved between the two species. However, there is a significant overlap (almost 800 genes) when zooming in to repeat-poor escapees involved in brain function and development.

Intriguingly, certain DMRs identified in hPGCs exhibited variable levels of DNA methylation in different embryos, sometimes showing a sex-dependent pattern (Tang et al., 2015). It is possible that these loci might be candidates for human TEI. However, to date, the evidence of these

DMRs changing their methylation levels in response to environmental cues is lacking and the direct involvement of the methylation status in disease progression remains unclear.

In mouse, a few studies reported that changes in paternal diet resulted in observable phenotypes in the progeny (F2, but not F3) (Carone et al., 2010; Radford et al., 2014). Importantly, however, even if methylation changes were detected in paternal sperm, they were not maintained in the offspring, suggesting an intergenerational effect (Shea et al., 2015). Alternative non-genetic mechanisms, such as transmission of noncoding RNAs, have been suggested as potential contributing factors to the heritability of these traits (Chen et al., 2016).

7.3 *Inheritance of Mitochondria Through the Female Germline*

Not only does the oocyte contribute genetic and potentially epigenetic information to the offspring, but it also passes on its cytoplasmic components. This accounts for the exclusively maternal inheritance of mitochondria, which possess their own genetic material (mtDNA) (Stewart & Larsson, 2014). Each cell contains multiple copies of mtDNA, ranging from around 100 molecules in sperm to over 100,000 in the egg (Stewart & Chinnery, 2015). Mutations in mtDNA are associated with mitochondrial diseases, which over the last years have emerged as a vast group of serious and incurable conditions, affecting approximately 1 in 5000 individuals (Chinnery et al., 2000; Gorman et al., 2015). Due to high energy demands of muscle, cardiac, and brain tissues, mitochondrial dysfunction usually leads to myopathies, cardiomyopathies, and encephalopathies (Tuppen, Blakely, Turnbull, & Taylor, 2010). The pattern of mitochondrial inheritance is complicated by the fact that most mitochondrial diseases only manifest themselves when the proportion of "faulty" mitochondria with mutant mtDNA exceeds a certain threshold (Durham, Samuels, Cree, & Chinnery, 2007; Stewart & Chinnery, 2015). Coexistence of mitochondria with mutant and wild-type DNA, i.e., with multiple haplotypes, is called heteroplasmy, as opposed to homoplasmy, when all the mitochondria in the cell possess identical genetic information (Stewart & Chinnery, 2015). Heteroplasmy levels can be significantly skewed within only one generation, whereby the proportion of mutant mtDNA inherited from a heteroplasmic mother varies between her children (Howell et al., 1992; Olivo, Van de Walle, Laipis, & Hauswirth, 1983). Therefore, healthy mothers with subclinical levels of heteroplasmy may have affected children with much higher content of mutant mtDNA (Chinnery & Hudson, 2013).

Statistical modeling shows that such behavior is consistent with rapid genetic drift, when random distribution of segregating units results in a rare genetic variant becoming quickly fixed or even predominant in a population (Birky, 1994; Wright, 1968). The rate of drift is inversely proportional to the population size and therefore drastic changes necessitate the existence of a bottleneck, a period, when the number of mitochondria is dramatically reduced (Chinnery & Hudson, 2013; Stewart & Chinnery, 2015). Experimental evidence from mice suggests that this bottleneck can occur in early PGCs, which indeed contain reduced levels of mtDNA (around 200 molecules/cell at E8.5) compared to later stages of PGCs (more than 2000 copies/cell at E10.5) and somatic cells (Cree et al., 2008; Stewart & Larsson, 2014; Wai, Teoli, & Shoubridge, 2008). There are however conflicting data, showing much higher content of mtDNA in early PGCs (Cao et al., 2007, 2009). The authors suggested that the bottleneck effect was due to the assembly of multiple mtDNA molecules into nucleoids that constitute new effective segregation units, which would result in the reduction of the population size (Cao et al., 2007, 2009). Of note, these studies used different markers to isolate PGCs, which could have confounded the results.

Alternatively, it is possible that some mtDNA variants are preferentially replicated in the developing oocyte before or after birth (Wai et al., 2008; Wolf, Hayama, & Mitalipov, 2017). This would have implications for mitochondrial replacement therapy (MRT) used to prevent the transmission of mtDNA-based disorders. Indeed, if a few defective mitochondria are left behind after MRT and proliferate extensively, it can undermine the efficacy of the procedure (Wolf et al., 2017).

Is there an evolutionary advantage of restricting the number of mitochondria in specifying PGCs? Even if the bottleneck indeed occurs at this stage, there is no evidence of active mtDNA reduction specifically in PGCs. Rather, mtDNA does not undergo gainful replication in early embryos before gastrulation (Piko & Taylor, 1987; St John, Facucho-Oliveira, Jiang, Kelly, & Salah, 2010). If the overall number of mtDNA is low, just a few copies of mutant mtDNA can severely compromise cellular functions leading to aberrant cell elimination. However, if the mutations lack immediate manifestation in this context, they would not be eliminated by purifying selection, which could explain a high incidence of mtDNA heteroplasmy (Stewart & Chinnery, 2015; Stewart & Larsson, 2014).

8. PERSPECTIVE

The germline transmits genetic information, epigenetic modifications, and cytoplasmic components from one generation to the next. In their developmental journey from PGCs to gametes, germ cells undergo a number of unique processes such as the mitochondrial bottleneck, global DNA demethylation, histone modification changes, and meiosis. All these events allow reacquisition of totipotency upon fertilization. Another significant feature of developing PGCs is their migration toward the developing gonads. It is a considerable challenge to fully recapitulate the elaborate germ cell development and induce functional human gametes *in vitro*. Fully defined conditions have not been achieved for mouse gametogenesis either: complete mouse oogenesis *in vitro* still requires co-culturing with embryonic gonadal cells. Further complication compared to the mouse is that the timescale of human germ cell development is much longer than that in mice (years versus months). Promisingly, however, recent years have seen the development of robust models for human PGC specification from pluripotent stem cells. The challenge remains to push their development further, to obtain the equivalent of migratory and gonadal PGCs, and, potentially, mature gametes. We envisage that better understanding of the signaling involved in later PGC development in non-rodent mammals would guide the stepwise approach toward human gamete generation *in vitro*. It would also be valuable to identify other cell types that support PGC development and could be used in reconstituted cell aggregates. hPSCs-based models of germ cell development can provide a useful platform to interrogate human germline biology, highlighting the molecular basis for normal development and the onset of disease. This could inform novel strategies for treatment of germ cell-related disorders.

ACKNOWLEDGMENTS

We thank members of the Surani lab for helpful discussions and especially Dr. Ufuk Günesdogan, Dr. Walfred W. C. Tang, Dr. Carlos le Sage, Dr. Christopher A. Penfold, Dr. Shienseog Kim, and Dr. Toshihiro Kobayashi for critical reading of the manuscript. The Gurdon Institute receives a core grant from the Wellcome Trust (092096) and a Cancer Research UK Grant (C6946/A14492). N.I. is supported by a Medical Research Council Research Grant and a Rosetrees Trust Research Funding. A.S. is supported by a Wellcome Trust 4-year PhD Studentship. M.A.S. is supported by a Wellcome Trust Investigator Award.

REFERENCES

Aksoy, I., Jauch, R., Chen, J., Dyla, M., Divakar, U., Bogu, G. K., et al. (2013). Oct4 switches partnering from Sox2 to Sox17 to reinterpret the enhancer code and specify endoderm. *The EMBO Journal, 32*, 938–953.

Almstrup, K., Nielsen, J. E., Mlynarska, O., Jansen, M. T., Jorgensen, A., Skakkebaek, N. E., et al. (2010). Carcinoma in situ testis displays permissive chromatin modifications similar to immature foetal germ cells. *British Journal of Cancer, 103*, 1269–1276.

Aramaki, S., Hayashi, K., Kurimoto, K., Ohta, H., Yabuta, Y., Iwanari, H., et al. (2013). A mesodermal factor, T, specifies mouse germ cell fate by directly activating germline determinants. *Developmental Cell, 27*, 516–529.

Barolo, S., & Posakony, J. W. (2002). Three habits of highly effective signaling pathways: Principles of transcriptional control by developmental cell signaling. *Genes & Development, 16*, 1167–1181.

Beddington, R. S., & Robertson, E. J. (1999). Axis development and early asymmetry in mammals. *Cell, 96*, 195–209.

Bedzhov, I., Leung, C. Y., Bialecka, M., & Zernicka-Goetz, M. (2014). In vitro culture of mouse blastocysts beyond the implantation stages. *Nature Protocols, 9*, 2732–2739.

Bedzhov, I., & Zernicka-Goetz, M. (2014). Self-organizing properties of mouse pluripotent cells initiate morphogenesis upon implantation. *Cell, 156*, 1032–1044.

Behringer, R. R., Wakamiya, M., Tsang, T. E., & Tam, P. P. (2000). A flattened mouse embryo: Leveling the playing field. *Genesis, 28*, 23–30.

Bellas, E., & Chen, C. S. (2014). Forms, forces, and stem cell fate. *Current Opinion in Cell Biology, 31*, 92–97.

Birky, C. W. (1994). Relaxed and stringent genomes: Why cytoplasmic genes don't obey Mendel's laws. *The Journal of Heredity, 85*, 355–365.

Boroviak, T., Loos, R., Bertone, P., Smith, A., & Nichols, J. (2014). The ability of inner-cell-mass cells to self-renew as embryonic stem cells is acquired following epiblast specification. *Nature Cell Biology, 16*, 516–528.

Boroviak, T., & Nichols, J. (2017). Primate embryogenesis predicts the hallmarks of human naive pluripotency. *Development (Cambridge, England), 144*, 175–186.

Broudy, V. C., Lin, N. L., Liles, W. C., Corey, S. J., O'Laughlin, B., Mou, S., et al. (1999). Signaling via Src family kinases is required for normal internalization of the receptor c-Kit. *Blood, 94*, 1979–1986.

Buecker, C., Srinivasan, R., Wu, Z., Calo, E., Acampora, D., Faial, T., et al. (2014). Reorganization of enhancer patterns in transition from naive to primed pluripotency. *Cell Stem Cell, 14*, 838–853.

Campolo, F., Gori, M., Favaro, R., Nicolis, S., Pellegrini, M., Botti, F., et al. (2013). Essential role of Sox2 for the establishment and maintenance of the germ cell line. *Stem Cells (Dayton, Ohio), 31*, 1408–1421.

Cantú, A. V., & Laird, D. J. (2017). A pilgrim's progress: Seeking meaning in primordial germ cell migration. *Stem Cell Research, 24*, 181–187.

Cao, L., Shitara, H., Horii, T., Nagao, Y., Imai, H., Abe, K., et al. (2007). The mitochondrial bottleneck occurs without reduction of mtDNA content in female mouse germ cells. *Nature Genetics, 39*, 386–390.

Cao, L., Shitara, H., Sugimoto, M., Hayashi, J., Abe, K., & Yonekawa, H. (2009). New evidence confirms that the mitochondrial bottleneck is generated without reduction of mitochondrial DNA content in early primordial germ cells of mice. *PLoS Genetics, 5*, e1000756.

Carone, B. R., Fauquier, L., Habib, N., Shea, J. M., Hart, C. E., Li, R., et al. (2010). Paternally induced transgenerational environmental reprogramming of metabolic gene expression in mammals. *Cell, 143*, 1084–1096.

Chambers, I., Silva, J., Colby, D., Nichols, J., Nijmeijer, B., Robertson, M., et al. (2007). Nanog safeguards pluripotency and mediates germline development. *Nature, 450*, 1230–1234.

Chatfield, J., O'Reilly, M. A., Bachvarova, R. F., Ferjentsik, Z., Redwood, C., Walmsley, M., et al. (2014). Stochastic specification of primordial germ cells from mesoderm precursors in axolotl embryos. *Development (Cambridge, England), 141*, 2429–2440.

Chen, Q., Yan, M., Cao, Z., Li, X., Zhang, Y., Shi, J., et al. (2016). Sperm tsRNAs contribute to intergenerational inheritance of an acquired metabolic disorder. *Science (New York, N.Y.), 351*, 397–400.

Chia, N. Y., Chan, Y. S., Feng, B., Lu, X., Orlov, Y. L., Moreau, D., et al. (2010). A genome-wide RNAi screen reveals determinants of human embryonic stem cell identity. *Nature*, *468*, 316–320.

Chinnery, P. F., & Hudson, G. (2013). Mitochondrial genetics. *British Medical Bulletin*, *106*, 135–159.

Chinnery, P. F., Johnson, M. A., Wardell, T. M., Singh-Kler, R., Hayes, C., Brown, D. T., et al. (2000). The epidemiology of pathogenic mitochondrial DNA mutations. *Annals of Neurology*, *48*, 188–193.

Chiquoine, A. D. (1954). The identification, origin, and migration of the primordial germ cells in the mouse embryo. *The Anatomical Record*, *118*, 135–146.

Clark, A. T., Bodnar, M. S., Fox, M., Rodriquez, R. T., Abeyta, M. J., Firpo, M. T., et al. (2004). Spontaneous differentiation of germ cells from human embryonic stem cells in vitro. *Human Molecular Genetics*, *13*, 727–739.

Cree, L. M., Samuels, D. C., de Sousa Lopes, S. C., Rajasimha, H. K., Wonnapinij, P., Mann, J. R., et al. (2008). A reduction of mitochondrial DNA molecules during embryogenesis explains the rapid segregation of genotypes. *Nature Genetics*, *40*, 249–254.

Cunningham, J. J., Ulbright, T. M., Pera, M. F., & Looijenga, L. H. (2012). Lessons from human teratomas to guide development of safe stem cell therapies. *Nature Biotechnology*, *30*, 849–857.

Damjanov, I., & Andrews, P. W. (2007). The terminology of teratocarcinomas and teratomas. *Nature Biotechnology*, *25*, 1212 [discussion 1212].

De Felici, M. (2013). Origin migration and proliferation of human primordial germ cells. In G. Coticchio (Ed.), *Oogenesis* (pp. 19–37). London: Springer.

de Jong, J., Stoop, H., Gillis, A. J., Hersmus, R., van Gurp, R. J., van de Geijn, G. J., et al. (2008). Further characterization of the first seminoma cell line TCam-2. *Genes, Chromosomes & Cancer*, *47*, 185–196.

de Jong, J., Stoop, H., Gillis, A. J., van Gurp, R. J., van de Geijn, G. J., Boer, M., et al. (2008). Differential expression of SOX17 and SOX2 in germ cells and stem cells has biological and clinical implications. *The Journal of Pathology*, *215*, 21–30.

Deglincerti, A., Croft, G. F., Pietila, L. N., Zernicka-Goetz, M., Siggia, E. D., & Brivanlou, A. H. (2016). Self-organization of the in vitro attached human embryo. *Nature*, *533*, 251–254.

Dobreva, M. P., Pereira, P. N., Deprest, J., & Zwijsen, A. (2010). On the origin of amniotic stem cells: Of mice and men. *The International Journal of Developmental Biology*, *54*, 761–777.

Durham, S. E., Samuels, D. C., Cree, L. M., & Chinnery, P. F. (2007). Normal levels of wild-type mitochondrial DNA maintain cytochrome c oxidase activity for two pathogenic mitochondrial DNA mutations but not for m.3243A→G. *American Journal of Human Genetics*, *81*, 189–195.

Easley, C. A., 4th, Phillips, B. T., McGuire, M. M., Barringer, J. M., Valli, H., Hermann, B. P., et al. (2012). Direct differentiation of human pluripotent stem cells into haploid spermatogenic cells. *Cell Reports*, *2*, 440–446.

Eckert, D., Biermann, K., Nettersheim, D., Gillis, A. J., Steger, K., Jack, H. M., et al. (2008). Expression of BLIMP1/PRMT5 and concurrent histone H2A/H4 arginine 3 dimethylation in fetal germ cells, CIS/IGCNU and germ cell tumors. *BMC Developmental Biology*, *8*, 106.

Economou, C., Tsakiridis, A., Wymeersch, F. J., Gordon-Keylock, S., Dewhurst, R. E., Fisher, D., et al. (2015). Intrinsic factors and the embryonic environment influence the formation of extragonadal teratomas during gestation. *BMC Developmental Biology*, *15*, 35.

Enders, A. C., & Lopata, A. (1999). Implantation in the marmoset monkey: Expansion of the early implantation site. *The Anatomical Record*, *256*, 279–299.

Extavour, C. G., & Akam, M. (2003). Mechanisms of germ cell specification across the metazoans: Epigenesis and preformation. *Development (Cambridge, England)*, *130*, 5869–5884.

Felix, W. (1912). *The development of the urinogenital organs, manual of human embryology II.* Philadelphia, USA: J. B. Lippincott Company, The Washington Square Press.

Fukushima, S., Yamashita, S., Kobayashi, H., Takami, H., Fukuoka, K., Nakamura, T., et al. (2017). Genome-wide methylation profiles in primary intracranial germ cell tumors indicate a primordial germ cell origin for germinomas. *Acta Neuropathologica*, *133*, 445–462.

Fuss, A. (1911). Über extraregionare Geschlechtszellen bei einem menschlichen Embryo von 4 Wochen. *Anatomischer Anzeiger*, *39*, 407–409.

Gafni, O., Weinberger, L., Mansour, A. A., Manor, Y. S., Chomsky, E., Ben-Yosef, D., et al. (2013). Derivation of novel human ground state naive pluripotent stem cells. *Nature*, *504*, 282–286.

Gillich, A., Bao, S., Grabole, N., Hayashi, K., Trotter, M. W., Pasque, V., et al. (2012). Epiblast stem cell-based system reveals reprogramming synergy of germline factors. *Cell Stem Cell*, *10*, 425–439.

Ginsburg, M., Snow, M. H., & McLaren, A. (1990). Primordial germ cells in the mouse embryo during gastrulation. *Development (Cambridge, England)*, *110*, 521–528.

Gkountela, S., Li, Z., Vincent, J. J., Zhang, K. X., Chen, A., Pellegrini, M., et al. (2013). The ontogeny of cKIT + human primordial germ cells proves to be a resource for human germ line reprogramming, imprint erasure and in vitro differentiation. *Nature Cell Biology*, *15*, 113–122.

Gkountela, S., Zhang, K. X., Shafiq, T. A., Liao, W. W., Hargan-Calvopina, J., Chen, P. Y., et al. (2015). DNA demethylation dynamics in the human prenatal germline. *Cell*, *161*, 1425–1436.

Gorman, G. S., Schaefer, A. M., Ng, Y., Gomez, N., Blakely, E. L., Alston, C. L., et al. (2015). Prevalence of nuclear and mitochondrial DNA mutations related to adult mitochondrial disease. *Annals of Neurology*, *77*, 753–759.

Grabole, N., Tischler, J., Hackett, J. A., Kim, S., Tang, F., Leitch, H. G., et al. (2013). Prdm14 promotes germline fate and naive pluripotency by repressing FGF signalling and DNA methylation. *EMBO Reports*, *14*, 629–637.

Guibert, S., Forne, T., & Weber, M. (2012). Global profiling of DNA methylation erasure in mouse primordial germ cells. *Genome Research*, *22*, 633–641.

Gunesdogan, U., Magnusdottir, E., & Surani, M. A. (2014). Primordial germ cell specification: A context-dependent cellular differentiation event [corrected]. *Philosophical Transactions of the Royal Society of London. Series B, Biological Sciences*, *369*.

Gunesdogan, U., & Surani, M. A. (2016). Developmental competence for primordial germ cell fate. *Current Topics in Developmental Biology*, *117*, 471–496.

Guo, F., Yan, L., Guo, H., Li, L., Hu, B., Zhao, Y., et al. (2015). The transcriptome and DNA methylome landscapes of human primordial germ cells. *Cell*, *161*, 1437–1452.

Hackett, J. A., Sengupta, R., Zylicz, J. J., Murakami, K., Lee, C., Down, T. A., et al. (2013). Germline DNA demethylation dynamics and imprint erasure through 5-hydroxymethylcytosine. *Science (New York, N.Y.)*, *339*, 448–452.

Hajkova, P., Ancelin, K., Waldmann, T., Lacoste, N., Lange, U. C., Cesari, F., et al. (2008). Chromatin dynamics during epigenetic reprogramming in the mouse germ line. *Nature*, *452*, 877–881.

Hara, K., Kanai-Azuma, M., Uemura, M., Shitara, H., Taya, C., Yonekawa, H., et al. (2009). Evidence for crucial role of hindgut expansion in directing proper migration of primordial germ cells in mouse early embryogenesis. *Developmental Biology*, *330*, 427–439.

Harrison, S. E., Sozen, B., Christodoulou, N., Kyprianou, C., & Zernicka-Goetz, M. (2017). Assembly of embryonic and extraembryonic stem cells to mimic embryogenesis in vitro. *Science (New York, N.Y.)*, *356*(6334), pii: eaal1810.

Hassoun, R., Püschel, B., & Viebahn, C. (2010). Sox17 expression patterns during gastrulation and early neurulation in the rabbit suggest two sources of endoderm formation. *Cells, Tissues, Organs*, *191*(2), 68–83.
Hayashi, K., Ohta, H., Kurimoto, K., Aramaki, S., & Saitou, M. (2011). Reconstitution of the mouse germ cell specification pathway in culture by pluripotent stem cells. *Cell*, *146*, 519–532.
Hayashi, K., & Saitou, M. (2013). Generation of eggs from mouse embryonic stem cells and induced pluripotent stem cells. *Nature Protocols*, *8*, 1513–1524.
Hayashi, K., & Saitou, M. (2014). Perspectives of germ cell development in vitro in mammals. *Animal Science Journal*, *85*(6), 617–626.
Hayashi, Y., Saitou, M., & Yamanaka, S. (2012). Germline development from human pluripotent stem cells toward disease modeling of infertility. *Fertility and Sterility*, *97*, 1250–1259.
Heaney, J. D., Anderson, E. L., Michelson, M. V., Zechel, J. L., Conrad, P. A., Page, D. C., et al. (2012). Germ cell pluripotency, premature differentiation and susceptibility to testicular teratomas in mice. *Development (Cambridge, England)*, *139*, 1577–1586.
Heard, E., & Martienssen, R. A. (2014). Transgenerational epigenetic inheritance: Myths and mechanisms. *Cell*, *157*, 95–109.
Hemberger, M., Dean, W., & Reik, W. (2009). Epigenetic dynamics of stem cells and cell lineage commitment: Digging Waddington's canal. *Nature Reviews. Molecular Cell Biology*, *10*, 526–537.
Hikabe, O., Hamazaki, N., Nagamatsu, G., Obata, Y., Hirao, Y., Hamada, N., et al. (2016). Reconstitution in vitro of the entire cycle of the mouse female germ line. *Nature*, *539*, 299–303.
Hopf, C., Viebahn, C., & Puschel, B. (2011). BMP signals and the transcriptional repressor BLIMP1 during germline segregation in the mammalian embryo. *Development Genes and Evolution*, *221*, 209–223.
Howell, N., Halvorson, S., Kubacka, I., McCullough, D. A., Bindoff, L. A., & Turnbull, D. M. (1992). Mitochondrial gene segregation in mammals: Is the bottleneck always narrow? *Human Genetics*, *90*, 117–120.
Idkowiak, J., Weisheit, G., Plitzner, J., & Viebahn, C. (2004). Hypoblast controls mesoderm generation and axial patterning in the gastrulating rabbit embryo. *Development Genes and Evolution*, *214*, 591–605.
Irie, N., Tang, W. W., & Azim Surani, M. (2014). Germ cell specification and pluripotency in mammals: A perspective from early embryogenesis. *Reproductive Medicine and Biology*, *13*, 203–215.
Irie, N., Weinberger, L., Tang, W. W. C., Kobayashi, T., Viukov, S., Manor, Y. S., et al. (2015). SOX17 is a critical specifier of human primordial germ cell fate. *Cell*, *160*, 253–268.
Ishikura, Y., Yabuta, Y., Ohta, H., Hayashi, K., Nakamura, T., Okamoto, I., et al. (2016). In vitro derivation and propagation of spermatogonial stem cell activity from mouse pluripotent stem cells. *Cell Reports*, *17*, 2789–2804.
Jackson, D. G., & Bell, J. I. (1990). Isolation of a cDNA encoding the human CD38 (T10) molecule, a cell surface glycoprotein with an unusual discontinuous pattern of expression during lymphocyte differentiation. *Journal of Immunology*, *144*, 2811–2815.
Jahn, T., Seipel, P., Coutinho, S., Urschel, S., Schwarz, K., Miething, C., et al. (2002). Analysing c-kit internalization using a functional c-kit-EGFP chimera containing the fluorochrome within the extracellular domain. *Oncogene*, *21*, 4508–4520.
Johnson, A. D., & Alberio, R. (2015). Primordial germ cells: The first cell lineage or the last cells standing? *Development (Cambridge, England)*, *142*, 2730–2739.
Jung, D., Xiong, J., Ye, M., Qin, X., Li, L., Cheng, S., et al. (2017). In vitro differentiation of human embryonic stem cells into ovarian follicle-like cells. *Nature Communications*, *8*, 15680.

Kagiwada, S., Kurimoto, K., Hirota, T., Yamaji, M., & Saitou, M. (2013). Replication-coupled passive DNA demethylation for the erasure of genome imprints in mice. *The EMBO Journal, 32*, 340–353.

Kanai-Azuma, M., Kanai, Y., Gad, J. M., Tajima, Y., Taya, C., Kurohmaru, M., et al. (2002). Depletion of definitive gut endoderm in Sox17-null mutant mice. *Development (Cambridge, England), 129*, 2367–2379.

Kanatsu-Shinohara, M., Inoue, K., Lee, J., Yoshimoto, M., Ogonuki, N., Miki, H., et al. (2004). Generation of pluripotent stem cells from neonatal mouse testis. *Cell, 119*, 1001–1012.

Kawakami, T., Zhang, C., Okada, Y., & Okamoto, K. (2006). Erasure of methylation imprint at the promoter and CTCF-binding site upstream of H19 in human testicular germ cell tumors of adolescents indicate their fetal germ cell origin. *Oncogene, 25*, 3225–3236.

Kaylan, K. B., Ermilova, V., Yada, R. C., & Underhill, G. H. (2016). Combinatorial micro-environmental regulation of liver progenitor differentiation by Notch ligands, TGFbeta, and extracellular matrix. *Scientific Reports, 6*, 23490.

Kee, K., Angeles, V. T., Flores, M., Nguyen, H. N., & Reijo Pera, R. A. (2009). Human DAZL, DAZ and BOULE genes modulate primordial germ-cell and haploid gamete formation. *Nature, 462*, 222–225.

Kehler, J., Tolkunova, E., Koschorz, B., Pesce, M., Gentile, L., Boiani, M., et al. (2004). Oct4 is required for primordial germ cell survival. *EMBO Reports, 5*, 1078–1083.

Kemp, C., Willems, E., Abdo, S., Lambiv, L., & Leyns, L. (2005). Expression of all Wnt genes and their secreted antagonists during mouse blastocyst and postimplantation development. *Developmental Dynamics, 233*, 1064–1075.

Kim, S., Gunesdogan, U., Zylicz, J. J., Hackett, J. A., Cougot, D., Bao, S. Q., et al. (2014). PRMT5 protects genomic integrity during global DNA demethylation in primordial germ cells and preimplantation embryos. *Molecular Cell, 56*, 564–579.

Ko, K., Tapia, N., Wu, G., Kim, J. B., Bravo, M. J., Sasse, P., et al. (2009). Induction of pluripotency in adult unipotent germline stem cells. *Cell Stem Cell, 5*, 87–96.

Kobayashi, T., Zhang, H., Tang, W. W. C., Irie, N., Withey, S., Klisch, D., et al. (2017). *Principles of early human development and germ cell program from conserved model systems.* Nature Advance Online publication.

Koshimizu, U., Watanabe, M., & Nakatsuji, N. (1995). Retinoic acid is a potent growth activator of mouse primordial germ cells in vitro. *Developmental Biology, 168*, 683–685.

Kurimoto, K., Yabuta, Y., Hayashi, K., Ohta, H., Kiyonari, H., Mitani, T., et al. (2015). Quantitative dynamics of chromatin remodeling during germ cell specification from mouse embryonic stem cells. *Cell Stem Cell, 16*, 517–532.

Labosky, P. A., Barlow, D. P., & Hogan, B. L. (1994). Mouse embryonic germ (EG) cell lines: Transmission through the germline and differences in the methylation imprint of insulin-like growth factor 2 receptor (Igf2r) gene compared with embryonic stem (ES) cell lines. *Development (Cambridge, England), 120*, 3197–3204.

Lawson, K. A., Dunn, N. R., Roelen, B. A., Zeinstra, L. M., Davis, A. M., Wright, C. V., et al. (1999). Bmp4 is required for the generation of primordial germ cells in the mouse embryo. *Genes & Development, 13*, 424–436.

Lawson, K. A., & Hage, W. J. (1994). Clonal analysis of the origin of primordial germ cells in the mouse. *Ciba Foundation Symposium, 182*, 68–84. discussion 84–91.

Leitch, H. G., Blair, K., Mansfield, W., Ayetey, H., Humphreys, P., Nichols, J., et al. (2010). Embryonic germ cells from mice and rats exhibit properties consistent with a generic pluripotent ground state. *Development (Cambridge, England), 137*, 2279–2287.

Leitch, H. G., Nichols, J., Humphreys, P., Mulas, C., Martello, G., Lee, C., et al. (2013). Rebuilding pluripotency from primordial germ cells. *Stem Cell Reports, 1*, 66–78.

Leitch, H. G., Okamura, D., Durcova-Hills, G., Stewart, C. L., Gardner, R. L., Matsui, Y., et al. (2014). On the fate of primordial germ cells injected into early mouse embryos. *Developmental Biology*, *385*, 155–159.

Leitch, H. G., & Smith, A. (2013). The mammalian germline as a pluripotency cycle. *Development (Cambridge, England)*, *140*, 2495–2501.

Leitch, H. G., Tang, W. W., & Surani, M. A. (2013). Primordial germ-cell development and epigenetic reprogramming in mammals. *Current Topics in Developmental Biology*, *104*, 149–187.

Li, L., Dong, J., Yan, L., Yong, J., Liu, X., Hu, Y., et al. (2017). Single-cell RNA-Seq analysis maps development of human germline cells and gonadal niche interactions. *Cell Stem Cell*, *20*, 891–892.

Litchfield, K., Holroyd, A., Lloyd, A., Broderick, P., Nsengimana, J., Eeles, R., et al. (2015). Identification of four new susceptibility loci for testicular germ cell tumour. *Nature Communications*, *6*, 8690.

Litchfield, K., Levy, M., Huddart, R. A., Shipley, J., & Turnbull, C. (2016). The genomic landscape of testicular germ cell tumours: From susceptibility to treatment. *Nature Reviews. Urology*, *13*, 409–419.

Litchfield, K., Levy, M., Orlando, G., Loveday, C., Law, P. J., Migliorini, G., et al. (2017). Identification of 19 new risk loci and potential regulatory mechanisms influencing susceptibility to testicular germ cell tumor. *Nature Genetics*, *49*, 1133–1140.

Litchfield, K., Thomsen, H., Mitchell, J. S., Sundquist, J., Houlston, R. S., Hemminki, K., et al. (2015). Quantifying the heritability of testicular germ cell tumour using both population-based and genomic approaches. *Scientific Reports*, *5*, 13889.

Liu, P., Wakamiya, M., Shea, M. J., Albrecht, U., Behringer, R. R., & Bradley, A. (1999). Requirement for Wnt3 in vertebrate axis formation. *Nature Genetics*, *22*, 361–365.

Long, H. K., Prescott, S. L., & Wysocka, J. (2016). Ever-changing landscapes: Transcriptional enhancers in development and evolution. *Cell*, *167*, 1170–1187.

Looijenga, L. H., Stoop, H., & Biermann, K. (2014). Testicular cancer: Biology and biomarkers. *Virchows Archiv: An International Journal of Pathology*, *464*, 301–313.

Lopes, S. M. C. D., van den Driesche, S., Carvalho, R. L. C., Larsson, J., Eggen, B., Surani, M. A., et al. (2005). Altered primordial germ cell migration in the absence of transforming growth factor beta signaling via ALK5. *Developmental Biology*, *284*, 194–203.

Luckett, W. P. (1975). The development of primordial and definitive amniotic cavities in early Rhesus monkey and human embryos. *The American Journal of Anatomy*, *144*, 149–167.

Luna, M. A., & Valenzuela-Tamariz, J. (1976). Germ-cell tumors of the mediastinum, postmortem findings. *American Journal of Clinical Pathology*, *65*, 450–454.

Magnúsdóttir, E., Dietmann, S., Murakami, K., Günesdogan, U., Tang, F., Bao, S., et al. (2013). A tripartite transcription factor network regulates primordial germ cell specification in mice. *Nature Cell Biology*, *15*(8), 905–915.

Magnúsdóttir, E., & Surani, M. A. (2014). How to make a primordial germ cell. *Development (Cambridge, England)*, *141*, 245–252.

Malavasi, F., Deaglio, S., Funaro, A., Ferrero, E., Horenstein, A. L., Ortolan, E., et al. (2008). Evolution and function of the ADP ribosyl cyclase/CD38 gene family in physiology and pathology. *Physiological Reviews*, *88*, 841–886.

Matsui, Y., Toksoz, D., Nishikawa, S., Nishikawa, S., Williams, D., Zsebo, K., et al. (1991). Effect of steel factor and leukaemia inhibitory factor on murine primordial germ cells in culture. *Nature*, *353*, 750–752.

Matsui, Y., Zsebo, K., & Hogan, B. L. (1992). Derivation of pluripotential embryonic stem cells from murine primordial germ cells in culture. *Cell*, *70*, 841–847.

McGrath, J., & Solter, D. (1984). Completion of mouse embryogenesis requires both the maternal and paternal genomes. *Cell*, *37*, 179–183.

Miska, E. A., & Ferguson-Smith, A. C. (2016). Transgenerational inheritance: Models and mechanisms of non-DNA sequence-based inheritance. *Science (New York, N.Y.)*, *354*, 59–63.

Murakami, K., Gunesdogan, U., Zylicz, J. J., Tang, W. W., Sengupta, R., Kobayashi, T., et al. (2016). NANOG alone induces germ cells in primed epiblast in vitro by activation of enhancers. *Nature*, *529*, 403–407.

Nakaki, F., Hayashi, K., Ohta, H., Kurimoto, K., Yabuta, Y., & Saitou, M. (2013). Induction of mouse germ-cell fate by transcription factors in vitro. *Nature*, *501*, 222–226.

Nakaki, F., & Saitou, M. (2014). PRDM14: A unique regulator for pluripotency and epigenetic reprogramming. *Trends in Biochemical Sciences*, *39*, 289–298.

Nettersheim, D., Westernstroer, B., Haas, N., Leinhaas, A., Brustle, O., Schlatt, S., et al. (2012). Establishment of a versatile seminoma model indicates cellular plasticity of germ cell tumor cells. *Genes, Chromosomes & Cancer*, *51*, 717–726.

Oestrup, O., Hall, V., Petkov, S. G., Wolf, X. A., Hyldig, S., & Hyttel, P. (2009). From zygote to implantation: Morphological and molecular dynamics during embryo development in the pig. *Reproduction in Domestic Animals*, *44*(Suppl. 3), 39–49.

Ohinata, Y., Ohta, H., Shigeta, M., Yamanaka, K., Wakayama, T., & Saitou, M. (2009). A signaling principle for the specification of the germ cell lineage in mice. *Cell*, *137*, 571–584.

Ohta, H., Kurimoto, K., Okamoto, I., Nakamura, T., Yabuta, Y., Miyauchi, H., et al. (2017). In vitro expansion of mouse primordial germ cell-like cells recapitulates an epigenetic blank slate. *The EMBO Journal*, *36*, 1888–1907.

Olivo, P. D., Van de Walle, M. J., Laipis, P. J., & Hauswirth, W. W. (1983). Nucleotide sequence evidence for rapid genotypic shifts in the bovine mitochondrial DNA D-loop. *Nature*, *306*, 400–402.

Oosterhuis, J. W., & Looijenga, L. H. (2005). Testicular germ-cell tumours in a broader perspective. *Nature Reviews. Cancer*, *5*, 210–222.

Oosterhuis, J. W., Stoop, H., Honecker, F., & Looijenga, L. H. (2007). Why human extragonadal germ cell tumours occur in the midline of the body: Old concepts, new perspectives. *International Journal of Andrology*, *30*, 256–263. [discussion 263–254].

Pallesen, G., & Hamilton-Dutoit, S. J. (1988). Ki-1 (CD30) antigen is regularly expressed by tumor cells of embryonal carcinoma. *The American Journal of Pathology*, *133*, 446–450.

Papp, B., & Plath, K. (2013). Epigenetics of reprogramming to induced pluripotency. *Cell*, *152*, 1324–1343.

Pauklin, S., & Vallier, L. (2013). The cell-cycle state of stem cells determines cell fate propensity. *Cell*, *155*, 135–147.

Payer, B., Rosenberg, M., Yamaji, M., Yabuta, Y., Koyanagi-Aoi, M., Hayashi, K., et al. (2013). Tsix RNA and the germline factor, PRDM14, link X reactivation and stem cell reprogramming. *Molecular Cell*, *52*, 805–818.

Pereira, P. N., Dobreva, M. P., Graham, L., Huylebroeck, D., Lawson, K. A., & Zwijsen, A. N. (2011). Amnion formation in the mouse embryo: The single amniochorionic fold model. *BMC Developmental Biology*, *11*, 48.

Phillips, B. T., Gassei, K., & Orwig, K. E. (2010). Spermatogonial stem cell regulation and spermatogenesis. *Philosophical Transactions of the Royal Society of London. Series B, Biological Sciences*, *365*, 1663–1678.

Piko, L., & Taylor, K. D. (1987). Amounts of mitochondrial DNA and abundance of some mitochondrial gene transcripts in early mouse embryos. *Developmental Biology*, *123*, 364–374.

Politzer, G. (1930). Uber einen menschlichen Embryo mit 7 Urwirbelpaaren. 3°. *Zeitschrift für Anatomie und Entwicklungsgeschichte*, *93*, 386–428.

Politzer, G. (1933). Die Keimbahn des Menschen. *Zeitschrift für Anatomie und Entwicklungsgeschichte, 100*, 331–361.
Rada-Iglesias, A., Bajpai, R., Swigut, T., Brugmann, S. A., Flynn, R. A., & Wysocka, J. (2011). A unique chromatin signature uncovers early developmental enhancers in humans. *Nature, 470*, 279–283.
Radford, E. J., Ito, M., Shi, H., Corish, J. A., Yamazawa, K., Isganaitis, E., et al. (2014). In utero effects. In utero undernourishment perturbs the adult sperm methylome and intergenerational metabolism. *Science (New York, N.Y.), 345*, 1255903.
Rajpert-De Meyts, E., McGlynn, K. A., Okamoto, K., Jewett, M. A., & Bokemeyer, C. (2016). Testicular germ cell tumours. *The Lancet, 387*, 1762–1774.
Rajpert-De Meyts, E., & Skakkebaek, N. E. (1993). The possible role of sex hormones in the development of testicular cancer. *European Urology, 23*, 54–59, discussion 60-51.
Ruark, E., Seal, S., McDonald, H., Zhang, F., Elliot, A., Lau, K., et al. (2013). Identification of nine new susceptibility loci for testicular cancer, including variants near DAZL and PRDM14. *Nature Genetics, 45*, 686–689.
Runyan, C., Gu, Y., Shoemaker, A., Looijenga, L., & Wylie, C. (2008). The distribution and behavior of extragonadal primordial germ cells in Bax mutant mice suggest a novel origin for sacrococcygeal germ cell tumors. *The International Journal of Developmental Biology, 52*, 333–344.
Saitou, M., Barton, S. C., & Surani, M. A. (2002). A molecular programme for the specification of germ cell fate in mice. *Nature, 418*(6895), 293–300.
Saitou, M., & Yamaji, M. (2012). Primordial germ cells in mice. *Cold Spring Harbor Perspectives in Biology, 4*(11), pii: a008375.
Sasaki, K., Nakamura, T., Okamoto, I., Yabuta, Y., Iwatani, C., Tsuchiya, H., et al. (2016). The germ cell fate of cynomolgus monkeys is specified in the nascent amnion. *Developmental Cell, 39*, 169–185.
Sasaki, K., Yokobayashi, S., Nakamura, T., Okamoto, I., Yabuta, Y., Kurimoto, K., et al. (2015). Robust in vitro induction of human germ cell fate from pluripotent stem cells. *Cell Stem Cell, 17*(2), 178–194.
Schneider, D. T., Schuster, A. E., Fritsch, M. K., Hu, J., Olson, T., Lauer, S., et al. (2001). Multipoint imprinting analysis indicates a common precursor cell for gonadal and nongonadal pediatric germ cell tumors. *Cancer Research, 61*, 7268–7276.
Scholz, M., Zehender, M., Thalmann, G. N., Borner, M., Thoni, H., & Studer, U. E. (2002). Extragonadal retroperitoneal germ cell tumor: Evidence of origin in the testis. *Annals of Oncology: Official Journal of the European Society for Medical Oncology, 13*, 121–124.
Seguin, C. A., Draper, J. S., Nagy, A., & Rossant, J. (2008). Establishment of endoderm progenitors by SOX transcription factor expression in human embryonic stem cells. *Cell Stem Cell, 3*, 182–195.
Seisenberger, S., Andrews, S., Krueger, F., Arand, J., Walter, J., Santos, F., et al. (2012). The dynamics of genome-wide DNA methylation reprogramming in mouse primordial germ cells. *Molecular Cell, 48*, 849–862.
Shahbazi, M. N., Jedrusik, A., Vuoristo, S., Recher, G., Hupalowska, A., Bolton, V., et al. (2016). Self-organization of the human embryo in the absence of maternal tissues. *Nature Cell Biology, 18*, 700–708.
Shamblott, M. J., Axelman, J., Wang, S., Bugg, E. M., Littlefield, J. W., Donovan, P. J., et al. (1998). Derivation of pluripotent stem cells from cultured human primordial germ cells. *Proceedings of the National Academy of Sciences of the United States of America, 95*, 13726–13731.
Shao, Y., Taniguchi, K., Townshend, R. F., Miki, T., Gumucio, D. L., & Fu, J. (2017). A pluripotent stem cell-based model for post-implantation human amniotic sac development. *Nature Communications, 8*, 208.

Shea, J. M., Serra, R. W., Carone, B. R., Shulha, H. P., Kucukural, A., Ziller, M. J., et al. (2015). Genetic and epigenetic variation, but not diet, shape the sperm methylome. *Developmental Cell*, *35*, 750–758.

Silvan, U., Diez-Torre, A., Moreno, P., Arluzea, J., Andrade, R., Silio, M., et al. (2013). The spermatogonial stem cell niche in testicular germ cell tumors. *The International Journal of Developmental Biology*, *57*, 185–195.

Singh, S. R., Burnicka-Turek, O., Chauhan, C., & Hou, S. X. (2011). Spermatogonial stem cells, infertility and testicular cancer. *Journal of Cellular and Molecular Medicine*, *15*, 468–483.

St John, J. C., Facucho-Oliveira, J., Jiang, Y., Kelly, R., & Salah, R. (2010). Mitochondrial DNA transmission, replication and inheritance: A journey from the gamete through the embryo and into offspring and embryonic stem cells. *Human Reproduction Update*, *16*, 488–509.

Stallock, J., Molyneaux, K., Schaible, K., Knudson, C. M., & Wylie, C. (2003). The pro-apoptotic gene Bax is required for the death of ectopic primordial germ cells during their migration in the mouse embryo. *Development (Cambridge, England)*, *130*, 6589–6597.

Stevens, L. C. (1967). Origin of testicular teratomas from primordial germ cells in mice. *Journal of the National Cancer Institute*, *38*, 549–552.

Stewart, J. B., & Chinnery, P. F. (2015). The dynamics of mitochondrial DNA heteroplasmy: Implications for human health and disease. *Nature Reviews. Genetics*, *16*, 530–542.

Stewart, J. B., & Larsson, N. G. (2014). Keeping mtDNA in shape between generations. *PLoS Genetics*, *10*, e1004670.

Sugawa, F., Arauzo-Bravo, M. J., Yoon, J., Kim, K. P., Aramaki, S., Wu, G., et al. (2015). Human primordial germ cell commitment in vitro associates with a unique PRDM14 expression profile. *The EMBO Journal*, *34*, 1009–1024.

Surani, M. A., Barton, S. C., & Norris, M. L. (1984). Development of reconstituted mouse eggs suggests imprinting of the genome during gametogenesis. *Nature*, *308*, 548–550.

Takashima, Y., Guo, G., Loos, R., Nichols, J., Ficz, G., Krueger, F., et al. (2014). Resetting transcription factor control circuitry toward ground-state pluripotency in human. *Cell*, *158*, 1254–1269.

Tanaka, T., Kanatsu-Shinohara, M., Hirose, M., Ogura, A., & Shinohara, T. (2015). Pluripotent cell derivation from male germline cells by suppression of Dmrt1 and Trp53. *The Journal of Reproduction and Development*, *61*, 473–484.

Tang, W. W., Dietmann, S., Irie, N., Leitch, H. G., Floros, V. I., Bradshaw, C. R., et al. (2015). A unique gene regulatory network resets the human germline epigenome for development. *Cell*, *161*, 1453–1467.

Tang, W. W., Kobayashi, T., Irie, N., Dietmann, S., & Surani, M. A. (2016). Specification and epigenetic programming of the human germ line. *Nature Reviews. Genetics*, *17*, 585–600.

Theunissen, T. W., Friedli, M., He, Y., Planet, E., O'Neil, R. C., Markoulaki, S., et al. (2016). Molecular criteria for defining the naive human pluripotent state. *Cell Stem Cell*, *19*, 502–515.

Theunissen, T. W., Powell, B. E., Wang, H., Mitalipova, M., Faddah, D. A., Reddy, J., et al. (2014). Systematic identification of culture conditions for induction and maintenance of naive human pluripotency. *Cell Stem Cell*, *15*, 471–487.

Tuppen, H. A., Blakely, E. L., Turnbull, D. M., & Taylor, R. W. (2010). Mitochondrial DNA mutations and human disease. *Biochimica et Biophysica Acta*, *1797*, 113–128.

Ushida, H., Kawakami, T., Minami, K., Chano, T., Okabe, H., Okada, Y., et al. (2012). Methylation profile of DNA repetitive elements in human testicular germ cell tumor. *Molecular Carcinogenesis*, *51*, 711–722.

Valdez Magana, G., Rodriguez, A., Zhang, H., Webb, R., & Alberio, R. (2014). Paracrine effects of embryo-derived FGF4 and BMP4 during pig trophoblast elongation. *Developmental Biology*, *387*, 15–27.

Viebahn, C., Stortz, C., Mitchell, S. A., & Blum, M. (2002). Low proliferative and high migratory activity in the area of Brachyury expressing mesoderm progenitor cells in the gastrulating rabbit embryo. *Development*, *129*(10), 2355–2365.

von Meyenn, F., Berrens, R. V., Andrews, S., Santos, F., Collier, A. J., Krueger, F., et al. (2016). Comparative principles of DNA methylation reprogramming during human and mouse in vitro primordial germ cell specification. *Developmental Cell*, *39*, 104–115.

Wai, T., Teoli, D., & Shoubridge, E. A. (2008). The mitochondrial DNA genetic bottleneck results from replication of a subpopulation of genomes. *Nature Genetics*, *40*, 1484–1488.

Wang, R. N., Green, J., Wang, Z., Deng, Y., Qiao, M., Peabody, M., et al. (2014). Bone morphogenetic protein (BMP) signaling in development and human diseases. *Genes & Diseases*, *1*, 87–105.

Wang, A., Yue, F., Li, Y., Xie, R., Harper, T., Patel, N. A., et al. (2015). Epigenetic priming of enhancers predicts developmental competence of hESC-derived endodermal lineage intermediates. *Cell Stem Cell*, *16*, 386–399.

Warrier, S., Van der Jeught, M., Duggal, G., Tilleman, L., Sutherland, E., Taelman, J., et al. (2017). Direct comparison of distinct naive pluripotent states in human embryonic stem cells. *Nature Communications*, *8*, 15055.

Weick, E. M., & Miska, E. A. (2014). piRNAs: From biogenesis to function. *Development (Cambridge, England)*, *141*, 3458–3471.

Witschi, E. (1948). Migration of germ cells of human embryos from the yolk sac to the primitive gonadal folds. In *Vol. 209. Contributions to embryology* (pp. 67–80): Carnegie Institution of Washington Publication.

Wolf, D. P., Hayama, T., & Mitalipov, S. (2017). Mitochondrial genome inheritance and replacement in the human germline. *The EMBO Journal*, *36*, 2177–2181.

Wright, S. (1968). *Evolution and the genetics of populations. A treatise*. Chicago: University of Chicago Press.

Yabuta, Y., Kurimoto, K., Ohinata, Y., Seki, Y., & Saitou, M. (2006). Gene expression dynamics during germline specification in mice identified by quantitative single-cell gene expression profiling. *Biology of Reproduction*, *75*, 705–716.

Yamaguchi, S., Kurimoto, K., Yabuta, Y., Sasaki, H., Nakatsuji, N., Saitou, M., et al. (2009). Conditional knockdown of Nanog induces apoptotic cell death in mouse migrating primordial germ cells. *Development (Cambridge, England)*, *136*, 4011–4020.

Yamaguchi, T. P., Takada, S., Yoshikawa, Y., Wu, N., & McMahon, A. P. (1999). T (Brachyury) is a direct target of Wnt3a during paraxial mesoderm specification. *Genes & Development*, *13*, 3185–3190.

Yamaji, M., Seki, Y., Kurimoto, K., Yabuta, Y., Yuasa, M., Shigeta, M., et al. (2008). Critical function of Prdm14 for the establishment of the germ cell lineage in mice. *Nature Genetics*, *40*, 1016–1022.

Yamaji, M., Ueda, J., Hayashi, K., Ohta, H., Yabuta, Y., Kurimoto, K., et al. (2013). PRDM14 ensures naive pluripotency through dual regulation of signaling and epigenetic pathways in mouse embryonic stem cells. *Cell Stem Cell*, *12*, 368–382.

Ying, Y., Liu, X. M., Marble, A., Lawson, K. A., & Zhao, G. Q. (2000). Requirement of Bmp8b for the generation of primordial germ cells in the mouse. *Molecular Endocrinology*, *14*, 1053–1063.

Ying, Y., Qi, X., & Zhao, G. Q. (2001). Induction of primordial germ cells from murine epiblasts by synergistic action of BMP4 and BMP8B signaling pathways. *Proceedings of the National Academy of Sciences of the United States of America*, *98*, 7858–7862.

Ying, Q. L., Wray, J., Nichols, J., Batlle-Morera, L., Doble, B., Woodgett, J., et al. (2008). The ground state of embryonic stem cell self-renewal. *Nature*, *453*, 519–523.

Ying, Y., & Zhao, G. Q. (2001). Cooperation of endoderm-derived BMP2 and extraembryonic ectoderm-derived BMP4 in primordial germ cell generation in the mouse. *Developmental Biology*, *232*, 484–492.

Yoshida, M., Kajikawa, E., Kurokawa, D., Tokunaga, T., Onishi, A., Yonemura, S., et al. (2016). Conserved and divergent expression patterns of markers of axial development in eutherian mammals. *Developmental Dynamics*, *245*, 67–86.

Zhou, Q., Wang, M., Yuan, Y., Wang, X., Fu, R., Wan, H., et al. (2016). Complete meiosis from embryonic stem cell-derived germ cells in vitro. *Cell Stem Cell*, *18*, 330–340.

FURTHER READING

Gamage, T. K. J. B., Chamley, L. W., & James, J. L. (2016). Stem cell insights into human trophoblast lineage differentiation. *Human Reproduction Update*, *23*, 77–103.

CHAPTER THREE

From Human Pluripotent Stem Cells to Cortical Circuits

Marc Astick[*,†,‡], **Pierre Vanderhaeghen**[*,†,‡,§,1]

[*]Université Libre de Bruxelles (U.L.B.), Institut de Recherches en Biologie Humaine et Moléculaire (IRIBHM), and ULB Neuroscience Institute (UNI), Brussels, Belgium
[†]VIB-KU Leuven Center for Brain & Disease Research, Leuven, Belgium
[‡]Department of Neurosciences, Leuven Brain Institute, KUL, Leuven, Belgium
[§]Welbio, Université Libre de Bruxelles (U.L.B.), Brussels, Belgium
[1]Corresponding author: e-mail address: pierre.vanderhaeghen@kuleuven.vib.be

Contents

Abstract

Understanding the development of the human brain in relation with evolution is an important frontier field in developmental biology. In particular, investigating the mechanisms underlying the greatly increased relative size and complexity of the cerebral

Current Topics in Developmental Biology, Volume 129
ISSN 0070-2153
https://doi.org/10.1016/bs.ctdb.2018.02.011

cortex, the seat of our enhanced cognitive abilities, remains a fascinating yet largely unsolved question. Though many advances in our understanding have been gained from the study of animal models, as well as human genetics and embryology, large gaps remain in our knowledge of the molecular mechanisms that control human cortical development. Interestingly, many aspects of corticogenesis can be recapitulated in vitro from mouse and human embryonic or induced pluripotent stem cells (PSCs), using a variety of experimental systems from 2D models to organoids to xenotransplantation. This has provided the opportunity to study these processes in an accessible and physiologically relevant setting. In this chapter, we will discuss how conserved and divergent features of primate/human corticogenesis can be modeled and studied mechanistically using PSC-based models of corticogenesis. We will also review what has been learned through these approaches about pathological defects of human corticogenesis, from early neurogenesis to late neuronal maturation and connectivity.

1. INTRODUCTION

The development of the human brain, unique in its relative size and complexity, remains one of the most challenging topics in developmental biology. In particular the cerebral cortex, the seat of higher cognitive function has strikingly evolved toward increased complexity in primate and hominid lineages. The links between development and evolution of the human cortex have emerged recently as a major topic of investigation, seeking to identify the genes, pathways, cell types, and neuronal circuits, which have equipped our species with such expanded cognitive abilities.

The mammalian cerebral cortex consists of a vast repertoire of excitatory and inhibitory neurons as well as glial cells, which are organized into a six-layer structure. Each layer is specifically defined by the neurons contained within, and while the overall structure is retained, cortical layers differ by composition across different areas of the cortex (Greig, Woodworth, Galazo, Padmanabhan, & Macklis, 2013). How the dozens of cortical neuronal types are generated in a highly coordinated fashion, both in time and space, and then connect with each other and the rest of the brain with exquisite specificity and plasticity constitute the major questions of cortical development.

A wealth of knowledge on cortical development has been gained through investigating features conserved throughout evolution. That is, the common mechanisms driving nervous system development in simple organisms such as fruit flys and nematode worms through to more complex vertebrate brains and ultimately primate species, including humans. However, there has been a recent alteration in focus, moving away from common toward divergent mechanisms of human brain development, that is, the divergence of the human brain from other species over recent hominid evolution. Specific insights into human brain development have come from various technological and conceptual advances, including human genetics centered on diseases presenting brain malformations and/or cognitive dysfunction, and comparative genomics unraveling the evolution of the human/hominid genome (Bae, Jayaraman, & Walsh, 2015; Varki, Geschwind, & Eichler, 2008). The access to human fetal brain material, although limited, has also proved invaluable to reveal key molecular and cellular features of human corticogenesis (Lein, Belgard, Hawrylycz, & Molnar, 2017; Lui, Hansen, & Kriegstein, 2011; Sousa, Meyer, Santpere, Gulden, & Sestan, 2017). However, it has remained challenging to integrate these findings, and those coming from animal models, especially when related to human-specific aspects of cortical development. This necessary synthesis would greatly benefit from a faithful and versatile experimental model where specific molecular and cellular processes could be studied in a systematic way in a human physiological context. This is where the recent burst in human embryonic/induced pluripotent stem cell (ESC/iPSC) technologies toward cortical differentiation brings great potential, by providing novel tools to study human developmental neurobiology.

In this chapter, we will first summarize the basic principles of cortical development that appear to be largely conserved among mammals, and then those that appear to be divergent and prominent in primate and human species. We will in parallel illustrate how human pluripotent stem cells (PSCs) can be used to study conserved and divergent features of human cortical development, as well as their pathology, from early cortical stem cells to late neuronal maturation and circuit assembly.

2. CONSERVED FEATURES OF MAMMALIAN CORTICOGENESIS

2.1 Early Cortical/Telencephalic Induction: It Is All in the Beginning

Neural development begins with the commitment of ectodermal cells to the neural lineage: importantly, this neural induction process proceeds mostly along a "default pathway." This was first revealed through pioneering studies in vertebrate models, which demonstrated that neural fate was mostly acquired through active inhibition of morphogen cues (reviewed in Munoz-Sanjuan & Brivanlou, 2002). In addition the first neural cells generated adopt a mostly anterior/rostral forebrain-like identity, which is converted to more caudal fates by embryonic cues. Thus, forebrain identity, from which the cortex emerges during embryogenesis essentially through inhibition of extrinsic cues, in particular of the BMP/Nodal and Wnt families (Wilson & Houart, 2004). Following this initial rostrocaudal patterning, the telencephalic vesicles emerge from the anterior forebrain, partly through induction by FGF8, and differentiate into two main germinal zones along the dorsoventral axis, leading to the formation of dorsal and ventral (D–V) telencephalon, mostly through the action of classical D–V morphogens such as Sonic Hedgehog (SHH) and BMP/Wnts signaling dorsally (Wilson & Rubenstein, 2000). This process of primitive neural/forebrain induction and subsequent D–V patterning is largely conserved across vertebrates, and elucidating the underlying mechanisms has opened the door to the current modeling of cortical development in vitro using PSC.

Indeed pioneering studies showed that when mouse ESCs are cultured in media lacking morphogens, they will rapidly express neural fate markers and convert into neural progenitor cells (NPCs) (Tropepe et al., 2001; Ying, Stavridis, Griffiths, Li, & Smith, 2003). Elaborating on this, most currently used protocols of neural induction from PSC, whether murine or human, essentially rely on culturing cells in the absence of serum and any morphogen, and supplementing the medium with active inhibitors of BMP/Tgf-β and/or Wnt signaling (Chambers et al., 2009; Eiraku et al., 2008; Gaspard et al., 2008; Pera et al., 2004; Watanabe et al., 2005).

Most importantly, neural cells induced by "default" adopt a forebrain-like identity, as in vivo (Eiraku et al., 2008; Gaspard et al., 2008; Watanabe et al., 2005). Their identity appears to mainly correspond to telencephalon, both ventral and dorsal from mouse ESC, which when cultured in the presence of cyclopamine, an inhibitor of the SHH pathway (normally expressed in the ventral neural tube), can then lead to a massive conversion to dorsal forebrain/telencephalic progenitors that mostly correspond to a cortical identity (Gaspard et al., 2008). Similarly, human PSC cultured through default pathways essentially lead to a collection of telencephalic cells, which interestingly in the human species have mostly a dorsal fate even when cultured without SHH inhibition (Espuny-camacho et al., 2013; Shi, Kirwan, Smith, Robinson, & Livesey, 2012).

Overall these data led to the surprising finding that cortical identity can be acquired mostly through a default pathway from PSCs and has provided the basis of most described protocols of corticogenesis, whether in two (2D) or three dimensions (3D). Interestingly, the intrinsic emergence of corticogenesis is not restricted to its initial steps, as we shall see in the next sections.

2.2 Cortical Neurogenesis: In Vitro Generation of a Complex Repertoire of Cortical Neurons

2.2.1 Pyramidal Neurons

The result of the early process of corticogenesis is the foundation of the neuroepithelial cell (NEC) population, from which the cerebral cortex will arise in the dorsal telencephalon (Fig. 1). The NEC founder population forms a pseudostratified monolayer, where each cell is attached via apical adherens and tight junctions. The NEC population divides symmetrically and undergoes a stereotyped interkinetic nuclear migration (IKNM), a

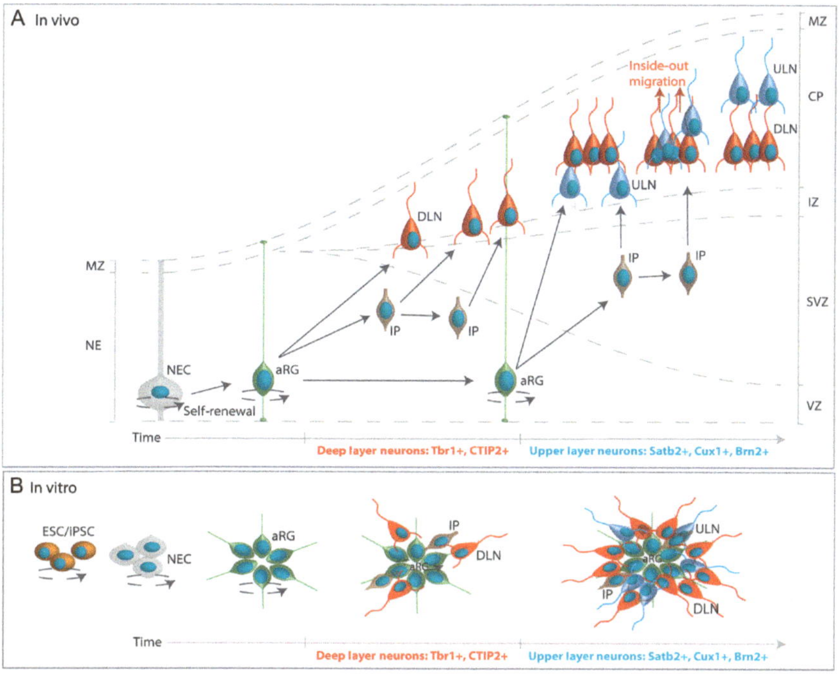

Fig. 1 (A) The founder NEC population gives rise to aRG from which cortical neurons are produced in temporal manner. (B) Corticogenesis modeled in vitro from PSCs follows the same time course of neuronal production as in vivo. *aRG*, apical radial glia; *CP*, cortical plate; *DLN*, deep layer neuron; *IP*, intermediate progenitor; *IZ*, intermediate zone; *MZ*, marginal zone; *NE*, neuroepithelium; *NEC*, neuroepithelial cell; *SVZ*, subventricular zone; *ULN*, upper layer neuron; *VZ*, ventricular zone.

process where the position of the cell nucleus is altered in line with the cell cycle, basally during G1 and S-phase, then directed apically during G2 phase with mitosis occurring at the VZ surface. This process gives the neuro-epithelium a pseudostratified appearance. In addition recent imaging studies using human fetal cortical samples have shown that NEC cells retract their basal process during mitosis, with both daughter cells regrowing this process after cell division (Subramanian, Bershteyn, Paredes, & Kriegstein, 2017). Following expansion of the pool of NEC, these will then convert to apical radial glial cells (aRGCs), which unlike NEC display a high neurogenic capacity. Indeed, all pyramidal cortical neurons are thought to be generated from the aRGCs, directly or indirectly (Kriegstein & Alvarez-Buylla, 2009). Characterized by a unique epithelial morphology and partially glial identity (Fig. 1), the aRGCs maintain a process with both the basal and apical surface of the developing cortex. The basal projection of the aRGC is thin and terminates by a bulb-like protrusion at the level of the basal lamina, a region rich in extracellular matrix proteins (Takahashi, Misson, & Caviness, 1990). In contrast to NECs, aRGCs retain their basal process during mitosis, which will constitute a stable radial scaffold essential for neuronal migration.

The aRGCs typically divide at the ventricular surface, undergoing IKNM, and divide either symmetrically to produce two progenitors or asymmetrically, regenerating progenitor aRGCs and generating neurons, either directly or indirectly via intermediate progenitors (IPCs) (Haubensak, Attardo, Denk, & Huttner, 2004; Mihalas et al., 2016; Noctor, Martinez-Cerdeno, Ivic, & Kriegstein, 2004). IPCs are a specific class of progenitor cell, which unlike aRGCs do not exhibit epithelial polarity. They divide in the subventricular zone (typically once in the mouse, with an increased number of divisions in primates) before generating neurons, and thereby enable an increase in the neurogenic output of aRGC.

As discussed previously, the cortex is characterized by a wide diversity of neurons that populate six distinct layers. A good part of this diversity is achieved by a fascinating mechanism of temporal patterning (Rossi, Fernandes, & Desplan, 2017), whereby progenitors sequentially generate distinct types of neurons that will settle in specific layers, from deep to superficial, according to the so-called inside out model (Gaspard & Vanderhaeghen, 2011). Specifically, cortical progenitors display a sequential shift in competence for the generation of different neuronal subtypes, such that the laminar identity and location of cortical neurons are tightly linked to neuronal birthdate. As neurogenesis occurs, early born neurons will occupy deep layers of the cortex, while late born neurons will migrate through these neurons to settle in the upper or superficial cortical layers (Fig. 1A). The mechanisms underlying temporal patterning of corticogenesis remain largely unknown, but are thought to involve progressive restriction of the competence of progenitors, as well as extrinsic influence from the cortical environment (Franco et al., 2012; Frantz & McConnell, 1996; McConnell & Kaznowski, 1991), although timely generation of a clone of distinct types of cortical neurons from single NEC can also occur in purely in vitro settings (Shen et al., 2006).

The capacity to generate neurons from six layers identities through the same sequence as in vivo is robustly retained in cortical progenitors derived from ESC (Eiraku et al., 2008; Espuny-camacho et al., 2013; Gaspard et al., 2008; Mariani et al., 2012; Shi et al., 2012). Indeed, birth dating and specific marker analyses have revealed that the first neurons generated during in vitro corticogenesis from ESC correspond to pioneer neurons such as Cajal–Retzius cells, followed by neurons expressing deep layer markers and finally upper layer markers (Fig. 1B). Moreover, single cell clonal analyses performed on ESC-derived cortical progenitors have demonstrated their multipotence, i.e., their ability to sequentially generate distinct layer-specific classes of neurons, as well as their time-dependent restriction of competence (Gaspard et al., 2008).

The PSC-derived neurons then mature into neurons displaying pyramidal morphology, including a large "apical" dendrite, and functional features of excitatory glutamatergic neurons. Microarray transcriptome profiling confirmed that in vitro corticogenesis from human PSC up to 3 months corresponds to midgestation cortical development (Espuny-camacho et al., 2013), while single cell RNA-seq analysis recapitulates the single cell transcriptomic signature of human midfetal cortical neurons (Handel et al., 2016). Thus, although the cortical neurons generated in the dish do not display obvious cytoarchitectural features such as layer organization, they display specific and diverse laminar identities as well as typical morphological and functional features of native pyramidal neurons. Importantly, when grafted into the mouse embryonic or neonatal cerebral cortex the neurons generated in vitro, whether from mouse or human ESC, were able to migrate, integrate, displaying axonal projections typical of various layers of the cerebral cortex and morphological and functional features of pyramidal neurons (Espuny-camacho et al., 2013; Gaspard et al., 2008; Nagashima et al., 2014).

2.2.2 Cortical Interneurons

While the above mechanisms account for the generation of pyramidal neurons, inhibitory interneurons on the other hand are generated in a separate ventral telencephalic niche (Wilson & Rubenstein, 2000), the medial and caudal ganglionic eminences (MGE/CGE), from which they undertake a highly stereotyped tangential migration to the dorsal telencephalon, where they will integrate into cortical circuits (Bandler, Mayer, & Fishell, 2017; Bartolini, Ciceri, & Marin, 2013; Southwell et al., 2014). Accordingly, ESC-derived cortical interneurons are not generated in vitro through the same pathway as pyramidal neurons, as they require the use of ventralizing cues (SHH, Wnt inhibitors, and FGF). In this way several classes of interneurons have been successfully generated from human PSC that display functional features of native interneurons and are able to integrate into the mouse cortex following xenotransplantation (Maroof et al., 2013; Nicholas et al., 2013).

3. CORTICAL ORGANOIDS: ADDING A NEW DIMENSION IN CORTICAL MODELING

The above-described models have revealed how much ESC-derived neural cells can recapitulate, in a seemingly simplistic way, several key features of the complex regional and temporal patterning leading to corticogenesis, despite little spatial organization. Indeed in 2D cultures rosettes of neural progenitors can display spatial features such as IKNM and apicobasal polarity, and neurons tend to segregate from rosettes spatially; however, these models do not reconstitute the developing cortical cytoarchitecture. This is why recent developments in 3D or so-called organoid modeling technologies, taking advantage of self-organizing properties of developing cells, hold much promise for the study of human corticogenesis events that depend on defined cytoarchitecture. Several recent reviews have covered this topic extensively from a conceptual and technological viewpoint (Huch, Knoblich, Lutolf, & Martinez-Arias, 2017; Lancaster & Knoblich, 2014; Quadrato, Brown, & Arlotta, 2016), so we will only summarize what is directly relevant to corticogenesis here. The basis of the technology comes from the seminal observation that ESC cultured in suspension as balls of cells and then differentiated into cortical cells, using in this case Wnt inhibitors, can spontaneously adopt a defined cytoarchitecture resembling the cortical anlage (Eiraku et al., 2008). Progenitors are found in the interior of the structure where they adopt a neuroepithelial cytoarchitecture highly reminiscent of the ventricular zone from which differentiating neurons, corresponding to various cortical layer neuronal identities, can segregate. Such an approach was modified to promote self-organization of a variety of brain structures, most strikingly the retina (Eiraku & Sasai, 2012). Since then such 3D models were generated in the absence of patterned molecules, first embedded in collagen matrix, leading to self-organized "cerebral" organoids that contained tissue representing many regions of the brain, which could be maintained in long-term cultures (up to 1 year in vitro), thus achieving an unprecedented level of cellular and architectural complexity (Lancaster et al., 2013). This cellular diversity was further shown to be remarkably similar to in vivo patterns using single cell RNA-seq (Camp et al., 2015; Quadrato et al., 2017). In parallel, the combination of cultured suspensions and morphogen inhibition led to "cortical" organoid generation (Kadoshima et al., 2013), which can recapitulate the distinctive cytoarchitecture of the developing cortex, i.e., distinct ventricular and subventricular zone-like neuroepithelium containing progenitors that generate cortical neurons. These neurons then delaminate outside of the VZ to form a mantle-like structure containing only postmitotic neurons, following a similar time line as in 2D systems. Subsequently there have been multiple different approaches used to generate organoids including cortical tissue (Pasca et al., 2015; Qian et al., 2016) as researchers seek to refine, improve, and standardize the production of cortical tissue (Iefremova et al., 2017; Lancaster et al., 2017; Mariani et al., 2015).

Another important aspect of cortical development for which organoid and 3D modeling could provide key insights is neuronal migration, whether radial for cortical pyramidal neurons or tangential for interneurons. As for radial migration, the emergence of structures that contain neurons of distinct layer identities suggests that radial migration can occur in organoids. However, it could also reflect cell sorting between neurons and progenitors as observed in neural rosettes in 2D culture, especially as these structures do not appear to form the sharp boundaries that distinguish the six cortical layers in vivo. Recently more direct evidence for radial migration in conjunction with radial glial cells was provided using an improved model of brain organoid (Lancaster et al., 2017), including direct time-lapse imaging of neurons leaving the ventricular zone. Conversely, ventral forebrain-like organoids are derived from ESCs utilizing small molecules that modulate the SHH and Wnt pathways. These organoids show high expression of the transcription factor Nkx2.1 and are found to contain extensive populations of GABAergic interneurons. When fused together these neurons migrate to and integrate with the dorsal cortical organoids, in patterns that resemble the characteristic tangential migration of cortical interneurons (Bagley, Reumann, Bian, Levi-Strauss, & Knoblich, 2017; Birey et al., 2017).

4. DIVERGENT FEATURES OF CORTICOGENESIS

Overall the studies above have revealed that a great number of conserved features of cortical development could be recapitulated in a similar fashion from PSCs of various species. In the next section, we will examine whether and how corticogenesis from PSCs also displays divergent features that are directly relevant to the species-specific features of human and primate corticogenesis.

Indeed, over the last several million years of hominid evolution the human cerebral cortex has undergone a significant increase in relative size and complexity. The developmental processes, which govern human cortex development, are similarly complex, involving the expansion of a limited number of an initial neural stem cell population to generate the myriad neurons and glial cells that make up the cortex. Here, we will summarize some of the specific features of corticogenesis that are thought to be altered between humans, other primates, and rodents, with potentially direct relevance to human brain evolution.

4.1 Temporal Dynamics of Corticogenesis

A first striking difference between human, primate, and mouse corticogenesis is the timing of NEC expansion prior to neurogenesis and thus number of cell divisions that NECs undergo to generate the initial pool of cortical progenitors. It was long proposed that extending the amplification period and thus the total derived number of the founder population of NECs prior to neurogenesis will lead to a greater size of cortex, in line with the radial unit model of corticogenesis (Rakic, 1995). This model proposes that neurons generated in the VZ will undergo a radial migration to settle in the cortical plate, constituting a radial column, positional information therefore being retained from the VZ.

Additional rounds of cell division in NECs could therefore lead to an exponential increase in column number, each radial column adding to the surface area of the cortex. While some evidence for this model has been obtained in mouse models (Chenn & Walsh, 2002; Sahara & O'Leary, 2009), it has been difficult to study given the very short period of amplification of NEC in this species. It is therefore quite interesting that PSC-derived models of corticogenesis actually recapitulate at least in part some of the temporal features of NEC, even in 2D (Fig. 2). Indeed, direct

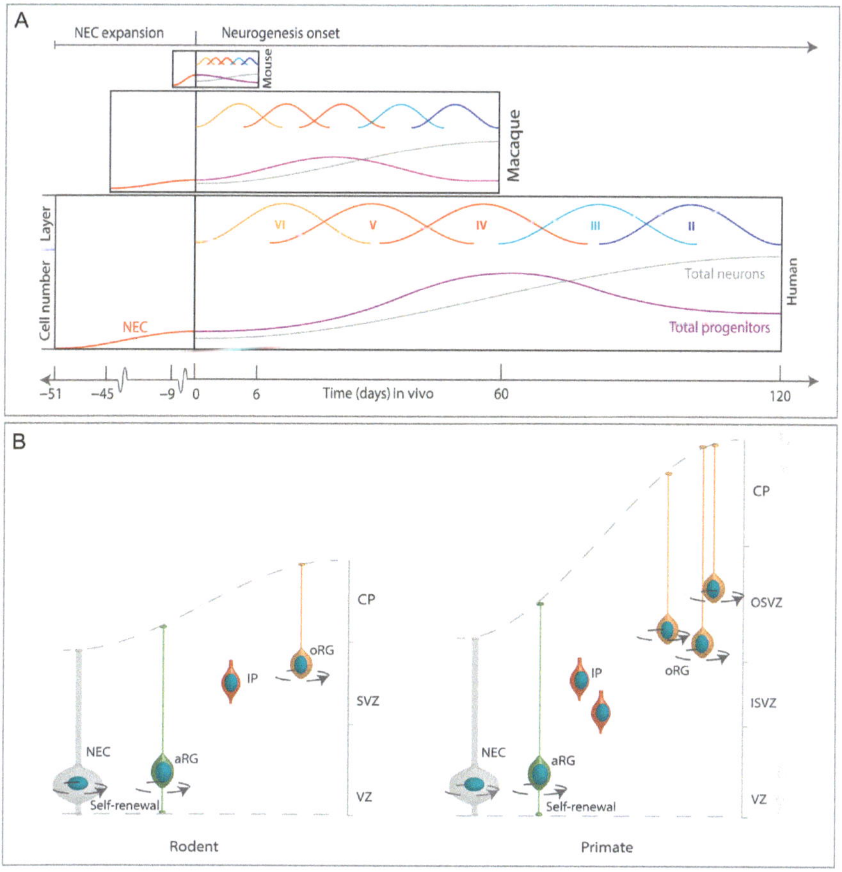

Fig. 2 (A) The temporal dynamics of corticogenesis differ across species. Human corticogenesis has a greatly extended period of NEC amplification prior to neurogenesis, while neurogenesis lasts for 120 days in human. (B) Progenitor diversity in rodents and primates. The developing human cortex contains a greatly expanded outer subventricular zone (OSVZ), which contains large numbers of outer radial glia (oRG).

comparison of human and mouse ESC-derived NEC using the exact same culture conditions revealed that while the first cortical neurons are generated after a few days following neural induction with mouse cells, it takes up to 3 weeks for the first neurons to be produced from human cells (Espuny-camacho et al., 2013). While the mechanisms underlying the higher amplification capacity of human/primate cells vs rodent NEC remain unknown, PSC models may prove to be very useful to discover the underlying mechanisms, inasmuch that NEC amplification is also hard to study in ex vivo human or primate material as it corresponds to very early stages of embryogenesis.

Once cortical neurogenesis starts, temporal differences can be further observed. For example, neurogenesis lasts around 120 days in humans, compared with about a week in the mouse (Sousa et al., 2017). As for NECs these timing differences are also observed during in vitro corticogenesis, with corticogenesis taking about 1 week with mouse ESC (Gaspard et al., 2008) and more than 3 months with human ESC (Espuny-camacho et al., 2013; Shi et al., 2012). The length of in vitro corticogenesis and temporal patterning of generation of layer-specific subtypes was directly compared between several primates using similar corticogenesis conditions from human, chimpanzee, and macaque ESC (Otani, Marchetto, Gage, Simons, & Livesey, 2016). While in each species neurogenesis appeared to occur at similar stages, the total length of neurogenic period was greatly protracted in human and chimpanzee cells compared with macaque cells, well in line with in vivo corticogenesis. Furthermore, clonal analyses of cultures of macaque and human cortical cells identified a critical phase during which the aRGC population continues to expand while producing neurons, which was prolonged in human cells, leading to higher neuronal output, as would be predicted in vivo. This behavior was found to be unaltered when human and macaque cells were cocultured, indicating a degree of autonomous regulation of the species-specific properties of cortical progenitors (Otani et al., 2016).

On the other hand, few differences were found overall between human and chimpanzee in vitro corticogenesis (Mora-Bermúdez et al., 2016; Otani et al., 2016). Comparative analyses in 2D and 3D systems confirmed previous findings that the cytoarchitecture, cellular diversity, and temporal gene expression are remarkably similar in human and chimpanzee developing cortex. However, live imaging performed on organoids at similar stages of corticogenesis revealed a significant lengthening in the prometaphase–metaphase transition in human aRGCs compared to chimpanzee and further between primates and mouse, while there is no significant difference in other phases of the cell cycle. Importantly this difference was found only in cortical aRGCs and not in nonneural cell types, while single cell RNA-seq analyses of the aRGCs revealed a human enrichment for several genes linked to proliferation (Mora-Bermúdez et al., 2016).

Overall these studies confirm that in vitro PSC-based models can model evolutionarily relevant differences and provide several experimental paradigms to identify the underlying molecular mechanisms.

4.2 Progenitor Diversity

Another key divergent feature of cortical development is the diversity of neuronal progenitors and their niche. Indeed, in particular in primates, at late stages of cortical neurogenesis the proliferative compartment expands basally into a so-called outer subventricular zone (OSVZ) that is much larger than the VZ (Dehay, Kennedy, & Kosik, 2015). The OSVZ contains specialized basal progenitors, the outer radial glia cells (oRGCs) (Fietz et al., 2010; Hansen, Lui, Parker, & Kriegstein, 2010; Reillo, de Juan Romero, Garcia-Cabezas, & Borrell, 2011), which display a distinctive morphology distinct from aRGCs, as they display a basal process but lack an apical process (Fig. 2B). The oRGCs display a specific gene expression profile particularly distinct from aRGCs, which may be linked in part to their different environmental niche (Pollen et al., 2015). They also display a specific and stereotyped movement at mitosis, that is a sharp apicobasal motion, known as mitotic somal translocation. Based on their strong enrichment in primate/human cortex and proliferative capacities, it has been proposed that oRGCs may contribute in a species-specific way to the expansion of the primate cortex as well as its gyrification (Fietz et al., 2010; Hansen et al., 2010; Reillo et al., 2011). The oRGCs can be found at later stages of corticogenesis from PSC-derived cortical cells, in 2D cultures (Shi et al., 2012) but more significantly in 3D organoid models (Kadoshima et al., 2013; Lancaster et al., 2013; Qian et al., 2016). As they settle in a more basal location than aRGCs, and thus may be able to develop in a distinct niche similar to that observed in vivo. Interestingly oRGCs could not be detected in brain organoids derived from mouse cells (Lancaster et al., 2013), thus recapitulating the species-specific relative abundance of apical vs basal progenitors like in vivo.

5. MODELING EARLY NEURODEVELOPMENTAL DISORDERS STRIKING THE HUMAN CORTEX

Overall, the studies detailed earlier demonstrate the suitability of various 2D and 3D PSC-based models to investigate distinct-specific features of human corticogenesis and complement the studies using existing animal models or human fetal brain material. In the next section, we will focus upon recent work combining these models with genetic or environmental manipulations that aim at modeling specific neurodevelopmental diseases and provide novel insight to the intrinsic and extrinsic mechanisms influencing human cortex ontogeny.

Many neurodevelopmental disorders affect the growth and development of the cerebral cortex and include a wide range of known conditions from structural cortical malformations such as lissencephaly and microcephaly to dysfunctions such as ID (intellectual deficiency) and ASD (autism spectrum disorder). Many types of genetic mutations or exposure to toxins or pathogens are all known to result in such disorders. The study of these conditions has historically been limited to animal models and patient tissue, but now PSC technologies have presented new opportunities for studying human developmental disorders in a more physiologically relevant setting, from early defects in neurogenesis to later disruption of neuronal connectivity.

5.1 Primary Microcephaly

Primary microcephaly (MCPH) is an ND in which the overall size of the brain is dramatically reduced, while overall brain structure is retained, with the most acute differences observable in the cerebral cortex. Its genetic form is a rare autosomal recessive disorder, and to date there have been over 16 genes identified in which mutation results in MCPH, predominantly encoding proteins important for mitotic function (Gilmore & Walsh, 2013). Identified mutations affect key processes underlying progenitor proliferation, differentiation, and survival, but the exact mechanisms leading to decreased cortical size, and whether they are common to all forms of genetic microcephaly, remain unclear. The study of the mechanisms of MCPH is particularly interesting in the context of human cortex development, since mouse models have revealed a usually milder phenotype, such as for ASPM, the most frequently mutated gene in human MCPH (Pulvers et al., 2010), suggesting that at least some of the MCPH-linked genes may bear some species specificity in their function.

MCPH was first studied in a human context using iPSC from a patient presenting a mutation in *CDK5RAP2* (Lancaster et al., 2013). Previously the effect of *CDK5RAP2* mutation had been investigated in the mouse, in which alterations were found in aRGC mitotic spindle orientation, together with precocious neuronal differentiation and increased progenitor cell death (Lizarraga et al., 2010). Cerebral organoids generated from the *CDK5RAP2* mutated patient cells showed overall reduction in organoid size when compared with control, together with increased neuronal differentiation and randomization of mitotic spindle orientation during neural progenitor division (Fig. 3B; Lancaster et al., 2013). Although the exact link between these phenotypes remains unclear (Delaunay, Kawaguchi,

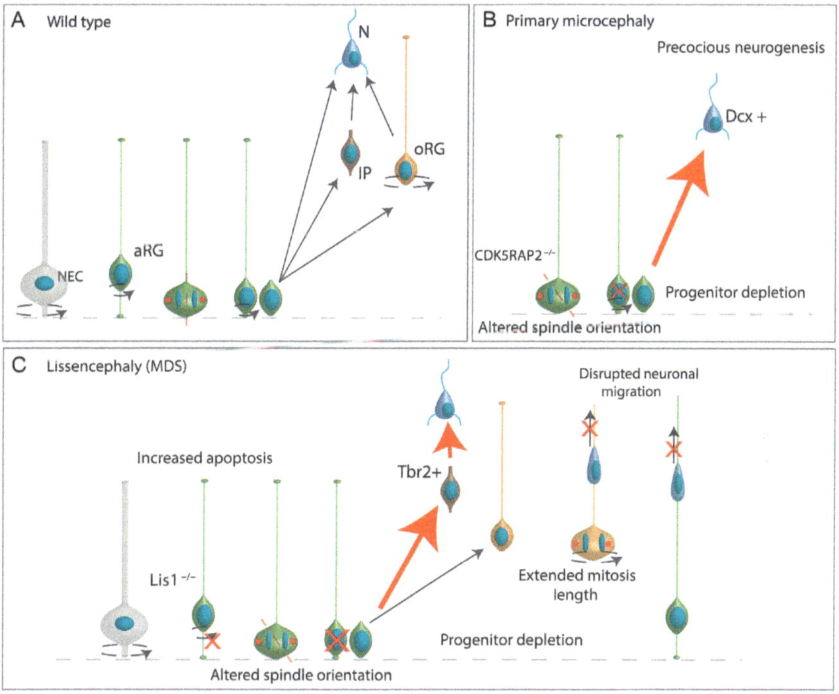

Fig. 3 (A) Wild-type corticogenesis. (B) Primary microcephaly results from mutations in *CDK5RAP2*; aRGCs exhibit altered spindle orientation and precocious neurogenesis resulting in progenitor pool depletion. (C) Lissencephaly/Miller–Dieker syndrome (MDS) is caused by mutations in *Lis1*; aRGCs exhibit altered spindle orientation concomitant with an increase in the number of Tbr2+ intermediate progenitors. oRGCs exhibit extended mitotic length, while neuronal migration is inhibited.

Dehay, & Matsuzaki, 2016), these data provide first proof of principle that a condition such as MCPH can be studied in a human PSC-derived model.

In line with defects in early neurogenesis underlying MCPH, models using iPSCs derived from Seckel syndrome patients demonstrated the importance of the primary cilium in neural progenitor maintenance (Gabriel et al., 2016). Seckel syndrome patients present with MCPH, but also have characteristic reduced body size; it is the result of a mutation in the *CPAP* gene, which codes for Centrosomal-P4.1-associated protein (CPAP). Studies in mice have previously revealed that loss of CPAP in aRGCs results in progressive loss of centrioles and centrosomes, defects in centriole genesis, and randomization of the mitotic spindle. As a result aRGCs delaminate from the apical surface, leading to apoptosis, reduction of the progenitor pool and microcephaly (Garcez et al., 2015; Insolera, Bazzi, Shao, Anderson, & Shi, 2014). On the other hand, CPAP is involved in maintenance of the centrosome and microtubule disassembly at the primary cilium that occurs during mitosis (Klingseisen & Jackson, 2011) the latter role was revealed by studying the neural differentiation from Seckel cell patient iPSCs (Gabriel et al., 2016). Patient-derived rosettes were observed to be less organized and smaller than controls; furthermore, aRGC mitotic spindle orientation was disrupted as observed in mouse models. Moreover, patient-derived NPCs lacked CPAP expression at the base of the cilium and failed to recruit cilium disassembly complex proteins. Consequently NPCs displayed elongated cilia and exhibited reduced proliferation, coupled with increased neurogenesis when compared with controls (Gabriel et al., 2016). The authors also report delayed G1–S transition in Seckel syndrome NPCs, which may be linked to the observed increase in neurogenesis (Lange, Huttner, & Calegari, 2009). Importantly these patients do not display null mutations but instead express CPAP lacking the microtubule interacting domain, thus defining a specific component of CPAP function at the level of cilia dynamics, which could directly underlie microcephaly in the patient context.

As well as investigating genetic causes of NDs, in vitro cortical models also provide a model to investigate other causes of cortical malformation, for example, fetal insults as a result of infections such as the zika virus. Zika virus infection is indeed linked to MCPH (Mlakar et al., 2016), which prompted the use of in vitro human ESC-based models to explore its neurotoxic nature and dissect the underlying pathogenic mechanisms (Qian, Nguyen, Jacob, Song, & Ming, 2017). It was demonstrated initially in 2D human ESC-derived neuronal cultures that the zika virus preferentially infects progenitor cells (Tang et al., 2016). Once infected, NPCs exhibit defects in proliferation concomitant with increased cell death in both neurons and progenitors (Qian et al., 2016), leading to defective growth of brain organoids (Garcez et al., 2016). Together with work in mouse models, the human PSC-based models are now being used to try and identify its cellular targets and potential inhibitors (Retallack et al., 2016; Yoon et al., 2017).

These data constitute collectively a convincing proof of principle that human PSC-based models can provide insights into the mechanisms controlling normal and abnormal size of the human cerebral cortex. It will be particularly interesting to expand these findings to other defects in brain size that have proven more difficult to model in the mouse such as ASPM deficiency.

5.2 Lissencephaly

Lissencephaly is characterized by a smooth and smaller brain surface and defects in the normal layer structure of the cortex. Deletions in the *PAFAH1B1* or *Lis1* gene located on chromosome 17p13.3, which codes for the protein Lis1, are the most common cause of lissencephaly (Reiner & Sapir, 2013). Miller–Dieker syndrome (MDS) is a lissencephaly form caused by large heterozygous deletions of 17p13.3, which encompasses not only *Lis1*, but also other genes including *YWHAE* coding for 14.3.3ε protein. Lis1 is essential for brain development and acts in a dose-dependent manner such that hemizygotic deletion at 17p13.3 results in lissencephaly, whereas increased dosage can result in developmental delay (Bi et al., 2009).

Studies using mouse models have revealed key functions for Lis1, demonstrating defects in neural proliferation, interkinetic motility, and neuronal migration among others (for review, see Reiner & Sapir, 2013). However, a major limitation of the mouse model is the lack of cortical gyrification and progenitor diversity observed in primates. Recent studies utilizing human iPSCs derived from Miller–Dieker syndrome patients, however, are beginning to shed light on the mechanisms underlying lissencephaly in a human-specific context (Bershteyn et al., 2017; Iefremova et al., 2017). These studies identified a number of critical phenotypes that are likely to contribute to MDS, including increased aRGC apoptosis in MDS patient-derived organoids, concomitant with decreased overall size of organoids when compared to controls. Further the cleavage orientation of aRGCs in the VZ is altered, with a shift from vertical to horizontal cleavage with respect to the apical surface associated with an increase in IPC number (Fig. 3C). Moreover, affixed organoids to culture plates allowing the observation of neurite and radial glia fiber extension and later characteristic saltatory neuronal migration revealed a reduction in overall migration speed in MDS patient-derived neurons, as well as increased time in the static position (Bershteyn et al., 2017). Such proliferative and migration defects are also found in mouse models of Lis1 mutations; however, one exciting aspect of cortical organoid development is the ability to image, in real time and in a physiologically relevant setting, the behavior of oRGCs, which are much less prevalent in the mouse. The presence of oRGCs in human cortical organoids was demonstrated by comparison of marker expression, morphology, and through single cell transcriptome analysis compared with human fetal cortical samples (Bershteyn et al., 2017). Real-time imaging of the cells revealed alteration in the length of mitosis specifically in MDS oRGC, while interestingly IPCs and aRGCs appeared to be unaffected, with MDS oRGCs remaining in mitosis for extended periods of time prior to cytokinesis. It will be interesting in future work to assess the relative impact of oRGC proliferation and neuron migration defects to the lack of gyrification in MDS patients, which may require studies in gyrencephalic animals such as ferret since cortical organoids do not actually undergo gyrification using current methods.

5.3 Macrocephaly

Other brain diseases of genetic origin are characterized by increased brain size, or macrocephaly, caused by mutations such as those of *PTEN*, implicated in hemimegancephaly and macrocephaly (Lee et al., 2012).

The underlying mechanisms were recently investigated in neural organoids using ESC bearing CRISPR–cas9-targeted mutations in *PTEN* (Li et al., 2017). Interestingly, an increase in surface area and volume of $PTEN^{-/-}$ organoids was observed when compared with isogenic controls, associated with sustained cell cycle reentry and delayed neurogenesis, thus leading to lateral expansion of early neural progenitors and subsequent increase in neuronal production, in line with the radial unit model of cortical expansion. Intriguingly an early increase in surface folding of the organoids, including VZ-like structures, was also observed in human PTEN mutant cells. It remains unclear how this folding may relate to normal gyrification of the cortex in vivo, which occurs much later in development and only in the neuronal compartment, while the ventricular compartment remains smooth. Interestingly similar mouse $PTEN^{-/-}$ ESC-derived organoids showed no alteration in surface folding, suggesting species-specific properties of mouse vs human neural organoids in the *PTEN* mutant background (Li et al., 2017).

In many cases macrocephaly is associated with cognitive defects, most strikingly symptoms related to ASD, including sporadic cases for which the genetic origin remains undetermined. To explore the potential links between sporadic ASD and brain size alterations, one recent study utilized iPSC derived from patients presenting with macrocephaly coupled with ASD (Mariani et al., 2015). Critical differences were observed in cortical organoids derived from patient vs unaffected parent iPSCs, in particular increased production of GABAergic neurons. This appeared to be caused by early alteration in the regional fate of the ASD patient-derived cells, which displayed increased expression of markers of the ventral telencephalon Dlx1/2 as well as FoxG1 telencephalic master gene. Remarkably, knockdown reduction of FoxG1 expression in ASD organoids normalized the organoids for GABAergic fate production. A similar study using cortical cells derived from another cohort of sporadic cases iPSC also revealed early defects, including overproliferation of neural precursors and deregulation of FoxG1 and other markers of ventral telencephalon (Marchetto et al., 2017). In this case, however, the phenotypes were primarily linked to Wnt pathway downregulation, which could be consistent with the known cross-inhibition of FoxG1 and Wnts in early vertebrate telencephalic development (Danesin et al., 2009). On the other hand, upregulation of GABAergic neuronal production was recently linked to some forms of ASD linked to CHD8 mutations, in both human and mouse models, but associated in this case with decreased Wnt signaling emphasizing the complexity of the topic (Katayama et al., 2016; Wang et al., 2017).

Intriguingly, a study investigating mutations in *DISC1* (disrupted in schizophrenia 1), thought to be causative of a rare form of familial schizophrenia (SCZ), has also identified similar defects during in vitro corticogenesis. Mutations in *DISC1* were shown to result in alterations in neuronal progenitor specification in a similar manner to those discussed for ASD, including increased Wnt signaling and reduced FoxG1 (Srikanth et al., 2015). Similarly, early alterations in apical–basal polarity of neural progenitors were detected in cells presenting chromosomal abnormalities linked to SCZ (Yoon et al., 2014).

Overall these data suggest a surprising model whereby some cases of complex cognitive disorders such as ASD or even SCZ could result partially from early alterations in regional patterning and fate that may eventually lead to an imbalance of neuronal subtype production, and ultimately circuit dysfunction. Future work combining human and animal models and targeting multiple forms of ASD/SCZ should aim to determine to which extent these early defects found in human in vitro models of ASD and SCZ do contribute to the disease physiopathology in vivo, or correspond to endophenotypes that would be only revealed in a sensitized in vitro situation.

6. MODELING LATE NEURODEVELOPMENT OF THE HUMAN CORTEX

In addition to modeling the first steps of corticogenesis, PSC models provide an opportunity to explore later aspects such as neuronal morphogenesis and synaptogenesis. In this frame several studies using 2D models of cortical neuron differentiation have focused on modeling human diseases thought to correspond to defects in synapse formation and/or function. Their exhaustive description goes beyond the scope of the present review focusing on corticogenesis, but to name a few, here we discuss some illustrative examples of such studies focused on cortical neurons.

6.1 Diseases Affecting Neuronal Maturation and Synapse Formation

One such condition for which iPSC-derived culture has been utilized is Timothy syndrome, a rare disease caused by a gain-of-function mutation in the *CACNA1C* gene, which encodes the voltage-gated calcium channel Cav1.2 alpha subunit, where patients present with arrhythmia, ID, and ASD. Neurons derived from patient iPSCs were shown to have defects in calcium signaling and activity-dependent gene expression (Pasca et al., 2011). Interestingly patient-derived neuronal cultures contain fewer CTIP2+ positive deep layer neurons and fewer Satb2+ callosal projection neurons, suggesting additional defects in neuronal specification, which were confirmed a posteriori by examining corresponding mouse mutant brain. A subsequent analysis of the transcriptome of patient- and control-derived progenitors and neurons, utilizing a combined gene coexpression network analysis with predicted transcription factor binding analysis, determined that the alterations in TS-associated gene expression were predicted to be coregulated by specific calcium-dependent transcription regulators (Tian et al., 2014). More recently, the migration behavior of neurons in TS patients was also explored using fused ventral–dorsal forebrain organoid system: this revealed potential defects in the tangential migration of mutant interneurons (Birey et al., 2017), consistent with previous data implicating calcium signaling in cortical interneuron positioning (Bortone & Polleux, 2009).

An important form of cognitive defect of genetic origin is Rett syndrome (RS) a progressive neurodevelopmental disease in which patients display initial normal cognitive and physical development, followed by severe regression with ID and ASD. RS is most frequently caused by an X-linked mutation in the *MECP2* gene and has been extensively studied in mouse models, revealing several key pathogenic mechanisms (Lombardi, Baker, & Zoghbi, 2015). The study of RS patient iPSC-derived neurons enabled the detection of alterations in early synapse development and subsequently function (Marchetto et al., 2010). Interestingly, increasing gene dosage of MeCP2, which is also pathological, was shown to result in the opposite phenotype, including increased synaptogenesis and dendritic complexity (Nageshappa et al., 2016). These studies point to a deregulation relatively early in cortical development, which may have long-lasting consequences for later cortical function (Ardhanareeswaran, Mariani, Coppola, Abyzov, & Vaccarino, 2017; Ben-Reuven & Reiner, 2016; Marchetto et al., 2010).

SHANK3 is a scaffold protein, which is enriched in the postsynaptic densities of excitatory cells and has been implicated in Phelan–McDermid syndrome (PMDS) (which is caused by heterozygous deletions of chromosome 22q13.3, a region containing the *SHANK3* gene) characterized by ID and ASD. Neurons derived from iPSCs of PMDS patients show reduced levels of SHANK3 expression and exhibit defects in excitatory synaptic transmission (Shcheglovitov et al., 2013). Recently, Yi and colleagues investigated the function of SHANK3 utilizing a heterozygous conditional knockout of *SHANK3* in human ESCs. Neurons derived from these mutant ESCs showed a range of anatomical and electrophysiological abnormalities including decreased neurite complexity, but also more surprising abnormalities in intrinsic properties such as increased input resistance and excitability (Yi et al., 2016). Further, this study demonstrated a direct interaction of SHANK3 with hyperpolarization-activated cyclic nucleotide-gated channels, suggesting that SHANK3 mutation results in a channelopathy, which could shed interesting light on ASD pathogenesis (Yi et al., 2016).

6.2 Evolution and Neoteny in the Human Cortex

Besides investigating the pathology of human NDs neurons, PSC-derived corticogenesis offers the possibility to explore potentially uniquely or preferentially human traits in cortical neurons. Indeed, the comparison of cortical neurons from human and nonhuman primates in several areas has revealed significant differences in their morphological properties, in particular increased levels of dendrite complexity and dendritic spine size and density (Benavides-Piccione, Ballesteros-Yanez, DeFelipe, & Yuste, 2002; Defelipe, 2011). In addition, the human neocortex displays a striking neoteny of its neuronal development compared to the other primates, i.e., the retention of juvenile properties even at mature stages, because of protracted neuronal maturation (Bufill, Agusti, & Blesa, 2011; Defelipe, 2011; Petanjek et al., 2011). Specifically, human cortical neurons, in particular at the level of the prefrontal cortex, display greatly protracted patterns of dendrite outgrowth, dendritic spine maturation, synapse formation, and pruning. They thus take years to reach mature adult-like stages, while macaque neurons take only a few months (Liu et al., 2016, 2012; Petanjek et al., 2011; Somel et al., 2009). Interestingly, many genes that display human-specific "neotenic" patterns of expression in the human cortex compared with nonhuman primates appear to encode for synaptic proteins, and are downregulated in the brain of ASD patients (Liu et al., 2016). This extended "critical period" of neuronal maturation could have significant implications for the unique differences in human cognitive abilities and developmental disorders when compared with other mammalian species, although it should be noted that the unusually slow rate maturation of human neurons is particularly striking for cortical neurons but is not unique to them, as many other types of human PSC-derived neurons also display slow maturation compared with their mouse counterparts (Zeltner & Studer, 2015).

Despite their potential importance, the molecular and cellular mechanisms underlying human neuronal neoteny remain essentially unknown, with the notable exception of the role of human-specific paralogues of *srGAP2* (Charrier et al., 2012). PSC-derived models may prove to be useful to elucidate the developmental mechanisms that underlie the unusually slow speed of human neuronal maturation. Indeed, the protracted maturation of human neurons was demonstrated in vitro, since from mouse ESC the generation of functional and synaptically connected pyramidal neurons takes about 3 weeks, while from human PSC similarly fated cortical neurons require several months (Espuny-camacho et al., 2013; Shi et al., 2012). Moreover, macaque cortical neurons also mature more quickly than both chimpanzee and human cortical neurons in vitro, following time lines that strongly resemble the in vivo situation (Otani et al., 2016).

These data strongly suggest a species-specific regulation of the rate of maturation of cortical neurons. This is probably best illustrated by transplantation studies, where mouse or human PSC-derived cortical cells are similarly transplanted into the mouse neonatal cortex.

Indeed, while transplanted mouse ESC-derived cortical neurons take about 4 weeks to develop fully, send long-range axonal projections, and display complex dendritic arbors, similarly transplanted human ESC-derived pyramidal neurons take up to 9–12 months posttransplantation to start to display mature patterns of morphology and physiology, including spine formation and robust synaptic connectivity (Fig. 4A–C) (Espuny-camacho et al., 2013), and this protracted development is also observed for transplanted interneurons (Maroof et al., 2013; Nicholas et al., 2013). These data thus may suggest that a timer mechanism, largely intrinsic to the human neurons, somehow controls their rate of maturation in a species-specific manner. Recent studies have started to explore how to speed up cortical neuronal fate acquisition and maturation (Qi et al., 2017), which will be

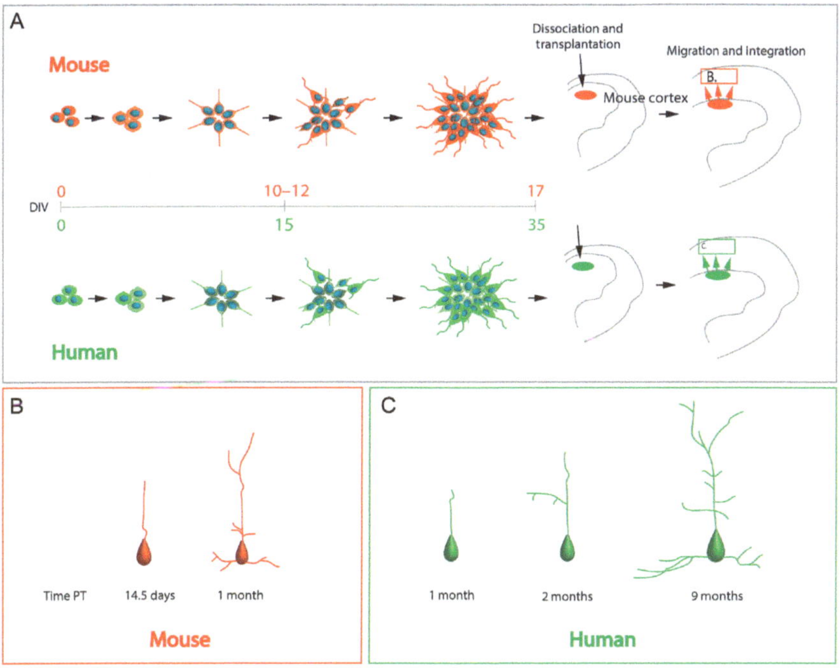

Fig. 4 Mouse and human cortical neurons derived in vitro are transplanted into the ventricle of mouse hosts (A). Transplanted neurons migrate and integrate into the host cortex, following their own species-specific time line (B and C).

interesting to explore further, in vitro and in vivo. Another interesting implication of the xenotransplantation findings is that transplanted juvenile neurons are capable of connecting with the adult cortex even a late stages of adulthood (new synapses between host and transplant are formed in 12-month-old mice), which bears interesting implications for the prospect of brain repair through transplantation of PSC-derived neurons (Michelsen et al., 2015; Tabar & Studer, 2014).

Moreover, the mouse–human neuron chimeric transplantation approach, though artificial, may constitute a fruitful paradigm to study, in an in vivo setting, human programs of neuronal development and connectivity, as well as their alterations in disease. The latter has been explored already in the context of neurodegenerative conditions, where human vs mouse cortical neurons were transplanted into transgenic Alzheimer's disease (AD) model mice overproducing amyloidogenic peptides (Espuny-Camacho et al., 2017). This revealed a specific loss of human neurons in the mouse AD model, several months posttransplantation, while similarly transplanted mouse ESC-derived neurons appeared largely unaffected, suggesting a species-specific sensitivity of human neurons. These data thus constitute a first proof of principle of the use of mouse–human transplantation to study complex brain diseases, in particular in the cases where species-specific mechanisms may underlie their origin or course.

7. CONCLUSION AND PERSPECTIVE

Overall the data discussed earlier illustrate a strikingly high level of self-organization properties of PSC-derived cortical cells, allowing the coordinated generation of a rich repertoire of cortical neuronal subtype, as well as a distinctive cortical-like cytoarchitecture to emerge de novo, in vitro. Beyond these exciting results more work is still clearly needed to make the 2D and 3D models accurate and robust enough to allow studies to be performed with the same mechanistic accuracy as with animal models, which will remain absolutely crucial for in vivo analyses of corticogenesis.

It is clear that there is no single human-specific difference occurring during corticogenesis that will account for the unique properties of the human brain, hence the need for innovative experimental models, in conjunction with the well-established ones, to tackle the complexity of the links between the evolution of the human genome, neural development, and brain structure and function, from stem cells to circuits.

ACKNOWLEDGMENTS

We apologize to the authors whose work could not be cited due to space constraints. The work from the authors presented in the review was funded by grants from the Belgian FNRS, the Belgian Queen Elizabeth Medical Foundation, the Interuniversity Attraction Poles Program (IUAP), the WELBIO Program of the Walloon Region, the European Research Council (ERC-2013-AG-340020), the AXA Research Fund, and Fondation ULB (to P.V.).

REFERENCES

Ardhanareeswaran, K., Mariani, J., Coppola, G., Abyzov, A., & Vaccarino, F. M. (2017). Human induced pluripotent stem cells for modelling neurodevelopmental disorders. *Nature Reviews. Neurology*, *13*(5), 265–278. https://doi.org/10.1038/nrneurol.2017.45.

Bae, B. I., Jayaraman, D., & Walsh, C. A. (2015). Genetic changes shaping the human brain. *Developmental Cell*, *32*(4), 423–434. https://doi.org/10.1016/j.devcel.2015.01.035.

Bagley, J. A., Reumann, D., Bian, S., Levi-Strauss, J., & Knoblich, J. A. (2017). Fused cerebral organoids model interactions between brain regions. *Nature Methods*, *14*(7), 743–751. https://doi.org/10.1038/nmeth.4304. http://www.nature.com/nmeth/journal/v14/n7/abs/nmeth.4304.html-supplementary-information.

Bandler, R. C., Mayer, C., & Fishell, G. (2017). Cortical interneuron specification: The juncture of genes, time and geometry. *Current Opinion in Neurobiology*, *42*, 17–24. https://doi.org/10.1016/j.conb.2016.10.003.

Bartolini, G., Ciceri, G., & Marin, O. (2013). Integration of GABAergic interneurons into cortical cell assemblies: Lessons from embryos and adults. *Neuron*, *79*(5), 849–864. https://doi.org/10.1016/j.neuron.2013.08.014.

Benavides-Piccione, R., Ballesteros-Yanez, I., DeFelipe, J., & Yuste, R. (2002). Cortical area and species differences in dendritic spine morphology. *Journal of Neurocytology*, *31*(3–5), 337–346. doi:5123787 [pii].

Ben-Reuven, L., & Reiner, O. (2016). Modeling the autistic cell: iPSCs recapitulate developmental principles of syndromic and nonsyndromic ASD. *Development, Growth & Differentiation*, *58*(5), 481–491. https://doi.org/10.1111/dgd.12280.

Bershteyn, M., Nowakowski, T. J., Pollen, A. A., Di Lullo, E., Nene, A., Wynshaw-Boris, A., et al. (2017). Human iPSC-derived cerebral organoids model cellular features of lissencephaly and reveal prolonged mitosis of outer radial glia. *Cell Stem Cell*, *20*(4), 435–449. e434. https://doi.org/10.1016/j.stem.2016.12.007.

Bi, W., Sapir, T., Shchelochkov, O. A., Zhang, F., Withers, M. A., Hunter, J. V., et al. (2009). Increased LIS1 expression affects human and mouse brain development. *Nature Genetics*, *41*(2), 168–177. http://www.nature.com/ng/journal/v41/n2/suppinfo/ng.302_S1.html.

Birey, F., Andersen, J., Makinson, C. D., Islam, S., Wei, W., Huber, N., et al. (2017). Assembly of functionally integrated human forebrain spheroids. *Nature*, *545*(7652), 54–59. https://doi.org/10.1038/nature22330.

Bortone, D., & Polleux, F. (2009). KCC2 expression promotes the termination of cortical interneuron migration in a voltage-sensitive calcium-dependent manner. *Neuron*, *62*(1), 53–71. https://doi.org/10.1016/j.neuron.2009.01.034.

Bufill, E., Agusti, J., & Blesa, R. (2011). Human neoteny revisited: The case of synaptic plasticity. *American Journal of Human Biology*, *23*(6), 729–739. https://doi.org/10.1002/ajhb.21225.

Camp, J. G., Badsha, F., Florio, M., Kanton, S., Gerber, T., Wilsch-Bräuninger, M., et al. (2015). Human cerebral organoids recapitulate gene expression programs of fetal neocortex development. *Proceedings of the National Academy of Sciences of the United States of America*, *112*(51), 15672–15677. https://doi.org/10.1073/pnas.1520760112.

Chambers, S. M., Fasano, C. A., Papapetrou, E. P., Tomishima, M., Sadelain, M., & Studer, L. (2009). Highly efficient neural conversion of human ES and iPS cells by dual inhibition of SMAD signaling. *Nature Biotechnology*, *27*(3), 275–280. https://doi.org/10.1038/nbt.1529.

Charrier, C., Joshi, K., Coutinho-Budd, J., Kim, J. E., Lambert, N., de Marchena, J., et al. (2012). Inhibition of SRGAP2 function by its human-specific paralogs induces neoteny during spine maturation. *Cell*, *149*(4), 923–935. https://doi.org/10.1016/j.cell.2012.03.034.

Chenn, A., & Walsh, C. A. (2002). Regulation of cerebral cortical size by control of cell cycle exit in neural precursors. *Science*, *297*(5580), 365–369. https://doi.org/10.1126/science.1074192.

Danesin, C., Peres, J. N., Johansson, M., Snowden, V., Cording, A., Papalopulu, N., et al. (2009). Integration of telencephalic Wnt and hedgehog signaling center activities by Foxg1. *Developmental Cell*, *16*(4), 576–587. https://doi.org/10.1016/j.devcel.2009.03.007.

Defelipe, J. (2011). The evolution of the brain, the human nature of cortical circuits, and intellectual creativity. *Frontiers in Neuroanatomy*, *5*, 29. https://doi.org/10.3389/fnana.2011.00029.

Dehay, C., Kennedy, H., & Kosik, K. S. (2015). The outer subventricular zone and primate-specific cortical complexification. *Neuron*, *85*(4), 683–694. https://doi.org/10.1016/j.neuron.2014.12.060.

Delaunay, D., Kawaguchi, A., Dehay, C., & Matsuzaki, F. (2016). Division modes and physical asymmetry in cerebral cortex progenitors. *Current Opinion in Neurobiology*, *42*, 75–83. https://doi.org/10.1016/j.conb.2016.11.009.

Eiraku, M., & Sasai, Y. (2012). Self-formation of layered neural structures in three-dimensional culture of ES cells. *Current Opinion in Neurobiology*, *22*(5), 768–777. https://doi.org/10.1016/j.conb.2012.02.005.

Eiraku, M., Watanabe, K., Matsuo-Takasaki, M., Kawada, M., Yonemura, S., Matsumura, M., et al. (2008). Self-organized formation of polarized cortical tissues from ESCs and its active manipulation by extrinsic signals. *Cell Stem Cell*, *3*(5), 519–532. https://doi.org/10.1016/j.stem.2008.09.002.

Espuny-Camacho, I., Arranz, A. M., Fiers, M., Snellinx, A., Ando, K., Munck, S., et al. (2017). Hallmarks of Alzheimer's disease in stem-cell-derived human neurons transplanted into mouse brain. *Neuron*, *93*(5) 1066–1081.e1068. https://doi.org/10.1016/j.neuron.2017.02.001.

Espuny-camacho, I., Michelsen, K. A., Gall, D., Linaro, D., Hasche, A., Lambert, N., et al. (2013). Pyramidal neurons derived from human pluripotent stem cells integrate efficiently into mouse brain circuits in vivo. *Neuron*, *77*, 440–456. https://doi.org/10.1016/j.neuron.2012.12.011.

Fietz, S. A., Kelava, I., Vogt, J., Wilsch-Brauninger, M., Stenzel, D., Fish, J. L., et al. (2010). OSVZ progenitors of human and ferret neocortex are epithelial-like and expand by integrin signaling. *Nature Neuroscience*, *13*(6), 690–699. https://doi.org/10.1038/nn.2553.

Franco, S. J., Gil-Sanz, C., Martinez-Garay, I., Espinosa, A., Harkins-Perry, S. R., Ramos, C., et al. (2012). Fate-restricted neural progenitors in the mammalian cerebral cortex. *Science*, *337*(6095), 746. https://doi.org/10.1126/science.1223616.

Frantz, G. D., & McConnell, S. K. (1996). Restriction of late cerebral cortical progenitors to an upper-layer fate. *Neuron*, *17*(1), 55–61. https://doi.org/10.1016/S0896-6273(00)80280-9.

Gabriel, E., Wason, A., Ramani, A., Gooi, L. M., Keller, P., Pozniakovsky, A., et al. (2016). CPAP promotes timely cilium disassembly to maintain neural progenitor pool. *The EMBO Journal*, *35*(8), 803–819. https://doi.org/10.15252/embj.201593679.

Garcez, P. P., Diaz-Alonso, J., Crespo-Enriquez, I., Castro, D., Bell, D., & Guillemot, F. (2015). Cenpj/CPAP regulates progenitor divisions and neuronal migration in the cerebral cortex downstream of Ascl1. *Nature Communications*, *6*, 6474. https://doi.org/10.1038/ncomms7474.

Garcez, P. P., Loiola, E. C., Madeiro da Costa, R., Higa, L. M., Trindade, P., Delvecchio, R., et al. (2016). Zika virus impairs growth in human neurospheres and brain organoids. *Science*, *352*(6287), 816. https://doi.org/10.1126/science.aaf6116.

Gaspard, N., Bouschet, T., Hourez, R., Dimidschstein, J., Naeije, G., van den Ameele, J., et al. (2008). An intrinsic mechanism of corticogenesis from embryonic stem cells. *Nature*, *455*(7211), 351–357. https://doi.org/10.1038/nature07287.

Gaspard, N., & Vanderhaeghen, P. (2011). Laminar fate specification in the cerebral cortex. *F1000 Biology Reports*, *3*, 6. https://doi.org/10.3410/B3-6.

Gilmore, E. C., & Walsh, C. A. (2013). Genetic causes of microcephaly and lessons for neuronal development. *Wiley Interdisciplinary Reviews: Developmental Biology*, *2*(4), 461–478. https://doi.org/10.1002/wdev.89.

Greig, L. C., Woodworth, M. B., Galazo, M. J., Padmanabhan, H., & Macklis, J. D. (2013). Molecular logic of neocortical projection neuron specification, development and diversity. *Nature Reviews Neuroscience*, *14*(11), 755–769. https://doi.org/10.1038/nrn3586.

Handel, A. E., Chintawar, S., Lalic, T., Whiteley, E., Vowles, J., Giustacchini, A., et al. (2016). Assessing similarity to primary tissue and cortical layer identity in induced pluripotent stem cell-derived cortical neurons through single-cell transcriptomics. *Human Molecular Genetics*, *25*(5), 989–1000. https://doi.org/10.1093/hmg/ddv637.

Hansen, D. V., Lui, J. H., Parker, P. R., & Kriegstein, A. R. (2010). Neurogenic radial glia in the outer subventricular zone of human neocortex. *Nature*, *464*(7288), 554–561. https://doi.org/10.1038/nature08845.

Haubensak, W., Attardo, A., Denk, W., & Huttner, W. B. (2004). Neurons arise in the basal neuroepithelium of the early mammalian telencephalon: A major site of neurogenesis. *Proceedings of the National Academy of Sciences of the United States of America*, *101*(9), 3196–3201. https://doi.org/10.1073/pnas.0308600100.

Huch, M., Knoblich, J. A., Lutolf, M. P., & Martinez-Arias, A. (2017). The hope and the hype of organoid research. *Development*, *144*(6), 938–941. https://doi.org/10.1242/dev.150201.

Iefremova, V., Manikakis, G., Krefft, O., Jabali, A., Weynans, K., Wilkens, R., et al. (2017). An organoid-based model of cortical development identifies non-cell-autonomous defects in Wnt signaling contributing to Miller-Dieker syndrome. *Cell Reports*, *19*(1), 50–59. https://doi.org/10.1016/j.celrep.2017.03.047.

Insolera, R., Bazzi, H., Shao, W., Anderson, K. V., & Shi, S. H. (2014). Cortical neurogenesis in the absence of centrioles. *Nature Neuroscience*, *17*(11), 1528–1535. https://doi.org/10.1038/nn.3831.

Kadoshima, T., Sakaguchi, H., Nakano, T., Soen, M., Ando, S., Eiraku, M., et al. (2013). Self-organization of axial polarity, inside-out layer pattern, and species-specific progenitor dynamics in human ES cell-derived neocortex. *Proceedings of the National Academy of Sciences of the United States of America*, *110*, 20284–20289. https://doi.org/10.1073/pnas.1315710110.

Katayama, Y., Nishiyama, M., Shoji, H., Ohkawa, Y., Kawamura, A., Sato, T., et al. (2016). CHD8 haploinsufficiency results in autistic-like phenotypes in mice. *Nature*, *537*(7622), 675–679. https://doi.org/10.1038/nature19357. http://www.nature.com/nature/journal/v537/n7622/abs/nature19357.html-supplementary-information.

Klingseisen, A., & Jackson, A. P. (2011). Mechanisms and pathways of growth failure in primordial dwarfism. *Genes & Development*, *25*(19), 2011–2024. https://doi.org/10.1101/gad.169037.

Kriegstein, A., & Alvarez-Buylla, A. (2009). The glial nature of embryonic and adult neural stem cells. *Annual Review of Neuroscience*, *32*, 149–184. https://doi.org/10.1146/annurev.neuro.051508.135600.

Lancaster, M. A., Corsini, N. S., Wolfinger, S., Gustafson, E. H., Phillips, A. W., Burkard, T. R., et al. (2017). Guided self-organization and cortical plate formation in human brain organoids. *Nature Biotechnology*, *35*(7), 659–666. https://doi.org/10.1038/nbt.3906.

Lancaster, M. A., & Knoblich, J. A. (2014). Organogenesis in a dish: Modeling development and disease using organoid technologies. *Science (New York, N.Y.)*, *345*, 1247125. https://doi.org/10.1126/science.1247125.

Lancaster, M. A., Renner, M., Martin, C. A., Wenzel, D., Bicknell, L. S., Hurles, M. E., et al. (2013). Cerebral organoids model human brain development and microcephaly. *Nature*, *501*(7467), 373–379. https://doi.org/10.1038/nature12517.

Lange, C., Huttner, W. B., & Calegari, F. (2009). Cdk4/cyclinD1 overexpression in neural stem cells shortens G1, delays neurogenesis, and promotes the generation and expansion of basal progenitors. *Cell Stem Cell*, *5*(3), 320–331. https://doi.org/10.1016/j.stem.2009.05.026.

Lee, J. H., Huynh, M., Silhavy, J. L., Kim, S., Dixon-Salazar, T., Heiberg, A., et al. (2012). De novo somatic mutations in components of the PI3K-AKT3-mTOR pathway cause hemimegalencephaly. *Nature Genetics*, *44*(8), 941–945. http://www.nature.com/ng/journal/v44/n8/abs/ng.2329.html-supplementary-information.

Lein, E. S., Belgard, T. G., Hawrylycz, M., & Molnar, Z. (2017). Transcriptomic perspectives on neocortical structure, development, evolution, and disease. *Annual Review of Neuroscience*, *40*, 629–652. https://doi.org/10.1146/annurev-neuro-070815-013858.

Li, Y., Muffat, J., Omer, A., Bosch, I., Lancaster, M. A., Sur, M., et al. (2017). Induction of expansion and folding in human cerebral organoids. *Cell Stem Cell*, *20*(3), 385–396. e383. https://doi.org/10.1016/j.stem.2016.11.017.

Liu, X., Han, D., Somel, M., Jiang, X., Hu, H., Guijarro, P., et al. (2016). Disruption of an evolutionarily novel synaptic expression pattern in autism. *PLoS Biology*, *14*(9), e1002558, https://doi.org/10.1371/journal.pbio.1002558.

Liu, X., Somel, M., Tang, L., Yan, Z., Jiang, X., Guo, S., et al. (2012). Extension of cortical synaptic development distinguishes humans from chimpanzees and macaques. *Genome Research*, *22*(4), 611–622. https://doi.org/10.1101/gr.127324.111.

Lizarraga, S. B., Margossian, S. P., Harris, M. H., Campagna, D. R., Han, A. P., Blevins, S., et al. (2010). Cdk5rap2 regulates centrosome function and chromosome segregation in neuronal progenitors. *Development*, *137*(11), 1907–1917. https://doi.org/10.1242/dev.040410.

Lombardi, L. M., Baker, S. A., & Zoghbi, H. Y. (2015). MECP2 disorders: From the clinic to mice and back. *The Journal of Clinical Investigation*, *125*(8), 2914–2923. https://doi.org/10.1172/jci78167.

Lui, J. H., Hansen, D. V., & Kriegstein, A. R. (2011). Development and evolution of the human neocortex. *Cell*, *146*(1), 18–36. https://doi.org/10.1016/j.cell.2011.06.030.

Marchetto, M. C., Belinson, H., Tian, Y., Freitas, B. C., Fu, C., Vadodaria, K., et al. (2017). Altered proliferation and networks in neural cells derived from idiopathic autistic individuals. *Molecular Psychiatry*, *22*(6), 820–835. https://doi.org/10.1038/mp.2016.95.

Marchetto, M. C., Carromeu, C., Acab, A., Yu, D., Yeo, G. W., Mu, Y., et al. (2010). A model for neural development and treatment of Rett syndrome using human induced pluripotent stem cells. *Cell*, *143*(4), 527–539. https://doi.org/10.1016/j.cell.2010.10.016.

Mariani, J., Coppola, G., Zhang, P., Abyzov, A., Provini, L., Tomasini, L., et al. (2015). FOXG1-dependent dysregulation of GABA/glutamate neuron differentiation in autism spectrum disorders. *Cell*, *162*(2), 375–390. https://doi.org/10.1016/j.cell.2015.06.034.

Mariani, J., Simonini, M. V., Palejev, D., Tomasini, L., Coppola, G., Szekely, A. M., et al. (2012). Modeling human cortical development in vitro using induced pluripotent stem cells. *Proceedings of the National Academy of Sciences of the United States of America*, *109*(31), 12770–12775. https://doi.org/10.1073/pnas.1202944109.

Maroof, A. M., Keros, S., Tyson, J. A., Ying, S. W., Ganat, Y. M., Merkle, F. T., et al. (2013). Directed differentiation and functional maturation of cortical interneurons from human embryonic stem cells. *Cell Stem Cell*, *12*(5), 559–572. https://doi.org/10.1016/j.stem.2013.04.008.

McConnell, S. K., & Kaznowski, C. E. (1991). Cell cycle dependence of laminar determination in developing neocortex. *Science*, *254*(5029), 282–285. https://doi.org/10.1126/science.1925583.

Michelsen, K. A., Acosta-Verdugo, S., Benoit-Marand, M., Espuny-Camacho, I., Gaspard, N., Saha, B., et al. (2015). Area-specific reestablishment of damaged circuits in the adult cerebral cortex by cortical neurons derived from mouse embryonic stem cells. *Neuron*, *85*(5), 982–997. https://doi.org/10.1016/j.neuron.2015.02.001.

Mihalas, A. B., Elsen, G. E., Bedogni, F., Daza, R. A., Ramos-Laguna, K. A., Arnold, S. J., et al. (2016). Intermediate progenitor cohorts differentially generate cortical layers and require Tbr2 for timely acquisition of neuronal subtype identity. *Cell Reports*, *16*(1), 92–105. https://doi.org/10.1016/j.celrep.2016.05.072.

Mlakar, J., Korva, M., Tul, N., Popović, M., Poljšak-Prijatelj, M., Mraz, J., et al. (2016). Zika virus associated with microcephaly. *New England Journal of Medicine*, *374*(10), 951–958. https://doi.org/10.1056/NEJMoa1600651.

Mora-Bermúdez, F., Badsha, F., Kanton, S., Camp, J. G., Vernot, B., Köhler, K., et al. (2016). Differences and similarities between human and chimpanzee neural progenitors during cerebral cortex development. *eLife*, *5*, e18683. https://doi.org/10.7554/eLife.18683.

Munoz-Sanjuan, I., & Brivanlou, A. H. (2002). Neural induction, the default model and embryonic stem cells. *Nature Reviews. Neuroscience*, *3*(4), 271–280. https://doi.org/10.1038/nrn786.

Nagashima, F., Suzuki, I. K., Shitamukai, A., Sakaguchi, H., Iwashita, M., Kobayashi, T., et al. (2014). Novel and robust transplantation reveals the acquisition of polarized processes by cortical cells derived from mouse and human pluripotent stem cells. *Stem Cells and Development*, *23*, 2129–2142. https://doi.org/10.1089/scd.2013.0251.

Nageshappa, S., Carromeu, C., Trujillo, C. A., Mesci, P., Espuny-Camacho, I., Pasciuto, E., et al. (2016). Altered neuronal network and rescue in a human MECP2 duplication model. *Molecular Psychiatry*, *21*(2), 178–188. https://doi.org/10.1038/mp.2015.128.

Nicholas, C. R., Chen, J., Tang, Y., Southwell, D. G., Chalmers, N., Vogt, D., et al. (2013). Functional maturation of hPSC-derived forebrain interneurons requires an extended timeline and mimics human neural development. *Cell Stem Cell*, *12*(5), 573–586. https://doi.org/10.1016/j.stem.2013.04.005.

Noctor, S. C., Martinez-Cerdeno, V., Ivic, L., & Kriegstein, A. R. (2004). Cortical neurons arise in symmetric and asymmetric division zones and migrate through specific phases. *Nature Neuroscience*, 7(2), 136–144. https://doi.org/10.1038/nn1172.

Otani, T., Marchetto, M. C., Gage, F. H., Simons, B. D., & Livesey, F. J. (2016). 2D and 3D stem cell models of primate cortical development identify species-specific differences in progenitor behavior contributing to brain size. *Cell Stem Cell*, *18*(4), 467–480. https://doi.org/10.1016/j.stem.2016.03.003.

Pasca, S. P., Portmann, T., Voineagu, I., Yazawa, M., Shcheglovitov, A., Pasca, A. M., et al. (2011). Using iPSC-derived neurons to uncover cellular phenotypes associated with Timothy syndrome. *Nature Medicine*, *17*(12), 1657–1662. https://doi.org/10.1038/nm.2576.

Pasca, A. M., Sloan, S. A., Clarke, L. E., Tian, Y., Makinson, C. D., Huber, N., et al. (2015). Functional cortical neurons and astrocytes from human pluripotent stem cells in 3D culture. *Nature Methods*, *12*(7), 671–678. https://doi.org/10.1038/nmeth.3415.

Pera, M. F., Andrade, J., Houssami, S., Reubinoff, B., Trounson, A., Stanley, E. G., et al. (2004). Regulation of human embryonic stem cell differentiation by BMP-2 and its antagonist noggin. *Journal of Cell Science*, *117*(Pt. 7), 1269–1280. https://doi.org/10.1242/jcs.00970.

Petanjek, Z., Judas, M., Simic, G., Rasin, M. R., Uylings, H. B., Rakic, P., et al. (2011). Extraordinary neoteny of synaptic spines in the human prefrontal cortex. *Proceedings of the National Academy of Sciences of the United States of America*, *108*(32), 13281–13286. https://doi.org/10.1073/pnas.1105108108.

Pollen, A. A., Nowakowski, T. J., Chen, J., Retallack, H., Sandoval-Espinosa, C., Nicholas, C. R., et al. (2015). Molecular identity of human outer radial glia during cortical development. *Cell*, *163*(1), 55–67. https://doi.org/10.1016/j.cell.2015.09.004.

Pulvers, J. N., Bryk, J., Fish, J. L., Wilsch-Brauninger, M., Arai, Y., Schreier, D., et al. (2010). Mutations in mouse Aspm (abnormal spindle-like microcephaly associated) cause not only microcephaly but also major defects in the germline. *Proceedings of the National Academy of Sciences of the United States of America*, *107*(38), 16595–16600. https://doi.org/10.1073/pnas.1010494107.

Qi, Y., Zhang, X. J., Renier, N., Wu, Z., Atkin, T., Sun, Z., et al. (2017). Combined small-molecule inhibition accelerates the derivation of functional cortical neurons from human pluripotent stem cells. *Nature Biotechnology*, *35*(2), 154–163. https://doi.org/10.1038/nbt.3777.

Qian, X., Nguyen, H. N., Jacob, F., Song, H., & Ming, G. L. (2017). Using brain organoids to understand Zika virus-induced microcephaly. *Development*, *144*(6), 952–957. https://doi.org/10.1242/dev.140707.

Qian, X., Nguyen, H. N., Song, M. M., Hadiono, C., Ogden, S. C., Hammack, C., et al. (2016). Brain-region-specific organoids using mini-bioreactors for modeling ZIKV exposure. *Cell*, *165*(5), 1238–1254. https://doi.org/10.1016/j.cell.2016.04.032.

Quadrato, G., Brown, J., & Arlotta, P. (2016). The promises and challenges of human brain organoids as models of neuropsychiatric disease. *Nature Medicine*, *22*(11), 1220–1228. https://doi.org/10.1038/nm.4214.

Quadrato, G., Nguyen, T., Macosko, E. Z., Sherwood, J. L., Min Yang, S., Berger, D. R., et al. (2017). Cell diversity and network dynamics in photosensitive human brain organoids. *Nature*, *545*(7652), 48–53. https://doi.org/10.1038/nature22047.

Rakic, P. (1995). A small step for the cell, a giant leap for mankind: A hypothesis of neocortical expansion during evolution. *Trends in Neurosciences*, *18*(9), 383–388. https://doi.org/10.1016/0166-2236(95)93934-P.

Reillo, I., de Juan Romero, C., Garcia-Cabezas, M. A., & Borrell, V. (2011). A role for intermediate radial glia in the tangential expansion of the mammalian cerebral cortex. *Cerebral Cortex*, *21*(7), 1674–1694. https://doi.org/10.1093/cercor/bhq238.

Reiner, O., & Sapir, T. (2013). LIS1 functions in normal development and disease. *Current Opinion in Neurobiology*, *23*(6), 951–956. https://doi.org/10.1016/j.conb.2013.08.001.

Retallack, H., Di Lullo, E., Arias, C., Knopp, K. A., Laurie, M. T., Sandoval-Espinosa, C., et al. (2016). Zika virus cell tropism in the developing human brain and inhibition by azithromycin. *Proceedings of the National Academy of Sciences of the United States of America*, *113*(50), 14408–14413.

Rossi, A. M., Fernandes, V. M., & Desplan, C. (2017). Timing temporal transitions during brain development. *Current Opinion in Neurobiology*, *42*, 84–92. https://doi.org/10.1016/j.conb.2016.11.010.

Sahara, S., & O'Leary, D. D. (2009). Fgf10 regulates transition period of cortical stem cell differentiation to radial glia controlling generation of neurons and basal progenitors. *Neuron*, *63*(1), 48–62. https://doi.org/10.1016/j.neuron.2009.06.006.

Shcheglovitov, A., Shcheglovitova, O., Yazawa, M., Portmann, T., Shu, R., Sebastiano, V., et al. (2013). SHANK3 and IGF1 restore synaptic deficits in neurons from 22q13 deletion syndrome patients. *Nature*, *503*(7475), 267–271. https://doi.org/10.1038/nature12618.

Shen, Q., Wang, Y., Dimos, J. T., Fasano, C. A., Phoenix, T. N., Lemischka, I. R., et al. (2006). The timing of cortical neurogenesis is encoded within lineages of individual progenitor cells. *Nature Neuroscience*, *9*(6), 743–751. https://doi.org/10.1038/nn1694.

Shi, Y., Kirwan, P., Smith, J., Robinson, H. P. C., & Livesey, F. J. (2012). Human cerebral cortex development from pluripotent stem cells to functional excitatory synapses. *Nature Neuroscience*, *15*, 477–486. https://doi.org/10.1038/nn.3041.

Somel, M., Franz, H., Yan, Z., Lorenc, A., Guo, S., Giger, T., et al. (2009). Transcriptional neoteny in the human brain. *Proceedings of the National Academy of Sciences of the United States of America*, *106*(14), 5743–5748. https://doi.org/10.1073/pnas.0900544106.

Sousa, A. M. M., Meyer, K. A., Santpere, G., Gulden, F. O., & Sestan, N. (2017). Evolution of the human nervous system function, structure, and development. *Cell*, *170*(2), 226–247. https://doi.org/10.1016/j.cell.2017.06.036.

Southwell, D. G., Nicholas, C. R., Basbaum, A. I., Stryker, M. P., Kriegstein, A. R., Rubenstein, J. L., et al. (2014). Interneurons from embryonic development to cell-based therapy. *Science*, *344*(6180), 1240622. https://doi.org/10.1126/science.1240622.

Srikanth, P., Han, K., Callahan, D. G., Makovkina, E., Muratore, C. R., Lalli, M. A., et al. (2015). Genomic DISC1 disruption in hiPSCs alters Wnt signaling and neural cell fate. *Cell Reports*, *12*(9), 1414–1429. https://doi.org/10.1016/j.celrep. 2015.07.061.

Subramanian, L., Bershteyn, M., Paredes, M. F., & Kriegstein, A. R. (2017). Dynamic behaviour of human neuroepithelial cells in the developing forebrain. *Nature Communications, 8,* 14167. https://doi.org/10.1038/ncomms14167.

Tabar, V., & Studer, L. (2014). Pluripotent stem cells in regenerative medicine: Challenges and recent progress. *Nature Reviews. Genetics, 15*(2), 82–92. https://doi.org/10.1038/nrg3563.

Takahashi, T., Misson, J.-P., & Caviness, V. S. (1990). Glial process elongation and branching in the developing murine neocortex: A qualitative and quantitative immunohistochemical analysis. *The Journal of Comparative Neurology, 302*(1), 15–28. https://doi.org/10.1002/cne.903020103.

Tang, H., Hammack, C., Ogden, S. C., Wen, Z., Qian, X., Li, Y., et al. (2016). Zika virus infects human cortical neural progenitors and attenuates their growth. *Cell Stem Cell, 18*(5), 587–590. https://doi.org/10.1016/j.stem.2016.02.016.

Tian, Y., Voineagu, I., Paşca, S. P., Won, H., Chandran, V., Horvath, S., et al. (2014). Alteration in basal and depolarization induced transcriptional network in iPSC derived neurons from Timothy syndrome. *Genome Medicine, 6*(10), 75. https://doi.org/10.1186/s13073-014-0075-5.

Tropepe, V., Hitoshi, S., Sirard, C., Mak, T. W., Rossant, J., & van der Kooy, D. (2001). Direct neural fate specification from embryonic stem cells: A primitive mammalian neural stem cell stage acquired through a default mechanism. *Neuron, 30*(1), 65–78. doi:S0896-6273(01)00263-X [pii].

Varki, A., Geschwind, D. H., & Eichler, E. E. (2008). Explaining human uniqueness: Genome interactions with environment, behaviour and culture. *Nature Reviews. Genetics, 9*(10), 749–763. https://doi.org/10.1038/nrg2428.

Wang, P., Mokhtari, R., Pedrosa, E., Kirschenbaum, M., Bayrak, C., Zheng, D., et al. (2017). CRISPR/Cas9-mediated heterozygous knockout of the autism gene CHD8 and characterization of its transcriptional networks in cerebral organoids derived from iPS cells. *Molecular Autism, 8,* 11. https://doi.org/10.1186/s13229-017-0124-1.

Watanabe, K., Kamiya, D., Nishiyama, A., Katayama, T., Nozaki, S., Kawasaki, H., et al. (2005). Directed differentiation of telencephalic precursors from embryonic stem cells. *Nature Neuroscience, 8*(3), 288–296. https://doi.org/10.1038/nn1402.

Wilson, S. W., & Houart, C. (2004). Early steps in the development of the forebrain. *Developmental Cell, 6*(2), 167–181. https://doi.org/10.1016/S1534-5807(04)00027-9.

Wilson, S. W., & Rubenstein, J. L. (2000). Induction and dorsoventral patterning of the telencephalon. *Neuron, 28*(3), 641–651.

Yi, F., Danko, T., Botelho, S. C., Patzke, C., Pak, C., Wernig, M., et al. (2016). Autism-associated SHANK3 haploinsufficiency causes Ih channelopathy in human neurons. *Science, 352*(6286), aaf2669. https://doi.org/10.1126/science.aaf2669.

Ying, Q. L., Stavridis, M., Griffiths, D., Li, M., & Smith, A. (2003). Conversion of embryonic stem cells into neuroectodermal precursors in adherent monoculture. *Nature Biotechnology, 21*(2), 183–186. https://doi.org/10.1038/nbt780.

Yoon, K. J., Nguyen, H. N., Ursini, G., Zhang, F., Kim, N. S., Wen, Z., et al. (2014). Modeling a genetic risk for schizophrenia in iPSCs and mice reveals neural stem cell deficits associated with adherens junctions and polarity. *Cell Stem Cell, 15*(1), 79–91. https://doi.org/10.1016/j.stem.2014.05.003.

Yoon, K.-J., Song, G., Qian, X., Pan, J., Xu, D., Rho, H.-S., et al. (2017). Zika-virus-encoded NS2A disrupts mammalian cortical neurogenesis by degrading adherens junction proteins. *Cell Stem Cell, 21*(3), 349–358.e346. https://doi.org/10.1016/j.stem.2017.07.014.

Zeltner, N., & Studer, L. (2015). Pluripotent stem cell-based disease modeling: Current hurdles and future promise. *Current Opinion in Cell Biology, 37,* 102–110. https://doi.org/10.1016/j.ceb.2015.10.008.

FURTHER READING

Gotz, M., & Huttner, W. B. (2005). The cell biology of neurogenesis. *Nature Reviews. Molecular Cell Biology*, *6*(10), 777–788. https://doi.org/10.1038/nrm1739.

Lehtinen, M. K., & Walsh, C. A. (2011). Neurogenesis at the brain–cerebrospinal fluid interface. *Annual Review of Cell and Developmental Biology*, *27*(1), 653–679. https://doi.org/10.1146/annurev-cellbio-092910-154026.

Taverna, E., Gotz, M., & Huttner, W. B. (2014). The cell biology of neurogenesis: Toward an understanding of the development and evolution of the neocortex. *Annual Review of Cell and Developmental Biology*, *30*, 465–502. https://doi.org/10.1146/annurev-cellbio-101011-155801.

CHAPTER FOUR

Studying the Brain in a Dish: 3D Cell Culture Models of Human Brain Development and Disease

Juliana Brown, Giorgia Quadrato, Paola Arlotta[1]
Harvard University, Cambridge, MA, United States
[1]Corresponding author: e-mail address: paola_arlotta@harvard.edu

Contents

Abstract

The study of the cellular and molecular processes of the developing human brain has been hindered by access to suitable models of living human brain tissue. Recently developed 3D cell culture models offer the promise of studying fundamental brain processes in the context of human genetic background and species-specific developmental mechanisms. Here, we review the current state of 3D human brain organoid models and consider their potential to enable investigation of complex aspects of human brain development and the underpinning of human neurological disease.

1. INTRODUCTION

Recent progress has enabled the development of cellular models of the developing human brain via the generation of 3D brain organoids derived from human pluripotent stem cells (hPSCs). Although reductionist in nature, brain organoids have several advantages that complement 2D cell culture models and animal studies to investigate aspects of human brain development

Current Topics in Developmental Biology, Volume 129
ISSN 0070-2153
https://doi.org/10.1016/bs.ctdb.2018.03.002

and pathology. In this chapter, we will focus on hPSC-derived in vitro models of brain development. Among these, emerging 3D models, in the form of brain organoids and spheroids, offer great promise for generating the complex cellular environment and tissue maturation necessary to study fundamental developmental processes, as well as the possibility of high-throughput phenotypic analysis and therapeutic agent screening for disease research.

2. DIRECTED DIFFERENTIATION OF HUMAN CNS CELL TYPES FROM iPSCs

With the advent of technology to reprogram somatic cells into induced pluripotent stem cells (iPSCs) has come the unique opportunity to direct the differentiation of a variety of cells of interest in the dish. Great progress has been made in this direction, including the development of a variety of protocols to direct the differentiation of hPSCs into several cell types of the human CNS (see Di Lullo & Kriegstein, 2017; Kelava & Lancaster, 2016; Quadrato, Brown, & Arlotta, 2016 for comprehensive reviews) (Fig. 1). Although all are reductionist in nature, these models differ in the types of cells produced and the complexity of the tissue made; the decision of which model to employ must ultimately be driven by the specific biological question.

2.1 Making Brain Cells in 2D

Early in vitro models of neuronal development from hPSCs involved adherent culture, producing 2D cell systems. Neural rosettes, which produce self-organizing neural progenitors around ventricle-like cavities (Chambers et al., 2009; Elkabetz et al., 2008; Shi et al., 2012; Zhang et al., 2001), have been a particularly widely employed 2D system. 2D culture models have the advantage that they can be used to generate relatively homogeneous cell populations, such as lower motor neurons, midbrain dopaminergic neurons, or excitatory neurons of the dorsal telencephalon, with acceptable purity and by using protocols that are scalable, owing to their comparatively easy implementation. 2D models, however, are largely not designed to generate extensive cellular diversity and, by definition, cannot recapitulate endogenous tissue architecture in three dimensions, which limits their ability to foster the complex cell interactions found in vivo. This is a disadvantage in the context of modeling developmental processes that involve multiple cell types across different brain regions and the functionality of the neural networks connecting them. It is also possible that the lack of 3D structure may impede cell–cell interactions that normally foster cellular maturation in vivo.

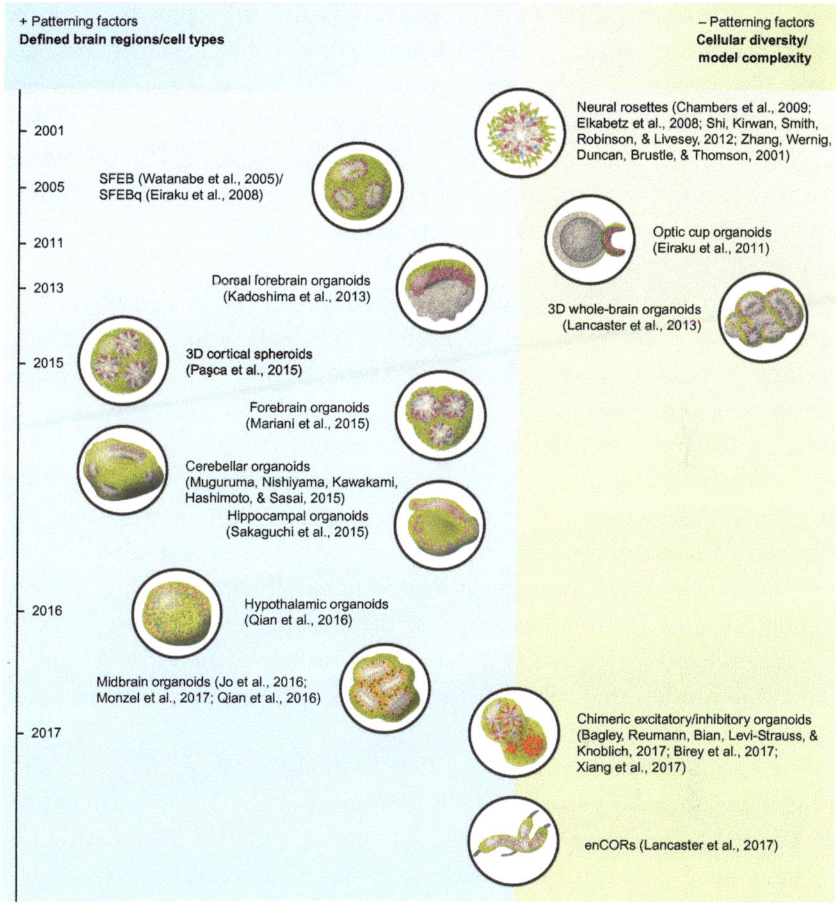

Fig. 1 hPSC-based cell culture models of human brain tissue. Selected examples of hPSC-derived 2D and 3D neural culture models, broadly divided into models that employ relatively greater (*left, blue background*) or less (*right, green background*) extrinsic patterning (such as the addition of regionalizing factors). Neural rosettes, fused chimeric organoids, and enCORs have been used in systems of both types. *enCORs*, engineered cerebral organoids; *SFEB*, serum-free culture of embryoid body-like aggregates; *SFEBq*, serum-free culture of embryoid body-like aggregates with quick reaggregation. *Modified from Quadrato, G., Brown, J., & Arlotta, P. (2016). The promises and challenges of human brain organoids as models of neuropsychiatric disease.* Nature Medicine, 22, *1220–1228.*

2.2 3D Human Brain Models: Patterned and Unpatterned Models

Recently developed 3D culture models, in principle, hold the potential of producing anatomically relevant spatial organization with cellular connections that might more closely recapitulate those found in vivo. Indeed, 3D brain tissue models in the form of brain organoids and spheroids have already been shown to recapitulate broad features of the developing brain, such as radial organization of cell types around ventricles as found in the early neural tube (Kadoshima et al., 2013; Lancaster et al., 2013; Otani, Marchetto, Gage, Simons, & Livesey, 2016; Qian et al., 2016). 3D models have also demonstrated promise for elucidating the mechanisms underlying neurodevelopmental disorders such as microcephaly (Cugola et al., 2016; Dang et al., 2016; Garcez et al., 2016; Lancaster et al., 2013; Qian et al., 2016). This field is still young; however, some predictions can be made both about the potential of the 3D systems and about future improvements.

Current 3D human organoid/spheroid models come in a variety of flavors, each with specific advantages and disadvantages. Broadly speaking, 3D protocols that rely on inducing early patterning of specific brain regions (the forebrain, for example) tend to generate tissue that is more homogenous in cellular composition, whereas omitting exogenous morphogenetic and fate-specifying cues to favor self-organization and self-patterning theoretically allows for the generation of multiple brain regions and diverse cell populations. The nomenclature for these models is currently somewhat inconsistent; in this chapter, we will preferentially use the term "organoid" to describe models that rely at least in part on principles of self-instructed organization and morphogenesis.

The serum-free embryoid bodies (SFEB) method (Watanabe et al., 2005), and its later "quick" version (SFEBq) (Eiraku et al., 2008), a particularly influential PSC-derived 3D model of the developing dorsal forebrain, was developed through modifications to the earlier rosette culture paradigm (Zhang et al., 2001) and the serum-free neural induction protocol (Ying, Stavridis, Griffiths, Li, & Smith, 2003). This protocol combines self-organization with patterning factors to favor the development of dorsal forebrain structures, including a polarized dorsal neuroepithelium, putative intermediate progenitors of the cerebral cortex, and distinct types of cortical excitatory neurons. Further improvements to this protocol led to the generation of dorsal forebrain (cortical) organoids that were grown exclusively in suspension to promote more protracted development (Kadoshima et al., 2013). This system successfully generates more mature cortical tissue that, in addition to containing a dorsalized neuroepithelium and subtypes of excitatory neurons that are correctly organized into a cortical plate-like structure, also contains a more diverse pool of progenitors that notably include outer radial glia cells typical of the human, but not the rodent, cortex.

Further protocols have been developed to streamline production and increase the reproducibility of 3D models. A method employing simplified culture conditions and scaffold-free growth was developed to generate cortical spheroids containing dorsal forebrain excitatory neurons (Paşca et al., 2015). This entirely 3D culture model includes a stage of neural progenitor expansion after patterning that favors the production of nonreactive astrocytes, cells that normally develop after the bulk of neurogenesis is complete, which perhaps suggests increased tissue maturation. Methods starting with 3D suspension cultures of initial hiPSCs have also been shown to improve control over the size and shape of neuronal spheroids (Rigamonti et al., 2016). This was applied to the production of a variety of neuronal types, including dorsal forebrain neurons and spinal cord motor neurons. Notably, arrays of miniaturized spinning bioreactors have recently been designed that allow for the culture of forebrain organoids in a small volume of media (Qian et al., 2016). This approach improved oxygen diffusion through the tissue, resulting in reduced apoptosis as compared to stationary culture, while also permitting the standardization of culture conditions and enhanced reproducibility. Because of their enhanced reproducibility, their greater control over organoid size and shape, and their scalability, these culture models are substantial steps toward downstream applications such as phenotypic and therapeutic screening using 3D systems.

Although many models have focused on generating tissue of the cerebral cortex, organoid models of other regions of the human brain have also been developed. 3D models of developing hippocampus (Sakaguchi et al., 2015), midbrain (Jo et al., 2016; Monzel et al., 2017; Qian et al., 2016), hypothalamus (Qian et al., 2016), and cerebellum (Muguruma et al., 2015) have also been generated. These systems may prove valuable to model development and dysfunction across different brain regions.

In all of these models, external pattering signals are provided to the cultures to favor the formation of specific brain regions, increasing reproducibility and providing better control over the types of cells produced, both of which are critical for future application of these systems to modeling development and disease. A major limitation of these systems, however, is evident when trying to model parts of the brain that contain cells that originate from different brain regions. For example, the excitatory and inhibitory neurons of the cerebral cortex originate from germinal zones of the dorsal and ventral telencephalon, respectively; thus, 3D models where "dorsalizing" signals are used to favor the formation of the dorsal telencephalon lead to the production of an "all excitatory" cortex, where inhibitory interneurons are produced in reduced numbers. Several recent approaches have addressed this problem through the generation of 3D chimeric systems, where spheroids or organoids patterned to a dorsal forebrain fate (in which neurons are predominantly excitatory) are fused together with others patterned to a ventral forebrain fate (predominantly inhibitory), to enable the intermigration and network integration of inhibitory and excitatory neurons (Bagley et al., 2017; Birey et al., 2017; Xiang et al., 2017). Future developments may enable similar approaches toward incorporating additional brain regions and more complex combinations; progress here will of course be affected by the rate at which patterning signals for distinct brain regions are identified, such that the different "building blocks" desired can be made and then engineered together.

A conceptually different approach has been used to generate 3D brain organoids that have the potential to form multiple brain regions. These models develop largely via self-patterning and self-organization, using intrinsic patterning principles similar to those that regulate the development of the embryonic brain. The first example of a model of neural tissue designed on these principles was described in a seminal study in which 3D optic cup organoids self-formed through morphogenetic and cellular differentiation events replicating those of the mammalian eye (Eiraku et al., 2011). This pioneering work paved the way for the generation of 3D, self-organizing, whole-brain organoids (Lancaster et al., 2013). Because in this system neural progenitors are not directed to acquire any specific regional fate, whole-brain organoids have in principle the potential to generate the largest diversity of cell types and brain regions among all 3D models. In addition, codevelopment of multiple brain regions within a single organoid might favor protracted maturation of the cells produced, which in turn would increase their ability to form robust neural networks. Cellular diversity (both neuronal and glial), enhanced maturation, and formation of active circuits are key requirements for modeling complex pathologies such as those of mental illnesses.

However, current self-patterning 3D brain organoids have substantial drawbacks compared to extrinsically patterned organoids, including reduced organoid-to-organoid reproducibility, unwanted differentiation into nonectodermal-derived cell types (Camp et al., 2015), and longer periods of development, with consequently higher production costs. Recent studies have attempted to reduce organoid-to-organoid variability by influencing their architecture. In one approach, individual neural rosettes were microdissected from adherent culture to seed organoid cultures, resulting in organoids with a single cortex-like structure (Lee et al., 2017). In another, providing microfilament scaffolds at the embryoid body stage promoted elongated structures and increased surface-area-to-volume ratio, increasing the fraction of embryoid bodies that successfully formed neuroectoderm, and the amount of neuroectoderm generated, in subsequent organoid culture (Lancaster et al., 2017).

Future advances in bioengineering might allow for spatially and temporally restricted delivery of exogenous patterning signals to foster the reliable generation of selected brain regions and to control their relative position within the organoid (Yin et al., 2016). It is of course important not to eliminate self-patterning entirely, as a major advantage of self-patterned brain organoids lies in their potential to generate a much larger diversity of cells, as compared to extensively patterned systems. Recapitulating the development of brain-like tissue in culture is, naturally, extremely difficult not only because of technical challenges but also because we lack a comprehensive picture of the complex patterning and morphogenesis events underlying normal embryonic development of the brain, especially in humans. It is therefore important that future protocol developments preserve some of the intrinsic capacity to generate regional and cellular complexity that whole-brain organoids have, while exerting more control over regional topography and overall tissue architecture.

Studies of basic developmental processes in brain organoid models have to date primarily examined early stages of development, in particular progenitor dynamics, such as effects of Shh signaling on the generation of human basal radial glia (Wang, Hou, & Han, 2016), and a role of retinoblastoma protein in the regulation of neuroblast proliferation (Matsui, Nieto-Estevez, Kyrychenko, Schneider, & Hsieh, 2017). Brain organoid systems have also facilitated interspecies comparisons of progenitor biology. An analysis of brain organoids derived from human, chimpanzee, and orangutan iPSCs demonstrated that, while human and chimpanzee organoids were grossly similar in cell composition and architecture, human organoids showed species-specific aspects of progenitor replication that may be related to increased brain size in humans (Mora-Bermudez et al., 2016). Deletion of the tumor suppressor PTEN led to increased organoid size in both human and mouse organoids, but resulted in surface folding only in the human organoids (consistent with the phenotypes of PTEN mutations in each species), apparently due to species-specific effects on proliferation of human outer radial glia (Li et al., 2017).

The nonneuronal cells in brain organoids may also deserve attention; two recent studies have described the generation of mature astrocytes from organoid cultures, which represent a valuable alternative to cadaveric primary human astrocytes for studies of human glial function (Dezonne et al., 2017; Sloan et al., 2017).

3. FUTURE PROSPECTS FOR IMPROVED 3D BRAIN MODELS

3D in vitro models of the human brain carry great potential to expand the range of features that can be established in culture, enabling higher-order investigation of brain development and functionality. Although this is speculative at this stage, self-patterning 3D systems have the greatest potential to generate broad cellular diversity, and 3D architecture may enable cell–cell interactions that are known to foster tissue maturation in the embryo, such as neuron–glia interactions that lead to synaptogenesis, myelination, and circuit maturation. By generating distinct regions of the brain and their cellular constituents, 3D systems have the potential to model long-distance connectivity among different parts of the brain.

The complexity of the processes that can be modeled in organoids will doubtlessly grow along with improved understanding of this relatively new in vitro system and further protocol development. Although it will certainly take time to move these models forward, some requirements are already clear.

Naturally, the biggest challenge of using iPSC technology to model human brain development and disease is to faithfully and reproducibly generate authentic cell types and to mimic the cell–cell interactions and circuit connectivity of the human CNS. There is an urgent need to map the cellular composition of and the diversity of cell types present within the various 3D systems. Some of the cell types present have been identified, largely through the use of single-gene markers, and more recently through single-cell profiling (Camp et al., 2015; Quadrato et al., 2017); however, comprehensive characterization over different models and protocols is critically lacking. Such characterization will be essential for assessing the potential applications of organoids as model systems, particularly for modeling neuropsychiatric disease, which reflects altered interactions and communication among a variety of cell types from multiple brain regions.

Conventional methods used in neurobiology to identify cell types (for example, immunohistochemistry) might not apply well to the resolution of cellular composition of organoids, owing to both their inconsistent neuroanatomy and the fact that analysis of single marker genes cannot conclusively identify cell types. High-throughput RNA-sequencing analysis of hundreds of thousands of single cells at comparatively low cost, such as Drop-seq (Macosko et al., 2015) and inDrop (Klein et al., 2015), has recently become a reality, and this technology is helping to enable, for the first time, a fine molecular mapping of the cellular composition of brain organoids. In parallel, in situ sequencing technologies such as fluorescent in situ RNA sequencing (FISSEQ) (Lee et al., 2014) will also be extremely valuable for identifying not only cell types but also their architectural organization. As, currently, our knowledge of the molecular profiles of individual cell types along the course of normal human brain development is still limited, this work must be accompanied by similar investigations of the endogenous developing brain to provide the benchmark for the fidelity of cells generated in vitro.

The field has already begun efforts to define the cell types present within these brain models and compare them to endogenous cells. Transcriptomic and epigenetic profiling of whole organoids vs other in vitro brain tissue models, while not able to resolve individual cell types, has suggested that organoid protocols produce tissue more closely resembling human brain, and that extending the organoid culture period improves this correlation (Luo et al., 2016; Xiang et al., 2017). An initial single-cell comparison using a relatively small number of cells demonstrated molecular similarities between cell types in early-stage organoids and those of the fetal human brain (Camp et al., 2015). A more extensive analysis profiled 80,000 cells from organoids at a more advanced stage of development, uncovering an extensive range of brain and retinal cell types, with very close correspondence to the gene expression profiles of the corresponding endogenous cell types (Quadrato et al., 2017).

This type of large-scale molecular profiling will allow for direct comparison of organoids with the endogenous human brain to define the similarities and differences of the various cell types to their in vivo counterparts, and provide essential information on the fidelity with which brain organoids resemble the endogenous tissue. In addition, it will provide the means for comparison of the cells and features produced across distinct 3D brain models, to define their advantages and disadvantages for current and future use. Although comprehensive atlases of single-cell transcriptomes from both endogenous developing brain and 3D systems will take time to generate, these studies will be important to identify cell types and developmental events that current models do not replicate, and are thus targets for future efforts to improve these model systems.

Furthermore, it is unclear how far the cells and circuits present within 3D organoids can develop in culture, and whether they can acquire structural and physiological features of the postnatal brain that can enable modeling to go beyond early embryonic events. Currently available culture techniques have constraints that are, in part, due to the challenges of maintaining tissue viability in long-term culture, but also the fact that protracted development, vs tissue survival, cannot easily be achieved in vitro (Hockemeyer & Jaenisch, 2016). It is difficult to associate particular in vitro models with an exact human developmental stage because of their relatively disorganized and unsynchronized development and the—so far—limited number of studies that have characterized the molecular and cellular features of the embryonic human brain at different ages.

Thus far, most models have been cultured for a relatively short period of time and have been used largely to study early developmental events of the embryonic brain, such as neural progenitor biology and dysfunction (Cugola et al., 2016; Dang et al., 2016; Garcez et al., 2016; Lancaster et al., 2013; Mariani et al., 2015; Otani et al., 2016; Qian et al., 2016; Tieng et al., 2014). These studies have highlighted the immense value of these systems to the investigation of endogenous developmental events. However, neuronal networks develop relatively late in embryogenesis and take substantial time to mature. Similarly, glia (both astrocytes and oligodendrocytes) are generated after the main wave of neurogenesis is complete. Myelination, in particular, is a canonical postnatal event, spanning years of postnatal life in humans (Tomassy, Dershowitz, & Arlotta, 2016). The field will need to develop robust methods for the production of more mature tissue in order to model later events such as circuit function or neuron–glia interactions.

There is no obvious, trivial solution to the problem of organoid maturation, but it is likely that longer culturing periods, incorporation of later patterning steps, and the induction of neuronal activity will foster this process. Protocol modifications that enable the development of more mature organoids are also likely to enrich the diversity of cell types produced and, possibly, favor connectivity among distinct brain regions, which in turn, much as in the embryo, could favor further fate specification and maturation. Engineered systems for perfusion of media throughout the organoid to facilitate oxygenation and nutrient distribution will also be of prime importance in achieving greater maturity, by preventing cell death at the center of older organoids and enabling size increases.

Modeling later stages of brain development will also require generating cell types of extracerebral origin. The key developmental roles known for microglia, particularly in synaptic pruning (Bilimoria & Stevens, 2015), indicate a need to incorporate this cell type into future organoid models. In the embryo, microglia originate from the yolk sac and invade the developing central nervous system as early as embryonic day 9.5 in the mouse (Ginhoux & Prinz, 2015). Because of their different embryonic origins, microglia are missing from most iPSC-derived brain models (Young-Pearse & Morrow, 2016). Protocols to generate microglia from human iPSCs have recently been developed (Muffat et al., 2016), which could potentially enable human microglia (with or without risk variants of interest) to be incorporated into organoids at early stages of differentiation. Recent studies support the feasibility of this type of approach (Abud et al., 2017; Schwartz et al., 2015).

Enhanced maturation of neurons in 3D may also favor the formation of mature neuronal traits such as dendritic spines, areas of synaptic contact that are a structural trait of more mature neurons, which have been difficult to generate in culture. Indeed, a recent study has demonstrated that extended organoid culture under optimized conditions allows development of neurons bearing dendritic spines (Quadrato et al., 2017). Thus, incorporating microglia into developing organoids might enable first-time modeling of events such as abnormal pruning of synapses and dendritic spines.

Another important issue that must be addressed is the high variability between organoids, even those derived from the same protocol and within the same batch. It is assumed that protocols that generate exogenously patterned 3D spheroids or organoids might allow for higher organoid-to-organoid reproducibility than 3D self-patterned whole-brain organoids, although there are as of yet no experimental data that compare organoids generated by each method in detail (i.e., beyond tissue histology and immunohistochemical analysis of selected marker genes). It is, however, safe to assume that reproducibility is a current limitation of all 3D models, and that increased regional and cellular diversity and longer periods of growth and development will be inversely correlated with reproducibility. Addressing this central problem will be fundamental for the downstream application of organoids to high-throughput phenotypic and therapeutic screening. Next-generation protocols that spatially and temporally control the application of patterning signals might lead to more reproducible tissue architecture and cellular composition. Future models might also incorporate fluorescent reporters, introduced into the starting iPSCs, that mark specific regions in organoids for ease of identification and selection.

As with all pluripotent stem cell-derived culture models, organoids can be produced in large quantities and are amenable to high-throughput downstream analysis. Advances in technology suggest that future avenues of research could include imaging-based screens, large-scale molecular characterization, and high-throughput electrophysiological analysis. The implementation of reporter lines in which specific cell types are fluorescently labeled will facilitate cell isolation, molecular profiling of specific classes of cells, and imaging of cells and circuits. The physiology of neurons and circuits can also be studied using techniques amenable to high-throughput pipelines, such as all-optical electrophysiology (Hochbaum et al., 2014), calcium imaging (Tsien, 1980), and high-density, multielectrode array recording (Scholvin et al., 2016). Combination of these multifaceted analyses will generate an unprecedented wealth of data from both control organoids and those derived from psychiatric patients or engineered to carry disease variants. Despite the real challenges involved in the collection, mining, and storage of "big data," it is likely that a systems-level analysis of datasets spanning multiple modes of analysis will be required to understand the neurobiology of neuropsychiatric diseases and to inform next-generation therapeutics.

4. MODELING NEUROPSYCHIATRIC DISEASE IN VITRO

Neuropsychiatric disorders, such as autism spectrum disorder (ASD), schizophrenia (SCZ), and bipolar disorder (BPD), have a devastating impact on patients' lives and represent a heavy social and economic burden for societies worldwide (Collins et al., 2011; World Health Organization, 2008). The complex symptomatology of these diseases, which affect uniquely human cognitive and behavioral traits, combined with their polygenic etiology, the lack of objective biomarkers for diagnosis, and the limitations of current research models, has made these disorders difficult to classify diagnostically and to study experimentally. Our current understanding of the structural, cellular, and circuit-level abnormalities affecting the brains of neuropsychiatric patients remains very superficial, and development of modern therapeutics has stalled (Hyman, 2012).

That neuropsychiatric disorders have a substantial genetic contribution has been known for some time. In the last several years, genomic analyses have provided a wealth of information regarding genes and pathways that are probably affected in these disorders, and have identified a large variety of both common genetic variants of small effect and rare mutations that lead to more severe effects (Gratten, Wray, Keller, & Visscher, 2014; McCarroll, Feng, & Hyman, 2014). However, many risk variants are also found in unaffected individuals (Doherty & Owen, 2014), and studies comparing loci across multiple disorders have identified a striking degree of overlap (Bulik-Sullivan et al., 2015; Cross-Disorder Group of the Psychiatric Genomics Consortium, 2013a, 2013b). These findings suggest that these disorders might have overlapping causative mechanisms, and that variation at additional risk-associated genes or exposure to sensitizing environmental factors divides this shared pathology into a spectrum of clinical presentations. As such, satisfactory model systems for these diseases must incorporate the ability to study complex genetic states, ideally including idiosyncratic genetic background.

Thus far, conventional experimental models, such as genetically modified mice and zebrafish, have been of great use to elucidate the role of individual mutant genes in brain development and dysfunction. The ability to engineer precise mutations in a controlled genetic background, the invariant and well-studied developmental trajectories of their nervous systems, and the variety of well-established experimental tools make these robust systems to examine the effect of a limited number of risk loci in great detail. However, commonly used animal models also have important limitations, owing to inherent differences in the development, architecture, and function of their brains as compared to humans. In addition, many genetic risk states identified in human patients are difficult to engineer into animal models—for example, large structural rearrangements, or polygenic states that resemble those found in the majority of patients, which would require multiple targeted mutations.

These limitations can be overcome in part with in vitro systems, which can incorporate patient-derived PSCs with idiosyncratic genetic composition and are amenable to large-scale genetic engineering and detailed molecular and functional analysis of living human tissue. An ideal in vitro model would also include enough cellular and regional diversity to enable the investigation of abnormalities that in patients involve multiple cell types, their interactions, and, ultimately, the functionality of both local and long-distance neural networks. Among existing models, brain organoids and spheroids offer great promise for generating the complex cellular environment and tissue maturation necessary to study fundamental disease processes and offer the possibility of high-throughput phenotypic analysis and therapeutic screening.

Patient-derived iPSCs can be used in in vitro systems to produce cultures of the particular cell type of interest with the patient's individual, idiopathic genetic composition, which is particularly important for modeling complex polygenic disorders for which risk loci may not yet have been identified (Wen, Christian, Song, & Ming, 2016). The possibilities for in vitro disease modeling have also greatly expanded with the coupling of PSC technology to genetic-engineering techniques, such as zinc-finger nucleases (Urnov et al., 2005), transcription activator-like effector nucleases (Miller et al., 2011), and CRISPR/Cas9 (Cong et al., 2013; Mali et al., 2013). These techniques allow for the introduction of precise, targeted mutations for analysis of the resulting cellular phenotype. Targeted gene repair of individual risk alleles allows the derivation of isogenic control cell lines, with otherwise identical genetic backgrounds, to provide the most perfectly analogous control samples or to parse polygenic states. iPSC-derived systems are also amenable to insertion of a given risk variant (or set of variants) into a panel of iPSC lines of different genetic backgrounds, which enables the investigation of how genetic composition modulates the ultimate phenotypic expression of a defined genetic perturbation.

Genetically engineered correction of the mutant allele at one or more loci in patient-derived cells may also help identify the most potent causal loci among the broader pool of risk alleles identified by gene association studies. This approach may assist in pinpointing causal genes within heterogeneous genetic backgrounds, and in cases in which the emergent phenotype may be influenced by multiple genetic variables (Wen et al., 2016). A recently reported technique has also raised the possibility of introducing multiple modifications at different loci simultaneously in the same cell (Gonzalez et al., 2014), which would be particularly useful for modeling the effects of multiple common variants in complex polygenic diseases.

5. 2D AND 3D MODELS OF NEUROPSYCHIATRIC DISEASE

Until recently, PSC-based psychiatric disease modeling has relied almost completely on 2D cellular models. These models have proven useful for measuring important parameters like gene expression, cell morphology, neuronal excitability, and synapse formation (Flaherty & Brennand, 2015; Wen et al., 2016).

Several studies have used neurons differentiated from patient-derived iPSCs to model psychiatric disease (reviewed in Quadrato et al., 2016). In one early study, neurons were differentiated from four patient iPSC lines to model schizophrenia (Brennand et al., 2011). These neurons exhibited various abnormalities in culture, including decreased neurite number and decreased expression of glutamate receptors and of the synaptic scaffolding protein PSD95. Patient-derived neurons also showed reduced connectivity, probably related to synaptic dysfunction. This same approach has also been applied to modeling abnormalities in neurons derived from patients with other neuropsychiatric diseases, such as BPD and Rett syndrome, a severe developmental disorder that includes ASD symptoms. A first example came from a study where neural progenitor cells and neurons were differentiated both from patients with Rett syndrome who presented with early brain overgrowth and from healthy control individuals with normal brain size. In addition to increased proliferation of neural progenitors, Rett syndrome neurons displayed reduced synaptogenesis and altered neuronal activity (Marchetto et al., 2010). A possible link between BPD and early neurodevelopmental defects has been suggested by studies in which neuronal cultures differentiated from a small number of patients with BPD exhibited abnormalities in the early stages of neural progenitor proliferation and neuronal fate specification as compared to unaffected controls (Chen et al., 2014; Madison et al., 2015). One study examined putative hippocampal neurons from BPD and healthy control cell lines and reported neuronal hyperexcitability and mitochondrial abnormalities (Mertens et al., 2015). A notable finding from this work was that the excitability of the patient-derived neurons decreased to control levels after treatment with lithium. This was true, however, only for neurons from patients who were responsive to lithium treatment. Neurons derived from patients who were unresponsive to lithium remained hyperexcitable after lithium treatment in vitro, which suggests a possible application of this model for the prediction of drug response and the development of therapeutics. However, for practical and technical reasons, studies examining neurons differentiated from patient-derived iPSCs have to date employed a limited set of patient-derived lines. Given the variability of human genetic background, the complex effects of the genetic context on both cellular phenotypes and disease manifestation, and the intrinsic variability of the directed differentiation protocols, it will be important to clarify the robustness of the phenotypes observed thus far by replicating these findings in larger cohorts of patient-derived lines and in other model systems.

In vitro models of single, highly penetrant mutations are perhaps more experimentally tractable. iPSCs can be derived from donor patients who carry a shared, known mutation, and strategies applied to enable the comparison of mutant and wild-type alleles within the same parental iPSC line. Prominent examples include a study that found synaptic pathology in iPSC-derived cortical forebrain neurons derived from patients with schizophrenia carrying a shared mutation in the disrupted in schizophrenia 1 (DISC1) gene (Wen et al., 2014); in this study, isogenic control lines were generated by correcting the mutant locus via genome editing. Other studies have taken advantage of X-inactivation in iPSCs to study iPSC-derived neurons from patients with Rett syndrome bearing heterozygous mutations in the X-linked genes methyl-CpG-binding protein 2 (MECP2) or cyclin-dependent kinase-like 5 (CDKL5) (associated with classical and Hanefeld variant Rett syndrome, respectively) (Ananiev, Williams, Li, & Chang, 2011; Cheung et al., 2011; Ricciardi et al., 2012); here, clones of the parental lines were screened to identify those expressing only the mutant or wild-type alleles. The use of such strategies to generate isogenic control cells is of critical importance for parsing the effect of genetic background from that of specific mutations on disease phenotypes.

Finally, iPSC-derived neurons have also been used to model the effects of large CNVs. Neurons derived from patients with a chromosomal deletion in 15q11.2, a region associated with multiple neuropsychiatric disorders, exhibited defects in cell polarity and adherens junctions, which perhaps suggests abnormal neuronal maturation and connectivity (Yoon et al., 2014). Putative cortical neurons derived from Phelan–McDermid Syndrome (PMDS)-associated iPSCs have also exhibited deficits in excitatory synaptic transmission owing to reduced expression of SH3 and multiple ankyrin repeat domains 3 (SHANK3), a gene located within the PMDS-causative 22q13.3 deletion (Shcheglovitov et al., 2013).

Molecular comparative analysis of hiPSC-derived neurons from patients with CNVs at 7q11.23 (resulting in Williams–Beuren syndrome and the autism spectrum disorder 7dupASD) found that 7q11.23 dosage imbalance disrupts disease-relevant programs of gene expression largely connected to the dysregulation of the transcription factor GTF2I (Adamo et al., 2015), which demonstrates the value of these models for identification of the specific contribution of individual genes within large CNVs.

These types of studies provide evidence that 2D culture systems can be used successfully to model some aspects of the complex pathology of these illnesses, most notably, the disruption of synaptic transmission and abnormal neuronal differentiation. However, these models have limited cellular complexity and are not intended to recapitulate the endogenous cellular diversity of and complex cell–cell interactions in the human brain.

Currently, 3D systems are beginning to be applied to the study of neuropsychiatric disease. In a first example, 3D forebrain (telencephalic) organoids were generated from four patients with idiopathic ASD and their unaffected first-degree relatives (Mariani et al., 2015). Patient-derived organoids exhibited dysregulation of forkhead box G1 (FOXG1) expression and increased production of GABAergic neurons. This work highlights the potential opportunity that organoid systems offer for identifying the cellular substrates of neuropsychiatric illnesses among the many cell types cogenerated within organoids.

As with 2D systems, organoids from hPSC lines bearing engineered mutations have been used to examine the effect of mutating single genes, including transcriptomic differences in CHD8 mutations (linked to autism) (Wang et al., 2017), defects in neurogenesis and neuronal differentiation in MECP2 mutations (linked to Rett syndrome) (Mellios et al., 2017). Several disorders with more complex genetic contributions have been investigated using patient-derived lines. Disorders relating to progenitor biology have been particularly amenable to modeling with current organoid protocols; recent studies have used patient-derived organoids to examine the effects of lissencephaly disorders (characterized by a reduction in brain folding) on proliferation of outer radial glia (Bershteyn et al., 2017) and neural migration (Bamba, Kanemura, Okano, & Yamasaki, 2017; Jin et al., 2017), of microcephaly disorders on neural progenitor differentiation (Gabriel & Gopalakrishnan, 2017), and of Timothy syndrome on interneuron migratory dynamics (Birey et al., 2017). However, patient-derived organoids and spheroids have also been used to examine early correlates of neurodegenerative disorders with onset in later life, including familial Alzheimer's disease (Lee et al., 2016; Raja et al., 2016) and familial Parkinson's disease (Son et al., 2017).

The approaches for improving 3D organoid models discussed earlier will also be of great importance to disease modeling. Neuropsychiatric disorders affect complex higher-order brain functions that involve multiple cell types across different brain regions, such as the prefrontal cortex, the thalamic reticular nucleus, the thalamus, and the basal ganglia, among others. Disease phenotypes have been reported at scales ranging from gross architectural changes and abnormal activity patterns that may reflect altered circuit function (Ecker, Bookheimer, & Murphy, 2015; Mouchlianitis, McCutcheon, & Howes, 2016), to abnormal functioning of individual ion channels (Heyes et al., 2015). In addition, many neuropsychiatric diseases manifest defects in later developmental and postnatal processes, such as altered circuit functionality, defective glia–neuron interaction, and abnormalities of myelin (Falk et al., 2016). Studies in patients and in rodent models have indicated that synaptic dysfunction and abnormal synaptic pruning are associated with psychiatric illness, especially schizophrenia (Hong, Dissing-Olesen, & Stevens, 2016).

Studying these processes in organoids will require methods to generate more complex brain architecture, improve maturation, and generate and incorporate the corresponding glial and extracerebral cells.

High-throughput single-cell molecular profiling will also be of great utility to disease studies. Profiling cultures from patient-derived or genetically engineered mutant iPSC lines would enable simultaneous analysis of the effect of genetic variants across multiple cell types; a particular strength of 3D brain organoid models, especially those that are largely self-patterned, as they have the greatest potential to produce diverse classes of cells. Conversely, a comprehensive analysis of the expression patterns of risk-associated genes across the spectrum of organoid brain cell types in healthy control organoids might help to identify the cellular substrates of individual disorders, as cell types that express multiple risk genes for a given disease may be more likely to be affected in that disorder.

6. CONCLUSIONS

It is not surprising that much excitement is currently being generated around the development of next-generation in vitro models of the human brain with the promise of exploring both normal development and the complex neurobiology of neuropsychiatric disorders in the dish. Although much work remains to be done to understand the value of 3D cellular systems, it is likely that the over the next few years, we will witness an increase in the use of 3D brain organoid and spheroid models. Protocol improvements that enable greater maturation and diversity of cell types will enable the analysis of previously experimentally inaccessible processes, such as pruning of dendritic spines and myelination. In particular, these models will be of great use to analyze the effects of specific risk alleles and human genetic backgrounds on the manifestations of neuropsychiatric and neurodegenerative disease.

REFERENCES

Abud, E. M., Ramirez, R. N., Martinez, E. S., Healy, L. M., Nguyen, C. H. H., Newman, S. A., et al. (2017). iPSC-derived human microglia-like cells to study neurological diseases. *Neuron, 94*, 278–293.e9.

Adamo, A., Atashpaz, S., Germain, P. L., Zanella, M., D'Agostino, G., Albertin, V., et al. (2015). 7q11.23 Dosage-dependent dysregulation in human pluripotent stem cells affects transcriptional programs in disease-relevant lineages. *Nature Genetics, 47*, 132–141.

Ananiev, G., Williams, E. C., Li, H., & Chang, Q. (2011). Isogenic pairs of wild type and mutant induced pluripotent stem cell (iPSC) lines from Rett syndrome patients as in vitro disease model. *PLoS One, 6*, e25255.

Bagley, J. A., Reumann, D., Bian, S., Levi-Strauss, J., & Knoblich, J. A. (2017). Fused cerebral organoids model interactions between brain regions. *Nature Methods*, *14*, 743–751.
Bamba, Y., Kanemura, Y., Okano, H., & Yamasaki, M. (2017). Visualization of migration of human cortical neurons generated from induced pluripotent stem cells. *Journal of Neuroscience Methods*, *289*, 57–63.
Bershteyn, M., Nowakowski, T. J., Pollen, A. A., Di Lullo, E., Nene, A., Wynshaw-Boris, A., et al. (2017). Human iPSC-derived cerebral organoids model cellular features of lissencephaly and reveal prolonged mitosis of outer radial glia. *Cell Stem Cell*, *20*, 435–449.e4.
Bilimoria, P. M., & Stevens, B. (2015). Microglia function during brain development: New insights from animal models. *Brain Research*, *1617*, 7–17.
Birey, F., Andersen, J., Makinson, C. D., Islam, S., Wei, W., Huber, N., et al. (2017). Assembly of functionally integrated human forebrain spheroids. *Nature*, *545*, 54–59.
Brennand, K. J., Simone, A., Jou, J., Gelboin-Burkhart, C., Tran, N., Sangar, S., et al. (2011). Modelling schizophrenia using human induced pluripotent stem cells. *Nature*, *473*, 221–225.
Bulik-Sullivan, B., Finucane, H. K., Anttila, V., Gusev, A., Day, F. R., Loh, P. R., et al. (2015). An atlas of genetic correlations across human diseases and traits. *Nature Genetics*, *47*, 1236–1241.
Camp, J. G., Badsha, F., Florio, M., Kanton, S., Gerber, T., Wilsch-Brauninger, M., et al. (2015). Human cerebral organoids recapitulate gene expression programs of fetal neocortex development. *Proceedings of the National Academy of Sciences of the United States of America*, *112*, 15672–15677.
Chambers, S. M., Fasano, C. A., Papapetrou, E. P., Tomishima, M., Sadelain, M., & Studer, L. (2009). Highly efficient neural conversion of human ES and iPS cells by dual inhibition of SMAD signaling. *Nature Biotechnology*, *27*, 275–280.
Chen, H. M., DeLong, C. J., Bame, M., Rajapakse, I., Herron, T. J., McInnis, M. G., et al. (2014). Transcripts involved in calcium signaling and telencephalic neuronal fate are altered in induced pluripotent stem cells from bipolar disorder patients. *Translational Psychiatry*, *4*, e375.
Cheung, A. Y., Horvath, L. M., Grafodatskaya, D., Pasceri, P., Weksberg, R., Hotta, A., et al. (2011). Isolation of MECP2-null Rett syndrome patient hiPS cells and isogenic controls through X-chromosome inactivation. *Human Molecular Genetics*, *20*, 2103–2115.
Collins, P. Y., Patel, V., Joestl, S. S., March, D., Insel, T. R., Daar, A. S., et al. (2011). Grand challenges in global mental health. *Nature*, *475*, 27–30.
Cong, L., Ran, F. A., Cox, D., Lin, S., Barretto, R., Habib, N., et al. (2013). Multiplex genome engineering using CRISPR/Cas systems. *Science*, *339*, 819–823.
Cross-Disorder Group of the Psychiatric Genomics Consortium. (2013a). Genetic relationship between five psychiatric disorders estimated from genome-wide SNPs. *Nature Genetics*, *45*, 984–994.
Cross-Disorder Group of the Psychiatric Genomics Consortium. (2013b). Identification of risk loci with shared effects on five major psychiatric disorders: A genome-wide analysis. *Lancet*, *381*, 1371–1379.
Cugola, F. R., Fernandes, I. R., Russo, F. B., Freitas, B. C., Dias, J. L., Guimaraes, K. P., et al. (2016). The Brazilian Zika virus strain causes birth defects in experimental models. *Nature*, *534*, 267–271.
Dang, J., Tiwari, S. K., Lichinchi, G., Qin, Y., Patil, V. S., Eroshkin, A. M., et al. (2016). Zika virus depletes neural progenitors in human cerebral organoids through activation of the innate immune receptor TLR3. *Cell Stem Cell*, *19*, 258–265.
Dezonne, R. S., Sartore, R. C., Nascimento, J. M., Saia-Cereda, V. M., Romao, L. F., Alves-Leon, S. V., et al. (2017). Derivation of functional human astrocytes from cerebral organoids. *Scientific Reports*, *7*, 45091.

Di Lullo, E., & Kriegstein, A. R. (2017). The use of brain organoids to investigate neural development and disease. *Nature Reviews. Neuroscience, 18*, 573–584.
Doherty, J. L., & Owen, M. J. (2014). Genomic insights into the overlap between psychiatric disorders: Implications for research and clinical practice. *Genome Medicine, 6*, 29.
Ecker, C., Bookheimer, S. Y., & Murphy, D. G. (2015). Neuroimaging in autism spectrum disorder: Brain structure and function across the lifespan. *Lancet Neurology, 14*, 1121–1134.
Eiraku, M., Takata, N., Ishibashi, H., Kawada, M., Sakakura, E., Okuda, S., et al. (2011). Self-organizing optic-cup morphogenesis in three-dimensional culture. *Nature, 472*, 51–56.
Eiraku, M., Watanabe, K., Matsuo-Takasaki, M., Kawada, M., Yonemura, S., Matsumura, M., et al. (2008). Self-organized formation of polarized cortical tissues from ESCs and its active manipulation by extrinsic signals. *Cell Stem Cell, 3*, 519–532.
Elkabetz, Y., Panagiotakos, G., Al Shamy, G., Socci, N. D., Tabar, V., & Studer, L. (2008). Human ES cell-derived neural rosettes reveal a functionally distinct early neural stem cell stage. *Genes & Development, 22*, 152–165.
Falk, A., Heine, V. M., Harwood, A. J., Sullivan, P. F., Peitz, M., Brustle, O., et al. (2016). Modeling psychiatric disorders: From genomic findings to cellular phenotypes. *Molecular Psychiatry, 21*, 1167–1179.
Flaherty, E. K., & Brennand, K. J. (2015). Using hiPSCs to model neuropsychiatric copy number variations (CNVs) has potential to reveal underlying disease mechanisms. *Brain Research, 1655*, 283–293.
Gabriel, E., & Gopalakrishnan, J. (2017). Generation of iPSC-derived human brain organoids to model early neurodevelopmental disorders. *Journal of Visualized Experiments, 122*. https://doi.org/10.3791/55372.
Garcez, P. P., Loiola, E. C., Madeiro da Costa, R., Higa, L. M., Trindade, P., Delvecchio, R., et al. (2016). Zika virus impairs growth in human neurospheres and brain organoids. *Science, 352*, 816–818.
Ginhoux, F., & Prinz, M. (2015). Origin of microglia: Current concepts and past controversies. *Cold Spring Harbor Perspectives in Biology, 7*, a020537.
Gonzalez, F., Zhu, Z., Shi, Z. D., Lelli, K., Verma, N., Li, Q. V., et al. (2014). An iCRISPR platform for rapid, multiplexable, and inducible genome editing in human pluripotent stem cells. *Cell Stem Cell, 15*, 215–226.
Gratten, J., Wray, N. R., Keller, M. C., & Visscher, P. M. (2014). Large-scale genomics unveils the genetic architecture of psychiatric disorders. *Nature Neuroscience, 17*, 782–790.
Heyes, S., Pratt, W. S., Rees, E., Dahimene, S., Ferron, L., Owen, M. J., et al. (2015). Genetic disruption of voltage-gated calcium channels in psychiatric and neurological disorders. *Progress in Neurobiology, 134*, 36–54.
Hochbaum, D. R., Zhao, Y., Farhi, S. L., Klapoetke, N., Werley, C. A., Kapoor, V., et al. (2014). All-optical electrophysiology in mammalian neurons using engineered microbial rhodopsins. *Nature Methods, 11*, 825–833.
Hockemeyer, D., & Jaenisch, R. (2016). Induced pluripotent stem cells meet genome editing. *Cell Stem Cell, 18*, 573–586.
Hong, S., Dissing-Olesen, L., & Stevens, B. (2016). New insights on the role of microglia in synaptic pruning in health and disease. *Current Opinion in Neurobiology, 36*, 128–134.
Hyman, S. E. (2012). Revolution stalled. *Science Translational Medicine, 4*, 155cm11.
Jin, M., Pomp, O., Shinoda, T., Toba, S., Torisawa, T., Furuta, K., et al. (2017). Katanin p80, NuMA and cytoplasmic dynein cooperate to control microtubule dynamics. *Scientific Reports, 7*, 39902.
Jo, J., Xiao, Y., Sun, A. X., Cukuroglu, E., Tran, H. D., Goke, J., et al. (2016). Midbrain-like organoids from human pluripotent stem cells contain functional dopaminergic and neuromelanin-producing neurons. *Cell Stem Cell, 19*, 248–257.

Kadoshima, T., Sakaguchi, H., Nakano, T., Soen, M., Ando, S., Eiraku, M., et al. (2013). Self-organization of axial polarity, inside-out layer pattern, and species-specific progenitor dynamics in human ES cell-derived neocortex. *Proceedings of the National Academy of Sciences of the United States of America*, *110*, 20284–20289.

Kelava, I., & Lancaster, M. A. (2016). Stem cell models of human brain development. *Cell Stem Cell*, *18*, 736–748.

Klein, A. M., Mazutis, L., Akartuna, I., Tallapragada, N., Veres, A., Li, V., et al. (2015). Droplet barcoding for single-cell transcriptomics applied to embryonic stem cells. *Cell*, *161*, 1187–1201.

Lancaster, M. A., Corsini, N. S., Wolfinger, S., Gustafson, E. H., Phillips, A. W., Burkard, T. R., et al. (2017). Guided self-organization and cortical plate formation in human brain organoids. *Nature Biotechnology*, *35*, 659–666.

Lancaster, M. A., Renner, M., Martin, C. A., Wenzel, D., Bicknell, L. S., Hurles, M. E., et al. (2013). Cerebral organoids model human brain development and microcephaly. *Nature*, *501*, 373–379.

Lee, C. T., Chen, J., Kindberg, A. A., Bendriem, R. M., Spivak, C. E., Williams, M. P., et al. (2017). CYP3A5 mediates effects of cocaine on human neocorticogenesis: Studies using an in vitro 3D self-organized hPSC model with a single cortex-like unit. *Neuropsychopharmacology*, *42*, 774–784.

Lee, J. H., Daugharthy, E. R., Scheiman, J., Kalhor, R., Yang, J. L., Ferrante, T. C., et al. (2014). Highly multiplexed subcellular RNA sequencing in situ. *Science*, *343*, 1360–1363.

Lee, H. K., Velazquez Sanchez, C., Chen, M., Morin, P. J., Wells, J. M., Hanlon, E. B., et al. (2016). Three dimensional human neuro-spheroid model of Alzheimer's disease based on differentiated induced pluripotent stem cells. *PLoS One*, *11*, e0163072.

Li, Y., Muffat, J., Omer, A., Bosch, I., Lancaster, M. A., Sur, M., et al. (2017). Induction of expansion and folding in human cerebral organoids. *Cell Stem Cell*, *20*, 385–396.e3.

Luo, C., Lancaster, M. A., Castanon, R., Nery, J. R., Knoblich, J. A., & Ecker, J. R. (2016). Cerebral organoids recapitulate epigenomic signatures of the human fetal brain. *Cell Reports*, *17*, 3369–3384.

Macosko, E. Z., Basu, A., Satija, R., Nemesh, J., Shekhar, K., Goldman, M., et al. (2015). Highly parallel genome-wide expression profiling of individual cells using nanoliter droplets. *Cell*, *161*, 1202–1214.

Madison, J. M., Zhou, F., Nigam, A., Hussain, A., Barker, D. D., Nehme, R., et al. (2015). Characterization of bipolar disorder patient-specific induced pluripotent stem cells from a family reveals neurodevelopmental and mRNA expression abnormalities. *Molecular Psychiatry*, *20*, 703–717.

Mali, P., Yang, L., Esvelt, K. M., Aach, J., Guell, M., DiCarlo, J. E., et al. (2013). RNA-guided human genome engineering via Cas9. *Science*, *339*, 823–826.

Marchetto, M. C., Carromeu, C., Acab, A., Yu, D., Yeo, G. W., Mu, Y., et al. (2010). A model for neural development and treatment of Rett syndrome using human induced pluripotent stem cells. *Cell*, *143*, 527–539.

Mariani, J., Coppola, G., Zhang, P., Abyzov, A., Provini, L., Tomasini, L., et al. (2015). FOXG1-dependent dysregulation of GABA/glutamate neuron differentiation in autism spectrum disorders. *Cell*, *162*, 375–390.

Matsui, T., Nieto-Estevez, V., Kyrychenko, S., Schneider, J. W., & Hsieh, J. (2017). Retinoblastoma protein controls growth, survival and neuronal migration in human cerebral organoids. *Development*, *144*, 1025–1034.

McCarroll, S. A., Feng, G., & Hyman, S. E. (2014). Genome-scale neurogenetics: Methodology and meaning. *Nature Neuroscience*, *17*, 756–763.

Mellios, N., Feldman, D. A., Sheridan, S. D., Ip, J. P. K., Kwok, S., Amoah, S. K., et al. (2017). MeCP2-regulated miRNAs control early human neurogenesis through differential effects on ERK and AKT signaling. *Molecular Psychiatry*, *23*, 1051–1065.

Mertens, J., Wang, Q. W., Kim, Y., Yu, D. X., Pham, S., Yang, B., et al. (2015). Differential responses to lithium in hyperexcitable neurons from patients with bipolar disorder. *Nature, 527*, 95–99.

Miller, J. C., Tan, S., Qiao, G., Barlow, K. A., Wang, J., Xia, D. F., et al. (2011). A TALE nuclease architecture for efficient genome editing. *Nature Biotechnology, 29*, 143–148.

Monzel, A. S., Smits, L. M., Hemmer, K., Hachi, S., Moreno, E. L., van Wuellen, T., et al. (2017). Derivation of human midbrain-specific organoids from neuroepithelial stem cells. *Stem Cell Reports, 8*, 1144–1154.

Mora-Bermudez, F., Badsha, F., Kanton, S., Camp, J. G., Vernot, B., Kohler, K., et al. (2016). Differences and similarities between human and chimpanzee neural progenitors during cerebral cortex development. *eLife, 5*. https://doi.org/10.7554/eLife.18683.

Mouchlianitis, E., McCutcheon, R., & Howes, O. D. (2016). Brain-imaging studies of treatment-resistant schizophrenia: A systematic review. *Lancet Psychiatry, 3*, 451–463.

Muffat, J., Li, Y., Yuan, B., Mitalipova, M., Omer, A., Corcoran, S., et al. (2016). Efficient derivation of microglia-like cells from human pluripotent stem cells. *Nature Medicine, 22*, 1358–1367.

Muguruma, K., Nishiyama, A., Kawakami, H., Hashimoto, K., & Sasai, Y. (2015). Self-organization of polarized cerebellar tissue in 3D culture of human pluripotent stem cells. *Cell Reports, 10*, 537–550.

Otani, T., Marchetto, M. C., Gage, F. H., Simons, B. D., & Livesey, F. J. (2016). 2D and 3D stem cell models of primate cortical development identify species-specific differences in progenitor behavior contributing to brain size. *Cell Stem Cell, 18*, 467–480.

Paşca, A. M., Sloan, S. A., Clarke, L. E., Tian, Y., Makinson, C. D., Huber, N., et al. (2015). Functional cortical neurons and astrocytes from human pluripotent stem cells in 3D culture. *Nature Methods, 12*, 671–678.

Qian, X., Nguyen, H. N., Song, M. M., Hadiono, C., Ogden, S. C., Hammack, C., et al. (2016). Brain-region-specific organoids using mini-bioreactors for modeling ZIKV exposure. *Cell, 165*, 1238–1254.

Quadrato, G., Brown, J., & Arlotta, P. (2016). The promises and challenges of human brain organoids as models of neuropsychiatric disease. *Nature Medicine, 22*, 1220–1228.

Quadrato, G., Nguyen, T., Macosko, E. Z., Sherwood, J. L., Min Yang, S., Berger, D. R., et al. (2017). Cell diversity and network dynamics in photosensitive human brain organoids. *Nature, 545*, 48–53.

Raja, W. K., Mungenast, A. E., Lin, Y. T., Ko, T., Abdurrob, F., Seo, J., et al. (2016). Self-organizing 3D human neural tissue derived from induced pluripotent stem cells recapitulate Alzheimer's disease phenotypes. *PLoS One, 11*, e0161969.

Ricciardi, S., Ungaro, F., Hambrock, M., Rademacher, N., Stefanelli, G., Brambilla, D., et al. (2012). CDKL5 ensures excitatory synapse stability by reinforcing NGL-1-PSD95 interaction in the postsynaptic compartment and is impaired in patient iPSC-derived neurons. *Nature Cell Biology, 14*, 911–923.

Rigamonti, A., Repetti, G. G., Sun, C., Price, F. D., Reny, D. C., Rapino, F., et al. (2016). Large-scale production of mature neurons from human pluripotent stem cells in a three-dimensional suspension culture system. *Stem Cell Reports, 6*, 993–1008.

Sakaguchi, H., Kadoshima, T., Soen, M., Narii, N., Ishida, Y., Ohgushi, M., et al. (2015). Generation of functional hippocampal neurons from self-organizing human embryonic stem cell-derived dorsomedial telencephalic tissue. *Nature Communications, 6*, 8896.

Scholvin, J., Kinney, J. P., Bernstein, J. G., Moore-Kochlacs, C., Kopell, N., Fonstad, C. G., et al. (2016). Close-packed silicon microelectrodes for scalable spatially oversampled neural recording. *IEEE Transactions on Biomedical Engineering, 63*, 120–130.

Schwartz, M. P., Hou, Z., Propson, N. E., Zhang, J., Engstrom, C. J., Santos Costa, V., et al. (2015). Human pluripotent stem cell-derived neural constructs for predicting neural toxicity. *Proceedings of the National Academy of Sciences of the United States of America, 112*, 12516–12521.

Shcheglovitov, A., Shcheglovitova, O., Yazawa, M., Portmann, T., Shu, R., Sebastiano, V., et al. (2013). SHANK3 and IGF1 restore synaptic deficits in neurons from 22q13 deletion syndrome patients. *Nature, 503*, 267–271.

Shi, Y., Kirwan, P., Smith, J., Robinson, H. P., & Livesey, F. J. (2012). Human cerebral cortex development from pluripotent stem cells to functional excitatory synapses. *Nature Neuroscience, 15*, 477–486. s1.

Sloan, S. A., Darmanis, S., Huber, N., Khan, T. A., Birey, F., Caneda, C., et al. (2017). Human astrocyte maturation captured in 3D cerebral cortical spheroids derived from pluripotent stem cells. *Neuron, 95*, 779–790.e6.

Son, M. Y., Sim, H., Son, Y. S., Jung, K. B., Lee, M. O., Oh, J. H., et al. (2017). Distinctive genomic signature of neural and intestinal organoids from familial Parkinson's disease patient-derived induced pluripotent stem cells. *Neuropathology and Applied Neurobiology, 43*, 584–603.

Tieng, V., Stoppini, L., Villy, S., Fathi, M., Dubois-Dauphin, M., & Krause, K. H. (2014). Engineering of midbrain organoids containing long-lived dopaminergic neurons. *Stem Cells and Development, 23*, 1535–1547.

Tomassy, G. S., Dershowitz, L. B., & Arlotta, P. (2016). Diversity matters: A revised guide to myelination. *Trends in Cell Biology, 26*, 135–147.

Tsien, R. Y. (1980). New calcium indicators and buffers with high selectivity against magnesium and protons: Design, synthesis, and properties of prototype structures. *Biochemistry, 19*, 2396–2404.

Urnov, F. D., Miller, J. C., Lee, Y. L., Beausejour, C. M., Rock, J. M., Augustus, S., et al. (2005). Highly efficient endogenous human gene correction using designed zinc-finger nucleases. *Nature, 435*, 646–651.

Wang, L., Hou, S., & Han, Y. G. (2016). Hedgehog signaling promotes basal progenitor expansion and the growth and folding of the neocortex. *Nature Neuroscience, 19*, 888–896.

Wang, P., Mokhtari, R., Pedrosa, E., Kirschenbaum, M., Bayrak, C., Zheng, D., et al. (2017). CRISPR/Cas9-mediated heterozygous knockout of the autism gene CHD8 and characterization of its transcriptional networks in cerebral organoids derived from iPS cells. *Molecular Autism, 8*, 11.

Watanabe, K., Kamiya, D., Nishiyama, A., Katayama, T., Nozaki, S., Kawasaki, H., et al. (2005). Directed differentiation of telencephalic precursors from embryonic stem cells. *Nature Neuroscience, 8*, 288–296.

Wen, Z., Christian, K. M., Song, H., & Ming, G. L. (2016). Modeling psychiatric disorders with patient-derived iPSCs. *Current Opinion in Neurobiology, 36*, 118–127.

Wen, Z., Nguyen, H. N., Guo, Z., Lalli, M. A., Wang, X., Su, Y., et al. (2014). Synaptic dysregulation in a human iPS cell model of mental disorders. *Nature, 515*, 414–418.

World Health Organization. (2008). *The global burden of disease: 2004 Update*. WHO Press.

Xiang, Y., Tanaka, Y., Patterson, B., Kang, Y. J., Govindaiah, G., Roselaar, N., et al. (2017). Fusion of regionally specified hPSC-derived organoids models human brain development and interneuron migration. *Cell Stem Cell, 21*, 383–398.e7.

Yin, X., Mead, B. E., Safaee, H., Langer, R., Karp, J. M., & Levy, O. (2016). Engineering stem cell organoids. *Cell Stem Cell, 18*, 25–38.

Ying, Q. L., Stavridis, M., Griffiths, D., Li, M., & Smith, A. (2003). Conversion of embryonic stem cells into neuroectodermal precursors in adherent monoculture. *Nature Biotechnology, 21*, 183–186.

Yoon, K. J., Nguyen, H. N., Ursini, G., Zhang, F., Kim, N. S., Wen, Z., et al. (2014). Modeling a genetic risk for schizophrenia in iPSCs and mice reveals neural stem cell deficits associated with adherens junctions and polarity. *Cell Stem Cell*, *15*, 79–91.

Young-Pearse, T. L., & Morrow, E. M. (2016). Modeling developmental neuropsychiatric disorders with iPSC technology: Challenges and opportunities. *Current Opinion in Neurobiology*, *36*, 66–73.

Zhang, S. C., Wernig, M., Duncan, I. D., Brustle, O., & Thomson, J. A. (2001). In vitro differentiation of transplantable neural precursors from human embryonic stem cells. *Nature Biotechnology*, *19*, 1129–1133.

CHAPTER FIVE

The Long Road to Making Muscle In Vitro

Olivier Pourquié*[,†,‡,1], Ziad Al Tanoury*[,†,‡], Jérome Chal*[,†,‡]
*Brigham and Women's Hospital, Boston, MA, United States
†Harvard Medical School, Boston, MA, United States
‡Harvard Stem Cell Institute, Boston, MA, United States
[1]Corresponding author: e-mail address: pourquie@genetics.med.harvard.edu

Contents

Abstract

The skeletal muscle lineage derives from the embryonic paraxial mesoderm (PM) which also gives rise to the axial skeleton, the dermis of the back, brown fat, meninges, and endothelial cells. Direct conversion was pioneered in skeletal muscle with overexpression of the transcription factor MyoD which can convert fibroblasts to a muscle fate. In contrast, directed differentiation of skeletal muscle from pluripotent cells (PC) in vitro has proven to be very difficult compared to other lineages and has only been achieved recently. Experimental strategies recapitulating myogenesis in vitro from mouse and human PC (ES/iPS) have now been reported and all rely on early activation of Wnt signaling at the epiblast stage. This leads to induction of neuromesodermal progenitors that can subsequently be induced to a PM fate and to skeletal muscle. These protocols can efficiently produce fetal muscle fibers and immature satellite cells. These new in vitro systems now open the possibility to better understand human myogenesis and to develop in vitro disease models as well as cell therapy approaches.

Current Topics in Developmental Biology, Volume 129
ISSN 0070-2153
https://doi.org/10.1016/bs.ctdb.2018.03.003

The production of skeletal muscle in vitro from mouse or human ES or iPS cells (collectively referred to as pluripotent cells (PC) in the text) has proven to be much more difficult than for other important lineages such as neurons or cardiomyocytes. Until recently, the only efficient protocols allowing differentiation of reasonably mature muscle cells relied on overexpression of transcription factors such as MyoD or Pax3/7 followed by cell sorting of the induced progenitors (Magli & Perlingeiro, 2017). Recently, several protocols have reported efficient directed differentiation of myogenic cells by recapitulating signaling cues from embryonic myogenesis. These approaches have been extensively reviewed recently (Chal & Pourquie, 2017; Magli & Perlingeiro, 2017). In this review, we will only discuss the most recent progress and open questions in this field, focusing exclusively on directed myogenic differentiation in monolayer cultures.

1. OVERVIEW OF MUSCLE DEVELOPMENT

The paraxial mesoderm (PM) forms during gastrulation and first constitutes the bilateral strips of presomitic mesoderm (PSM) which flank the forming neural tube at the posterior end of the embryo (Chal & Pourquie, 2009). Cells in the posterior PSM express a set of specific genes such as *Mesogenin1* (*Msgn1*) and they experience oscillations of the segmentation clock which drives the rhythmic production of the embryonic segments called somites (Hubaud & Pourquie, 2014). The segmental pattern is established as a stripe of gene expression which prefigures the future segment as cells enter the anterior third of the PSM. Concomitant with segment determination, PSM cells activate the transcription factor *Pax3* and begin to adopt an epithelial organization. Somites form as epithelial spheres at the tip of the anterior PSM and they define the characteristic periodic pattern of the vertebral column and its associated muscles. The somites soon become subdivided into a ventral compartment (the sclerotome) that contributes to the axial skeleton and a dorsal epithelial compartment called dermomyotome which maintains expression of Pax3 and contains the progenitors of the skeletal muscles, dermis, and brown fat (Chal & Pourquie, 2009).

Initiation of myogenesis is first observed in cells of the dermomyotome when the transcription factor Myf5 becomes expressed (Chal & Pourquie, 2017). Myogenic cells delaminate under the dermomyotome and differentiate into muscle fibers forming the first embryonic muscles called myotomes.

Myotomes eventually contribute to the paraxial muscles controlling the movements of the spine. At the same time, myogenic progenitors delaminate from the lateral edge of the dermomyotome and migrate to the lateral plate to form the rudiments of the limb and girdle muscles. Muscles in the head derive from the anterior part of the PM, which does not form somites but migrates to establish the head muscles primordia. Myogenesis in the trunk and in the head relies on different genetic programs involving a distinct set of transcription factors (Tzahor, 2009).

The first phase of myogenesis is called embryonic or primary myogenesis, and it generates a first set of muscle fibers that serve as a scaffold for the secondary or fetal myogenesis (Biressi, Molinaro, & Cossu, 2007). This next phase is initiated by a population of myogenic progenitors expressing Pax7, which originate from the Pax3+ dermomyotomal progenitors and are first found intermingled with primary muscle fibers (Gros, Manceau, Thome, & Marcelle, 2005; Hutcheson, Zhao, Merrell, Haldar, & Kardon, 2009; Kassar-Duchossoy et al., 2005; Relaix, Rocancourt, Mansouri, & Buckingham, 2005). These progenitors proliferate and generate new fibers and also fuse to existing fibers. A subset of these Pax7+ cells is set aside to form the satellite cells. These cells lie quiescent below the basal lamina of muscle fibers, and they represent the muscle stem cells able to regenerate damaged fibers in the adult (Brack & Rando, 2012; Dumont, Wang, & Rudnicki, 2015). This fetal phase results in the establishment of the definitive pattern of individual muscles which arise by splitting initial muscle masses into definitive muscle groups. Subsequent phases of myogenesis involve growth, first by addition of new nuclei by fusion of progenitors, and then by hypertrophy of the myofibers (White, Bierinx, Gnocchi, & Zammit, 2010). In parallel, muscles fibers become innervated by motoneurons and they mature into fully functional contractile fibers. Maturation of the muscle fiber involves a switch from embryonic to adult myosin heavy chain types and the adoption of either a slow or fast fiber phenotype. This choice also conditions the future metabolism of the fiber (glycolytic vs oxidative). Formation of the muscle fiber also implies the assembly of actin and myosin into sarcomeres, the basic contractile unit of muscle, followed by the organization of sarcomeres into myofibrils (Lemke & Schnorrer, 2017). In parallel, the membrane (sarcolemma) of the muscle fiber forms elaborate postsynaptic elements (the neuromuscular junction) and a complex network of tubules leading to the formation of triads which are important for calcium handling during excitation contraction.

2. Sox2+/T+ NEUROMESODERMAL PROGENITORS (NMPs) REPRESENT THE FIRST STEP IN MYOGENIC DIFFERENTIATION IN VITRO

In amniotes, the skeletal muscles derive from the PM which originates from cells of the anterior region of the epithelial epiblast (Pourquie, 2004). These cells undergo an epithelium-to-mesenchyme transition (EMT) to ingress into the anterior primitive streak during gastrulation and they ultimately produce the muscles of the neck, limbs, and trunk. PM progenitors remain as a self-renewing population in the tail bud to contribute to more posterior regions of the body (Wilson, Olivera-Martinez, & Storey, 2009). Recently, a population of NMPs able to contribute to both the PM and to the spinal cord has been identified in mouse by retrospective clonal analysis (Tzouanacou, Wegener, Wymeersch, Wilson, & Nicolas, 2009). In vivo, these cells can contribute to clones spanning large domains along the antero-posterior (AP) axis. While there is still no good direct lineage tracing of these NMP progenitors at the single-cell level in vivo, indirect evidence suggest that they correspond to cells expressing both the neural marker Sox2 and the mesodermal marker T/Brachyury (Henrique, Abranches, Verrier, & Storey, 2015). These cells are found in the anterior-most primitive streak region (the Node Streak Border) and in the caudo-lateral epiblast adjacent to the PS in the early amniote embryo and in the chordo-neural hinge of the tail bud at later stages (Garriock et al., 2015; Olivera-Martinez, Harada, Halley, & Storey, 2012; Wymeersch et al., 2016). In mouse, NMPs are first detected as a Sox2+/T+ cell population in the anterior PS at day 7.5 and they remain as a self-renewing population in the epiblast and in the tail bud until the completion of body axis extension (Gouti et al., 2017; Wymeersch et al., 2016). Such Sox2/T-expressing cells have been identified in the epiblast of all vertebrate species examined from fish to humans indicating that they represent a conserved population of PM progenitors (Kimelman, 2016; Olivera-Martinez et al., 2012).

In amniotes, the PM fate is specified in the anterior PS and adjacent epiblast in response to TGF beta/Nodal and Wnt signaling, while more lateral fates (including lateral plate and intermediate mesoderm) depend on BMP signaling (Robertson, 2014; Zinski, Tajer, & Mullins, 2017). Wnt and FGF signaling control several key aspects of PM formation during gastrulation, including fate specification, ingression, proliferation, and maintenance of the precursor cell populations (Boulet & Capecchi, 2012;

Ciruna & Rossant, 2001; Ciruna, Schwartz, Harpal, Yamaguchi, & Rossant, 1997; Naiche, Holder, & Lewandoski, 2011; Yamaguchi, Harpal, Henkemeyer, & Rossant, 1994). FGF signaling controls cells' glycolytic activity, which peaks in the tail bud and in turn regulates Wnt signaling (Oginuma et al., 2017). FGF signaling is downstream of Wnt signaling in the tail bud thus closing the signaling loop (Aulehla et al., 2003). Inhibition of Wnt, FGF, or glycolysis signaling breaks this loop, leading to the arrest of PM production and to differentiation of the NMPs to a Sox2+ neural fate. This results in the formation of ectopic neural tissue instead of PM (Chapman & Papaioannou, 1998; Oginuma et al., 2017; Yamaguchi, Takada, Yoshikawa, Wu, & McMahon, 1999). As the PM plays a key role in elongation of the posterior body axis (Benazeraf et al., 2010), this ultimately leads to an arrest of elongation and a truncation of the body axis. Such axis truncation phenotypes are also observed when mutating the Wnt downstream targets, T or Tbx6 (Chapman & Papaioannou, 1998; Javali et al., 2017; Yamaguchi et al., 1999). Thus, a right balance of Wnt and FGF signaling is necessary for NMP maintenance in the tail bud in vivo.

In vitro, production of NMPs can be efficiently achieved by treating mouse and human PSCs with Wnt activators such as the GSK3 beta inhibitor Chiron (Chir) (reviewed in Henrique et al., 2015). Treatment of PC at the epiblast stage, i.e., mouse ES cells differentiated for 2 days with FGF or N2B27, or of human ES/iPS cells (which are already at an epiblast stage) for 1 day with Chir in vitro is enough to obtain a very high percentage (>80%) of Sox2+/T+ cells in vitro. NMP induction efficiency can be further improved by adding FGF to the medium in addition to Chir (Gouti et al., 2014; Lippmann et al., 2015; Tsakiridis et al., 2014; Turner, Hayward, et al., 2014). However, NMP production can also be efficiently achieved without FGF addition (Denham et al., 2015) probably because FGF is autonomously produced by the cells downstream of Wnt signaling (Aulehla et al., 2003). In addition, maintenance of Sox2 required for NMP induction in vitro requires low doses of RA, which can be autonomously produced in cells by the retinaldehyde dehydrogenase Aldh1a2 as long as vitamin A is present in the culture medium (Gouti et al., 2017). Cdx genes play an important role in controlling RA levels thus allowing NMP induction and PM differentiation (Gouti et al., 2017). These NMP cells induced in vitro can differentiate to the paraxial and the neural lineage and their bipotentiality has been demonstrated by clonal analysis (Tsakiridis & Wilson, 2015) and by grafting in the chicken and mouse embryos (Edri, Hayward, Baillie-Johnson, Steventon, & Martinez Arias, 2018; Gouti et al., 2014).

While Wnt is absolutely required for PM differentiation from NMPs in vitro, the neural fate can be efficiently induced by RA treatment (Gouti et al., 2017). In vitro, identity of the NMPs progressively changes with age as exemplified by the progressive expression of more posterior Hox genes (Gouti et al., 2017; Lippmann et al., 2015).

In vivo, whether all the body PM and spinal cord derives from NMPs is currently unclear. Fate mapping studies indicate that cells located in the anterior PS (excluding the node region) mostly contribute to anterior somites or to anterior somites and lateral plate and not to the spinal cord (Henrique et al., 2015; Iimura, Yang, Weijer, & Pourquie, 2007; Psychoyos & Stern, 1996; Wymeersch et al., 2016). Conversely, labeling T-expressing cells of the mouse primitive streak using a T-Cre labeling strategy shows that while the entire PM is labeled, cells of the anterior spinal cord do not express the T gene (Timofeeva et al., 2005). These experiments argue that the anterior region of the PM and of the spinal cord derive from different pools of progenitors. Several observations suggest that NMPs rather contribute to cells located posteriorly in the embryo. While a few clones contributing to the spinal cord and the PM along the entire axis were reported by Tzouanacou et al. (2009), several of these clones also had progeny in other tissues such as notochord. In fact most of the NMP clones identified in this study resided posterior to the thoracic region. Furthermore, in mouse mutants for genes of the Wnt pathway, ectopic neural tubes form in place of the PM only at posterior levels (Chapman & Papaioannou, 1998; Galceran, Farinas, Depew, Clevers, & Grosschedl, 1999; Yamaguchi et al., 1999). Together, these data argue that, in vivo, NMP actively contribute mostly to the posterior PM and spinal cord.

Also, the PM derivatives were shown to originate from two distinct sets of progenitors (Cambray & Wilson, 2007; Iimura et al., 2007; Selleck & Stern, 1991). The first one is located in the anterior PS/node region. It gives rise to the medial somite and its descendants are able to autonomously segment (Freitas, Rodrigues, Charrier, Teillet, & Palmeirim, 2001). Its cells can self-renew and contribute to long clones along the AP axis (Cambray & Wilson, 2002, 2007). The second pool is located in a more posterior territory of the anterior PS and it gives rise to cells of the lateral somite (Cambray & Wilson, 2007; Iimura et al., 2007; Psychoyos & Stern, 1996), which depend on an axial signal for segmentation (Freitas et al., 2001). These cells do not self-renew and only produce shorter clones (Cambray & Wilson, 2007; Iimura et al., 2007; Psychoyos & Stern, 1996). These observations contrast with the recent reports discussed earlier which show that most human and mouse PCs differentiated in Chir-containing medium acquire what appears to be a unique NMP Sox2+/T+ fate. It is nevertheless possible that the NMP population represents the only PM population in the embryo and that its fate (neural or medial or lateral PM) is dictated by the local signaling environment. Accordingly, markers have yet to be identified that distinguish the different progenitors pools.

In vivo, Nodal has been shown to play an important role in the induction of the anterior primitive streak which gives rise to NMPs (Robertson, 2014). Hence, some protocols include an early treatment with TGF beta activators (activin) to first induce the anterior PS/NMP fate (Chalamalasetty et al., 2014; Loh et al., 2016; Tsakiridis et al., 2014; Turner, Hayward, et al., 2014). Differentiation of mouse Nodal−/− epi-SCs with Chir and FGF results in an increase of Sox2 and a decrease of the NMP markers Fgf8 and Cyp26a1 suggesting that NMPs are poorly induced in these conditions (Edri et al., 2018). However, NMP induction can be achieved very efficiently in mouse and human cells without Nodal/Activin treatment, arguing that these factors are produced by the cells themselves in response to Wnt activation (Henrique et al., 2015, our own observations). In vivo, Nodal is produced downstream of Wnt/beta-catenin signaling (Robertson, 2014), suggesting that the Wnt activators added to the culture in vitro could be sufficient to trigger endogenous Nodal activation.

Thus, the only conserved factor between all the published protocols reporting efficient induction of Sox2+/T+ NMPs in vitro is Wnt signaling which needs to be activated at the equivalent of the epiblast stage for cells to differentiate to the NMP fate. As discussed earlier, Wnt activation appears to be sufficient to trigger activation of other pathways such as Nodal and FGF, whose addition does not seem to be critical for NMP differentiation in vitro.

3. PSM FORMATION FROM NMPs

The first step in the differentiation of NMPs toward the PM fate is the downregulation of Sox2 and the activation of the transcription factors *Tbx6* and *Msgn1*. Cells acquire an identity of mesoderm progenitors cells (MPCs) characterized by the expression of *T*, *Msgn1*, and *Tbx6* (Chalamalasetty et al., 2014). In vivo, MPCs are located in the posterior-most PSM, in the tail bud. Acquisition of the MPC fate is paralleled by an EMT which requires the transcription factors T, Snai1, and Msgn1 (Turner, Rue, Mackenzie, Davies, & Martinez Arias, 2014). Once ingressed in the PSM, these cells are highly motile, a feature that plays an important role in the control of axis elongation (Benazeraf et al., 2010). Both MPC differentiation and EMT/motility have been recapitulated in vitro during differentiation of the NMPs to a PM fate with mouse ES cells (Chalamalasetty et al., 2014; Turner, Rue, et al., 2014). In the mouse embryo, *Tbx6* is already expressed in the primitive streak, i.e., slightly earlier than *Msgn1* (Javali et al., 2017; Nowotschin, Ferrer-Vaquer, Concepcion, Papaioannou, & Hadjantonakis, 2012). In vitro studies of differentiating mouse ES lines in which *Tbx6* was mutated showed that *Tbx6−/−* mouse ES cells remain in an NMP-like state but acquire a progressively more posterior identity (Gouti et al., 2017). Thus, *Tbx6* is required for cells to exit the NMP state and acquire an MPC fate. Similar studies established that *Msgn1* controls the next step of PSM differentiation, i.e., the exit from the MPC state and acquisition of the posterior PSM (pPSM) fate (Chalamalasetty et al., 2014; Gouti et al., 2017). This stage is characterized by the expression of *Msgn1* and *Tbx6* and the downregulation of *T* (Chalamalasetty et al., 2014). The signals controlling the transition from NMP to MPC have not been identified and in vitro, MPCs and pPSM spontaneously differentiate when maintaining ES cells in Chir-containing medium, suggesting that these signals are autonomously produced by the cells. FGF signaling which is strongly activated in the posterior PSM in vivo, was reported to maintain cells in a more undifferentiated PSM state in mouse ES cells differentiating in Chir-containing medium in vitro (Sudheer et al., 2016). Both MPC and pPSM cells experience periodic signaling involving Notch, Wnt, and FGF pathways in vivo reflecting activity of the segmentation clock (Hubaud & Pourquie, 2014). Expression of so-called cyclic genes such as *Hes7* or *Lunatic Fringe* is detected in both mouse and human PSM-like cells differentiated in vitro (Chal et al., 2018, 2015; Loh et al., 2016), and oscillations of *Hes7* have recently been reported in differentiated mouse ES cells (Matsumiya, Tomita, Yoshioka-Kobayashi, Isomura, & Kageyama, 2018).

Mouse and human PC lines harboring fluorescent reporters driven by *T*, *Msgn1*, or *Tbx6* regulatory sequences have been generated and used to optimize induction of PSM-like cells (Chal et al., 2018, 2015; Choi et al., 2016; Sudheer et al., 2016; Turner, Rue, et al., 2014). Several studies have shown that activating Wnt signaling in mouse or human PC is sufficient to induce expression of *Msgn1* and *Tbx6*, although the production of NMPs or MPCs has not been specifically evaluated in these studies (Chal et al., 2015; Choi et al., 2016; Henrique et al., 2015; Loh et al., 2016; Shelton et al., 2014; Sudheer et al., 2016; Xi et al., 2017). However, all these protocols involve a similar step of Wnt activation at the epiblast stage, suggesting that NMPs are also produced in these conditions. Additional treatments with other compounds interfering with other signaling pathways such as PI3K have been reported to further stimulate PSM production in vitro (Choi et al., 2016; Loh et al., 2016). However, the biological rationale for these treatments is not obvious and very high level of PSM induction can be achieved without them suggesting that they are dispensable. One of the surprising conclusions of all these different studies is that while in vivo, the precise spatiotemporal deployment of various signaling activities involving the BMP, Nodal, Wnt, and FGF signaling pathways is critical to achieve efficient PM fate induction, such precision is not necessary to obtain the correct fates in vitro. This could suggest that much of the developmental processes, including activation of the appropriate signaling molecules are spontaneously recapitulated in the differentiating monolayers of PCs, requiring only an initial trigger (Wnt activation in this case). Such a situation is observed in organoid cultures which can recapitulate the development of complex embryonic structures in vitro largely based on self-organization rather than directed administration of signaling cues (Sasai, Eiraku, & Suga, 2012).

4. CROSSING THE DETERMINATION FRONT

The next critical stage in PM cells differentiation takes place in the anterior PSM, when cells reach the so-called determination front level (Dubrulle, McGrew, & Pourquie, 2001). At this level, oscillations of the segmentation clock arrest (Hubaud & Pourquie, 2014). The segmental identity is first established at the front level in response to the activation of genes such as *Mesp2*, which are expressed in stripes marking the future segment boundaries (Morimoto, Takahashi, Endo, & Saga, 2005). *Msgn1* expression is downregulated at this position, soon followed by *Tbx6* while *Pax3* becomes first expressed (Chal & Pourquie, 2009). In vivo, this transition is controlled by the posterior FGF/Wnt gradient system (Aulehla & Pourquie, 2009). Wnt and FGF signaling levels control the AP position where *Mesp2* and *Pax3* are activated (Aulehla et al., 2003, 2008; Delfini, Dubrulle, Malapert, Chal, & Pourquie, 2005; Dubrulle et al., 2001; Naiche et al., 2011). Some PM differentiation protocols are recapitulating this signaling transition by removing the Wnt activator from the medium and adding FGF/ERK inhibitors to simulate the downregulation of FGF and Wnt signaling that occurs at the determination front (Chal et al., 2018; Loh et al., 2016). However, other studies have reported that maintenance in

Chir-containing medium (with or without FGF) is sufficient to efficiently recapitulate this transition both in mouse and in human PCs (reviewed in Chal & Pourquie, 2017). In both sets of conditions, *Msgn1* is down-regulated, soon followed by *Mesp2* transient upregulation and *Pax3* activation as observed in vivo. This suggests that the response to cell signaling is largely controlled autonomously in these 2D culture conditions.

Most of the characterization of the PM differentiation in vivo has been performed in model organisms such as mouse, chicken, or fish embryos. Very little is known about this process in human embryos due to the difficulty to access these early developmental stages. Comparison of the transcriptome of FACS-sorted mouse and human Msgn1-YFP+ reporter PCs differentiated in vitro demonstrated that the same set of key transcription factors and signaling cues are activated in PSM-like cells of the two species (Chal et al., 2018). A transcriptomic analysis of the differentiating PM at 4.5/5 weeks in human embryos was also recently performed, revealing many similarities with mouse PSM development (Xi et al., 2017). One difference though was the down-regulation of the TGF beta pathway taking place in the PSM. Treating human ES or iPS cells, differentiated to PSM stage in Chir-containing medium, with a TGF beta inhibitor in addition to a BMP inhibitor was shown to improve differentiation toward an anterior PSM/somite fate (Loh et al., 2016; Xi et al., 2017).

5. INDUCED PM CELLS DRIFT TO A LATERAL PLATE FATE IN ABSENCE OF BMP INHIBITION

BMP signaling represents another important signaling pathway which antagonizes the PM fate to promote the lateral plate identity in vivo. The lateral plate mesoderm is the tissue adjacent to the PM which gives rise to the limbs and to the body wall. In vivo, inhibition of BMP signaling is necessary for PM formation and exposing posterior PSM to BMP4 can convert the cells to a lateral plate fate (Tonegawa, Funayama, Ueno, & Takahashi, 1997). Conversely graft of a Noggin-producing bead adjacent to the territory of the PS fated to give rise to the lateral plate, converts these cells to a PM fate (Tonegawa & Takahashi, 1998). Treatment of both mouse and human Msgn1-YFP ES/iPS reporter cells with Wnt activators results in a large number of fluorescent cells induced after 2 days in vitro (Chal et al., 2018, 2015; Choi et al., 2016; Sudheer et al., 2016). Examination of the transcriptome of these cells revealed strong activation of *BMP4* and lateral plate markers such as *FoxF1* and *Hand1* (Chal et al., 2018). In contrast, when cells were treated with BMP inhibitors in addition to the Wnt activator, *BMP4* was downregulated and strong upregulation of anterior PSM markers was observed (Chal et al., 2018, 2015; Loh et al., 2016). These observations are surprising as *Msgn1* is specifically expressed in the posterior PSM, suggesting that cells first acquire a pPSM fate and then drift to a lateral plate fate. In vivo, bipotential progenitors located in the posterior PSM and able to give rise to both PM and lateral plates have been identified (Stern, Fraser, Keynes, & Primmett, 1988), supporting the idea that Msgn1+ cells could bifurcate to a lateral plate fate in response to BMP4. Thus, addition of BMP inhibitors together with Wnt agonists to PSC cultures can stabilize the PM fate. This would argue that the BMP inhibitors act after cells activate *Msgn1*, i.e., after day 2 in both mouse and human cultures. Efficient induction of anterior PSM markers was indeed observed when BMP inhibitors were added after day 2 (Xi et al., 2017), although similar results were observed when cells are cultured with the BMP inhibitor throughout PSM induction (Chal et al., 2015). While some protocols reporting muscle production in vitro in absence of BMP inhibitors have been reported, the induction of lateral plate markers has not been analyzed in these cultures and how the lateral plate drift is overcome remains unclear (Borchin, Chen, & Barberi, 2013; Shelton et al., 2014).

6. RECAPITULATION OF MYOGENESIS IN VITRO

In vitro, once PM cells differentiated as described earlier have reached the somite stage characterized by Pax3 expression, they can subsequently be induced to a myogenic fate using various strategies (reviewed in Chal & Pourquie, 2017; Magli & Perlingeiro, 2017). When cultured in medium containing HGF, FGF, and IGF, the complete myogenesis program can be recapitulated with mouse ES cells (Chal et al., 2015). Formation of mononucleated myocytes resembling myotomal cells produced during primary myogenesis is seen starting around day 9. This phase is followed by massive production of Pax7+ myogenic progenitors peaking around 2–3 weeks, giving rise to muscle fibers expressing more mature sets of myosin heavy chain such as MYH2 or markers of secondary fibers such as NFIX. This eventually leads to the production of millimeter long muscle fibers containing up to 100 nuclei showing organized myofibrils and contractile activity (Chal et al., 2015). Based on morphological criteria, these fibers resemble early postnatal mouse fibers (White et al., 2010). Remarkably, such linear long fibers are rarely observed when differentiating primary myoblasts or cell lines in culture which tend to form bags of nuclei (myosacs), and thus PSC myogenic cultures provide a new model better approximating the cell biology of muscle (Fig. 1).

While the developmental myogenic sequence is reasonably well described in mouse, little is known about prenatal human myogenesis. Transposition of the mouse differentiation protocols to human PCs was shown to recapitulate the overall differentiation sequence in several cases (reviewed in Chal & Pourquie, 2017; Magli & Perlingeiro, 2017). The fusion efficiency of myogenic cells obtained from human PC differentiated in vitro as described in Chal et al. (2015) and Shelton et al. (2014) was recently compared to human

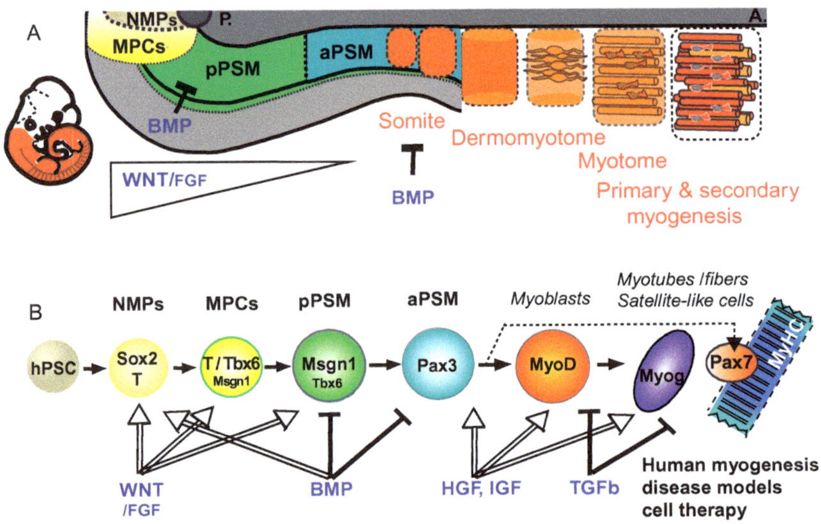

Fig. 1 Myogenesis in vivo (A) and in vitro (B). (A) Trunk skeletal myogenesis (*far left*). The anatomical domain corresponding to trunk myogenesis is depicted in *red* on a E10.5 mouse embryo (*center*). Schematic of paraxial mesoderm (PM) differentiation and skeletal myogenesis progression in the posterior domain of a developing embryo. Key developmental stages and spatial map of the PM are shown. The mesoderm is patterned by specific signaling pathways in particular Wnt, FGF, and BMP pathways. Dorsal view, posterior (P.) to the *left*, anterior (A.) to the *right*. For clarity, only one side of the bilateral embryo is depicted, *midline on top*. (B) Diagram recapitulating the differentiation of human pluripotent stem cells (hPSC) into PM toward somitic progenitors and skeletal myogenesis. From *left* to *right*, identified intermediate cell types and their key marker genes are shown. Cell types are color coded according to the fate map shown in (A). Key signaling events identified during the in vitro differentiation of PM and skeletal myogenesis are shown. *hPSCs*: human pluripotent stem cells; *NMPs*: neuromesodermal progenitors, *MPCs*: mesodermal progenitor cells; *pPSM*: posterior presomitic mesoderm; *aPSM*: anterior presomitic mesoderm.

fetal muscle at weeks 9, 14, 17, and adult muscle (Hicks et al., 2018). The fusion indices and nuclei per myotube were found similar between the myogenic cells differentiated in vitro and week 9 embryonic muscles while they were significantly higher in older samples. RNAseq was further used to compare these different samples and confirmed that the myogenic cells differentiated in vitro are more closely related to the week 9 cells than the later ones (Hicks et al., 2018). This stage corresponds to the late primary myogenesis. Spectacular tridimensional analyses of muscle development in cleared human embryos ranging from 8 to 14 weeks of gestation with antibodies against PAX7, MYOG, and myosin heavy chain have been recently reported (Belle et al., 2017). Myosin heavy chain was already detected at week 8 in muscles that were colonized with motoneurons. At 9 weeks, massive amounts of PAX7 and MYOG were detected in the developing muscles resembling the situation in mouse at late fetal stages (Belle et al., 2017). Although the data are quite fragmentary, it appears that the prenatal stages of muscle development in humans resemble those of mice.

An alternative protocol for myogenesis induction of human iPS cells recently described involves inhibition of Notch signaling with DAPT from day 4 to 12 following an initial Chir treatment (Choi et al., 2016). This also results in efficient differentiation of contractile muscle fibers exhibiting sarcomeric organization after 30 days. However, the progression of the myogenic sequence and whether or not PAX7+ cells differentiate in these conditions has not been studied. This protocol was used to produce fusion competent myoblasts from wild type and mutant iPS lines from DMD patients to study the phenotype induced by absence of dystrophin in vitro. This led to the identification of an upregulation of BMP and TGF beta signaling in the mutant fibers that were correlated with a fusion defect (Choi et al., 2016). Interestingly, TGF beta signaling was also identified in the study described earlier comparing myotubes differentiated in vitro from human PCs to human myotubes derived from primary cultures (Hicks et al., 2018). TGF beta1 signaling was enriched in the myotubes differentiated in vitro and shown to inhibit myotube formation. Treating differentiating iPS cells with the TGF beta inhibitors SB431542 or A83-01 greatly improves the maturation of myotubes which differentiated to a late fetal stage in vitro. MYH1 and MYH8 were increased and better organized sarcomeres were observed with electron microscopy. Importantly, this treatment does not deplete PAX7+ cells in the culture. Strikingly, treatment with TGF beta inhibitors in vivo also stimulated engraftment of myogenic cells differentiated from iPS in vitro (Hicks et al., 2018).

So far, as seen for many other cell types differentiated in vitro from PCs, muscle fibers generated in vitro from mouse or human PCs exhibit a fetal/perinatal phenotype. Differentiation of myofibers to an adult-like phenotype from PCs has not been reported. Even muscle fibers differentiated in vitro from PCs by forced expression of the transcription factor Pax7 were shown to reach a significant level of maturation, but retained calcium and voltage properties characteristic of immature myofibers (Skoglund et al., 2014).

7. GENERATION OF THE Pax7+ MYOGENIC LINEAGE IN VITRO

Adult skeletal muscles are endowed with significant regenerative capacity upon injury. This regeneration is made possible thanks to a small subpopulation of quiescent adult stem cells called satellite cells (Brack & Rando, 2012; Dumont et al., 2015). These cells are small cells which lie under the basal lamina of muscle fibers and which express the transcription factor Pax7. Upon injury, satellite cells become activated and proliferate extensively to generate a population of myoblasts which fuse to form postmitotic myotubes, reconstructing the damaged muscle. Satellite cells are also able to self-renew and a small population of Pax7+ cells is maintained during the regeneration process to recreate the quiescent muscle stem cell compartment. Much is known about the regenerative properties of adult satellite cells, but their developmental trajectory is poorly understood. Most of our understanding of satellite cell development comes from studies performed in mouse, where satellite cells derive mostly from the pool of Pax7+ myogenic progenitors that generate the fetal muscle fibers (Gros et al., 2005; Kassar-Duchossoy et al., 2005; Relaix et al., 2005). Access to in vitro systems recapitulating satellite cells differentiation in mouse and human would be extremely useful to understand the segregation of this lineage during myogenesis. Satellite cells represent the ideal candidate for cell therapy approaches aiming at reconstructing muscles. However, these cells are scarce in adult muscles and they are difficult to amplify in vitro, as they tend to lose their regenerative properties in culture. Thus, access to an unlimited source of satellite cells differentiated from PCs would represent a true breakthrough and could open the possibility to develop cell therapy protocols for muscular dystrophies.

In mouse in vitro cultures, the long muscle fibers generated from PCs are flanked by a limited number of quiescent Pax7+ cells sometimes found inside the forming basal lamina suggesting that these cells correspond to immature satellite cells (Chal et al., 2015). When these Pax7+ cells differentiated from an ES Pax7-GFP reporter line were sorted by FACS and transplanted into the *Tibialis Anterior* of *mdx* mice (Chal et al., 2015), they contributed to the generation of muscle fibers and also to satellite cells. This demonstrates their capacity to self-renew in vivo suggesting that these cells exhibit properties of satellite cells. Since the mouse fibers produced in vitro correspond to perinatal fibers, this suggests that the Pax7+ cells also correspond to the equivalent stage. Recently, mouse perinatal satellite cells were shown to exhibit a stronger regenerative potential than adult cells (Tierney et al., 2016). Thus, the fact that Pax7+ satellite-like cells generated from PC in vitro exhibit a perinatal phenotype might be an advantage for the development of cell therapy approaches for muscular dystrophies.

Differentiation of human PCs to a myogenic fate with the protocols described earlier can also lead to efficient induction of Pax7+ cells (around 25% after 3 weeks) (reviewed in Chal & Pourquie, 2017). However, these PAX7+ cells have not yet been tested for their regenerative potential in vivo. While human PAX7-GFP iPS reporter lines have been reported, their use for producing Pax7+ cells in vitro has not been published yet (Wu, Hunt, Xue, Liu, & Darabi, 2016). An alternative to such reporters is the identification of cell surface markers specifically expressed in satellite cells. Using a weighted gene coexpression analysis approach comparing the transcriptome human fetal muscle, cartilage, tendon, bone, and ligament, ERBB3 and NGFR were identified as interesting markers able to enrich in human PAX7+ progenitors (Hicks et al., 2018). When grafted in *mdx* mouse muscles, cells from human myogenic iPS cultures according to Chal et al. (2016) or to Shelton et al. (2014) FACS-sorted with these surface markers were able to generate significant amounts of human myofibers indicating that they share properties with satellite cells.

8. CONCLUSION

The late stages of muscle development are poorly understood mostly due to the difficulty to follow cell fates in large tissues. Thus development of in vitro models of myogenesis will certainly shed light on processes difficult to study in vivo such as cell fusion or satellite cell differentiation. Moreover, it will lead to a better understanding of human myogenesis which has been very poorly studied so far due to the difficulties to access embryos. Beyond these expected contributions to the development of skeletal muscle, the development of in vitro differentiation protocols will contribute to the development of better disease models and to cell therapy of muscular diseases.

Maturing muscle fibers in vitro to the adult stage remains also a major challenge. In vivo, innervation plays an important role in fiber maturation and thus coculture of PSC-derived muscle fibers and motoneurons could help fiber maturation. Deciphering methods to improve maturation of these cells toward an adult-like phenotype is therefore an important goal for the field. Access to in vitro models of mature muscle fibers will also provide unique opportunities to study fiber aging and explore therapeutic avenues for muscle wasting diseases such as Cachexia and Sarcopenia.

ACKNOWLEDGMENTS

We thank members of the Pourquié lab for discussions and James Briscoe for comments. Research in the Pourquié lab was funded by the French Muscular Dystrophy Association (AFM), by an advanced grant from the European Research Council and from the National Institute of Health (R01HD085121).

REFERENCES

Aulehla, A., & Pourquie, O. (2009). Signaling gradient during paraxial mesoderm development. In J. Briscoe, P. Lawrence, & J. Vincent (Eds.), *Reading and interpreting gradients during development*: Cold Spring Harbor Laboratory Press.

Aulehla, A., Wehrle, C., Brand-Saberi, B., Kemler, R., Gossler, A., Kanzler, B., et al. (2003). Wnt3a plays a major role in the segmentation clock controlling somitogenesis. *Developmental Cell*, *4*, 395–406.

Aulehla, A., Wiegraebe, W., Baubet, V., Wahl, M. B., Deng, C., Taketo, M., et al. (2008). A beta-catenin gradient links the clock and wavefront systems in mouse embryo segmentation. *Nature Cell Biology*, *10*, 186–193.

Belle, M., Godefroy, D., Couly, G., Malone, S. A., Collier, F., Giacobini, P., et al. (2017). Tridimensional visualization and analysis of early human development. *Cell*, *169*. 161–173. e112.

Benazeraf, B., Francois, P., Baker, R. E., Denans, N., Little, C. D., & Pourquie, O. (2010). A random cell motility gradient downstream of FGF controls elongation of an amniote embryo. *Nature*, *466*, 248–252.

Biressi, S., Molinaro, M., & Cossu, G. (2007). Cellular heterogeneity during vertebrate skeletal muscle development. *Developmental Biology*, *308*, 281–293.

Borchin, B., Chen, J., & Barberi, T. (2013). Derivation and FACS-mediated purification of PAX3+/PAX7+ skeletal muscle precursors from human pluripotent stem cells. *Stem Cell Reports*, *1*, 620–631.

Boulet, A. M., & Capecchi, M. R. (2012). Signaling by FGF4 and FGF8 is required for axial elongation of the mouse embryo. *Developmental Biology*, *371*, 235–245.

Brack, A. S., & Rando, T. A. (2012). Tissue-specific stem cells: Lessons from the skeletal muscle satellite cell. *Cell Stem Cell*, *10*, 504–514.

Cambray, N., & Wilson, V. (2002). Axial progenitors with extensive potency are localised to the mouse chordoneural hinge. *Development*, *129*, 4855–4866.

Cambray, N., & Wilson, V. (2007). Two distinct sources for a population of maturing axial progenitors. *Development, 134*, 2829–2840.
Chal, J., Al Tanoury, Z., Hestin, M., Gobert, B., Aivio, S., Hick, A., Cherrier, T., Nesmith, A. P., Parker, K. K., & Pourquié, O. (2016). Generation of human muscle fibers and satellite-like cells from human pluripotent stem cells in vitro. *Nature Protocols, 11*, 1833–1850. https://doi.org/10.1038/nprot.2016.110.
Chal, J., Al Tanoury, Z., Oginuma, M., Moncuquet, P., Gobert, B., Miyanari, A., et al. (2018). Recapitulating early development of mouse musculoskeletal precursors of the paraxial mesoderm in vitro. *Development, 145*(6). https://doi.org/10.1242/dev.157339.
Chal, J., Oginuma, M., Al Tanoury, Z., Gobert, B., Sumara, O., Hick, A., et al. (2015). Differentiation of pluripotent stem cells to muscle fiber to model Duchenne muscular dystrophy. *Nature Biotechnology, 33*, 962–969.
Chal, J., & Pourquie, O. (2009). Patterning and differentiation of the vertebrate spine. In *The Skeletal System* (pp. 41–116): Cold Spring Harbor Monograph.
Chal, J., & Pourquie, O. (2017). Making muscle: Skeletal myogenesis in vivo and in vitro. *Development, 144*, 2104–2122.
Chalamalasetty, R. B., Garriock, R. J., Dunty, W. C., Jr., Kennedy, M. W., Jailwala, P., Si, H., et al. (2014). Mesogenin 1 is a master regulator of paraxial presomitic mesoderm differentiation. *Development, 141*, 4285–4297.
Chapman, D. L., & Papaioannou, V. E. (1998). Three neural tubes in mouse embryos with mutations in the T-box gene Tbx6. *Nature, 391*, 695–697.
Choi, I. Y., Lim, H., Estrellas, K., Mula, J., Cohen, T. V., Zhang, Y., et al. (2016). Concordant but varied phenotypes among Duchenne muscular dystrophy patient-specific myoblasts derived using a human iPSC-based model. *Cell Reports, 15*, 2301–2312.
Ciruna, B., & Rossant, J. (2001). FGF signaling regulates mesoderm cell fate specification and morphogenetic movement at the primitive streak. *Developmental Cell, 1*, 37–49.
Ciruna, B. G., Schwartz, L., Harpal, K., Yamaguchi, T. P., & Rossant, J. (1997). Chimeric analysis of fibroblast growth factor receptor-1 (Fgfr1) function: A role for FGFR1 in morphogenetic movement through the primitive streak. *Development, 124*, 2829–2841.
Delfini, M. C., Dubrulle, J., Malapert, P., Chal, J., & Pourquie, O. (2005). Control of the segmentation process by graded MAPK/ERK activation in the chick embryo. *Proceedings of the National Academy of Sciences of the United States of America, 102*, 11343–11348.
Denham, M., Hasegawa, K., Menheniott, T., Rollo, B., Zhang, D., Hough, S., et al. (2015). Multipotent caudal neural progenitors derived from human pluripotent stem cells that give rise to lineages of the central and peripheral nervous system. *Stem Cells, 33*, 1759–1770.
Dubrulle, J., McGrew, M. J., & Pourquie, O. (2001). FGF signaling controls somite boundary position and regulates segmentation clock control of spatiotemporal Hox gene activation. *Cell, 106*, 219–232.
Dumont, N. A., Wang, Y. X., & Rudnicki, M. A. (2015). Intrinsic and extrinsic mechanisms regulating satellite cell function. *Development, 142*, 1572–1581.
Edri, S., Hayward, P., Baillie-Johnson, P., Steventon, B., & Martinez Arias, A. (2018). An epiblast stem cell derived multipotent progenitor population for axial extension, bioRxiv 242461. https://doi.org/10.1101/242461.
Freitas, C., Rodrigues, S., Charrier, J. B., Teillet, M. A., & Palmeirim, I. (2001). Evidence for medial/lateral specification and positional information within the presomitic mesoderm. *Development, 128*, 5139–5147.
Galceran, J., Farinas, I., Depew, M. J., Clevers, H., & Grosschedl, R. (1999). Wnt3a-/–like phenotype and limb deficiency in Lef1(-/-)Tcf1(-/-) mice. *Genes & Development, 13*, 709–717.

Garriock, R. J., Chalamalasetty, R. B., Kennedy, M. W., Canizales, L. C., Lewandoski, M., & Yamaguchi, T. P. (2015). Lineage tracing of neuromesodermal progenitors reveals novel Wnt-dependent roles in trunk progenitor cell maintenance and differentiation. *Development*, *142*, 1628–1638.

Gouti, M., Delile, J., Stamataki, D., Wymeersch, F. J., Huang, Y., Kleinjung, J., et al. (2017). A gene regulatory network balances neural and mesoderm specification during vertebrate trunk development. *Developmental Cell*, *41*, 243–261. e247.

Gouti, M., Tsakiridis, A., Wymeersch, F. J., Huang, Y., Kleinjung, J., Wilson, V., et al. (2014). In vitro generation of neuromesodermal progenitors reveals distinct roles for wnt signalling in the specification of spinal cord and paraxial mesoderm identity. *PLoS Biology*, *12*, e1001937.

Gros, J., Manceau, M., Thome, V., & Marcelle, C. (2005). A common somitic origin for embryonic muscle progenitors and satellite cells. *Nature*, *435*, 954–958.

Henrique, D., Abranches, E., Verrier, L., & Storey, K. G. (2015). Neuromesodermal progenitors and the making of the spinal cord. *Development*, *142*, 2864–2875.

Hicks, M. R., Hiserodt, J., Paras, K., Fujiwara, W., Eskin, A., Jan, M., et al. (2018). ERBB3 and NGFR mark a distinct skeletal muscle progenitor cell in human development and hPSCs. *Nature Cell Biology*, *20*, 46–57.

Hubaud, A., & Pourquie, O. (2014). Signalling dynamics in vertebrate segmentation. *Nature Reviews. Molecular Cell Biology*, *15*, 709–721.

Hutcheson, D. A., Zhao, J., Merrell, A., Haldar, M., & Kardon, G. (2009). Embryonic and fetal limb myogenic cells are derived from developmentally distinct progenitors and have different requirements for beta-catenin. *Genes & Development*, *23*, 997–1013.

Iimura, T., Yang, X., Weijer, C. J., & Pourquie, O. (2007). Dual mode of paraxial mesoderm formation during chick gastrulation. *Proceedings of the National Academy of Sciences of the United States of America*, *104*, 2744–2749.

Javali, A., Misra, A., Leonavicius, K., Acharyya, D., Vyas, B., & Sambasivan, R. (2017). Co-expression of Tbx6 and Sox2 identifies a novel transient neuromesoderm progenitor cell state. *Development*, *144*, 4522–4529.

Kassar-Duchossoy, L., Giacone, E., Gayraud-Morel, B., Jory, A., Gomes, D., & Tajbakhsh, S. (2005). Pax3/Pax7 mark a novel population of primitive myogenic cells during development. *Genes & Development*, *19*, 1426–1431.

Kimelman, D. (2016). Tales of tails (and trunks): Forming the posterior body in vertebrate embryos. *Current Topics in Developmental Biology*, *116*, 517–536.

Lemke, S. B., & Schnorrer, F. (2017). Mechanical forces during muscle development. *Mechanisms of Development*, *144*, 92–101.

Lippmann, E. S., Williams, C. E., Ruhl, D. A., Estevez-Silva, M. C., Chapman, E. R., Coon, J. J., et al. (2015). Deterministic HOX patterning in human pluripotent stem cell-derived neuroectoderm. *Stem Cell Reports*, *4*, 632–644.

Loh, K. M., Chen, A., Koh, P. W., Deng, T. Z., Sinha, R., Tsai, J. M., et al. (2016). Mapping the pairwise choices leading from pluripotency to human bone, heart, and other mesoderm cell types. *Cell*, *166*, 451–467.

Magli, A., & Perlingeiro, R. R. C. (2017). Myogenic progenitor specification from pluripotent stem cells. *Seminars in Cell & Developmental Biology*, *72*, 87–98.

Matsumiya, M., Tomita, T., Yoshioka-Kobayashi, K., Isomura, A., & Kageyama, R. (2018). ES cell-derived presomitic mesoderm-like tissues for analysis of synchronized oscillations in the segmentation clock. *Development*, *145*(4), 1–11. https://doi.org/10.1242/dev.156836.

Morimoto, M., Takahashi, Y., Endo, M., & Saga, Y. (2005). The Mesp2 transcription factor establishes segmental borders by suppressing Notch activity. *Nature*, *435*, 354–359.

Naiche, L. A., Holder, N., & Lewandoski, M. (2011). FGF4 and FGF8 comprise the wavefront activity that controls somitogenesis. *Proceedings of the National Academy of Sciences of the United States of America*, *108*, 4018–4023.

Nowotschin, S., Ferrer-Vaquer, A., Concepcion, D., Papaioannou, V. E., & Hadjantonakis, A. K. (2012). Interaction of Wnt3a, Msgn1 and Tbx6 in neural versus paraxial mesoderm lineage commitment and paraxial mesoderm differentiation in the mouse embryo. *Developmental Biology*, *367*, 1–14.

Oginuma, M., Moncuquet, P., Xiong, F., Karoly, E., Chal, J., Guevorkian, K., et al. (2017). A gradient of glycolytic activity coordinates FGF and Wnt signaling during elongation of the body axis in Amniote embryos. *Developmental Cell*, *40*, 342–353. e310.

Olivera-Martinez, I., Harada, H., Halley, P. A., & Storey, K. G. (2012). Loss of FGF-dependent mesoderm identity and rise of endogenous retinoid signalling determine cessation of body axis elongation. *PLoS Biology*, *10*, e1001415.

Perantoni, A. O., Timofeeva, O., Naillat, F., Richman, C., Pajni-Underwood, S., Wilson, C., et al. (2005). Inactivation of FGF8 in early mesoderm reveals an essential role in kidney development. *Development*, *132*, 3859–3871.

Pourquie, O. (2004). The chick embryo: A leading model in somitogenesis studies. *Mechanisms of Development*, *121*, 1069–1079.

Psychoyos, D., & Stern, C. D. (1996). Fates and migratory routes of primitive streak cells in the chick embryo. *Development*, *122*, 1523–1534.

Relaix, F., Rocancourt, D., Mansouri, A., & Buckingham, M. (2005). A Pax3/Pax7-dependent population of skeletal muscle progenitor cells. *Nature*, *435*, 948–953.

Robertson, E. J. (2014). Dose-dependent nodal/Smad signals pattern the early mouse embryo. *Seminars in Cell & Developmental Biology*, *32*, 73–79.

Sasai, Y., Eiraku, M., & Suga, H. (2012). In vitro organogenesis in three dimensions: Self-organising stem cells. *Development*, *139*, 4111–4121.

Selleck, M. A., & Stern, C. D. (1991). Fate mapping and cell lineage analysis of Hensen's node in the chick embryo. *Development*, *112*, 615–626.

Shelton, M., Metz, J., Liu, J., Carpenedo, R. L., Demers, S. P., Stanford, W. L., et al. (2014). Derivation and expansion of PAX7-positive muscle progenitors from human and mouse embryonic stem cells. *Stem Cell Reports*, *3*, 516–529.

Skoglund, G., Laine, J., Darabi, R., Fournier, E., Perlingeiro, R., & Tabti, N. (2014). Physiological and ultrastructural features of human induced pluripotent and embryonic stem cell-derived skeletal myocytes in vitro. *Proceedings of the National Academy of Sciences of the United States of America*, *111*, 8275–8280.

Stern, C. D., Fraser, S. E., Keynes, R. J., & Primmett, D. R. (1988). A cell lineage analysis of segmentation in the chick embryo. *Development*, *104*, 231–244.

Sudheer, S., Liu, J., Marks, M., Koch, F., Anurin, A., Scholze, M., et al. (2016). Different concentrations of FGF ligands, FGF2 or FGF8 determine distinct states of WNT-induced presomitic mesoderm. *Stem Cells*, *34*, 1790–1800. https://doi.org/10.1002/stem.2371.

Tierney, M. T., Gromova, A., Boscolo Sesillo, F., Sala, D., Spenlé, C., Orend, G., et al. (2016). Autonomous extracellular matrix remodeling controls a progressive adaptation in muscle stem cell regenerative capacity during development. *Cell Reports*, *14*, 1940–1952.

Tonegawa, A., Funayama, N., Ueno, N., & Takahashi, Y. (1997). Mesodermal subdivision along the mediolateral axis in chicken controlled by different concentrations of BMP-4. *Development*, *124*, 1975–1984.

Tonegawa, A., & Takahashi, Y. (1998). Somitogenesis controlled by Noggin. *Developmental Biology*, *202*, 172–182.

Tsakiridis, A., Huang, Y., Blin, G., Skylaki, S., Wymeersch, F., Osorno, R., et al. (2014). Distinct Wnt-driven primitive streak-like populations reflect in vivo lineage precursors. *Development, 141*, 1209–1221.

Tsakiridis, A., & Wilson, V. (2015). Assessing the bipotency of in vitro-derived neuromesodermal progenitors. *F1000Research, 4*, 100.

Turner, D. A., Hayward, P. C., Baillie-Johnson, P., Rue, P., Broome, R., Faunes, F., et al. (2014). Wnt/beta-catenin and FGF signalling direct the specification and maintenance of a neuromesodermal axial progenitor in ensembles of mouse embryonic stem cells. *Development, 141*, 4243–4253.

Turner, D. A., Rue, P., Mackenzie, J. P., Davies, E., & Martinez Arias, A. (2014). Brachyury cooperates with Wnt/beta-catenin signalling to elicit primitive-streak-like behaviour in differentiating mouse embryonic stem cells. *BMC Biology, 12*, 63.

Tzahor, E. (2009). Heart and craniofacial muscle development: A new developmental theme of distinct myogenic fields. *Developmental Biology, 327*, 273–279.

Tzouanacou, E., Wegener, A., Wymeersch, F. J., Wilson, V., & Nicolas, J. F. (2009). Redefining the progression of lineage segregations during mammalian embryogenesis by clonal analysis. *Developmental Cell, 17*, 365–376.

White, R. B., Bierinx, A. S., Gnocchi, V. F., & Zammit, P. S. (2010). Dynamics of muscle fibre growth during postnatal mouse development. *BMC Developmental Biology, 10*, 21.

Wilson, V., Olivera-Martinez, I., & Storey, K. G. (2009). Stem cells, signals and vertebrate body axis extension. *Development, 136*, 1591–1604.

Wu, J., Hunt, S. D., Xue, H., Liu, Y., & Darabi, R. (2016). Generation and validation of PAX7 reporter lines from human iPS cells using CRISPR/Cas9 technology. *Stem Cell Research, 16*, 220–228.

Wymeersch, F. J., Huang, Y., Blin, G., Cambray, N., Wilkie, R., Wong, F. C., et al. (2016). Position-dependent plasticity of distinct progenitor types in the primitive streak. *eLife, 5*, e10042. https://doi.org/10.7554/eLife.10042.

Xi, H., Fujiwara, W., Gonzalez, K., Jan, M., Liebscher, S., Van Handel, B., et al. (2017). In vivo human somitogenesis guides somite development from hPSCs. *Cell Reports, 18*, 1573–1585.

Yamaguchi, T. P., Harpal, K., Henkemeyer, M., & Rossant, J. (1994). fgfr-1 is required for embryonic growth and mesodermal patterning during mouse gastrulation. *Genes & Development, 8*, 3032–3044.

Yamaguchi, T. P., Takada, S., Yoshikawa, Y., Wu, N., & McMahon, A. P. (1999). T (Brachyury) is a direct target of Wnt3a during paraxial mesoderm specification. *Genes & Development, 13*, 3185–3190.

Zinski, J., Tajer, B., & Mullins, M. C. (2017). TGF-beta family signaling in early vertebrate development. *Cold Spring Harbor Perspectives in Biology*. https://doi.org/10.1101/cshperspect.a033274.

CHAPTER SIX

Recapitulating and Deciphering Human Pancreas Development From Human Pluripotent Stem Cells in a Dish

Maja B.K. Petersen[*,†,2], **Carla A.C. Gonçalves**[*,2], **Yung Hae Kim**[*,1], **Anne Grapin-Botton**[*,1]
[*]Novo Nordisk Foundation Center for Stem Cell Biology (DanStem), University of Copenhagen, Copenhagen, Denmark
[†]Department of Islet and Stem Cell Biology, Novo Nordisk, Måløv, Denmark
[1]Corresponding authors: e-mail address: yung.kim@sund.ku.dk; anne.grapin-botton@sund.ku.dk

Contents

Abstract

Here, we review how human pluripotent stem cell models of pancreas development have emerged and became an important tool to study human development and disease. Initially developed toward the production of β cells for diabetes therapy, the protocols have been refined based on knowledge of pancreas development in model organisms. While the cells produced are closer and closer to the end goal of a functional β cell, these models have also been used to carry out functional experiments addressing

[2] Equal contribution.

Current Topics in Developmental Biology, Volume 129
ISSN 0070-2153
https://doi.org/10.1016/bs.ctdb.2018.02.009

gene function and expression as well as regulatory and epigenetic landscape changes during human pancreas development. They thereby complement model organisms and reports from human genetic variants predisposing to different forms of diabetes, as well as observations on human fetal tissue. In this review, we therefore compare these different sources of information and discuss how human stem cell models are evolving to inform us on pancreatic diseases and possible treatments.

1. INTRODUCTION

Human pluripotent stem cells (hPSCs) have long been used for the production of pancreatic β cells, with the aim of developing a cell therapy to treat patients with type 1 and possibly some forms of type 2 diabetes. Initial attempts to generate β cells from human embryonic stem cells (hESCs) demonstrated that insulin-positive cells could be derived in vitro from spontaneously differentiated hPSCs (Assady et al., 2001; Lumelsky et al., 2001). However, it was later demonstrated that the insulin immunoreactivity in some of these early studies could be attributed to the uptake of insulin from the medium by the cultured cells (Hansson et al., 2004; Rajagopal, Anderson, Kume, Martinez, & Melton, 2003). Thereafter, the field has been driven forward by continuous optimization of protocols that mimic pancreas development. Due to the limited availability of human studies, these protocols have been mostly informed by animal models. The field has progressed tremendously in the last 20 years and clinical trials are ongoing to test the safety and efficacy of hESC-derived pancreatic progenitor cells in patients with type I diabetes (ViaCyte Inc., https://clinicaltrials.gov, identifiers: NCT02239354, NCT02939118, NCT03162926, NCT03163511). The current trials generate pancreas progenitor implantation in protective devices and rely on their in vivo differentiation into β cells (Kroon et al., 2008), but the production of insulin-producing cells in vitro has also been reproducibly achieved in multiple laboratories, although these cells are not yet fully mature and functional (Johnson, 2016; Kushner, MacDonald, & Atkinson, 2014).

hPSCs are also now being used as a way to gain information on human pancreas development and pancreatic affections. The cell culture models are amenable to genetic manipulations, thereby enabling to mimic human polymorphisms, to create full gene inactivation and gain of function, and to make reporters marking how specific genes or pathways are activated during pancreatic differentiation. They also give access to regulatory landscape and epigenetic changes occurring during the organ-formation process. This opens opportunities to glimpse into the mechanisms of organ formation in human and complements what we know from model organisms, human genetics, and scarce static observations in human fetal samples.

We know relatively little about human pancreas development from observations in human. Research on human embryos (up to about 8 weeks' postconception, wpc) and fetuses (thereafter) is limited by ethical restrictions and availability. However, histological observations on normal fetal tissue recovered after termination of pregnancy have been done at nearly every stage permitted by regulations. The other major way in which we learn from human observations is through the study of pancreatic affections caused by mutations. Without the guidance of what we know of pancreas development in model organisms (reviewed in Larsen & Grapin-Botton, 2017), we would not be able to make much sense of these observations, but extrapolations from animal models have to be considered with prudence, as we already know that some processes are not fully conserved. More and more information can be gained from in vitro models derived from hPSCs. In the long term, in addition to providing models of diabetes and possibly pancreatic cancer, two-dimensional (2D) and three-dimensional (3D) stem cell-derived culture models may provide a platform to test gene corrections or treatments.

In this review we will summarize what is known of each sequential phase of human pancreatic development. We will scrutinize what is inferred from in vivo studies in animal models, how this has informed the development of in vitro models, and how these models have reciprocally contributed knowledge to better understand human development.

2. ENDODERM INDUCTION AND PATTERNING

The endoderm forms during gastrulation along with the other germ layers. The main inducer of definitive endoderm during development in vertebrate embryos is Nodal signaling (reviewed by Zorn & Wells, 2009). The gastrulation events have never been observed in human due to ethical guidelines that prevent culture of in vitro-fertilized human embryos past day 14 of development.[a] However, they are thought to be very similar, though they are deployed in embryos of diverse shapes such as a flat disc in human, chick, and rabbits or a cup in rodents.

[a] ISSCR Guidelines: http://www.isscr.org/home/about-us/news-press-releases/2016/2016/05/12/isscr-releases-updated-guidelines-for-stem-cell-research-and-clinical-translation. National Academies 2010: https://www.ncbi.nlm.nih.gov/books/NBK210068/.

Our understanding of endoderm induction and regionalization in model organisms has enabled the directed differentiation of hESCs into endoderm and several derived organs. The work of D'Amour et al. (2005) (NovoCell, currently ViaCyte) has provided the proof of concept that sequential steps of pancreatic development could be mimicked in vitro. They started with a protocol demonstrating efficient generation of definitive endoderm by using Activin A, a cheaper Nodal surrogate (D'Amour et al., 2005), and later improved it with combined activation of WNT signaling (D'Amour et al., 2006). In vivo, canonical WNTs are also required for endoderm formation and increase *Nodal* levels in vertebrates (reviewed by Zorn & Wells, 2009). Most protocols still use high doses of TGF-β family members such as Activin A or GDF8, though Nodal is optimal (Chen, Borowiak, Sherwood, Kweudjeu, & Melton, 2013), along with activation of β-catenin by WNT3A or chemical inhibition of GSK3β (using, e.g., the small molecule inhibitor CHIR99021) (Fig. 1) (see Cai et al., 2010; Jiang, Au, et al., 2007; Jiang, Shi, et al., 2007; Kroon et al., 2008; Loh et al., 2014; Nostro et al., 2011; Rezania et al., 2013; Rostovskaya, Bredenkamp, & Smith, 2015; Zhang et al., 2009 for a comparative study). The concentration of Activin A is important, as high concentration (100 ng/mL) ensures anterior endoderm induction. The combination of Activin A and WNT3A at the start of definitive endoderm induction enabled to shorten the first stage from 4 days (D'Amour et al., 2005) to 3 days (D'Amour et al., 2006). The resulting definitive endoderm is characterized by the expression of SRY-box 17 (SOX17), CXCR4, and Forkhead Box A2 (FOXA2) and can be expanded in vitro (Cheng et al., 2012).

In mouse, the first diversification of endoderm through patterning events occurs during its induction. The first cells migrate through the primitive streak and move anteriorly, while those migrating later locate posteriorly, thus giving rise to medial–lateral and more posterior endoderm, respectively (reviewed by Tam and Loebel, 2007). The endoderm is further patterned into regions marked by different transcription factors upon the influence of secreted signals from the WNT, BMP, FGF, and retinoid acid (RA) family (reviewed by Zorn & Wells, 2009). The earliest samples of human fetal tissue that have been collected correspond to a stage where this regionalization has started but is not completed. Such observations, staged as Carnegie stage (CS) 9, corresponding to 22–26 days postconception (dpc) (O'Rahilly & Muller, 2010), are rare. At this stage, the embryo has about four somites, and the endoderm is still a sheet of cells, which begins to invaginate anteriorly, to form a tube—the primitive gut (see Table 1 for

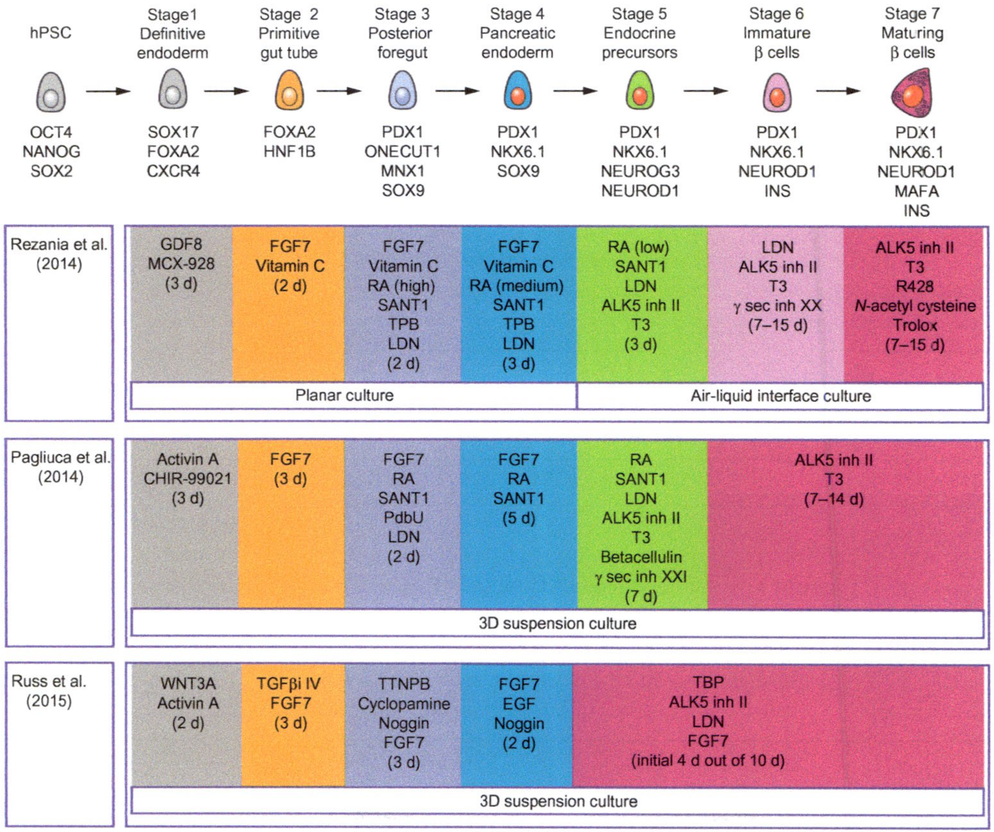

Fig. 1 Step-wise in vitro differentiation protocols of hPSCs toward pancreatic β cells. Schematic description of pancreatic development and desired markers of cell type at each stage is shown. Three recent differentiation protocols are compared according to defined stages including lists of components used and duration of each stage. *ALK5 inh II*, ALK5 (TGF-β receptor) inhibitor; *Betacellulin*, Erbb1 (EGFR) and Erbb4 agonists; *GDF8, Activin A*, TGF-β family member; *MCX-928, CHIR-99021*, GSK3β inhibitor; *LDN*, BMP receptor inhibitor; *N-acetyl cysteine, Trolox (Vit E analogue)*, antioxidant; *R428*, AXL inhibitor; *T3*, thyroid hormone; *TGF-βi IV*, TGF-β inhibitor; *TPB, PdbU*, pKC activator; *TTNPB*, RA agonist; *SANT1*, SHH inhibitor; *γ-sec inh XX, XXI*, Notch inhibitor (γ-secretase inhibitor); .

Table 1 Human Pancreas Development

Fetal Age	CS	dpc	Pancreas Development	References
First trimester				
4 wpc (22–28 dpc)	CS9	22–26	Endoderm has not yet folded and is in open communication with the visceral endoderm of the yolk sac	Jennings et al. (2013)
	CS10	25–27	Endoderm folds into gut tube with formation of the anterior intestinal port (AIP). Notochord is adjacent to dorsal foregut endoderm, from which SHH expression is excluded. SHH is detected in the ventral endoderm, which is adjacent to the lateral mesoderm. FOXA2 is detected in all epithelial cells. SOX17 is only expressed in the domain that lacks SHH. PDX1 and GATA4 are not detected	Jennings et al. (2013)
	CS11	27–29	Earliest detection of PDX1 reported at ~25 dpc	Piper et al. (2004)
5 wpc (29–35 dpc)	CS12	29–31	PDX1 detected in dorsal and ventral buds along with GATA4, FOXA2, and SOX17. Dorsal pancreatic endoderm is separated from the notochord and the dorsal aorta by mesenchyme	Jennings et al. (2013)
	CS13	30–33	Ventral and dorsal buds express SOX9, PDX1, and GATA4 and are in close proximity to the ventral and dorsal anastomoses of the left and right vitelline veins. Microlumens detected in dorsal bud. SOX9 transcripts (in situ hybridization) are detected at 32 dpc. NKX6.1 first detected, most readily in dorsal bud. NKX2.2 not detected. SOX17 expression lost	Piper et al. (2002), Piper et al. (2004), and Jennings et al. (2013)
	CS14	33–35	N/A	

6 wpc (36–42 dpc)	CS15	35–37	Gut rotation brings the pancreatic buds to each side of the portal vein. Presence of microlumens in both buds. Expression of PDX1, FOXA2, SOX9, and NKX6.1. GATA4 very weak/negative	Jennings et al. (2013)
	CS16	37–40	Epithelium 1–2 cells thick, clearly branched. Expression of PDX1, FOXA2, SOX9, and NKX6.1. NKX2.2 not detected	Jennings et al. (2013)
	CS17	39–42	Pancreatic buds fuse late in the 6th week	Jeon, Correa-Medina, Ricordi, Edlund, and Diez (2009), Bocian-Sobkowska, Zabel, Wozniak, and Surdyk-Zasada (1999), and Gittes (2009)
7 wpc (43–49 dpc)	CS18	42–45	SOX9 transcripts detected in the entire epithelium at ~52 dpc	Piper et al. (2002)
	CS19	45–47	GATA4 detected in some peripheral cells and excluded from trunk regions. NKX6.1 expressed with varying intensity, not restricted to trunk region yet. FOXA2 also expressed at varying intensity. SOX9 strongly expressed in all cells	Jennings et al. (2013) and Piper et al. (2002)
	CS20	47–50	N/A	
8 wpc (50–56 dpc)	CS21	49–52	Insulin$^+$ cells at ~7.5 wpc. GATA4 and CPA1 coexpressed in tip cells indicating acinar differentiation, whereas NKX6.1 and SOX9 are not excluded yet	Piper et al. (2004), Jeon et al. (2009), and Jennings et al. (2013)
	CS22	52–55	N/A	
	CS23	53–58	N/A	

Continued

Table 1 Human Pancreas Development—cont'd

Fetal Age	CS	dpc	Pancreas Development	References
9 wpc			Insulin, glucagon, and somatostatin$^+$ cells detected at 8.5 wpc. PP is not detected. Prevalence: insulin > glucagon > SST	Piper et al. (2004)
10 wpc			NKX6.1 excluded from GATA4 and CPA1 coexpressing tip cells. SOX9 transcript detected more weakly (in situ hybridization)	Jennings et al. (2013) and Piper et al. (2002)
Second trimester				
12–13 wpc			Islet formation and initiation of vascularization and innervation. Beta cells express PCSK1 but only low levels of GLUT2. Both β and α cells express IAPP, NKX2.2, and MAFB	Piper et al. (2004), Roost et al. (2014), and Proshchina, Krivova, Barabanov, and Saveliev (2014)
14 wpc			SOX9 excluded from tips. Reduced SOX9 transcript levels (in situ hybridization)	Jennings et al. (2013) and Piper et al. (2002)
20 wpc			Islets are completely vascularized	Jennings et al. (2013) and Piper et al. (2002)
Third trimester				
27–28 wpc			Islets are highly innervated (greatly reduced in adult)	Jennings et al. (2013) and Rodriguez-Diaz et al. (2011)

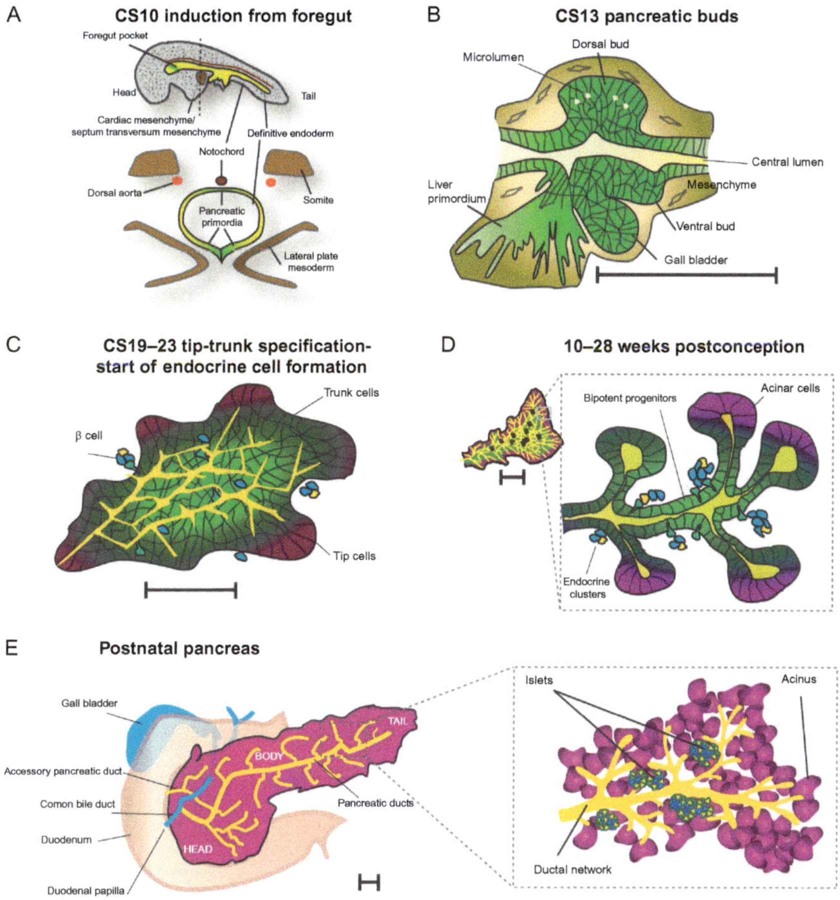

Fig. 2 Key events in pancreas development in human. (A) PDX1 is induced in the dorsal endoderm, it is in contact with the notochord, whereas the mesenchyme has separated the dorsal endoderm from the notochord and dorsal aorta when PDX1 is first detected. (B) The pancreas expands into the surrounding mesenchyme, and microlumens form, which later fuse to form a plexus. (C) Branching morphogenesis is initiated, during which the pancreatic progenitors are segregated into acinar-fated tip- and ducto-endocrine bipotent trunk domains. Endocrine differentiation in the mouse is biphasic, but there is likely only a single phase in humans starting at CS21 (6 wpc). (D) Islet formation, vascularization, and innervation are completed in humans before birth. (E) After birth, pancreas shows elaborated ductal network and islet of Langerhans. Scale bars indicate relative size of the developing pancreas.

an overview of human pancreas development and stages) (Jennings et al., 2013). Gut tube closure occurs at CS10, 25–27 dpc, corresponding to E8–E8.5 in the mouse (Fig. 2A). At this stage, the endoderm expresses FOXA2 and the dorsal endoderm expresses SOX17.

To establish foregut identity in hPSC-derived endoderm and subsequently generate pancreatic progenitors expressing pancreatic and duodenal homeobox 1 (PDX1), protocols use various combinations and timing of activation of FGF or EGF, RA, and WNT signaling, as well as inhibition of BMP and hedgehog signaling (Fig. 1) (see Rostovskaya et al., 2015 for a comparative study). In the initial protocols, primitive gut tube formation required the removal of Activin A and addition of FGF10, while posterior foregut was generated with sustained FGF signaling and addition of RA and an inhibitor of hedgehog signaling, which induced PDX1 expression (D'Amour et al., 2006). PDX1 expression could be obtained in the absence of RA, but endocrine differentiation was not induced at the subsequent stage. The recent protocols use FGF7, which is less expensive than FGF10 and does not require heparin sulfate proteoglycans for signaling, RA, a chemical SHH inhibitor, SANT1, and a BMP inhibitor, LDN, as well as vitamin C (Bruin et al., 2014; Rezania et al., 2012, 2014) (Table 2; Fig. 1). The BMP inhibitor presumably prevents ventral endoderm derivatives from forming, notably the liver, while the SHH inhibitor prevents lung tissue induction (Gouon-Evans et al., 2006; Green et al., 2011). Changes in the doses of some of these factors enable the production of more anterior (lungs, stomach) or more posterior (small intestine) endoderm derivatives (Ameri et al., 2010; Green et al., 2011; Spence et al., 2011; Touboul et al., 2010). In addition to that, the dosage and exposure time of some factors needs to be adjusted to different cell line requirements, notably the strength of BMP inhibition, to optimize later β-cell differentiation (Nostro et al., 2011).

3. PANCREATIC ENDODERM INDUCTION AND MAINTENANCE

The outgrowth of the dorsal pancreatic bud is detected at 26 dpc, followed by the ventral bud at around 30 dpc (Jennings et al., 2013; Piper et al., 2004) (Table 1; Fig. 2B). In vitro, these morphogenetic events are not recapitulated, but several of the transcription factors maintaining pancreatic identity and their triggers are conserved.

From what we know in the mouse, we expect the human pancreas to be made of pancreatic progenitors endowed with the potential to give rise to all pancreatic cells (Gu, Dubauskaite, & Melton, 2002). These cells are marked by a series of transcription factors that give them their identity and enable them to proliferate (Fig. 3). *PDX1* is an essential gene for pancreas formation, and *PDX1* mutations in humans are associated with pancreatic

Table 2 Genetic Mutations Causing Abnormal Pancreas Development

Gene	Mouse Null Mutation Pancreatic Phenotype	Human Mutation	Pancreatic Phenotype	hPSC Mutation Model	References
Transcription factors					
GATA4	KO: embryonic lethal before pancreas development	Haploinsufficiency	Pancreas agenesis/ hypoplasia and variable exocrine insufficiency	Hom: reduced PE specification and $PDX1^{+}NKX6.1^{+}$ cells	Carrasco, Delgado, Soria, Martin, and Rojas (2012), Xuan et al. (2012), Shaw-Smith et al. (2014), Tiyaboonchai et al. (2017), and Shi et al. (2017)
	cKO: normal/ euglycemic (Pdx1-cre)			Het: reduced $PDX1^{+}NKX6.1^{+}$ cells	
GATA6	KO: embryonic lethal before pancreas development	Haploinsufficiency	Pancreas agenesis/ hypoplasia	Het and Hom: reduced PE specification and $PDX1^{+}NKX6.1^{+}$ cells (dose dependent)	Carrasco et al. (2012), Xuan et al. (2012), Suzuki et al. (2014), Gong et al. (2013), De Franco et al. (2013), Yorifuji et al. (2012), Bonnefond et al. (2012), Allen et al. (2011), Tiyaboonchai et al. (2017), and Shi et al. (2017)
	cKO: normal/ euglycemic (Pdx1-cre)		Permanent neonatal diabetes		
GATA4/6	cKO: pancreatic agenesis (Pdx1-cre)			Het and Hom mutations: more severe (dose dependent)	Carrasco et al. (2012), Xuan et al. (2012), and Shi et al. (2017)
GLIS3	KO: neonatal diabetes	Homozygous or cHet mutations	Permanent neonatal diabetes and variable exocrine insufficiency (cystic dysplasia)	No phenotype	Kang et al. (2009), Senee et al. (2006), Dimitri et al. (2011), and Zhu et al. (2016)

Continued

Table 2 Genetic Mutations Causing Abnormal Pancreas Development—cont'd

Gene	Mouse Null Mutation Pancreatic Phenotype	Human Mutation	Pancreatic Phenotype	hPSC Mutation Model	References
HNF1B	KO: embryonic lethal before pancreas development	Heterozygous mutations	Pancreas hypoplasia Permanent neonatal diabetes MODY5	MODY5 hiPSC lines: compensatory increase in pancreatic transcription factors CXCR4, SOX17, FOXA2, GATA4/6, HNF1B, PDX1, MNX1, RFX6, etc. and decrease in PAX6	Haumaitre et al. (2005), De Vas et al. (2015), Horikawa et al. (1997), Lindner et al. (1999), Bellanne-Chantelot et al. (2004), Yorifuji et al. (2004), and Teo et al. (2016)
	cKO: pancreatic hypoplasia, cystic ducts, impaired acinar, and endocrine differentiation (Pdx1-cre)				
MNX1	KO: dorsal pancreatic agenesis	Homozygous mutations	Pancreas hypoplasia	Reduced $PDX1^+$ (PE) and Ins^+ cells	Harrison, Thaler, Pfaff, Gu, and Kehrl (1999), Li, Arber, Jessell, and Edlund (1999), Bonnefond et al. (2013), Flanagan et al. (2014), and Zhu et al. (2016)
			Permanent neonatal diabetes		
NEUROD1	KO: neonatal diabetes	Homozygous mutations	Permanent neonatal diabetes MODY6	NA	Naya et al. (1997) and Rubio-Cabezas et al. (2010)
		Heterozygous mutations			

NEUROG3	KO: neonatal diabetes	Homozygous or cHet mutations	Permanent neonatal diabetes or later onset diabetes	Reduced Ins^+ cells	Gradwohl, Dierich, LeMeur, and Guillemot (2000), Wang et al. (2010), Pinney et al. (2011), Rubio-Cabezas et al. (2011), and Zhu et al. (2016)
NKX2.2	KO: neonatal diabetes	Homozygous mutations	Permanent neonatal diabetes	NA	Sussel et al. (1998) and Flanagan et al. (2014)
PAX6	KO: neonatal diabetes	cHet mutations	Permanent neonatal diabetes or later onset diabetes	NA	St-Onge, Sosa-Pineda, Chowdhury, Mansouri, and Gruss (1997), and Solomon et al. (2009)
PDX1	KO: pancreatic agenesis	Homozygous or cHet mutations	Pancreas agenesis/ hypoplasia Permanent neonatal diabetes MODY4	Reduced PE and reduced Ins^+ cells	Jonsson, Carlsson, Edlund, and Edlund (1994), Stoffers, Zinkin, Stanojevic, Clarke, and Habener, (1997), Piper et al. (2002), Schwitzgebel et al. (2003), Thomas et al. (2009), and Zhu et al. (2016)
PTF1A	KO: pancreatic agenesis	Homozygous mutations	Pancreas agenesis Permanent neonatal diabetes	No phenotype	Kawaguchi et al. (2002), Sellick et al. (2004), and Zhu et al. (2016)

Continued

Table 2 Genetic Mutations Causing Abnormal Pancreas Development—cont'd

Gene	Mouse Null Mutation Pancreatic Phenotype	Human Mutation	Pancreatic Phenotype	hPSC Mutation Model	References
PTF1A enhancer		Homozygous or cHet mutations	Pancreas-specific agenesis	NA	Weedon et al. (2014)
RFX6	KO: neonatal diabetes	Homozygous mutations	Pancreas hypoplasia	Reduced $PDX1^+$ (PE) and Ins^+ cells	Smith et al. (2010), Concepcion et al. (2014), and Zhu et al. (2016)
			Permanent neonatal diabetes		
SOX9	KO: embryonic lethal (haploinsufficiency)	Haploinsufficiency	Pancreas hypoplasia, embryonic/fetal lethality	NA	Jennings et al. (2013) and Piper et al. (2002)
	cKO: pancreas hypoplasia (Pdx1-Cre)				
Other genes					
NEK8/ NPHP9	KO: no pancreas phenotype described (randomization of left-right asymmetry, cardiac anomalies, and glomerular kidney cysts)	cHet or homozygous mutations	Cystic dysplasia	NA	Manning et al. (2013) and Frank et al. (2013)
NPHP3	KO: no pancreas phenotype described (Situs inversus and cystic kidney disease)	Homozygous mutations	Cystic dysplasia	NA	Bergmann et al. (2008), Fiskerstrand et al. (2010), and Frank et al. (2013)
UBR1	KO: exocrine pancreatic insufficiency	cHet and homozygous mutations	Exocrine insufficiency	NA	Atik et al. (2015), Zenker et al. (2005), and Zenker, Mayerle, Reis, and Lerch (2006)

cHet, compound heterozygous mutation; *cKO*, conditional knockout; *NA*, not available; *KO*, knockout.

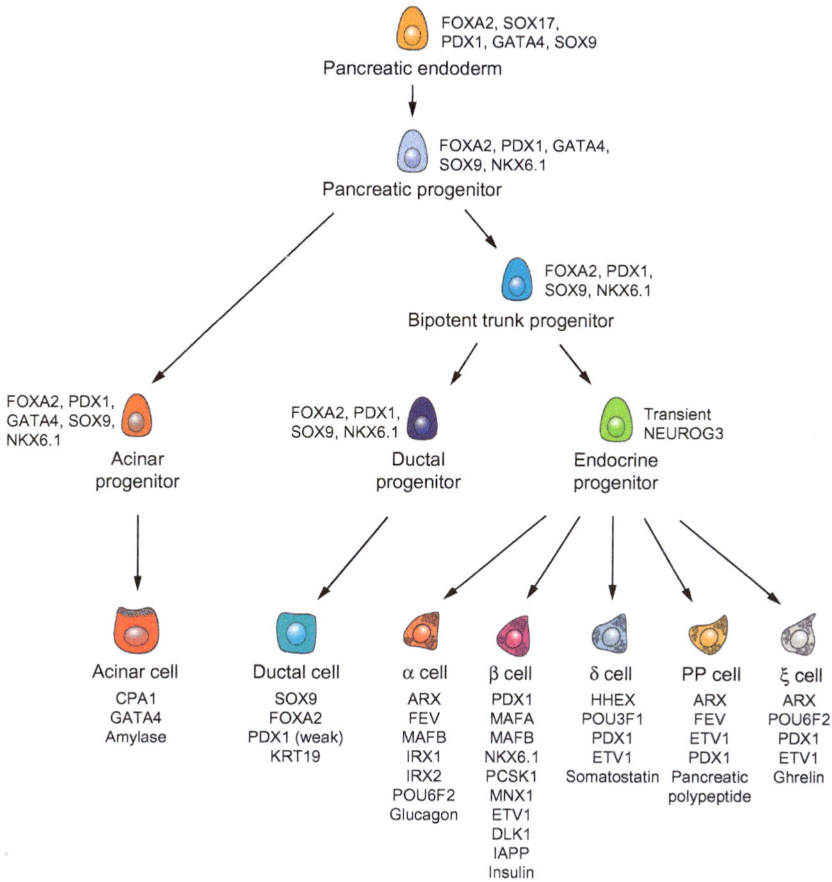

Fig. 3 Overview of lineage segregation in human pancreas development. Key transcription factors and other markers that identify different cell types are listed in each lineage. The indicated factors are based on data from immunohistochemical studies of human pancreas development, supplemented with data from single-cell gene expression analysis of adult human islet cells (Petersen et al., 2017).

agenesis/hypoplasia and neonatal diabetes (Piper et al., 2002; Stoffers et al., 1997; Thomas et al., 2009) (Table 2). PDX1 is detected early in the forming dorsal and ventral buds at 26 dpc (CS12) (Jennings et al., 2013; Piper et al., 2004). *Pdx1* is also essential for pancreas development in mice, as homozygous loss of *Pdx1* leads to pancreatic agenesis (Jonsson et al., 1994). Though *PDX1*$^{-/-}$ hESC lines have been generated, their effect on pancreas progenitors has not been studied in detail other than the obvious loss of PDX1 (Zhu et al., 2016). Little is known about what triggers PDX1 expression in human. In the mouse, the induction of *Pdx1* in the dorsal pancreatic anlage requires RA signaling (Molotkov, Molotkova, & Duester, 2005) and Activin from the overlying notochord, preventing *Shh* expression in the dorsal endoderm (Hebrok, Kim, & Melton, 1998; Hebrok, Kim, St Jacques, McMahon, & Melton, 2000) (Fig. 2A). Subsequent signaling from endothelial cells is essential for the maintenance of PDX1 expression in the dorsal pancreas (Yoshitomi & Zaret, 2004). In human, the dorsal endoderm is also in contact with the notochord at CS10, and it may be the reason why there is no SHH in the dorsal human foregut endoderm at this stage (Jennings et al., 2013).

In vitro, after posterior foregut induction, FGF7, vitamin C, the BMP inhibitor LDN, RA, the SHH inhibitor SANT1, and a PCK activator are maintained to produce pancreas progenitors expressing PDX1 (Fig. 1). The protocols thus partially mimic the mouse mechanisms of pancreas induction, directly repressing SHH rather than mimicking notochord signals. They appear to neglect the elusive endothelial signals. It is likely that the pancreas produced in vitro is similar to the dorsal pancreas, since the BMP inhibitor should prevent ventral endoderm from forming. There are not many markers discriminating the dorsal from the ventral pancreas in other species, but MNX1/HLXB9 is a general dorsal endoderm marker that would deserve further investigations in human systems. Mimicking dorsal pancreas generation may have secondary effects as the ventral pancreas generates more PP cells in mice and human (Brereton, Vergari, Zhang, & Clark, 2015; Fiocca et al., 1983). Vitamin C is sometimes added in these in vitro protocols, increasing total cell number and reducing endocrine progenitor specification at early stages in 2D culture but not in 3D suspension culture systems (Rezania et al., 2014).

Other transcription factors are known to maintain pancreas identity in human and their study in hESCs is starting to shed light on the conservation of their function between mouse and human (Table 2 and references therein). RFX6 is a transcription factor expressed in a wider endoderm area than the pancreas for which patient mutations, mouse inactivation, and the observed severe reduction in $PDX1^+$ cells in $RFX6^{-/-}$ hESCs are in agreement and indicate an early role in pancreas formation and a later role in pancreas and intestinal endocrine cell differentiation.

However, a few recent examples have revealed differences in activities between mice and humans. GATA4 and GATA6 form a redundant pair in mice and their double inactivation results in pancreatic agenesis, while heterozygous mutants are normal (Carrasco et al., 2012; Xuan et al., 2012) (Table 2). In contrast, both *GATA4* and *GATA6* haploinsufficient mutations lead to neonatal or childhood-onset diabetes, ranging from pancreatic agenesis to hypoplasia with variable exocrine insufficiency (Allen et al., 2011; Bonnefond et al., 2012; Cebola et al., 2015; Shaw-Smith et al., 2014; Xuan et al., 2012). This is in agreement with their expression in progenitors in both species, though *GATA4* expression is delayed in humans (CS12) compared to mice, where it is expressed before gut tube closure (Jennings et al., 2013; Kuo et al., 1997; Molkentin, Lin, Duncan, & Olson, 1997). The use of hPSC culture models bearing patient-relevant GATA6 haploinsufficient mutations (patient-derived induced PSCs (iPSCs) or engineered hESCs) has provided mechanistic insight into the role of GATA factors in human (Shi et al., 2017; Tiyaboonchai et al., 2017). They show that GATA6 is expressed earlier and more highly than GATA4 in human and that GATA6 has a regulatory role over GATA4. GATA6 haploinsufficiency leads to reduced endocrine cell formation, while pancreas progenitor formation is reduced in haploinsufficient cells if a suboptimal differentiation protocol using limiting RA is used (Tiyaboonchai et al., 2017). As in mice, additional reduction in *GATA4* gene dosage exacerbates the defect. These studies also highlight that optimized differentiation protocols can mask the role of a gene at a given stage and overcome the manifestation of genetic mutations by providing molecules triggering downstream biological pathways (here RA) and thus call for caution in interpretations.

MNX1 (motor neuron and pancreas homeobox 1) is another transcription factor that requires caution in the interpretation of phenotypes analyzed in hESCs in vitro. Data in mice show that this gene has two functions, one in dorsal—but not ventral bud formation and one in β-cell maturation (Harrison et al., 1999; Li et al., 1999; Pan, Brissova, Powers, Pfaff, & Wright, 2015; Thompson, Gésina, Scheinert, Bucher, & Grapin-Botton, 2012) (Table 2). Human homozygous mutations result in neonatal diabetes without developmental pancreatic defects (Bonnefond et al., 2013), indicating that MNX1 also has an important role in β-cell development in humans. It is thus not clear whether the early function in dorsal pancreas formation is conserved though MNX1 protein expression is detected at gestational week 7 (G7w) in $SOX9^+$ progenitor cells, and then in β cells and $SOX9^+$ ductal cells at G18w (Pan et al., 2015). Even more confusingly, inactivation of *MNX1* in hES cells has no effect on $PDX1^+$ or β-cell numbers (Zhu et al., 2016). It is possible that, as above, a component added in the culture medium bypasses the function of MNX1.

In the same spirit, no defect was detected in pancreatic transcription factor 1 subunit α $(PTF1A)^{-/-}$ hESC-derived progenitors and β cells

(Zhu et al., 2016), although it is known to be crucial from the earlier stages of human pancreas development, since mutations in *PTF1A* coding or regulatory sequences lead to pancreas agenesis and permanent neonatal diabetes mellitus (Sellick et al., 2004; Weedon et al., 2014) (Table 2). Expression of PTF1A protein during human fetal pancreas development has not been examined due to the lack of reliable antibodies (Jennings et al., 2013), but transcripts for *PTF1A* have been detected from 7 to 21 wpc, with a peak around 16 wpc (Jeon et al., 2009). The study of *Ptf1a*-null mice shows that Ptf1a maintains pancreatic identity, prevents pancreas progenitors from contributing to the duodenum, and is later required for acinar cell biogenesis and function (Burlison, Long, Fujitani, Wright, & Magnuson, 2008; Kawaguchi et al., 2002).

hPSC models have in other instances been useful to reveal the impact of specific patient mutations. Hepatic nuclear factor 1 homeobox B (*HNF1B*) heterozygous mutations can cause a spectrum of defects ranging from pancreatic agenesis or hypoplasia (associated with polycystic kidney disease and dysgenetic gonads), neonatal diabetes or maturity onset diabetes of the young (MODY5), in agreement with inactivation at different stages in mice (Bellanne-Chantelot et al., 2004; Haumaitre et al., 2006; Horikawa et al., 1997; Lindner et al., 1999; Yorifuji et al., 2004). Though little is known about HNF1B expression in human, a recent report shows *HNF1B* expression peaking at pancreatic endoderm stage during in vitro differentiation of hiPSCs (Teo et al., 2016). When MODY5 patient-derived iPSC lines harboring a point mutation causing pancreatic hypoplasia were differentiated into pancreatic endoderm, multiple markers for definitive endoderm and pancreatic endoderm were upregulated (Teo et al., 2016). This study revealed a direct upregulation of *PDX1* by mutant HNF1B that was not phenocopied by wild-type HNF1B. At the same time, the mutation caused an indirect decrease in $PAX6^+$ cells, indicating impaired endocrine differentiation. The interpretation of the in vitro results is somehow difficult to fit to the phenotype of patients with dorsal pancreatic agenesis, and this may be due to a long-term indirect effect of persistent progenitor state in conjunction with impaired β-cell differentiation.

Several other transcription factors are known to be expressed in human pancreas progenitors and hESC-derived progenitors, such as SOX9, although their function has not been studied in vitro (Table 2). Another essential pancreatic endoderm marker, NKX6.1, is detected from CS13 (30–33 dpc), is expressed throughout the pancreatic epithelium by 9 wpc, and becomes enriched in β cells at 13 wpc with weak expression in other cell types (Jennings et al., 2013; Riedel et al., 2012). In adult human islets, NKX6.1

is only detected in β cells (Riedel et al., 2012), which is similar in mouse (Sander et al., 2000). Neither inactivation in hESCs nor the occurrence of spontaneous mutations has informed its function in human. Additional factors are important for the maintenance and expansion of pancreatic identity in mouse, such as one cut homeobox genes (*ONECUT1/HNF6*, *ONECUT2*, and *ONECUT3*) and Prospero Homeobox 1 (*PROX1*) (Haumaitre et al., 2005; Jacquemin et al., 2000; Westmoreland et al., 2012). Though global profiling has revealed their expression in vitro at the stage when pancreatic progenitors peak (Xie et al., 2013), little is known about their role in human. Global transcriptome and epigenome signatures in hESC models of pancreas differentiation reveal that the pancreas progenitor program is induced progressively, associated with removal of histone H3 lysine 27 trimethylation marks triggered by Polycomb proteins, first on genes such as *ONECUT2*, then *SOX9* and *PDX1*, and later on *NKX6.1*, *HNF6*, and *PTF1A* (Xie et al., 2013). The enhancers regulating their transcription have been identified and are poised from the stage of gut tube induction (Wang et al., 2015). This analysis revealed enrichment for *FOXA*, *GATA*, *PDX1*, *HNF4A*, *HNF1*, and *RFX* motifs, and direct targets of *PDX1* and *FOXA2* were determined by Chip-Seq (Wang et al., 2015).

Recent protocols have aimed at inducing cells coexpressing PDX1 and NKX6.1, as a hallmark of pancreatic progenitors with a potency to differentiate into insulin-producing β cells. It is based on the observations that $PDX1^+$ cells that are not $NKX6.1^+$ may be duodenal (Kroon et al., 2008; Nostro et al., 2015; Russ et al., 2015) and NKX6.1-high pancreatic endoderm cells generate more mature β cells after transplantation (Rezania et al., 2013). However, it was recently shown that NKX6.1 expression in pancreas progenitors is not necessary to form β cells and expression at the endocrine progenitor stage is sufficient (Petersen et al., 2017). Recently, glycoprotein 2 (GP2), a cell surface protein specific for pancreatic progenitors was identified by transcriptional profiling as a marker of $PDX1^+/NKX6.1^+$ progenitors (Ameri et al., 2017; Cogger et al., 2017). It can also identify the same cells in the human fetal pancreas (at week 9.1) and is thus a convenient marker for cell sorting, more specific than the previously identified CD142 (Ameri et al., 2017; Kelly et al., 2011).

4. PANCREAS EXPANSION

Cell proliferation leads to a rapid expansion of the pancreatic buds into the surrounding mesenchyme. Lineage tracing in mice has shown that buds contain multipotent progenitors that give rise to the major lineages of the pancreas: ductal, acinar, and endocrine cells (Gu et al., 2002; Kopp et al., 2011; Larsen et al., 2017; Pan et al., 2013; Solar et al., 2009; Zhou et al., 2007) (Fig. 3). In human, pancreas progenitors are also very proliferative, apparently with a decrease in time and they remain so after transplantation of the 8 wpc pancreas under the kidney capsule in immunocompromised SCID mice (Castaing, Duvillie, Quemeneur, Basmaciogullari, & Scharfmann, 2005; Castaing et al., 2001; Piper et al., 2004; Ye, Duvillie, & Scharfmann, 2005). These grafts also revealed that they can generate endocrine and acinar cells.

Pancreas progenitors produced in vitro using different protocols have been shown in some instances to have the potential to generate acinar and endocrine cells. However, the majority of current protocols do not recapitulate a pancreatic progenitor expansion step, rather relying on hESC expansion to increase cell numbers. It was shown only recently that hESC-derived pancreatic progenitors can be expanded in vitro (Trott et al., 2017). In this protocol, a feeder layer reportedly prevents further differentiation into the exocrine lineage, although the mechanism responsible for this was not investigated. Small molecules and growth factors were used to maintain a strong adhesion of the pancreatic progenitors to the feeder layer (TGF-β inhibitor) or to ensure proliferation and maintenance of progenitor identity (RA, EGF, FGF10, Notch inhibitor). Though the role of the Notch pathway has not been tested thoroughly in human, another study showed that *HES1* inactivation in hESCs does not have an effect on the number of PDX1^{+} cells (Zhu et al., 2016). However, the assay was most likely carried out in confluent cultures where growth is largely inhibited and it would be worth testing this in a context that enables expansion.

In mouse, the FGF, EGF, and Notch pathways are also required for pancreatic expansion, as well as the WNT pathway (reviewed in Larsen & Grapin-Botton, 2017). In human, many FGF-family members are expressed in the pancreas at 6–9 wpc and at least FGF7 and FGF10 are known to be expressed in the mesenchyme. The addition of these factors to explanted human pancreas leads to a twofold increase of proliferation in vitro (Ye et al., 2005). WNT signaling is enriched in CS16–18 pancreatic progenitors (Cebola et al., 2015), and dissociated cells from early pancreas rudiments expand in culture in response to a culture medium containing WNT agonists, EGF and FGF10 (Bonfanti et al., 2015). Recently, TEAD and YAP were also found to activate the enhancer network of human embryonic pancreatic progenitors and regulate their expansion (Cebola et al., 2015). Both the YAP and WNT pathway would be worth testing in hPSC-derived progenitors.

5. EMERGENCE OF AN EXOCRINE GLAND: FORMATION OF BRANCHES AND SEGREGATION OF THE ACINAR LINEAGE

In mouse, the multipotent pancreatic progenitors soon segregate into distinct populations with restricted competence. While *Ptf1a, Gata4, Sox9,* and *Nkx6.1* are initially coexpressed, *Ptf1a* and *Gata4* become gradually restricted to the tip cells of the pancreatic epithelium, whereas *Nkx6.1* and *Sox9* are restricted to the central trunk domain (Schaffer, Freude, Nelson, & Sander, 2010). A similar process occurs in human. By CS19 (45–47 dpc), the central areas of the pancreas exhibit less GATA4, whereas carboxypeptidase A1 (CPA1), GATA4, NKX6.1, and SOX9 are expressed for an extended period in the tip cells (Jennings et al., 2013) (Table 1; Fig. 2C). About 2.5 weeks later, by 10 wpc, NKX6.1 becomes restricted to the central areas (presumably trunk progenitors), followed by SOX9 at 14 wpc (Jennings et al., 2013). By comparison to mice, we assume that the tips are made of acinar-committed cells, whereas the central trunk is made of bipotent endocrine–ductal progenitors (Kopinke et al., 2011; Kopp et al., 2011; Solar et al., 2009) (Fig. 2D). It is unclear whether a population with bipotent ducto-endocrine potential also exists in vitro, but by comparison with mice, it would be expected to lack PTF1A expression. Producing acinar cells from hPSCs has not been a major focus. However, the presence of Indolactam V or FGF7 in the culture medium reportedly promotes the formation of acinar cells (Chen et al., 2009; Hohwieler et al., 2017; Takizawa-Shirasawa et al., 2013), while a combination of BMP and valproic acid produces ductal cells (Simsek et al., 2016). In addition, even protocols designed for the production of endocrine cells enable the formation of exocrine cells or their progenitors (Ma, Wert, Shvartsman, Melton, & Jaenisch, 2018). However, to what extent these molecules mimic acinar and ductal commitment in vivo is unclear.

Notch signaling plays an important role in the tip–trunk decision, preventing differentiation of both acinar and endocrine cells at different time points (Hald et al., 2003; Horn et al., 2012; Murtaugh, Stanger, Kwan, & Melton, 2003). It would be interesting to investigate whether the *HES1* homozygous mutants made by Zhu et al. have an increased acinar differentiation in addition to the increase in endocrine differentiation (Zhu et al., 2016). The transplantation of the 8 wpc human pancreas under the mouse kidney capsule demonstrated that in human, as in mice, $CPA1^+$ acinar cells proliferate efficiently, but it is not known whether they do so after in vitro production (Castaing et al., 2005). Interactions with the basal lamina via integrins may also play a role in defining the outer cells and later formation of acini (Kesavan et al., 2009; Shih, Panlasigui, Cirulli, & Sander, 2016).

The segregation of acinar cells is preceded shortly by the initiation of pancreatic ducts and branches that will be needed in vivo to transport pancreatic enzymes to the duodenum. The ducts are initiated by the formation of rosettes of progenitors defining central microlumen that will connect to form a network (Fig. 2C and D) and remodel into branches (Fig. 2E) of single-layered epithelial ducts (Kesavan et al., 2009; Villasenor, Chong, Henkemeyer, & Cleaver, 2010). Likewise, microlumens are detected in the human developing pancreas around CS13 (30–33 dpc; Table 1; Fig. 2B) (Jennings et al., 2013). The 2D nature of most hESC-derived cultures limits the investigation of these mechanisms in vitro. However, the pancreatic endoderm cells form more than one layer, especially in areas expressing PDX1, and microlumens can be observed (Löf-Öhlin et al., 2017). When cultured in Matrigel, in 3D, hPSC-derived progenitors can form spheres with a monolayer of cells lining a large lumen (Boj et al., 2015; Hohwieler et al., 2017; Huang et al., 2015; Lee et al., 2013). Currently there is no model suitable for studying mutations that cause diabetes associated to morphological defects in human, such as HNF1B, NEK8, NPHP3, and GLIS3 (Table 2 and references therein).

6. ENDOCRINE SPECIFICATION THROUGH A TRANSIENT NEUROG3$^+$ STATE

Producing endocrine cells has been the main focus of the work using hESCs. In mice, Notch signaling activates the transcription factor *Hes1* that acts as a gatekeeper of endocrine commitment. The contribution of Notch signaling to human fetal pancreas development is largely unknown. However, *HES1*$^{-/-}$ hESC lines exhibit a significant increase in endocrine differentiation, in agreement with the function of *Hes1* in mice (Fujikura et al., 2006; Zhu et al., 2016) (Table 2). HES1 is a repressor of the bHLH transcription factor Neurogenin 3 (NEUROG3), which is transiently expressed to initiate endocrine cell specification (Gradwohl et al., 2000; Jensen et al., 2000) (Fig. 3).

During human fetal development, the expression of *NEUROG3* transcripts has been detected between 9 and 35 wpc in humans, peaking between 9 and 19 wpc (Salisbury et al., 2014) (Table 1; Fig. 2D). The presence of two peaks at 12 and 17 wpc is controversial (Jeon et al., 2009; Sarkar et al., 2008). Thus, there are no clear indications that *NEUROG3* expression and endocrine development occur in two waves in humans as in mice (Villasenor, Chong, & Cleaver, 2008). Studies showing reliable detection of NEUROG3 protein in human fetal samples are scarce, as nearly all studies have used an antibody raised against mouse NEUROG3, which displays very limited affinity toward human NEUROG3 (Honoré et al., 2016). The pancreatic program of endocrine differentiation occurs with a similar timing after transplanting human fetal pancreas of 7–9 wpc into mice, showing that pancreas development can progress autonomously in this supportive adult environment (Capito et al., 2013).

Compound heterozygous point mutations in *NEUROG3* in human can cause permanent neonatal diabetes with malabsorptive diarrhea, suggesting its role in pancreatic and intestinal endocrine cell production (Rubio-Cabezas et al., 2011). Functional studies on several patient-relevant NEUROG3 mutant forms suggest that only those with severe functional impairment lead to diabetes, while small loss of activity affects enteroendocrine differentiation (Pinney et al., 2011; Rubio-Cabezas et al., 2011; Wang et al., 2006). A dosage sensitivity has also been shown in mouse (Wang et al., 2010).

Two recent studies employed the CRISPR–Cas9 system to inactivate *NEUROG3* in hESC lines to address its role in human pancreatic endocrine development. *NEUROG3*$^{-/-}$ hESC lines fail to produce any endocrine cells (McGrath, Watson, Ingram, Helmrath, & Wells, 2015) or exhibit a severe reduction (Zhu et al., 2016) depending on the protocol used. A 75% reduction of endocrine cell numbers in *NEUROG3*$^{+/-}$ hESC lines was observed in the protocol the most sensitive to NEUROG3. In summary, the findings from patient-derived *NEUROG3* mutations and studies using hESC differentiation clearly demonstrate that NEUROG3 is important for human pancreatic endocrine cell specification but leave a possibility that other proteins may act redundantly in endocrine specification. This would not be entirely surprising as pancreatic endocrine cells are specified by other proneural bHLH factors rather than Neurog3 in zebrafish (Flasse et al., 2013).

Although human studies have focused on the role of NEUROG3 in the generation of β cells, lineage tracing studies in mice have demonstrated that the *Neurog3*$^{+}$ population represents a precursor state for all pancreatic endocrine lineages (Gu et al., 2002) (Fig. 3). The timing of NEUROG3 onset has been demonstrated to influence the fate choice of endocrine progenitors in mice: the early cells becoming polyhormonal coexpressing glucagon and low levels of insulin or glucagon only, while other endocrine cells form in the second wave of NEUROG3 expression indicating a dynamic competence of endocrine progenitors during pancreas development. A similar time-controlled NEUROG3 gain of function at different stages of hESC differentiation in a *NEUROG3*-null hESC line (Zhu et al., 2016) showed that endocrine cells expressing insulin, glucagon, or somatostatin formed at all stages of induction, but it is unclear which stages of human development are explored in these protocols, which are much shorter than the period of NEUROG3 cell formation in vivo. It is also possible that pancreatic progenitors are exposed to different niches or extracellular signals at different stages of development in vivo, which may lack in vitro.

The initiation of NEUROG3 expression from hESC-derived pancreas progenitors can be monitored in vitro using reporter lines (Löf-Öhlin et al., 2017; Petersen et al., 2017), but the process that triggers its activation in vitro is unclear. Though most of the protocols use a Notch inhibitor, it is marginally beneficial and not required for NEUROG3 induction (D'Amour et al., 2006; Russ et al., 2015) or used much later than the stage of NEUROG3 induction (Rezania et al., 2014). It is also unclear why Notch inhibition enriches the generation of insulin-producing PDX1^{+}/NKX6.1^{+}/NEUROD1^{+} cells, over other hormone-producing cells.

Instead, ALK5 inhibition may be sufficient to induce endocrine differentiation from progenitors in vitro (Fig. 1). ALK5 inhibitor reinforces NEUROG3 and its downstream endocrine marker expression. However, *NEUROG3* is induced already at pancreatic progenitor stage before the ALK5 inhibitor exposure (Petersen et al., 2017). The mechanisms by which ALK5 promotes NEUROG3 and more generally endocrine specification, as well as their in vivo relevance and evolutionary conservation remain unclear. Taken together, it would be interesting to investigate how *NEUROG3* is induced from pancreatic progenitors and which signaling or niche is important to enrich NEUROG3^{+} cells.

7. LAUNCHING THE ENDOCRINE PROGRAM DOWNSTREAM OF NEUROG3

Once cells reach the NEUROG3 expression threshold needed for endocrine progenitor commitment, they delaminate from the epithelium through a partial epithelial-to-mesenchymal transition dependent on the expression of snail family transcriptional repressor 2 (*Snail2*) and down-regulation of cadherin 1 (*Cdh1*; E-cadherin) (Gouzi, Kim, Katsumoto, Johansson, & Grapin-Botton, 2011; Rukstalis & Habener, 2007). Delaminating NEUROG3^{+} cells retains a narrow luminal tether for a while

(Bechard et al., 2016) before migrating to form clusters with other endocrine cells at the interface between the ductal epithelium and the mesenchyme. Further differentiation of these cells to hormone-producing endocrine cells occurs simultaneously (Gouzi et al., 2011). The process driving the migration and subsequent clustering of endocrine progenitors and their progeny is poorly understood but has been shown to involve integrins (Cirulli et al., 2000). These clusters represent the immature islets of Langerhans, which have been demonstrated to be polyclonal in both mice and humans (Deltour et al., 1991; Scharfmann, Xiao, Heimberg, Mallet, & Ravassard, 2008). The delamination process is likely to occur in human as well, but its steps and molecular control are not known. hESC-derived cells turning on *NEUROG3* expression in vitro also lose apicobasal polarity (Löf-Öhlin et al., 2017) and aggregate into delaminating clusters. Intriguing observations were recently made where inhibition of the rho kinase pathway promotes beta cell formation from hPSCs, though it remains unclear whether this affect the delamination process or another step (Ghazizadeh et al., 2017; Toyoda et al., 2017). The aggregation of endocrine cells is seen in some human stem cell models in vitro, especially in 3D (Pagliuca et al., 2014; Rezania et al., 2014; Russ et al., 2015).

NEUROG3^{+} endocrine progenitors and their progeny are largely postmitotic, as *Neurog3* expression and endocrine commitment induce the expression of cell-cycle inhibitor, cyclin-dependent kinase inhibitor 1a (*Cdkn1a*), and p21 protein (Cdc42/Rac1)-activated kinase 3 (*Pak3*) (Kim et al., 2015; Miyatsuka, Kosaka, Kim, & German, 2011; Piccand et al., 2014). However, it has recently been demonstrated that a population of lowly expressing *Neurog3*$^{+}$ cells with low or no detectable NEUROG3 protein expression represents a subpopulation of endocrine progenitors with proliferative capacity (Bechard et al., 2016; Larsen et al., 2017). Whether a similar population exists in human is not certain, as *NEUROG3*-expressing cells in 8 wpc pancreas neither incorporate BrdU nor express KI67 (Castaing et al., 2005). The terminally differentiated endocrine cells reenter the cell cycle in a transient burst of proliferation during the perinatal period, which is critical for establishing the adequate size of the β-cell population (Georgia & Bhushan, 2004). Proliferation in the human fetal pancreas has been examined using Ki67 staining in samples from 12 to 41 wpc, but did not indicate an increased proliferation rate in endocrine cells in late gestation (Bouwens, Lu, & De Krijger, 1997). On the contrary, the Ki67 labeling index decreased in endocrine cells throughout development. Instead, the primary β-cell expansion in humans appears to occur postnatally during infancy (Gregg et al., 2012;

Kassem, Ariel, Thornton, Scheimberg, & Glaser, 2000; Meier et al., 2008). A very low level of NEUROG3 is sustained in differentiated endocrine cells and has been demonstrated to be important for β-cell maturation and function (Wang et al., 2009). A similar activity has not been investigated in human or in hPSC-derived β cells.

Following their specification by a peak of high NEUROG3 expression, endocrine cells progressively activate the genes that will be necessary for their function. Some of these genes are common for all endocrine cells, while others are specific for the different types of endocrine cells (Fig. 3). The lineage trajectory of NEUROG3 targets has been studied in vitro at single-cell resolution, delineating steps of progression on the endocrine path (Petersen et al., 2017). Newborn endocrine progenitors expressing *NEUROG3* retain pancreatic progenitor markers (*SOX9*, *ONECUT1*, *HES1*, *NOTCH1/2*), while the first endocrine transcription factors initiated downstream of *NEUROG3* are *RUNX1T1* and *NKX2.2*, followed by *PAX4* and *NEUROD1* and subsequently functional genes such as *CHGA*, *GCK*, and the prohormone convertases subtilisin/kexin type 1/2 (*PCSK1/2*). These transcription factors are NEUROG3 targets in mice. It remains to be seen whether this sequence is conserved in vivo. There is some evidence that impaired function of several genes downstream of NEUROG3 can lead to diabetes due to the formation of a suboptimal or functionally mis-specified endocrine and β-cell mass. Hence mutations in *PAX6*, *GLIS3*, *RFX6*, *NEUROD*, *NKX2.2*, and *PDX1* lead to neonatal diabetes (Table 2 and references therein). GLIS3 also predisposes to type 1 and type 2 diabetes, though it is not clear whether this is due to its function in emerging endocrine cells (Nogueira et al., 2013; Wen & Yang, 2017). Several other polymorphisms increasing type 2 diabetes risk are close to the loci of *PROX1*, *HHEX*, *PDX1*, *HNF1B*, *PDX1*, and *NOTCH2*, and it would be interesting to know whether they affect endocrine cell development in these patients (Table 2). Human hPSC models have been valuable to assess the function of some genes functioning downstream of NEUROG3 and have shown that the disruption of *RFX6* and *ARX*, but not *MNX1* and *GLIS3*, affects β-cell production in vitro (Gage et al., 2015; Zhu et al., 2016). Exploration of other protocols may reveal functions of GLIS3 and MNX1, which inactivation leads to strong phenotypes in mice (Kang et al., 2009; Li & Edlund, 2001). It is known that impaired development can increase the risk of type 2 diabetes, notably from the Dutch winter famine at the end of World War 2. The offspring from women exposed to the famine up to 32 weeks of development, the window period during which NEUROG3 specifies endocrine cells, developed impairments in glucose secretion, glucose intolerance but normal insulin resistance unless individuals became diabetic (de Rooij et al., 2006).

Human endocrine cells are assembled into islet-like structures already at 12–14 wpc (Gregg et al., 2012; Jeon et al., 2009) (Table 1). At this time, β cells coexpress PCSK1 and islet amyloid polypeptide (IAPP) but have only low levels of GLUT2. α cells express IAPP at this stage but not PCSK1 (Piper et al., 2004). Both β and α cells express NKX2.2 and MAF BZIP transcription factor B (MAFB) during development and in adult human islets (Riedel et al., 2012). These clusters initially have β cells in the center and α cells in the periphery, like in mice, but after birth, islets with more intermingled α and β cells progressively emerge, possibly by expansion of islets leading to coalescence (Bosco et al., 2010; Brissova et al., 2005; Cabrera et al., 2006; Jeon et al., 2009; Steiner, Kim, Miller, & Hara, 2011). Human islet vascularization begins during early development with endothelial cells in close proximity to endocrine cell clusters at 10 wpc and blood vessels penetrating the forming islets at 14 wpc (Piper et al., 2004). Compared to the mouse, human islets appear to be fully vascularized long before birth around 20 wpc (Roost et al., 2014), though the adult human islets are less vascularized than mouse islets (Brissova et al., 2015) (Table 1). Human fetal islet innervation begins at around 12 wpc, and they are highly innervated in the third trimester (27–28 wpc) (Proshchina et al., 2014) (Table 1). In contrast, adult human islets appear less innervated, and neurons are closely associated with the islet microvasculature rather than the endocrine cells (Rodriguez-Diaz et al., 2011), whereas mouse adult islets are highly innervated with direct contact between endocrine cells and axons (Reinert et al., 2014). The protocols used to derive pancreatic endocrine cells do not produce endothelial and neural cells and the islets are therefore not vascularized and innervated in vitro. However, endothelial cells can be later assembled (Takebe et al., 2015).

8. DIFFERENTIATING THE FIVE ENDOCRINE SUBTYPES

While mouse islets consist of ~80% β cells, human islets have fewer β cells (~60%) and more α (~30%) and δ cells (~10%) (Brissova et al., 2005; Cabrera et al., 2006; Steiner et al., 2011). At birth, human islets have equal numbers of α, β, and δ cells, and the adult composition is established through proliferation of β cells predominantly within the first 2 years of life and a decrease in δ cells (Gregg et al., 2012). The differences in cell arrangement and proportion may cause functional differences between mouse and human (reviewed by Chandrasekera & Pippin, 2013; Islam, 2010). In vitro hPSC differentiation has so far emphasized the production of β cells, though all other endocrine types are also produced. Control over the ratio of different endocrine cell types has not been prioritized.

While in mouse there is a specific temporal order of endocrine subtype differentiation and cells formed in the primary transition express glucagon and low levels of insulin, this order is not conserved in human. As in fish, in humans, the first endocrine cells detected at 52 dpc (~7.5 wpc) are insulin$^+$ cells, while glucagon, somatostatin, and PP are not found (Jeon et al., 2009; Piper et al., 2004) (Table 1). At 9 wpc, all five endocrine cell types of the pancreas are present (Riedel et al., 2012). During human pancreatic development, bihormonal cells expressing both glucagon and insulin have been reported at various frequencies (Jeon et al., 2009; Piper et al., 2004; Polak, Bouchareb-Banaei, Scharfmann, & Czernichow, 2000; Riedel et al., 2012; Sarkar et al., 2008). Riedel et al. reported a progressive decline in the number of these bihormonal cells from the onset of endocrine differentiation at around 8 wpc and the absence in adult pancreas. The purpose and fate of polyhormonal cells in human development are unknown; however, their transcriptional profile is similar to α cells more than β cells, determined by their expression of ARX and the lack of PDX1, NKX6.1, and MAFA (Riedel et al., 2012). Likewise, in vitro hPSC-derived bihormonal cells have the same transcriptional profile, and when transplanted into mice, these cells predominantly form monohormonal glucagon$^+$ cells (Basford et al., 2012; Kelly et al., 2011; Rezania et al., 2011, 2013). These findings have led to the hypothesis that polyhormonal cells during human development potentially represent cells committed toward the α-cell lineage. Lineage tracing of the early cells in mouse using a *NEUROG3-CreER* driver has shown that the early cells contribute all types of endocrine cells to the adult islets (Gu et al., 2002). Although lineage tracing that makes use of transgenic lines by using hormone promoters suggests that cells expressing one hormone do not later express another hormone in mice (Herrera, 2000), the short promoter elements used in this study may not faithfully recapitulate expression of the two insulin genes present in rodents. In humans, ghrelin-producing ε cells are detected during development as early as 9 wpc, but, unlike in mice, ghrelin is not coproduced with glucagon or other hormones, except sporadically with PPY (Andralojc et al., 2009; Heller et al., 2005; Riedel et al., 2012). Although ε cells are more frequent in the embryonic pancreas, they are also present in adult human islets, where they are localized primarily to the periphery of islets (Andralojc et al., 2009; Wierup, Svensson, Mulder, & Sundler, 2002).

All endocrine cell types have been produced in vitro from hESCs. One of the first protocols used exendin-4, in addition to IGF1 and HGF, to produce endocrine cells though none of the components appeared essential on its own (D'Amour et al., 2006). The protocol generated ~7% C-peptide$^+$ cells, while glucagon$^+$, ghrelin$^+$, PPY$^+$, and somatostatin$^+$ cells were also produced (but not quantified). Coexpression of glucagon and C-peptide was also detected. While the insulin$^+$ cells were able to respond with insulin secretion to direct depolarization, they did not respond to high glucose levels (Basford et al., 2012; D'Amour et al., 2006). However, transplanting the pancreatic progenitors into mice allowed their maturation into insulin-secreting cells (Kelly et al., 2011; Kroon et al., 2008), which were able to reverse hyperglycemia in mouse models of type 1 diabetes after 12–16 weeks of maturation in vivo (Bruin et al., 2013; Rezania et al., 2012, 2013). In 2014, two protocols were published demonstrating the differentiation of insulin-secreting β-like cells in vitro (Fig. 1). Rezania et al. published a seven-stage protocol, which produced β-like cells that responded to elevated glucose levels with increased insulin secretion, although it was delayed and reduced compared with human islets (Fig. 1). Furthermore, a minority of the β-like cells exhibited calcium responses when stimulated with increased glucose levels, which were also reduced and were slower than what was measured in human islets. Finally, the cells were able to reverse diabetes after 40 days in a diabetic mouse model, which is faster than with pancreas progenitor transplantation (Rezania et al., 2014). This finding has, however, been proposed to rely mostly on basal insulin secretion from the transplanted cells, since the measured glucose-stimulated insulin secretion was still relatively low when diabetes had been fully reversed (Johnson, 2016). These cells have important transcriptional differences with adult β cells and retain transcriptional coexpression of hormones in all cells produced, though some are monohormonal at the protein level (Petersen et al., 2017). The other protocol was published by Melton and colleagues and was a six-stage protocol; however, here the cells were differentiated in suspension culture throughout the protocol (Pagliuca et al., 2014) (Fig. 1). The produced β-like cells displayed glucose-responsive insulin secretion 2 weeks after transplantation into immunocompromised mice and could prevent the progression of diabetes upon transplantation in a host mimicking type 1 diabetes (Pagliuca et al., 2014). The available data suggest that although the cells displayed some glucose-stimulated insulin secretion in vitro and calcium responses even before transplantation, none of the cells produced in these protocols are fully mature in vitro (Johnson, 2016; Pagliuca et al., 2014; Petersen et al., 2017). The authors later demonstrated that encapsulated β-like cells could reverse diabetes in STZ-induced diabetic mice. However, this study had one important limitation, as it was not shown that the reversed hyperglycemia was indeed due to the grafted cells by removal of the graft (Vegas et al., 2016).

Since these two publications, Hebrok and colleagues published a protocol in 2015, in which the number of polyhormonal cells (INS^+/GCG^+) was greatly reduced. The generated insulin$^+$ cells displayed modest glucose responsiveness upon transplantation into nondiabetic mice. Their transplantation into STZ-induced diabetic mice resulted in significantly reduced blood-glucose levels, but did not, however, reverse diabetes (Russ et al., 2015). An alternative protocol functioning over multiple lines was also proposed by Nostro et al (2015). It is likely that the maturation of β cells produced in vitro improves in the coming years. Moreover, it would be desirable to increase the emphasis on producing other endocrine cell types. For therapy, assembling different endocrine cells would enable the local cross talk between different endocrine cell types. It would also be useful for disease modeling to test the function of variants that predispose to diabetes (Beer & Gloyn, 2016). The most effective differentiation to the α-lineage obtained in vitro so far produces 65% glucagon- and ARX-expressing cells, which effectively secrete glucagon upon stimulation with arginin, KCl, or carbachol and also reduce secretion upon stimulation with an analogue of somatostatin (Rezania et al., 2011). The cells produced in vitro are not fully mature, and they secrete GLP1 and continue to mature upon in vitro transplantation.

We know little about the mechanisms that initiate the formation of specific endocrine subtypes. However, several transcription factors are differentially expressed among endocrine subtypes (Fig. 3). Most of the work has focused on NKX6.1, PAX4, and ARX. Paired box 4 (PAX4) is necessary for β- and δ-cell formation and knockout mice form more α cells, many of which coexpress ghrelin (Sosa-Pineda, Chowdhury, Torres, Oliver, & Gruss, 1997). Lineage tracing has revealed that *Pax4* labels endocrine progenitors that contribute equally to α, β, δ, and PP cells (Greenwood, Li, Jones, & Melton, 2007). It is unknown whether the protein is ever expressed in α cells though and in β cells, its expression must be transient, as forced expression in adults alters β-cell function (Collombat et al., 2009). In the developing human pancreas, PAX4 is detected from 8–9 wpc to 21 wpc, similar to NEUROG3, but similar to the mouse, gene expression is not detected in adult human islets (Dorrell et al., 2008; Sarkar et al., 2008).

Conversely, Aristaless-related homeobox (ARX) is needed for α- and PP-cell development in humans, as patients with *ARX*-null mutations have no α or PP cells in the pancreas (Itoh et al., 2010). Likewise, *Arx* is necessary for α-cell development in mice, though not the early forming α- and poly-hormonal cells and *Arx* deficiency leads to a compensatory increase of β and δ cells. In contrast to humans, *Arx* loss does not affect PP cell differentiation in mice (Collombat et al., 2003). However, forced expression of *Arx* in early pancreatic cells (driven by *Pdx1* or *Pax6*) drives endocrine progenitors toward either an α or a PP cell fate (Collombat et al., 2007). In contrast to the mouse, *ARX*-null mutations in humans were not accompanied by an obvious increase in β and δ cells (Itoh et al., 2010). Collombat and colleagues have further shown a reciprocal repression mechanism between ARX and PAX4, proposed to be important for the specification of the α vs β/δ lineages (Collombat et al., 2005). During human fetal development, ARX expression (increasing between 13 and 19 wpc) is detected slightly later than PAX4 (from 9–10 wpc to 19 wpc) (Jeon et al., 2009) and this is also observed at the single-cell level in hESC differentiation (Petersen et al., 2017). In support of a role in α-cell development in humans, hESC lines deficient for *ARX* fail to generate any glucagon$^+$ cells after in vitro differentiation toward the pancreatic lineage, and this is accompanied by an increase in somatostatin$^+$ cells (Gage et al., 2015; Zhu et al., 2016). Introducing PAX4 expression by adenovirus at the pancreatic progenitor stage of hESC differentiation further showed reduced ARX expression, reduced glucagon$^+$ cell numbers, and increased numbers of insulin monohormonal cells (Gage, Baker, & Kieffer, 2014).

A lot of emphasis has been put on NKX6.1 in human, arguing that it enables the selection of progenitors on the path to β cells. In human and mouse, NKX6.1 is expressed in β cells and in progenitors but not in the other endocrine cells. In mouse, Nkx6.1 is needed to maintain a subset of the β-cell transcriptional network and thereby β-cell function (Henseleit et al., 2005; Sander et al., 2000). Its inactivation in β cells after or before birth leads to a loss of insulin expression and expression of somatostatin instead (Schaffer et al., 2013; Taylor, Liu, & Sander, 2013). It is also needed in NEUROG3$^+$ cells to produce β cells expressing insulin. Its role in progenitors remains unclear as no experimental setup used so far modified its expression strictly in this population though global inactivation suggests that it is transiently required in progenitors to produce NEUROG3$^+$ cells until E12.5 (Nelson, Schaffer, & Sander, 2007).

The three transcription factors form a regulatory network where both NKX6.1 and PAX4 directly repress ARX in synergy (Nelson et al., 2007; Schaffer et al., 2013) and where NKX6.1 expression is partially dependent on PAX4 activity (Wang et al., 2004).

9. FROM THE PRODUCTION OF ENDOCRINE CELL TYPES FROM PSCS TO THERAPIES

An important goal of in vitro β-cell production from hPSCs is to transplant patients with type 1 and possibly advanced forms of type 2 diabetes that do not respond to other medications. The first ongoing human phase I/II trial for patients with type 1 diabetes (ViaCyte Inc. clinical trials identifier: NCT02239354) is based on the transplantation of progenitor cells in encapsulation devices (Bruin et al., 2013). These cells mature in the body in 16 weeks and by this time point can rescue diabetic mice. The encapsulation device harbors large numbers of cells, and their semipermeable membranes prevent both immune cell contact and the release of possibly tumorigenic cells that have not matured properly, while enabling diffusion of nutrients, oxygen, proteins, peptides, and metabolites regulating hormone secretion, as well as the diffusion of the hormones produced by the cells. Whether the diffusion exchanges are fast and efficient enough to treat patients will require further investigation. Other types of encapsulation materials have been envisaged, some of which reduce alloreaction against the graft (Vegas et al., 2016).

More recently, research has enabled the efficient production of β cells in vitro (Pagliuca et al., 2014; Rezania et al., 2014; Russ et al., 2015). Selectively sorting β-cell progenitors as they are expanding (Ameri et al., 2017) or more mature β cells can further optimize transplantations. Experiments in mice show that transplanting β cells rather than progenitors enables faster treatment. However, the high oxygen demands of β cells make them prone to cell death, especially during the time needed for revascularization. This can be reduced by an oxygen reservoir (Ludwig et al., 2013).

In spite of getting closer and closer to fully functional β cells, all cells produced in vitro still lack low glucose shut-off. Moreover, it is unclear whether all the cells produced respond to glucose, many functional genes are missing, and several indicate an immature state (Domínguez-Bendala, Lanzoni, Klein, Álvarez-Cubela, & Pastori, 2015; Johnson, 2016; Kieffer, 2016; Kushner et al., 2014; Petersen et al., 2017; Spagnoli, 2015). Though some patients with no other treatment option may benefit from the graft of suboptimal cells, efforts are ongoing to understand how maturation operates in vivo and how to promote it in a dish.

Another relevant step for a therapeutic product is the development of GMP protocols compatible with mass production. The use of 3D culture protocols in reactors is a first step in this direction (Pagliuca et al., 2014; Russ et al., 2015). Transposing protocols enabling proliferation of pancreas progenitors to 3D may offer other possibilities for mass production (Trott et al., 2017).

For economic reasons, the ongoing clinical trials are designed to produce progenitors from a single ES cell line for use in all patients and the encapsulation method should alleviate allo- and autoreaction. While it is technically feasible to produce personalized β cells using iPSCs derived from each patient (Millman et al., 2016), which would likely require protocol tailoring to cell line (Nostro et al., 2015), this would be a solution of extreme cost and would not solve autoimmune destruction.

10. OUTLOOK

With the merger of BetaLogics (now a Janssen subsidiary) with ViaCyte, and the creation of Semma Therapeutics to commercialize the progress made by the Melton group, as well as other initiatives, the translation of basic research to the clinic is expected to progress in the coming years. In the meantime academic laboratories continue to play active roles in protocol development and in the development of methods to produce other pancreatic cell types. Stem cells are also expected to provide an alternative to cell lines and to develop as disease models, as well as to provide models to investigate human pancreatic cell physiology, paving the way to future drug testing. This will also be important to understand genetic predisposition to diabetes with all the emerging gene association studies. Stem cell-derived models could also be useful in the field of pancreatic cancer if more emphasis is put on the exocrine pancreas.

ACKNOWLEDGMENTS

The authors' work on human stem cells is supported by grant NNF17CC0027852 of the Novo Nordisk Foundation to DanStem, Grant DFF-7016-00045 from the Danish Council for Independent Research to A.G.-B. and C.A.C.G. and an industrial PhD Grant 1355-00115B from the Innovation Fund Denmark to M.K.B.P.

CONFLICT OF INTEREST

M.B.K.P. was an employee of Novo Nordisk A/S and may hold shares in the company.

REFERENCES

Allen, H. L., Flanagan, S. E., Shaw-Smith, C., De Franco, E., Akerman, I., Caswell, R., et al. (2011). GATA6 haploinsufficiency causes pancreatic agenesis in humans. *Nature Genetics*, *44*(1), 20–22. https://doi.org/10.1038/ng.1035.

Ameri, J., Borup, R., Prawiro, C., Ramond, C., Schachter, K. A., Scharfmann, R., et al. (2017). Efficient generation of glucose-responsive beta cells from isolated GP2 + human pancreatic progenitors. *Cell Reports*, *19*, 36–49. https://doi.org/10.1016/j.celrep.2017.03.032.

Ameri, J., Ståhlberg, A., Pedersen, J., Johansson, J. K., Johannesson, M. M., Artner, I., et al. (2010). FGF2 specifies hESC-derived definitive endoderm into foregut/midgut cell lineages in a concentration-dependent manner. *Stem Cells (Dayton, Ohio)*, *28*, 45–56. https://doi.org/10.1002/stem.249.

Andralojc, K. M., Mercalli, A., Nowak, K. W., Albarello, L., Calcagno, R., Luzi, L., et al. (2009). Ghrelin-producing epsilon cells in the developing and adult human pancreas. *Diabetologia*, *52*, 486–493. https://doi.org/10.1007/s00125-008-1238-y.

Assady, S., Maor, G., Amit, M., Itskovitz-Eldor, J., Skorecki, K. L., & Tzukerman, M. (2001). Insulin production by human embryonic stem cells. *Diabetes*, *50*, 1691–1697. https://doi.org/10.2337/diabetes.50.8.1691.

Atik, T., Karakoyun, M., Sukalo, M., Zenker, M., Ozkinay, F., & Aydogdu, S. (2015). Two novel UBR1 gene mutations in a patient with johanson blizzard syndrome: A mild phenotype without mental retardation. *Gene*, *570*(1), 153–155. https://doi.org/10.1016/j.gene.2015.06.082.

Basford, C. L., Prentice, K. J., Hardy, A. B., Sarangi, F., Micallef, S. J., Li, X., et al. (2012). The functional and molecular characterisation of human embryonic stem cell-derived insulin-positive cells compared with adult pancreatic beta cells. *Diabetologia*, *55*(2), 358–371. https://doi.org/10.1007/s00125-011-2335-x.

Bechard, M. E., Bankaitis, E. D., Hipkens, S. B., Ustione, A., Piston, D. W., Yang, Y.-P., et al. (2016). Precommitment low-level Neurog3 expression defines a long-lived mitotic endocrine-biased progenitor pool that drives production of endocrine-committed cells. *Genes & Development*. https://doi.org/10.1101/gad.284729.116.

Beer, N. L., & Gloyn, A. L. (2016). Genome-edited human stem cell-derived beta cells: A powerful tool for drilling down on type 2 diabetes GWAS biology. *F1000Res*, *5*, https://doi.org/10.12688/f1000research.8682.1.

Bellanne-Chantelot, C., Chauveau, D., Gautier, J. F., Dubois-Laforgue, D., Clauin, S., Beaufils, S., et al. (2004). Clinical spectrum associated with hepatocyte nuclear factor-1beta mutations. *Annals of Internal Medicine*, *140*(7), 510–517.

Bergmann, C., Fliegauf, M., Bruchle, N. O., Frank, V., Olbrich, H., Kirschner, J., et al. (2008). Loss of nephrocystin-3 function can cause embryonic lethality, Meckel-Gruber-like syndrome, situs inversus, and renal-hepatic-pancreatic dysplasia. *American Journal of Human Genetics*, *82*(4), 959–970.

Bocian-Sobkowska, J., Zabel, M., Wozniak, W., & Surdyk-Zasada, J. (1999). Polyhormonal aspect of the endocrine cells of the human fetal pancreas. *Histochemistry and Cell Biology*, *112*, 147–153. https://doi.org/10.1007/s004180050401.

Boj, S. F., Hwang, C. I., Baker, L. A., Chio, I. I., Engle, D. D., Corbo, V., et al. (2015). Organoid models of human and mouse ductal pancreatic cancer. *Cell*, *160*(1-2), 324–338. https://doi.org/10.1016/j.cell.2014.12.021.

Bonfanti, P., Nobecourt, E., Oshima, M., Albagli-Curiel, O., Laurysens, V., Stange, G., et al. (2015). Ex vivo expansion and differentiation of human and mouse fetal pancreatic progenitors are modulated by epidermal growth factor. *Stem Cells and Development*, *24*(15), 1766–1778. https://doi.org/10.1089/scd.2014.0550.

Bonnefond, A., Sand, O., Guerin, B., Durand, E., De Graeve, F., Huyvaert, M., et al. (2012). GATA6 inactivating mutations are associated with heart defects and, inconsistently, with pancreatic agenesis and diabetes. *Diabetologia*, *55*(10), 2845–2847. https://doi.org/10.1007/s00125-012-2645-7.

Bonnefond, A., Vaillant, E., Philippe, J., Skrobek, B., Lobbens, S., Yengo, L., et al. (2013). Transcription factor gene MNX1 is a novel cause of permanent neonatal diabetes in a consanguineous family. *Diabetes & Metabolism*, *39*, 276–280. https://doi.org/10.1016/j.diabet.2013.02.007.

Bosco, D., Armanet, M., Morel, P., Niclauss, N., Sgroi, A., Muller, Y. D., et al. (2010). Unique arrangement of alpha- and beta-cells in human Islets of Langerhans. *Diabetes*, *59*, 1202–1210. https://doi.org/10.2337/db09-1177.

Bouwens, L., Lu, W. G., & De Krijger, R. (1997). Proliferation and differentiation in the human fetal endocrine pancreas. *Diabetologia*, *40*, 398–404. https://doi.org/10.1007/s001250050693.

Brereton, M. F., Vergari, E., Zhang, Q., & Clark, A. (2015). Alpha-, delta- and PP-cells: Are they the architectural cornerstones of islet structure and co-ordination? *The Journal of Histochemistry and Cytochemistry*, *63*(8), 575–591. https://doi.org/10.1369/0022155415583535.

Brissova, M., Fowler, M. J., Nicholson, W. E., Chu, A., Hirshberg, B., Harlan, D. M., et al. (2005). Assessment of human pancreatic islet architecture and composition by laser scanning confocal microscopy. *Journal of Histochemistry and Cytochemistry*, *53*, 1087–1097. https://doi.org/10.1369/jhc.5C6684.2005.

Brissova, M., Shostak, A., Fligner, C. L., Revetta, F. L., Washington, M. K., Powers, A. C., et al. (2015). Human islets have fewer blood vessels than mouse islets and the density of islet vascular structures is increased in type 2 diabetes. *The Journal of Histochemistry and Cytochemistry: Official Journal of the Histochemistry Society*, *63*, 637–645. https://doi.org/10.1369/0022155415573324.

Bruin, J. E., Erener, S., Vela, J., Hu, X., Johnson, J. D., Kurata, H. T., et al. (2014). Characterization of polyhormonal insulin-producing cells derived in vitro from human embryonic stem cells. *Stem Cell Research*, *12*, 194–208. https://doi.org/10.1016/j.scr.2013.10.003.

Bruin, J. E., Rezania, A., Xu, J., Narayan, K., Fox, J. K., O'Neil, J. J., et al. (2013). Maturation and function of human embryonic stem cell-derived pancreatic progenitors in macroencapsulation devices following transplant into mice. *Diabetologia*, *56*, 1987–1998. https://doi.org/10.1007/s00125-013-2955-4.

Burlison, J. S., Long, Q., Fujitani, Y., Wright, C. V., & Magnuson, M. A. (2008). Pdx-1 and Ptf1a concurrently determine fate specification of pancreatic multipotent progenitor cells. *Developmental Biology*, *316*(1), 74–86. https://doi.org/10.1016/j.ydbio.2008.01.011.

Cabrera, O., Berman, D. M., Kenyon, N. S., Ricordi, C., Berggren, P.-O., & Caicedo, A. (2006). The unique cytoarchitecture of human pancreatic islets has implications for islet cell function. *Proceedings of the National Academy of Sciences of the United States of America*, *103*, 2334–2339. https://doi.org/10.1073/pnas.0510790103.

Cai, J., Yu, C., Liu, Y., Chen, S., Guo, Y., Yong, J., et al. (2010). Generation of homogeneous PDX1(+) pancreatic progenitors from human ES cell-derived endoderm cells. *Journal of Molecular Cell Biology*, *2*, 50–60. https://doi.org/10.1093/jmcb/mjp037.

Capito, C., Simon, M. T., Aiello, V., Clark, A., Aigrain, Y., Ravassard, P., et al. (2013). Mouse muscle as an ectopic permissive site for human pancreatic development. *Diabetes*, *62*, 3479–3487. https://doi.org/10.2337/db13-0554.

Carrasco, M., Delgado, I., Soria, B., Martin, F., & Rojas, A. (2012). GATA4 and GATA6 control mouse pancreas organogenesis. *Journal of Clinical Investigation*, *122*, 3504–3515. https://doi.org/10.1172/JCI63240.

Castaing, M., Duvillie, B., Quemeneur, E., Basmaciogullari, A., & Scharfmann, R. (2005). Ex vivo analysis of acinar and endocrine cell development in the human embryonic pancreas. *Developmental Dynamics*, *234*(2), 339–345. https://doi.org/10.1002/dvdy.20547.

Castaing, M., Peault, B., Basmaciogullari, A., Casal, I., Czernichow, P., & Scharfmann, R. (2001). Blood glucose normalization upon transplantation of human embryonic pancreas into beta-cell-deficient SCID mice. *Diabetologia*, *44*(11), 2066–2076. https://doi.org/10.1007/s001250100012.

Cebola, I., Rodriguez-Segui, S. A., Cho, C. H., Bessa, J., Rovira, M., Luengo, M., et al. (2015). TEAD and YAP regulate the enhancer network of human embryonic pancreatic progenitors. *Nature Cell Biology*, *17*(5), 615–626. https://doi.org/10.1038/ncb3160.

Chandrasekera, P., & Pippin, J. (2013). Of rodents and men: Species-specific glucose regulation and type 2 diabetes research. *ALTEX*, *31*, 157–176. https://doi.org/10.14573/altex.1309231.

Chen, A. E., Borowiak, M., Sherwood, R. I., Kweudjeu, A., & Melton, D. A. (2013). Functional evaluation of ES cell-derived endodermal populations reveals differences between Nodal and Activin A-guided differentiation. *Development*, *140*(3), 675–686. https://doi.org/10.1242/dev.085431.

Chen, S., Borowiak, M., Fox, J. L., Maehr, R., Osafune, K., Davidow, L., et al. (2009). A small molecule that directs differentiation of human ESCs into the pancreatic lineage. *Nature Chemical Biology*, *5*(4), 258–265. https://doi.org/10.1038/nchembio.154.

Cheng, X., Ying, L., Lu, L., Galvão, A. M., Mills, J. A., Lin, H. C., et al. (2012). Self-renewing endodermal progenitor lines generated from human pluripotent stem cells figures and legends. *Cell Stem Cell*, *11*, 371–384. https://doi.org/10.1016/j.stem.2012.02.024.

Cirulli, V., Beattie, G. M., Klier, G., Ellisman, M., Ricordi, C., Quaranta, V., et al. (2000). Expression and function of αvβ3 and αvβ5 integrins in the developing pancreas. *The Journal of Cell Biology*, *150*, 1445–1459.

Cogger, K. F., Sinha, A., Sarangi, F., McGaugh, E. C., Saunders, D., Dorrell, C., et al. (2017). Glycoprotein 2 is a specific cell surface marker of human pancreatic progenitors. *Nature Communications*, *8*(1), 331. https://doi.org/10.1038/s41467-017-00561-0.

Collombat, P., Hecksher-Sorensen, J., Broccoli, V., Krull, J., Ponte, I., Mundiger, T., et al. (2005). The simultaneous loss of Arx and Pax4 genes promotes a somatostatin-producing cell fate specification at the expense of the alpha- and beta-cell lineages in the mouse endocrine pancreas. *Development*, *132*(13), 2969–2980. https://doi.org/10.1242/dev.01870.

Collombat, P., Hecksher-Sorensen, J., Krull, J., Berger, J., Riedel, D., Herrera, P. L., et al. (2007). Embryonic endocrine pancreas and mature beta cells acquire alpha and PP cell phenotypes upon Arx misexpression. *The Journal of Clinical Investigation*, *117*(4), 961–970. https://doi.org/10.1172/JCI29115.

Collombat, P., Mansouri, A., Hecksher-Sorensen, J., Serup, P., Krull, J., Gradwohl, G., et al. (2003). Opposing actions of Arx and Pax4 in endocrine pancreas development. *Genes & Development*, *17*(20), 2591–2603.

Collombat, P., Xu, X., Ravassard, P., Sosa-Pineda, B., Dussaud, S., Billestrup, N., et al. (2009). The ectopic expression of Pax4 in the mouse pancreas converts progenitor cells into alpha and subsequently beta cells. *Cell*, (3), 449–462. https://doi.org/10.1016/j.cell.2009.05.035.

Concepcion, J. P., Reh, C. S., Daniels, M., Liu, X., Paz, V. P., Ye, H., et al. (2014). Neonatal diabetes, gallbladder agenesis, duodenal atresia, and intestinal malrotation caused by a novel homozygous mutation in RFX6. *Pediatric Diabetes*, *15*(1), 67–72. https://doi.org/10.1111/pedi.12063.

D'Amour, K. A., Agulnick, A. D., Eliazer, S., Kelly, O. G., Kroon, E., & Baetge, E. E. (2005). Efficient differentiation of human embryonic stem cells to definitive endoderm. *Nature Biotechnology*, *23*(12), 1534–1541.

D'Amour, K. A., Bang, A. G., Eliazer, S., Kelly, O. G., Agulnick, A. D., Smart, N. G., et al. (2006). Production of pancreatic hormone-expressing endocrine cells from human embryonic stem cells. *Nature Biotechnology*, *24*(11), 1392–1401.

De Franco, E., Shaw-Smith, C., Flanagan, S. E., Shepherd, M. H., International NDM Consortium, Hattersley, A. T., et al. (2013). GATA6 mutations cause a broad phenotypic spectrum of diabetes from pancreatic agenesis to adult-onset diabetes without exocrine insufficiency. *Diabetes*, *62*(3), 993–997. https://doi.org/10.2337/db12-0885.

Deltour, L., Leduque, P., Paldi, A., Ripoche, M. A., Dubois, P., & Jami, J. (1991). Polyclonal origin of pancreatic islets in aggregation mouse chimaeras. *Development (Cambridge, England)*, *112*, 1115–1121.

de Rooij, S. R., Painter, R. C., Phillips, D. I., Osmond, C., Michels, R. P., Godsland, I. F., et al. (2006). Impaired insulin secretion after prenatal exposure to the Dutch famine. *Diabetes Care*, *29*(8), 1897–1901. https://doi.org/10.2337/dc06-0460.

De Vas, M. G., Kopp, J. L., Heliot, C., Sander, M., Cereghini, S., & Haumaitre, C. (2015). Hnf1b controls pancreas morphogenesis and the generation of Ngn3+ endocrine progenitors. *Development*, *142*(5), 871–882. https://doi.org/10.1242/dev.110759.

Dimitri, P., Warner, J. T., Minton, J. A., Patch, A. M., Ellard, S., Hattersley, A. T., et al. (2011). Novel GLIS3 mutations demonstrate an extended multisystem phenotype. *European Journal of Endocrinology*, *164*(3), 437–443. https://doi.org/10.1530/EJE-10-0893.

Domínguez-Bendala, J., Lanzoni, G., Klein, D., Álvarez-Cubela, S., & Pastori, R. L. (2015). The human endocrine pancreas: New insights on replacement and regeneration. *Trends in Endocrinology and Metabolism*, *27*(3), 153–162. https://doi.org/10.1016/j.tem.2015.12.003.

Dorrell, C., Abraham, S. L., Lanxon-Cookson, K. M., Canaday, P. S., Streeter, P. R., & Grompe, M. (2008). Isolation of major pancreatic cell types and long-term culture-initiating cells using novel human surface markers. *Stem Cell Research*, *1*, 183–194. https://doi.org/10.1016/j.scr.2008.04.001.

Fiocca, R., Sessa, F., Tenti, P., Usellini, L., Capella, C., O'Hare, M. M., et al. (1983). Pancreatic polypeptide (PP) cells in the PP-rich lobe of the human pancreas are identified ultrastructurally and immunocytochemically as F cells. *Histochemistry*, 77(4), 511–523.

Fiskerstrand, T., Houge, G., Sund, S., Scheie, D., Leh, S., Boman, H., et al. (2010). Identification of a gene for renal-hepatic-pancreatic dysplasia by microarray-based homozygosity mapping. *The Journal of Molecular Diagnostics*, *12*(1), 125–131. https://doi.org/10.2353/jmoldx.2010.090033.

Flanagan, S. E., De Franco, E., Lango Allen, H., Zerah, M., Abdul-Rasoul, M. M., Edge, J. A., et al. (2014). Analysis of transcription factors key for mouse pancreatic development establishes NKX2-2 and MNX1 mutations as causes of neonatal diabetes in man. *Cell Metabolism*, *19*, 146–154. https://doi.org/10.1016/j.cmet.2013.11.021.

Flasse, L. C., Stern, D. G., Pirson, J. L., Manfroid, I., Peers, B., & Voz, M. L. (2013). The bHLH transcription factor Ascl1a is essential for the specification of the intestinal secretory cells and mediates Notch signaling in the zebrafish intestine. *Developmental Biology*, *376*(2), 187–197. https://doi.org/10.1016/j.ydbio.2013.01.011.

Frank, V., Habbig, S., Bartram, M. P., Eisenberger, T., Veenstra-Knol, H. E., Decker, C., et al. (2013). Mutations in NEK8 link multiple organ dysplasia with altered Hippo signalling and increased c-MYC expression. *Human Molecular Genetics*, *22*(11), 2177–2185. https://doi.org/10.1093/hmg/ddt070.

Fujikura, J., Hosoda, K., Iwakura, H., Tomita, T., Noguchi, M., Masuzaki, H., et al. (2006). Notch/Rbp-j signaling prevents premature endocrine and ductal cell differentiation in the pancreas. *Cell Press, 3*(1), 59–65. https://doi.org/10.1016/j.cmet.2005.12.005.

Gage, B. K., Asadi, A., Baker, R. K., Webber, T. D., Wang, R., Itoh, M., et al. (2015). The role of ARX in human pancreatic endocrine specification. *PLoS One, 10,* e0144100. https://doi.org/10.1371/journal.pone.0144100.

Gage, B. K., Baker, R. K., & Kieffer, T. J. (2014). Overexpression of PAX4 reduces glucagon expression in differentiating hESCs. *Islets, 6.* https://doi.org/10.4161/isl.29236.

Georgia, S., & Bhushan, A. (2004). Beta cell replication is the primary mechanism for maintaining postnatal beta cell mass. *Journal of Clinical Investigation, 114,* 963–968. https://doi.org/10.1172/JCI200422098.

Ghazizadeh, Z., Kao, D. I., Amin, S., Cook, B., Rao, S., Zhou, T., et al. (2017). ROCKII inhibition promotes the maturation of human pancreatic beta-like cells. *Nature Communications, 8*(1), 298. https://doi.org/10.1038/s41467-017-00129-y.

Gittes, G. K. (2009). Developmental biology of the pancreas: A comprehensive review. *Developmental Biology, 326*(1), 4–35. https://doi.org/10.1016/j.ydbio.2008.10.024.

Gong, M., Simaite, D., Kuhnen, P., Heldmann, M., Spagnoli, F., Blankenstein, O., et al. (2013). Two novel GATA6 mutations cause childhood-onset diabetes mellitus, pancreas malformation and congenital heart disease. *Hormone Research in Pædiatrics, 79*(4), 250–256. https://doi.org/10.1159/000348844.

Gouon-Evans, V., Boussemart, L., Gadue, P., Nierhoff, D., Koehler, C. I., Kubo, A., et al. (2006). BMP-4 is required for hepatic specification of mouse embryonic stem cell-derived definitive endoderm. *Nature Biotechnology, 24*(11), 1402–1411.

Gouzi, M., Kim, Y. H., Katsumoto, K., Johansson, K., & Grapin-Botton, A. (2011). Neurogenin3 initiates stepwise delamination of differentiating endocrine cells during pancreas development. *Developmental Dynamics, 240*(3), 589–604. https://doi.org/10.1002/dvdy.22544.

Gradwohl, G., Dierich, A., LeMeur, M., & Guillemot, F. (2000). Neurogenin3 is required for the development of the four endocrine cell lineages of the pancreas. *Proceedings of the National Academy of Sciences of the United States of America, 97*(4), 1607–1611.

Green, M. D., Chen, A., Nostro, M. C., d'Souza, S. L., Schaniel, C., Lemischka, I. R., et al. (2011). Generation of anterior foregut endoderm from human embryonic and induced pluripotent stem cells. *Nature Biotechnology, 29*(3), 267–272. https://doi.org/10.1038/nbt.1788.

Greenwood, A. L., Li, S., Jones, K., & Melton, D. A. (2007). Notch signaling reveals developmental plasticity of Pax4+ pancreatic endocrine progenitors and shunts them to a duct fate. *Mechanisms of Development, 124,* 97–107. https://doi.org/10.1016/j.mod.2006.11.002.

Gregg, B. E., Moore, P. C., Demozay, D., Hall, B. A., Li, M., Husain, A., et al. (2012). Formation of a human β-cell population within pancreatic islets is set early in life. *The Journal of Clinical Endocrinology and Metabolism, 97,* 1–10. https://doi.org/10.1210/jc.2012-1206.

Gu, G., Dubauskaite, J., & Melton, D. A. (2002). Direct evidence for the pancreatic lineage: NGN3+ cells are islet progenitors and are distinct from duct progenitors. *Development, 129*(10), 2447–2457.

Hald, J., Hjorth, J. P., German, M. S., Madsen, O. D., Serup, P., & Jensen, J. (2003). Activated Notch1 prevents differentiation of pancreatic acinar cells and attenuate endocrine development. *Developmental Biology, 260*(2), 426–437.

Hansson, M., Tonning, A., Frandsen, U., Petri, A., Rajagopal, J., Englund, M. C., et al. (2004). Artifactual insulin release from differentiated embryonic stem cells. *Diabetes, 53*(10), 2603–2609.

Harrison, K. A., Thaler, J., Pfaff, S. L., Gu, H., & Kehrl, J. H. (1999). Pancreas dorsal lobe agenesis and abnormal islets of Langerhans in Hlxb9-deficient mice. *Nature Genetics, 23*(1), 71–75.

Haumaitre, C., Barbacci, E., Jenny, M., Ott, M. O., Gradwohl, G., & Cereghini, S. (2005). Lack of TCF2/vHNF1 in mice leads to pancreas agenesis. *Proceedings of the National Academy of Sciences of the United States of America*, *102*(5), 1490–1495. https://doi.org/10.1073/pnas.0405776102.

Haumaitre, C., Fabre, M., Cormier, S., Baumann, C., Delezoide, A. L., & Cereghini, S. (2006). Severe pancreas hypoplasia and multicystic renal dysplasia in two human fetuses carrying novel HNF1beta/MODY5 mutations. *Human Molecular Genetics*, *15*(15), 2363–2375.

Hebrok, M., Kim, S. K., & Melton, D. A. (1998). Notochord repression of endodermal Sonic hedgehog permits pancreas development. *Genes and Development*, *12*(11), 1705–1713.

Hebrok, M., Kim, S. K., St Jacques, B., McMahon, A. P., & Melton, D. A. (2000). Regulation of pancreas development by hedgehog signaling. *Development*, *127*(22), 4905–4913.

Heller, R. S., Jenny, M., Collombat, P., Mansouri, A., Tomasetto, C., Madsen, O. D., et al. (2005). Genetic determinants of pancreatic epsilon-cell development. *Developmental Biology*, *286*, 217–224. https://doi.org/10.1016/j.ydbio.2005.06.041.

Henseleit, K. D., Nelson, S. B., Kuhlbrodt, K., Hennings, J. C., Ericson, J., & Sander, M. (2005). NKX6 transcription factor activity is required for alpha- and beta-cell development in the pancreas. *Development (Cambridge, England)*, *132*, 3139–3149. https://doi.org/10.1242/dev.01875.

Herrera, P. L. (2000). Adult insulin- and glucagon-producing cells differentiate from two independent cell lineages. *Development*, *127*(11), 2317–2322.

Hohwieler, M., Illing, A., Hermann, P. C., Mayer, T., Stockmann, M., Perkhofer, L., et al. (2017). Human pluripotent stem cell-derived acinar/ductal organoids generate human pancreas upon orthotopic transplantation and allow disease modelling. *Gut*, *66*(3), 473–486. https://doi.org/10.1136/gutjnl-2016-312423.

Honoré, C., Rescan, C., Hald, J., McGrath, P. S., Petersen, M. B. K., Hansson, M., et al. (2016). Revisiting the immunocytochemical detection of Neurogenin 3 expression in mouse and man. *Diabetes, Obesity and Metabolism*, *18*, 10–22. https://doi.org/10.1111/dom.12718.

Horikawa, Y., Iwasaki, N., Hara, M., Furuta, H., Hinokio, Y., Cockburn, B. N., et al. (1997). Mutation in hepatocyte nuclear factor-1 beta gene (TCF2) associated with MODY. *Nature Genetics*, *17*(4), 384–385.

Horn, S., Kobberup, S., Jørgensen, M. C., Kalisz, M., Klein, T., Kageyama, R., et al. (2012). Mind bomb 1 is required for pancreatic β-cell formation. *Proceedings of the National Academy of Sciences of the United States of America*, *109*, 7356–7361. https://doi.org/10.1073/pnas.1203605109.

Huang, L., Holtzinger, A., Jagan, I., BeGora, M., Lohse, I., Ngai, N., et al. (2015). Ductal pancreatic cancer modeling and drug screening using human pluripotent stem cell- and patient-derived tumor organoids. *Nature Medicine*, *21*(11), 1364–1371. https://doi.org/10.1038/nm.3973.

Islam, M. S. (2010). Calcium signaling in the islets. In M. S. Islam (Ed.), *Islets of Langerhans: Vol. 654* (pp. 235–259) 20518156. Springer Science and Business Media.

Itoh, M., Takizawa, Y., Hanai, S., Okazaki, S., Miyata, R., Inoue, T., et al. (2010). Partial loss of pancreas endocrine and exocrine cells of human ARX-null mutation: Consideration of pancreas differentiation. *Differentiation*, *80*, 118–122. https://doi.org/10.1016/j.diff.2010.05.003.

Jacquemin, P., Durviaux, S. M., Jensen, J., Godfraind, C., Gradwohl, G., Guillemot, F., et al. (2000). Transcription factor hepatocyte nuclear factor 6 regulates pancreatic endocrine cell differentiation and controls expression of the proendocrine gene ngn3. *Molecular and Cell Biology*, *20*(12), 4445–4454.

Jennings, R. E., Berry, A. A., Kirkwood-Wilson, R., Roberts, N. A., Hearn, T., Salisbury, R. J., et al. (2013). Development of the human pancreas from foregut to endocrine commitment. *Diabetes*, *62*(10), 3514–3522. https://doi.org/10.2337/db12-1479.

Jensen, J., Pedersen, E. E., Galante, P., Hald, J., Heller, R. S., Ishibashi, M., et al. (2000). Control of endodermal endocrine development by Hes-1. *Nature Genetics*, *24*, 36–44.

Jeon, J., Correa-Medina, M., Ricordi, C., Edlund, H., & Diez, J. A. (2009). Endocrine cell clustering during human pancreas development. *The Journal of Histochemistry and Cytochemistry*, *57*(9), 811–824. https://doi.org/10.1369/jhc.2009.953307.

Jiang, J., Au, M., Lu, K., Eshpeter, A., Korbutt, G., Fisk, G., et al. (2007). Generation of insulin-producing islet-like clusters from human embryonic stem cells. *Stem Cells*, *25*(8), 1940–1953. https://doi.org/10.1634/stemcells.2006-0761.

Jiang, W., Shi, Y., Zhao, D., Chen, S., Yong, J., Zhang, J., et al. (2007). In vitro derivation of functional insulin-producing cells from human embryonic stem cells. *Cell Research*, *17*(4), 333–344. https://doi.org/10.1038/cr.2007.28.

Johnson, J. D. (2016). The quest to make fully functional human pancreatic beta cells from embryonic stem cells: Climbing a mountain in the clouds. *Diabetologia*, *59*(10), 2047–2057. https://doi.org/10.1007/s00125-016-4059-4.

Jonsson, J., Carlsson, L., Edlund, T., & Edlund, H. (1994). Insulin-promoter-factor 1 is required for pancreas development in mice. *Nature*, *371*(6498), 606–609.

Kang, H. S., Kim, Y. S., ZeRuth, G., Beak, J. Y., Gerrish, K., Kilic, G., et al. (2009). Transcription factor Glis3, a novel critical player in the regulation of pancreatic beta-cell development and insulin gene expression. *Molecular and Cellular Biology*, *29*(24), 6366–6379. https://doi.org/10.1128/MCB.01259-09.

Kassem, S. A., Ariel, I., Thornton, P. S., Scheimberg, I., & Glaser, B. (2000). Beta-cell proliferation and apoptosis in the developing normal human pancreas and in hyperinsulinism of infancy. *Diabetes*, *49*, 1325–1333. https://doi.org/10.2337/diabetes.49.8.1325.

Kawaguchi, Y., Cooper, B., Gannon, M., Ray, M., MacDonald, R. J., & Wright, C. V. (2002). The role of the transcriptional regulator Ptf1a in converting intestinal to pancreatic progenitors. *Nature Genetics*, *32*(1), 128–134.

Kelly, O. G., Chan, M. Y., Martinson, L. A., Kadoya, K., Ostertag, T. M., Ross, K. G., et al. (2011). Cell-surface markers for the isolation of pancreatic cell types derived from human embryonic stem cells. *Nature Biotechnology*, *29*(8), 750–756. https://doi.org/10.1038/nbt.1931.

Kesavan, G., Sand, F. W., Greiner, T. U., Johansson, J. K., Kobberup, S., Wu, X., et al. (2009). Cdc42-mediated tubulogenesis controls cell specification. *Cell*, *139*(4), 791–801. https://doi.org/10.1016/j.cell.2009.08.049.

Kieffer, T. J. (2016). Closing in on mass production of mature human beta cells. *Cell Stem Cell*, *18*, 699–702. https://doi.org/10.1016/j.stem.2016.05.014.

Kim, Y. H., Larsen, H. L., Rué, P., Lemaire, L. A., Ferrer, J., & Grapin-Botton, A. (2015). Cell cycle–dependent differentiation dynamics balances growth and endocrine differentiation in the pancreas. *PLoS Biology*, *13*, 1–25. https://doi.org/10.1371/journal.pbio.1002111.

Kopinke, D., Brailsford, M., Shea, J. E., Leavitt, R., Scaife, C. L., & Murtaugh, L. C. (2011). Lineage tracing reveals the dynamic contribution of Hes1 + cells to the developing and adult pancreas. *Development*, *138*, 431–441. https://doi.org/10.1242/dev.053843.

Kopp, J. L., Dubois, C. L., Schaffer, A. E., Hao, E., Shih, H. P., Seymour, P. A., et al. (2011). Sox9 + ductal cells are multipotent progenitors throughout development but do not produce new endocrine cells in the normal or injured adult pancreas. *Development*, *138*(4), 653–665. https://doi.org/10.1242/dev.056499.

Kroon, E., Martinson, L. A., Kadoya, K., Bang, A. G., Kelly, O. G., Eliazer, S., et al. (2008). Pancreatic endoderm derived from human embryonic stem cells generates glucose-responsive insulin-secreting cells in vivo. *Nature Biotechnology, 26*(4), 443–452. https://doi.org/10.1038/nbt1393.

Kuo, C. T., Morrisey, E. E., Anandappa, R., Sigrist, K., Lu, M. M., Parmacek, M. S., et al. (1997). GATA4 transcription factor is required for ventral morphogenesis and heart tube formation. *Genes and Development, 11*(8), 1048–1060.

Kushner, J. A., MacDonald, P. E., & Atkinson, M. A. (2014). Stem cells to insulin secreting cells: Two steps forward and now a time to pause? *Cell Stem Cell, 15*(5), 535–536. https://doi.org/10.1016/j.stem.2014.10.012.

Larsen, H. L., & Grapin-Botton, A. (2017). The molecular and morphogenetic basis of pancreas organogenesis. *Seminars in Cell & Developmental Biology, 66*, 51–68. https://doi.org/10.1016/j.semcdb.2017.01.005.

Larsen, H. L., Martín-Coll, L., Nielsen, A. V., Wright, C., Trusina, A., Kim, Y. H., et al. (2017). Stochastic priming and spatial cues orchestrate heterogeneous clonal contribution to pancreas organogenesis. *Nature Communications, 8*(1), 605. https://doi.org/10.1038/s41467-017-00258-4.

Lee, J., Sugiyama, T., Liu, Y., Wang, J., Gu, X., Lei, J., et al. (2013). Expansion and conversion of human pancreatic ductal cells into insulin-secreting endocrine cells. *eLife, 2*, e00940. https://doi.org/10.7554/eLife.00940.

Li, H., Arber, S., Jessell, T. M., & Edlund, H. (1999). Selective agenesis of the dorsal pancreas in mice lacking homeobox gene Hlxb9. *Nature Genetics, 23*(1), 67–70.

Li, H., & Edlund, H. (2001). Persistent expression of Hlxb9 in the pancreatic epithelium impairs pancreatic development. *Developmental Biology, 240*(1), 247–253.

Lindner, T. H., Njolstad, P. R., Horikawa, Y., Bostad, L., Bell, G. I., & Sovik, O. (1999). A novel syndrome of diabetes mellitus, renal dysfunction and genital malformation associated with a partial deletion of the pseudo-POU domain of hepatocyte nuclear factor-1beta. *Human Molecular Genetics, 8*(11), 2001–2008.

Löf-Öhlin, Z. M., Nyeng, P., Bechard, M. E., Hess, K., Bankaitis, E., Greiner, T. U., et al. (2017). EGFR signalling controls cellular fate and pancreatic organogenesis by regulating apicobasal polarity. *Nature Cell Biology, 19*(11), 1313–1325. https://doi.org/10.1038/ncb3628.

Loh, K. M., Ang, L. T., Zhang, J., Kumar, V., Ang, J., Auyeong, J. Q., et al. (2014). Efficient endoderm induction from human pluripotent stem cells by logically directing signals controlling lineage bifurcations. *Cell Stem Cell, 14*(2), 237–252. https://doi.org/10.1016/j.stem.2013.12.007.

Ludwig, B., Reichel, A., Steffen, A., Zimerman, B., Schally, A. V., Block, N. L., et al. (2013). Transplantation of human islets without immunosuppression. *Proceedings of the National Academy of Sciences of the United States of America, 110*(47), 19054–19058. https://doi.org/10.1073/pnas.1317561110.

Lumelsky, N., Blondel, O., Laeng, P., Velasco, I., Ravin, R., & McKay, R. (2001). Differentiation of embryonic stem cells to insulin-secreting structures similar to pancreatic islets. *Science, 292*(5520), 1389–1394.

Ma, H., Wert, K. J., Shvartsman, D., Melton, D. A., & Jaenisch, R. (2018). Establishment of human pluripotent stem cell-derived pancreatic β-like cells in the mouse pancreas. *Proceedings of the National Academy of Sciences of the United States of America*, 201702059, Epub ahead of print. https://doi.org/10.1073/pnas.1702059115.

Manning, D. K., Sergeev, M., van Heesbeen, R. G., Wong, M. D., Oh, J. H., Liu, Y., et al. (2013). Loss of the ciliary kinase Nek8 causes left-right asymmetry defects. *Journal of the American Society of Nephrology, 24*(1), 100–112. https://doi.org/10.1681/ASN.2012050490.

McGrath, P. S., Watson, C. L., Ingram, C., Helmrath, M. A., & Wells, J. M. (2015). The basic helix-loop-helix transcription factor NEUROG3 is required for development of the human endocrine pancreas. *Diabetes*. https://doi.org/10.2337/db14-1412.

Meier, J., Butler, A., Saisho, Y., Monchamp, T., Galasso, R., Bhushan, A., et al. (2008). Postnatal expansion of β-cell mass in humans. *Diabetes*, *57*, 1584–1594. https://doi.org/10.2337/db07-1369.

Millman, J. R., Xie, C., Van Dervort, A., Gurtler, M., Pagliuca, F. W., & Melton, D. A. (2016). Generation of stem cell-derived beta-cells from patients with type 1 diabetes. *Nature Communications*, 7, 11463. https://doi.org/10.1038/ncomms11463.

Miyatsuka, T., Kosaka, Y., Kim, H., & German, M. S. (2011). Neurogenin3 inhibits proliferation in endocrine progenitors by inducing Cdkn1a. *Proceedings of the National Academy of Sciences of the United States of America*, *108*(1), 185–190. https://doi.org/10.1073/pnas.1004842108.

Molkentin, J. D., Lin, Q., Duncan, S. A., & Olson, E. N. (1997). Requirement of the transcription factor GATA4 for heart tube formation and ventral morphogenesis. *Genes and Development*, *11*(8), 1061–1072.

Molotkov, A., Molotkova, N., & Duester, G. (2005). Retinoic acid generated by Raldh2 in mesoderm is required for mouse dorsal endodermal pancreas development. *Developmental Dynamics*, *232*(4), 950–957.

Murtaugh, L. C., Stanger, B. Z., Kwan, K. M., & Melton, D. A. (2003). Notch signaling controls multiple steps of pancreatic differentiation. *Proceedings of the National Academy of Sciences of the United States of America*, *100*(25), 14920–14925.

Naya, F. J., Huang, H.-P., Qiu, Y., Mutoh, H., DeMayo, F. J., Leiter, A. B., et al. (1997). Diabetes, defective pancreatic morphogenesis, and abnormal enteroendocrine differentiation in BETA2/NeuroD-deficient mice. *Genes & Development*, *11*, 2323–2334. https://doi.org/10.1101/gad.11.18.2323.

Nelson, S. B., Schaffer, A. E., & Sander, M. (2007). The transcription factors Nkx6.1 and Nkx6.2 possess equivalent activities in promoting beta-cell fate specification in Pdx1+ pancreatic progenitor cells. *Development (Cambridge, England)*, *134*, 2491–2500. https://doi.org/10.1242/dev.002691.

Nogueira, T. C., Paula, F. M., Villate, O., Colli, M. L., Moura, R. F., Cunha, D. A., et al. (2013). GLIS3, a susceptibility gene for type 1 and type 2 diabetes, modulates pancreatic beta cell apoptosis via regulation of a splice variant of the BH3-only protein Bim. *PLoS Genetics*, *9*(5). e1003532. https://doi.org/10.1371/journal.pgen.1003532.

Nostro, C. M., Sarangi, F., Yang, C., Holland, A., Elefanty, A. G., Stanley, E. G., et al. (2015). Efficient generation of NKX6-1+ pancreatic progenitors from multiple human pluripotent stem cell lines. *Stem Cell Reports*, *4*(4), 591–604. https://doi.org/10.1016/j.stemcr.2015.02.017.

Nostro, M. C., Sarangi, F., Ogawa, S., Holtzinger, A., Corneo, B., Li, X., et al. (2011). Stage-specific signaling through TGF family members and WNT regulates patterning and pancreatic specification of human pluripotent stem cells. *Development*, *138*, 1445. https://doi.org/10.1242/dev.065904.

O'Rahilly, R., & Muller, F. (2010). Developmental stages in human embryos: Revised and new measurements. *Cells, Tissues, Organs*, *192*(2), 73–84. https://doi.org/10.1159/000289817.

Pagliuca, F. W., Millman, J. R., Gürtler, M., Segel, M., Van Dervort, A., Ryu, J. H., et al. (2014). Generation of functional human pancreatic β cells in vitro. *Cell*, *159*, 428–439. https://doi.org/10.1016/j.cell.2014.09.040.

Pan, F. C., Bankaitis, E. D., Boyer, D., Xu, X., Van de Casteele, M., Magnuson, M. A., et al. (2013). Spatiotemporal patterns of multipotentiality in Ptf1a-expressing cells during pancreas organogenesis and injury-induced facultative restoration. *Development*, *140*(4), 751–764. https://doi.org/10.1242/dev.090159.

Pan, F. C., Brissova, M., Powers, A. C., Pfaff, S., & Wright, C. V. (2015). Inactivating the permanent neonatal diabetes gene Mnx1 switches insulin-producing beta-cells to a delta-like fate and reveals a facultative proliferative capacity in aged beta-cells. *Development*, *142*(21), 3637–3648. https://doi.org/10.1242/dev. 126011.

Petersen, M. B. K., Azad, A., Ingvorsen, C., Hess, K., Hansson, M., Grapin-Botton, A., et al. (2017). Single-cell gene expression analysis of a human ESC model of pancreatic endocrine development reveals different paths to β-cell differentiation. *Stem Cell Reports*, *9*, 1246–1261.

Piccand, J., Meunier, A., Merle, C., Jia, Z., Barnier, J.-V., & Gradwohl, G. (2014). Pak3 promotes cell cycle exit and differentiation of β-cells in the embryonic pancreas and is necessary to maintain glucose homeostasis in adult mice. *Diabetes*, *63*, 203–215. https://doi.org/10.2337/db13-0384.

Pinney, S. E., Oliver-Krasinski, J., Ernst, L., Hughes, N., Patel, P., Stoffers, D. A., et al. (2011). Neonatal diabetes and congenital malabsorptive diarrhea attributable to a novel mutation in the human neurogenin-3 gene coding sequence. *The Journal of Clinical Endocrinology and Metabolism*, *96*, 1960–1965. https://doi.org/10.1210/jc.2011-0029.

Piper, K., Ball, S. G., Keeling, J. W., Mansoor, S., Wilson, D. I., & Hanley, N. A. (2002). Novel SOX9 expression during human pancreas development correlates to abnormalities in Campomelic dysplasia. *Mechanisms of Development*, *116*(1–2), 223–226.

Piper, K., Brickwood, S., Turnpenny, L. W., Cameron, I. T., Ball, S. G., Wilson, D. I., et al. (2004). Beta cell differentiation during early human pancreas development. *The Journal of Endocrinology*, *181*(1), 11–23.

Polak, M., Bouchareb-Banaei, L., Scharfmann, R., & Czernichow, P. (2000). Early pattern of differentiation in the human pancreas. *Diabetes*, *49*, 225–232. https://doi.org/10.2337/diabetes.49.2.225.

Proshchina, A. E., Krivova, Y. S., Barabanov, V. M., & Saveliev, S. V. (2014). Ontogeny of neuro-insular complexes and islets innervation in the human pancreas. *Frontiers in Endocrinology*, *5*, 1–8. https://doi.org/10.3389/fendo.2014.00057.

Rajagopal, J., Anderson, W. J., Kume, S., Martinez, O. I., & Melton, D. A. (2003). Insulin staining of ES cell progeny from insulin uptake. *Science*, *299*(5605), 363.

Reinert, R. B., Cai, Q., Hong, J.-Y., Plank, J. L., Aamodt, K., Prasad, N., et al. (2014). Vascular endothelial growth factor coordinates islet innervation via vascular scaffolding. *Development (Cambridge, England)*, *141*, 1480–1491. https://doi.org/10.1242/dev.098657.

Rezania, A., Bruin, J. E., Arora, P., Rubin, A., Batushansky, I., Asadi, A., et al. (2014). Reversal of diabetes with insulin-producing cells derived in vitro from human pluripotent stem cells. *Nature Biotechnology*, *32*(11), 1121–1133. https://doi.org/10.1038/nbt.3033.

Rezania, A., Bruin, J. E., Riedel, M. J., Mojibian, M., Asadi, A., Xu, J., et al. (2012). Maturation of human embryonic stem cell-derived pancreatic progenitors into functional islets capable of treating pre-existing diabetes in mice. *Diabetes*, *61*, 2016–2029. https://doi.org/10.2337/db11-1711.

Rezania, A., Bruin, J. E., Xu, J., Narayan, K., Fox, J. K., O'Neil, J. J., et al. (2013). Enrichment of human embryonic stem cell-derived NKX6.1-expressing pancreatic progenitor cells accelerates the maturation of insulin-secreting cells in vivo. *Stem Cells*, *31*(11), 2432–2442. https://doi.org/10.1002/stem.1489.

Rezania, A., Riedel, M., Wideman, R., Karanu, F., Ao, Z., Warnock, G. L., et al. (2011). Production of functional glucagon-secreting α-cells from human embryonic stem cells. *Diabetes*, *60*, 239–247. https://doi.org/10.2337/db10-0573.A.R.

Riedel, M. J., Asadi, A., Wang, R., Ao, Z., Warnock, G. L., & Kieffer, T. J. (2012). Immunohistochemical characterisation of cells co-producing insulin and glucagon in the developing human pancreas. *Diabetologia*, *55*, 372–381. https://doi.org/10.1007/s00125-011-2344-9.

Rodriguez-Diaz, R., Abdulreda, M. H., Formoso, A. L., Gans, I., Ricordi, C., Berggren, P. O., et al. (2011). Innervation patterns of autonomic axons in the human endocrine pancreas. *Cell Metabolism*, *14*, 45–54. https://doi.org/10.1016/j.cmet.2011.05.008.

Roost, M. S., Van Iperen, L., De Melo Bernardo, A., Mummery, C. L., Carlotti, F., De Koning, E. J., et al. (2014). Lymphangiogenesis and angiogenesis during human fetal pancreas development. *Vascular Cell*, *6*, 1–11. https://doi.org/10.1186/2045-824X-6-22.

Rostovskaya, M., Bredenkamp, N., & Smith, A. (2015). Towards consistent generation of pancreatic lineage progenitors from human pluripotent stem cells. *Philosophical Transactions of the Royal Society B*, *370*, 20140365. https://doi.org/10.1098/rstb.2014.0365.

Rubio-Cabezas, O., Jensen, J. N., Hodgson, M. I., Codner, E., Ellard, S., Serup, P., et al. (2011). Permanent neonatal diabetes and enteric anendocrinosis associated with biallelic mutations in NEUROG3. *Diabetes*, *60*, 1349–1353. https://doi.org/10.2337/db10-1008.

Rubio-Cabezas, O., Minton, J. A., Kantor, I., Williams, D., Ellard, S., & Hattersley, A. T. (2010). Homozygous mutations in NEUROD1 are responsible for a novel syndrome of permanent neonatal diabetes and neurological abnormalities. *Diabetes*, *59*(9), 2326–2331. https://doi.org/10.2337/db10-0011.

Rukstalis, J. M., & Habener, J. F. (2007). Snail2, a mediator of epithelial-mesenchymal transitions, expressed in progenitor cells of the developing endocrine pancreas. *Gene Expression Patterns*, *7*, 471–479. https://doi.org/10.1016/j.modgep.2006.11.001.

Russ, H. A., Parent, A. V., Ringler, J. J., Hennings, T. G., Nair, G. G., Shveygert, M., et al. (2015). Controlled induction of human pancreatic progenitors produces functional beta-like cells in vitro. *The EMBO Journal*, *34*(13), 1759–1772. https://doi.org/10.15252/embj.201591058.

Salisbury, R. J., Blaylock, J., Berry, A. A., Jennings, R. E., De Krijger, R., Piper Hanley, K., et al. (2014). The window period of NEUROGENIN3 during human gestation. *Islets*, *6*, e954436. https://doi.org/10.4161/19382014.2014.954436.

Sander, M., Sussel, L., Conners, J., Scheel, D., Kalamaras, J., Dela Cruz, F., et al. (2000). Homeobox gene Nkx6.1 lies downstream of Nkx2.2 in the major pathway of beta-cell formation in the pancreas. *Development*, *127*(24), 5533–5540.

Sarkar, S. A., Kobberup, S., Wong, R., Lopez, A. D., Quayum, N., Still, T., et al. (2008). Global gene expression profiling and histochemical analysis of the developing human fetal pancreas. *Diabetologia*, *51*, 285–297. https://doi.org/10.1007/s00125-007-0880-0.

Schaffer, A. E., Freude, K. K., Nelson, S. B., & Sander, M. (2010). Nkx6 transcription factors and Ptf1a function as antagonistic lineage determinants in multipotent pancreatic progenitors. *Developmental Cell*, *18*(6), 1022–1029. https://doi.org/10.1016/j.devcel.2010.05.015.

Schaffer, A. E., Taylor, B. L., Benthuysen, J. R., Liu, J., Thorel, F., Yuan, W., et al. (2013). Nkx6.1 controls a gene regulatory network required for establishing and maintaining pancreatic beta cell identity. *PLoS Genetics*, *9*, e1003274. https://doi.org/10.1371/journal.pgen.1003274.

Scharfmann, R., Xiao, X., Heimberg, H., Mallet, J., & Ravassard, P. (2008). Beta cells within single human islets originate from multiple progenitors. *PLoS ONE*, *3*(10). e3559, https://doi.org/10.1371/journal.pone.0003559.

Schwitzgebel, V. M., Mamin, A., Brun, T., Ritz-Laser, B., Zaiko, M., Maret, A., et al. (2003). Agenesis of human pancreas due to decreased half-life of insulin promoter factor 1. *The Journal of Clinical Endocrinology and Metabolism*, *88*(9), 4398–4406. https://doi.org/10.1210/jc.2003-030046.

Sellick, G. S., Barker, K. T., Stolte-Dijkstra, I., Fleischmann, C., Coleman, R. J., Garrett, C., et al. (2004). Mutations in PTF1A cause pancreatic and cerebellar agenesis. *Nature Genetics*, *36*(12), 1301–1305.

Senee, V., Chelala, C., Duchatelet, S., Feng, D., Blanc, H., Cossec, J. C., et al. (2006). Mutations in GLIS3 are responsible for a rare syndrome with neonatal diabetes mellitus and

congenital hypothyroidism. *Nature Genetics*, *38*(6), 682–687. https://doi.org/10.1038/ng1802.
Shaw-Smith, C., De Franco, E., Lango Allen, H., Batlle, M., Flanagan, S. E., Borowiec, M., et al. (2014). GATA4 mutations are a cause of neonatal and childhood-onset diabetes. *Diabetes*, *63*(8), 2888–2894. https://doi.org/10.2337/db14-0061.
Shi, Z. D., Lee, K., Yang, D., Amin, S., Verma, N., Li, Q. V., et al. (2017). Genome editing in hPSCs reveals GATA6 haploinsufficiency and a genetic interaction with GATA4 in human pancreatic development. *Cell Stem Cell*, *20*(5), 675–688. e676. https://doi.org/10.1016/j.stem.2017.01.001.
Shih, H. P., Panlasigui, D., Cirulli, V., & Sander, M. (2016). ECM signaling regulates collective cellular dynamics to control pancreas branching morphogenesis. *Cell Reports*, *14*, 169–179.
Simsek, S., Zhou, T., Robinson, C. L., Tsai, S. Y., Crespo, M., Amin, S., et al. (2016). Modeling cystic fibrosis using pluripotent stem cell-derived human pancreatic ductal epithelial cells. *Stem Cells Translational Medicine*, *5*(5), 572–579. https://doi.org/10.5966/sctm.2015-0276.
Smith, S. B., Qu, H. Q., Taleb, N., Kishimoto, N. Y., Scheel, D. W., Lu, Y., et al. (2010). Rfx6 directs islet formation and insulin production in mice and humans. *Nature*, *463*(7282), 775–780. https://doi.org/10.1038/nature08748.
Solar, M., Cardalda, C., Houbracken, I., Martin, M., Maestro, M. A., De Medts, N., et al. (2009). Pancreatic exocrine duct cells give rise to insulin-producing beta cells during embryogenesis but not after birth. *Developmental Cell*, *17*(6), 849–860.
Solomon, B. D., Pineda-Alvarez, D. E., Balog, J. Z., Hadley, D., Gropman, A. L., Nandagopal, R., et al. (2009). Compound heterozygosity for mutations in PAX6 in a patient with complex brain anomaly, neonatal diabetes mellitus, and microophthalmia. *American Journal of Medical Genetics Part A*, *149A*(11), 2543–2546. https://doi.org/10.1002/ajmg.a.33081.
Sosa-Pineda, B., Chowdhury, K., Torres, M., Oliver, G., & Gruss, P. (1997). The Pax4 gene is essential for differentiation of insulin-producing beta cells in the mammalian pancreas. *Nature*, *386*, 399–402.
Spagnoli, F. M. (2015). Simply the right time to turn on insulin. *The EMBO Journal*, *34*, 2014–2016. https://doi.org/10.15252/embj.201591894.
Spence, J. R., Mayhew, C. N., Rankin, S. A., Kuhar, M. F., Vallance, J. E., Tolle, K., et al. (2011). Directed differentiation of human pluripotent stem cells into intestinal tissue in vitro. *Nature*, *470*(7332), 105–109. https://doi.org/10.1038/nature09691.
Steiner, D. J., Kim, A., Miller, K., & Hara, M. (2011). Pancreatic islet plasticity: Interspecies comparison of islet architecture and composition. *Islets*, *2*, 135–145.
Stoffers, D. A., Zinkin, N. T., Stanojevic, V., Clarke, W. L., & Habener, J. F. (1997). Pancreatic agenesis attributable to a single nucleotide deletion in the human IPF1 gene coding sequence. *Nature Genetics*, *15*(1), 106–110. https://doi.org/10.1038/ng0197-106.
St-Onge, L., Sosa-Pineda, B., Chowdhury, K., Mansouri, A., & Gruss, P. (1997). Pax6 is required for differentiation of glucagon-producing alpha-cells in mouse pancreas. *Nature*, *387*(6631), 406–409. https://doi.org/10.1038/387406a0.
Sussel, L., Kalamaras, J., Hartigan-O'Connor, D. J., Meneses, J. J., Pedersen, R. A., Rubenstein, J. L., et al. (1998). Mice lacking the homeodomain transcription factor Nkx2.2 have diabetes due to arrested differentiation of pancreatic beta cells. *Development*, *125*(12), 2213–2221.
Suzuki, S., Nakao, A., Sarhat, A. R., Furuya, A., Matsuo, K., Tanahashi, Y., et al. (2014). A case of pancreatic agenesis and congenital heart defects with a novel GATA6 nonsense mutation: Evidence of haploinsufficiency due to nonsense-mediated mRNA decay. *American Journal of Medical Genetics. Part A*, *164A*(2), 476–479. https://doi.org/10.1002/ajmg.a.36275.

Takebe, T., Enomura, M., Yoshizawa, E., Kimura, M., Koike, H., Ueno, Y., et al. (2015). Vascularized and complex organ buds from diverse tissues via mesenchymal cell-driven condensation. *Cell Stem Cell, 16*(5), 556–565. https://doi.org/10.1016/j.stem.2015.03.004.

Takizawa-Shirasawa, S., Yoshie, S., Yue, F., Mogi, A., Yokoyama, T., Tomotsune, D., et al. (2013). FGF7 and cell density are required for final differentiation of pancreatic amylase-positive cells from human ES cells. *Cell and Tissue Research, 354*(3), 751–759. https://doi.org/10.1007/s00441-013-1695-6.

Tam, P. P. L., & Loebel, D. A. F. (2007). Gene function in mouse embryogenesis: Get set for gastrulation. *Nature Reviews. Genetics, 8*, 368–381. https://doi.org/10.1038/nrg2084.

Taylor, B., Liu, F. F., & Sander, M. (2013). Nkx6.1 is essential for maintaining the functional state of pancreatic beta cells. *Cell Reports, 4*, 1262–1275. https://doi.org/10.1016/j.celrep.2013.08.010.

Teo, A. K. K., Lau, H. H., Valdez, I. A., Dirice, E., Tjora, E., Raeder, H., et al. (2016). Early developmental perturbations in a human stem cell model of MODY5/HNF1B pancreatic hypoplasia. *Stem Cell Reports, 6*, 357–367. https://doi.org/10.1016/j.stemcr.2016.01.007.

Thomas, I. H., Saini, N. K., Adhikari, A., Lee, J. M., Kasa-Vubu, J. Z., Vazquez, D. M., et al. (2009). Neonatal diabetes mellitus with pancreatic agenesis in an infant with homozygous IPF-1 Pro63fsX60 mutation. *Pediatric Diabetes, 10*(7), 492–496. https://doi.org/10.1111/j.1399-5448.2009.00526.x.

Thompson, N., Gésina, E., Scheinert, P., Bucher, P., & Grapin-Botton, A. (2012). RNA profiling and chromatin immunoprecipitation-sequencing reveal that PTF1a stabilizes pancreas progenitor identity via the control of MNX1/HLXB9 and a network of other transcription factors. *Molecular and Cellular Biology, 32*, 1189–1199. https://doi.org/10.1128/MCB.06318-11.

Tiyaboonchai, A., Cardenas-Diaz, F. L., Ying, L., Maguire, J. A., Sim, X., Jobaliya, C., et al. (2017). GATA6 plays an important role in the induction of human definitive endoderm, development of the pancreas, and functionality of pancreatic beta cells. *Stem Cell Reports, 8*(3), 589–604. https://doi.org/10.1016/j.stemcr.2016.12.026.

Touboul, T., Hannan, N. R., Corbineau, S., Martinez, A., Martinet, C., Branchereau, S., et al. (2010). Generation of functional hepatocytes from human embryonic stem cells under chemically defined conditions that recapitulate liver development. *Hepatology, 51*(5), 1754–1765. https://doi.org/10.1002/hep. 23506.

Toyoda, T., Kimura, A., Tanaka, H., Ameku, T., Mima, A., Hirose, Y., et al. (2017). Rho-associated kinases and non-muscle myosin IIs inhibit the differentiation of human iPSCs to pancreatic endoderm. *Stem Cell Reports, 9*(2), 419–428. https://doi.org/10.1016/j.stemcr.2017.07.005.

Trott, J., Tan, E. K., Ong, S., Titmarsh, D. M., Denil, S., Giam, M., et al. (2017). Long-term culture of self-renewing pancreatic progenitors derived from human pluripotent stem cells. *Stem Cell Reports, 8*(6), 1675–1688. https://doi.org/10.1016/j.stemcr.2017.05.019.

Vegas, A. J., Veiseh, O., Gürtler, M., Millman, J. R., Pagliuca, F. W., Bader, A. R., et al. (2016). Long-term glycemic control using polymer-encapsulated human stem cell–derived beta cells in immune-competent mice. *Nature Medicine, 22*, 306–311. https://doi.org/10.1038/nm.4030.

Villasenor, A., Chong, D. C., & Cleaver, O. (2008). Biphasic Ngn3 expression in the developing pancreas. *Developmental Dynamics, 237*(11), 3270–3279. https://doi.org/10.1002/dvdy.21740.

Villasenor, A., Chong, D. C., Henkemeyer, M., & Cleaver, O. (2010). Epithelial dynamics of pancreatic branching morphogenesis. *Development, 137*(24), 4295–4305. https://doi.org/10.1242/dev.052993.

Wang, A., Yue, F., Li, Y., Xie, R., Harper, T., Patel, N. A., et al. (2015). Epigenetic priming of enhancers predicts developmental competence of hESC-derived endodermal lineage intermediates. *Cell Stem Cell, 16*(4), 386–399. https://doi.org/10.1016/j.stem.2015.02.013.

Wang, J., Cortina, G., Wu, S., Tran, R., Cho, J.-H., Tsai, M.-J., et al. (2006). Mutant neurogenin-3 in congenital malabsorptive diarrhea. *New England Journal of Medicine, 355*, 270–280.

Wang, J., Elghazi, L., Parker, S. E., Kizilocak, H., Asano, M., Sussel, L., et al. (2004). The concerted activities of Pax4 and Nkx2.2 are essential to initiate pancreatic beta-cell differentiation. *Developmental Biology, 266*(1), 178–189.

Wang, S., Jensen, J. N., Seymour, P. A., Hsu, W., Dor, Y., Sander, M., et al. (2009). Sustained Neurog3 expression in hormone-expressing islet cells is required for endocrine maturation and function. *Proceedings of the National Academy of Sciences of the United States of America, 106*(24), 9715–9720. https://doi.org/10.1073/pnas.0904247106.

Wang, S., Yan, J., Anderson, D. A., Xu, Y., Kanal, M. C., Cao, Z., et al. (2010). Neurog3 gene dosage regulates allocation of endocrine and exocrine cell fates in the developing mouse pancreas. *Developmental Biology, 339*, 26–37. https://doi.org/10.1016/j.ydbio.2009.12.009.

Weedon, M. N., Cebola, I., Patch, A.-M., Flanagan, S. E., De Franco, E., Caswell, R., et al. (2014). Recessive mutations in a distal PTF1A enhancer cause isolated pancreatic agenesis. *Nature Genetics, 46*, 61–64. https://doi.org/10.1038/ng.2826.

Wen, X., & Yang, Y. (2017). Emerging roles of GLIS3 in neonatal diabetes, type 1 and type 2 diabetes. *Journal of Molecular Endocrinology, 58*(2), R73–R85. https://doi.org/10.1530/JME-16-0232.

Westmoreland, J. J., Kilic, G., Sartain, C., Sirma, S., Blain, J., Rehg, J., et al. (2012). Pancreas-specific deletion of Prox1 affects development and disrupts homeostasis of the exocrine pancreas. *Gastroenterology, 142*, 999–1009. e1006. https://doi.org/10.1053/j.gastro.2011.12.007.

Wierup, N., Svensson, H., Mulder, H., & Sundler, F. (2002). The ghrelin cell: A novel developmentally regulated islet cell in the human pancreas. *Regulatory Peptides, 107*, 63–69. https://doi.org/10.1016/S0167-0115(02)00067-8.

Xie, R., Everett, L. J., Lim, H. W., Patel, N. A., Schug, J., Kroon, E., et al. (2013). Dynamic chromatin remodeling mediated by polycomb proteins orchestrates pancreatic differentiation of human embryonic stem cells. *Cell Stem Cell, 12*(2), 224–237. https://doi.org/10.1016/j.stem.2012.11.023.

Xuan, S., Borok, M. J., Decker, K. J., Battle, M. A., Duncan, S. A., Hale, M. A., et al. (2012). Pancreas-specific deletion of mouse Gata4 and Gata6 causes pancreatic agenesis. *The Journal of Clinical Investigation, 122*(10), 3516–3528. https://doi.org/10.1172/JCI63352.

Ye, F., Duvillie, B., & Scharfmann, R. (2005). Fibroblast growth factors 7 and 10 are expressed in the human embryonic pancreatic mesenchyme and promote the proliferation of embryonic pancreatic epithelial cells. *Diabetologia, 48*(2), 277–281. https://doi.org/10.1007/s00125-004-1638-6.

Yorifuji, T., Kawakita, R., Hosokawa, Y., Fujimaru, R., Yamaguchi, E., & Tamagawa, N. (2012). Dominantly inherited diabetes mellitus caused by GATA6 haploinsufficiency: Variable intrafamilial presentation. *Journal of Medical Genetics, 49*(10), 642–643. https://doi.org/10.1136/jmedgenet-2012-101161.

Yorifuji, T., Kurokawa, K., Mamada, M., Imai, T., Kawai, M., Nishi, Y., et al. (2004). Neonatal diabetes mellitus and neonatal polycystic, dysplastic kidneys: Phenotypically discordant recurrence of a mutation in the hepatocyte nuclear factor-1beta gene due to germline mosaicism. *The Journal of Clinical Endocrinology and Metabolism, 89*(6), 2905–2908.

Yoshitomi, H., & Zaret, K. S. (2004). Endothelial cell interactions initiate dorsal pancreas development by selectively inducing the transcription factor Ptf1a. *Development*, *131*(4), 807–817.

Zenker, M., Mayerle, J., Lerch, M. M., Tagariello, A., Zerres, K., Durie, P. R., et al. (2005). Deficiency of UBR1, a ubiquitin ligase of the N-end rule pathway, causes pancreatic dysfunction, malformations and mental retardation (Johanson-Blizzard syndrome). *Nature Genetics*, *37*(12), 1345–1350. https://doi.org/10.1038/ng1681.

Zenker, M., Mayerle, J., Reis, A., & Lerch, M. M. (2006). Genetic basis and pancreatic biology of Johanson-Blizzard syndrome. *Endocrinology and Metabolism Clinics of North America*, *35*(2), 243–253. vii–viii. https://doi.org/10.1016/j.ecl.2006.02.013.

Zhang, D., Jiang, W., Liu, M., Sui, X., Yin, X., Chen, S., et al. (2009). Highly efficient differentiation of human ES cells and iPS cells into mature pancreatic insulin-producing cells. *Cell Research*, *19*(4), 429–438. https://doi.org/10.1038/cr.2009.28.

Zhou, Q., Law, A. C., Rajagopal, J., Anderson, W. J., Gray, P. A., & Melton, D. A. (2007). A multipotent progenitor domain guides pancreatic organogenesis. *Developmental Cell*, *13*(1), 103–114.

Zhu, Z., Li, Q. V., Lee, K., Rosen, B. P., González, F., Soh, C.-L., et al. (2016). Genome editing of lineage determinants in human pluripotent stem cells reveals mechanisms of pancreatic development and diabetes. *Cell Stem Cell*, *18*, 755–768. https://doi.org/10.1016/j.stem.2016.03.015.

Zorn, A. M., & Wells, J. M. (2009). Vertebrate endoderm development and organ formation. *Annual Review of Cell and Developmental Biology*, *25*, 221–251. https://doi.org/10.1146/annurev.cellbio.042308.113344.

MIX
Papier | Fördert gute Waldnutzung
FSC® C083411

Kolibri 360 GmbH
Ferdinand-Jühlke-Straße 7
99095 Erfurt, Deutschland
produktsicherheit@kolibri360.de